IRRIGATION SYSTEM DESIGN

An Engineering Approach

Richard H. Cuenca

Department of Agricultural Engineering
Oregon State University

PRENTICE HALL, Englewood Cliffs, New Jersey 07632

Library of Congress Cataloging-in-Publication Data

Cuenca, Richard H.
Irrigation system design: an engineering approach/ Richard H. Cuenca.
p. cm.
Includes index.
ISBN 0-13-506163-6
1. Irrigation engineering. 2. Irrigation—Equipment and supplies.
I. Title.
TC805.C84 1989
627′.52—dc19 88-18614
CIP

Editorial/production supervision
and interior design: Sophie Papanikolaou
Cover design: Ben Santora
Manufacturing buyer: Bob Anderson

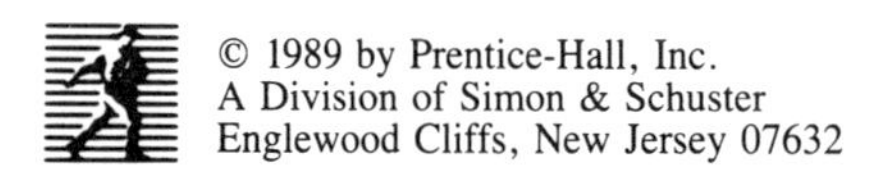

DEDICATION

This book is dedicated to Gino and Tranquilla Toschi and Hermenegildo and Pilar Cuenca, my grandparents, who had the courage to leave the Old World and give their families the opportunity to pursue their dreams in the New World.

Printed in the United States of America
10 9 8 7 6 5 4 3 2 1

ISBN 0-13-506163-6

Prentice-Hall International (UK) Limited, *London*
Prentice-Hall of Australia Pty. Limited, *Sydney*
Prentice-Hall Canada Inc., *Toronto*
Prentice-Hall Hispanoamericana, S.A., *Mexico*
Prentice-Hall of India Private Limited, *New Delhi*
Prentice-Hall of Japan, Inc., *Tokyo*
Simon & Schuster Asia Pte. Ltd., *Singapore*
Editora Prentice-Hall do Brasil, Ltda., *Rio de Janeiro*

Contents

Preface

The goal of this book is to demonstrate the quantification of parameters necessary for the design, installation, and operation of various types of irrigation systems. The book is broad in scope with different sections devoted to fundamental principles, water application systems, and water distribution systems. Complete books have been written covering only individual chapters in this manuscript. But the goal of this text is to quantitatively demonstrate in a single volume modern design concepts applied to optimizing the use of water resources for irrigation over a wide range of field conditions. The focus will be on the supply, distribution, and application of water for irrigation at the farm level.

This text is directed at upper-division engineering students who have had the normal complement of mathematics, physics, and chemistry courses required in an accredited lower-division engineering curriculum. Practicing engineers working in the area of irrigation system design and operations will consider this book a valuable reference due to the use of demonstrative example problems and to the wide range of subjects covered. Although this textbook demonstrates and encourages the use of computerized solutions to irrigation system design problems, no special computer programming skills are assumed at the outset.

The underlying concepts in the design of a physical system for irrigation are based on the physics and chemistry of soils, on crop water requirements, and on hydraulics and economics. This book will lay an adequate foundation in these principles so the final solution can be developed which is based on a sound understanding of subjects critical to proper design. To build this foundation, chapters 2 through 5 are devoted to economics, soil physics and chemistry, and crop water requirements.

Chapters 6 through 8 discuss the design of surface, sprinkler, and trickle water application systems for irrigation. System design examples are included in each chapter to demonstrate design principles. These examples will pull specific topics covered in the chapter together with the underlying principles discussed in the first five chapters. Many of the examples will be based on actual field cases and will therefore bring the reader closer to the realities of field situations.

The topics of the final four chapters are pump systems, pipelines, groundwater wells, and open channel flow. The emphasis in these final chapters is therefore the supply and distribution of water at the farm level. This material completes the requirements of the design engineer working with on-farm irrigation systems to respond to questions covering water supply, distribution, and application.

One of the advantages of this book is its liberal use of example problems to clarify concepts. Example problems offer readers the first opportunity to observe the application of possibly abstract concepts in arriving at the quantitative solution required for system design. Example problems are therefore an integral element of this textbook.

A modern engineering textbook would not be complete without an indication of the potential of computerized solutions to design problems and example applications. At the same time, it is felt that over-reliance on preprogrammed, or "canned," solutions must be avoided if the reader is to develop skills and confidence in computer programming. One must also consider that no computerized design program can efficiently handle every condition which may arise in the field.

We present certain programs and computerized spreadsheets which are considered to have general application in design of irrigation systems. The programs are written in BASIC and will be conveniently applicable to desk-top computer systems. In some cases, readers are asked to design their own program to accomplish a given task. The programming oriented questions are presented in a sequence so the reader may gradually develop confidence in programming skills.

Each chapter after the first concludes with a problem set designed to allow the reader to apply skills developed through the chapter. Problems will almost always be quantitative in nature since those are the types of problems engineers are expected to deal with. Problems may cover a single principle, parts of a system design, and possible computer applications. When applicable, suggestions will also be given for topics requiring further research.

As in many other cases, this text evolved out of the need to have adequate updated teaching material available for students, in this case upper-division engineering students in irrigation system design. It is anticipated that the author's practical experience in the design and operation of irrigation systems in the United States, Europe, and Africa will also make this book a valuable reference for practicing engineers.

Acknowledgments

This book is no different than other technical works in that it owes its existence to the strong personalities that shaped the author during his development. At the University of California at Davis, I am particularly indebted to the guidance and example of Jim Amorocho, Miguel Mariño, Bill Pruitt, and Don Nielsen. Thanks are due to Ian Stewart, who patiently taught me how to conduct and interpret field experiments, and to Marv Shearer at Oregon State University from whom I learned a lot by trying to shut up and listen in the desert of central Tunisia. Marvin Jensen of USDA contributed to critiques of early drafts of this text and set a tremendous example of productivity through ASCE committee work. A significant credit for the technical thoroughness of this book is owed to Rick Allen of Utah State University who did more to sharpen up this book than should be asked of any reviewer. His effort is especially appreciated. I must also sincerely thank my students of the past ten years at Oregon State University. They taught me more by their questions than I taught them by my answers. Don Nielsen once said to me that it is by teaching that you learn. As usual, he was right. And finally, all the above would have come to nothing if not for the patience of my family, my wife Shirley and especially my girls Theresa and Alicia, who put up with all those lost weekends and evenings with Daddy bashing away at the keyboard.

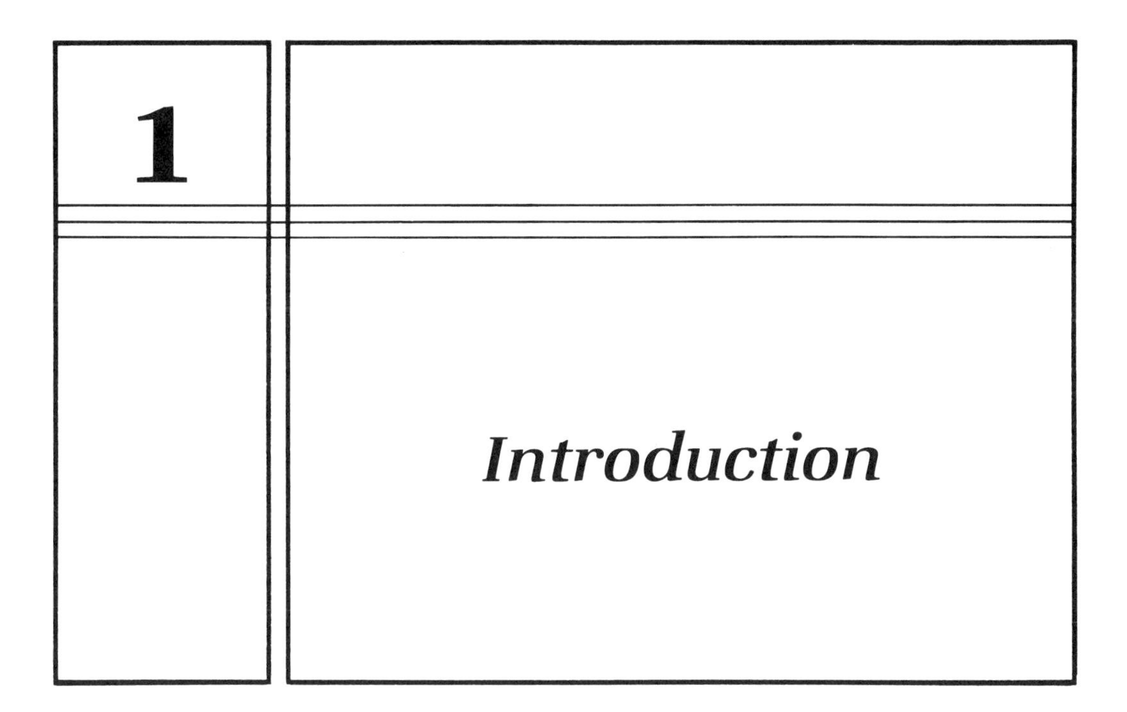

1.1 Fundamental Principles

Purpose of Irrigation

Irrigation is basically an attempt by man to locally alter the hydrologic cycle and to promote increased agricultural productivity. Irrigation in some form has been practiced since before recorded history. Irrigation system design combines elements of soil science, agronomy, social science, hydraulics, hydrology, and economic analysis. The science and art of irrigation system design integrates numerous disciplines and is a field in which the neglect of any one contributing discipline can preclude the final project from attaining its goals. An irrigation project cannot be successfully designed, constructed, and operated within the borders of a single scientific subject area. This book will stress the soil, hydrologic, hydraulic, and economic aspects of irrigation system design. However, the contributions of agronomy and social science can never be too far removed from the view of the competent design engineer.

In locally altering the hydrologic cycle and increasing agricultural productivity, the impact of irrigation on the development of modern civilization has been profound. Irrigated agriculture has both reacted to and participated in changes in rural and urban societies. The increased stability of food resources brought about through irrigation is one fundamental development which allowed early civilizations to gradually shift from nomadic, food gathering societies to groups with permanent or semi-permanent dwellings. The impact of production advantages due to irrigated agriculture remain tremendously important. Today's global economy causes reaction to a failed wheat crop in the Great Plains of the United States to be felt in bread supplies in urban centers in the Soviet Union. Conversely, mild winter conditions on the Russian Steppe are reflected in wheat and bread prices in the United States. A significant part of a country's balance of payments for buying and selling commodities in the world market may be made up of agricultural production. Due to the decrease in risk brought about by irrigation, countries with a larger proportion of land in irrigation tend to have increased stability in those sectors of the regional and national economy affected by agriculture.

Increased productivity in agriculture through irrigation also leads to increased opportunities in businesses which supply the agricultural sector. These effects are felt directly in sales of irrigation equipment and indirectly in sales of seed, fertilizer, pesticides, herbicides, and agricultural machinery. This expected increase in sales assumes that irrigated agriculture will lead to increased profits over dry-land or rain-fed agriculture. In the quasi-free agricultural market system in the United States and other western economies, the goal of increased and more stable profits is the principal reason for the design and installation of irrigation systems. In such economies, irrigation systems should not be designed which do not lead to increased profits over the long term. In other cases, irrigation systems may be thought of as a means of achieving a national strategic goal of more stable food supplies and less dependence on foreign agricultural resources. In any case, the goals of the irrigation system, whether a large regional system or a system to serve an individual farm, must be clearly defined and the ultimate design formulated to give the highest probability within cost constraints of meeting those goals.

Expected Benefits of Irrigation In Terms of Crop Yield

To quantify the economic benefit of irrigation, it is necessary to be able to quantify the expected increase in yield as a function of increasing amounts of water delivered by the irrigation system. A graphical representation of this relationship is termed a crop-water production function. The most straightforward case of a crop-water production function will be investigated in this section. Nutrient levels will be assumed adequate to produce maximum growth at any level of irrigation and the crop will be assumed free of disease, weeds, and pests. The crop response to irrigation will therefore represent a maximum at any given water level.

A question which arises is what measure of crop water use to apply in evaluating the crop-water production function. Figure 1-1 indicates a typical functional shape for depth of applied water (AW) to the crop over the growing season versus marketable yield (Y). This curve was developed by fitting a regression equation through experimental field data.

A number of interesting features may be determined from the yield versus applied water curve. The function starts with a relatively high slope indicating that water is efficiently used to increase production at low levels of irrigation. As applied water levels increase, the slope diminishes as demonstrated by ΔY_1 being greater

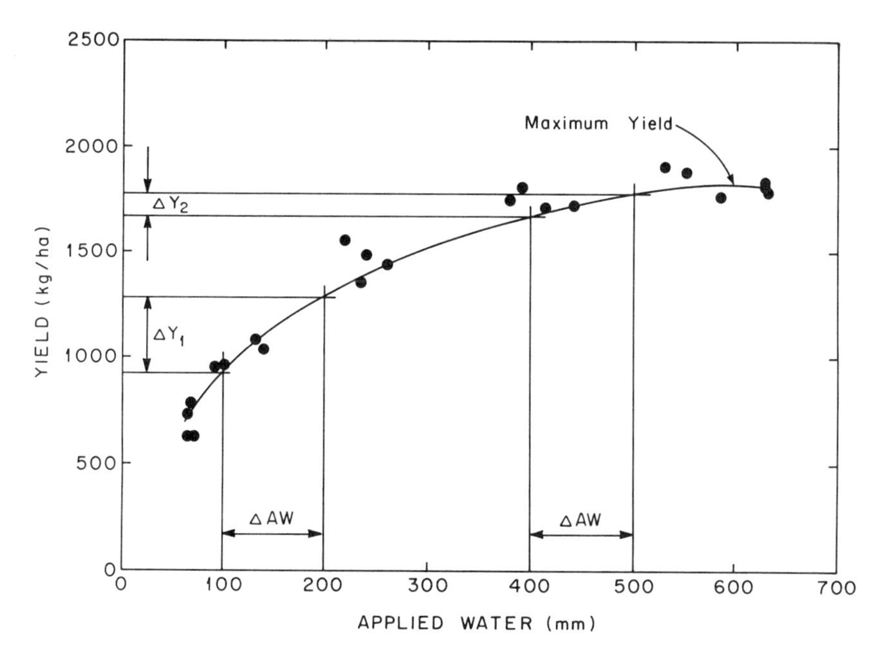

Figure 1-1 Yield as a function of applied water for SJ2 variety cotton. (Adapted from Cuenca, 1978.)

than ΔY_2 for the same incremental ΔAW. In fact, as the function approaches maximum yield, the slope goes to zero. Beyond this point, additional increments of applied water tend to decrease yield due to decreased aeration and impedance of necessary gas transfer within the root zone. Not all crops respond precisely as indicated in Fig. 1-1, but the response shown is typical.

An alternate way of expressing the crop-water production function is to plot consumptive use, or evapotranspiration, versus yield. Evapotranspiration will be the preferred term in this book. Evapotranspiration represents the combination of water evaporated from the plant and soil surface plus that amount of water which passes through the soil into the roots, through the stem of the plant, and to the leaves where it passes into the atmosphere through small pores termed stomates (singular stomata). Movement of water through the stomates is in response to the gradient of water vapor between the plant leaves and the atmosphere. The translocation of moisture from the root zone through the stomates to the atmosphere is termed transpiration. This process combined with evaporation represents the quantity of water consumptively used by the plant, or evapotranspiration.

Figure 1-2 indicates a plot of yield versus evapotranspiration for the same field data indicated in Fig. 1-1. It is interesting to note that this relationship can be described as a linear function. One does not generally expect nature to respond in a linear fashion, yet crop-water production functions such as that indicated in Fig. 1-2 have been noted by numerous researchers (Barrett, 1977; Vaux et al., 1981). In fact, current methodologies for quantifying crop response to irrigation rely implicitly on a linear relationship between evapotranspiration and yield (Doorenbos and Kassam, 1979). The maximum yield and maximum evapotranspiration points in Fig. 1-2 are

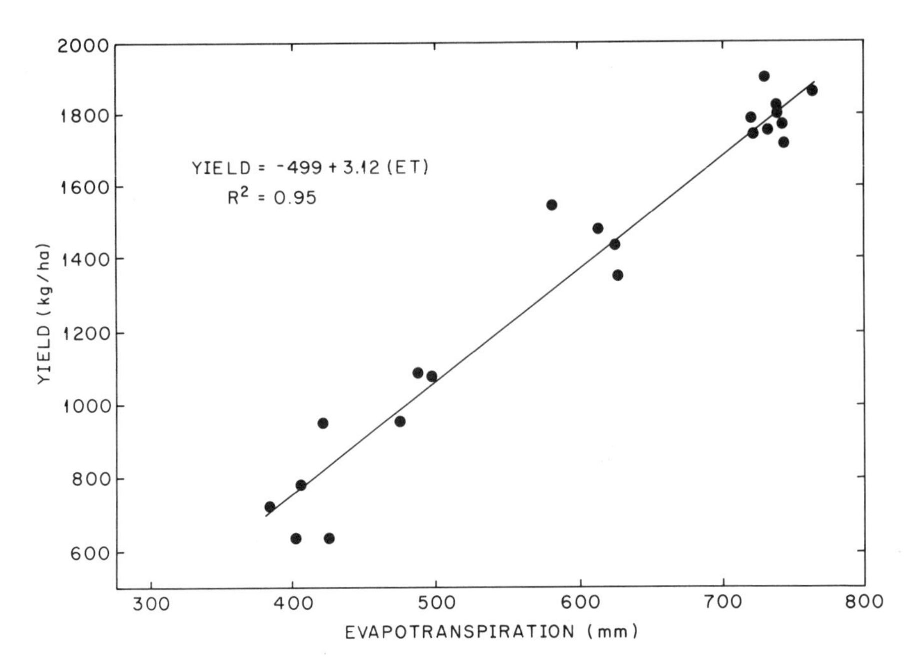

Figure 1-2 Yield as a function of evapotranspiration for SJ2 variety cotton. (Adapted from Cuenca, 1978.)

assumed to be varietal characteristics of the crop, as is the slope of the production function. It should be noted that the same growth conditions assumed to hold for Fig. 1-1 are also assumed to hold for Fig. 1-2, and no data reflect ill-timed irrigations. Both figures may therefore be interpreted as representing the maximum response to various levels of irrigation.

Figure 1-3 presents a combination of the type of data shown in the previous two figures. This figure indicates plots of yield versus applied water and yield versus that amount of evapotranspiration determined to be from irrigation. The lowest level of evapotranspiration is assumed due to available soil moisture at time of planting plus precipitation throughout the growing season. This evapotranspiration from available soil moisture at time of planting plus precipitation has been termed basal evapotranspiration (Stewart and Hagan, 1973).

The applied water in Fig. 1-3 includes soil moisture at time of planting plus growing season precipitation. The curves that fit through both characterizations of the crop-water production function in Fig. 1-3 start at the same point indicating that all the applied water is used as evapotranspiration at low levels of irrigation. As the irrigation level increases, less applied water is directly converted to evapotranspiration—that is, more percolates past the root zone, runs off the field, or is stored in the root zone at the time of harvest. Stewart and Hagan (1973) were the first to term the difference in water use between the two functions in Fig. 1-3 at a given level of yield as non-evapotranspiration irrigation.

The nonlinearity in the difference between the two functions in Fig. 1-3 leads to a subsequent nonlinearity in some measures of irrigation efficiency as the irriga-

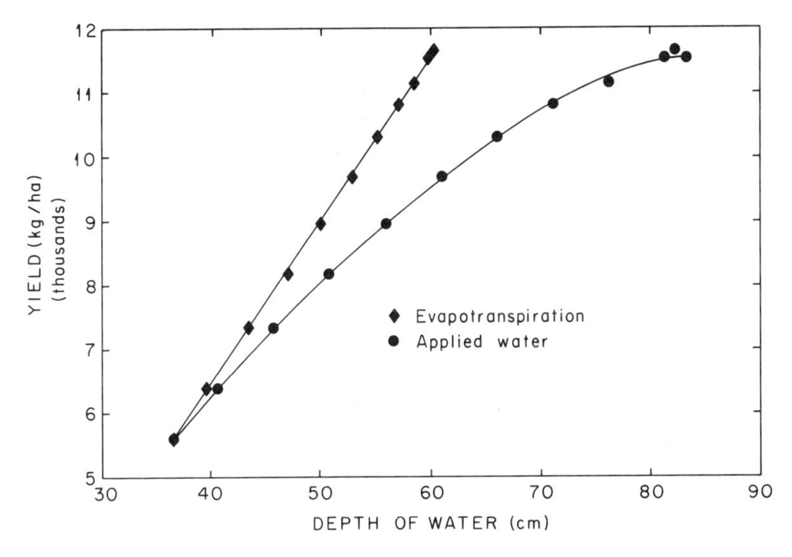

Figure 1-3 Idealized water production function for yield versus applied water and evapotranspiration.

tion level increases. This nonlinearity in efficiency has not traditionally been accounted for in system design. Application of crop-water production functions will be discussed further in Chapter 2. Of particular importance will be relative crop-production functions for yield versus evapotranspiration. Such functions range from zero to 100 percent on both axis by dividing measured values by the maximum yield and maximum evapotranspiration. These quantities are assumed to be varietal characteristics of the crop. Such relative production functions avoid some of the problems associated with application of the more site-specific production functions indicated in Figs. 1-1 through 1-3.

Example Problem 1-1

Assume that the applied water for Fig. 1-3 includes the available soil water at time of planting plus precipitation plus irrigation. The yield versus applied water relationship may be described by the following quadratic equation:

$$Y = -4941 + 35.83(AW) - 0.0195(AW)^2$$

where

$$Y = \text{yield, kg/ha}$$
$$AW = \text{applied water, mm}$$

The yield versus evapotranspiration relationship may be described by the following linear equation:

$$Y = -3783 + 25.65(ET)$$

where

$$ET = \text{evapotranspiration, mm}$$

(a) Compute the ratio of the change in yield to the change in applied water when the applied water is increased from 400 mm to 500 mm and from 700 mm to 800 mm.

(b) Define the efficiency as the amount of evapotranspiration produced for a given level of applied water expressed as a percent. Compute the efficiency at an applied water level of 450 mm and at a level of 750 mm.

Solution

(a) For AW = 400 mm

$$Y = -4941 + 35.83(400) - 0.0195(400)^2$$
$$Y = 6275 \text{ kg/ha}$$

For AW = 500 mm

$$Y = -4941 + 35.85(500) - 0.0195(500)^2$$
$$Y = 8105 \text{ kg/ha}$$
$$\frac{\Delta Y}{\Delta AW} = \frac{8105 - 6275}{500 - 400} = 18.3 \frac{\text{kg/ha}}{\text{mm}}$$

For AW = 700 mm

$$Y = -4941 + 35.83(700) - 0.0195(700)^2$$
$$Y = 10{,}596 \text{ kg/ha}$$

For AW = 800 mm

$$Y = -4941 + 35.83(800) - 0.0195(800)^2$$
$$Y = 11{,}258 \text{ kg/ha}$$
$$\frac{\Delta Y}{\Delta AW} = \frac{11{,}258 - 10{,}596}{800 - 700} = 6.6 \frac{\text{kg/ha}}{\text{mm}}$$

(b) For AW = 450 mm

$$Y = -4941 + 35.83(450) - 0.195(450)^2$$
$$Y = 7238 \text{ kg/ha}$$

Rearranging the equation for Y vs ET

$$ET = \frac{Y + 3783}{25.65}$$
$$ET = \frac{7238 + 3783}{25.65}$$
$$ET = 430 \text{ mm}$$
$$\text{Eff} = \frac{430 \text{ mm}}{450 \text{ mm}} \times 100 = 95.6\%$$

For AW = 750 mm

$$Y = 4941 + 35.83(750) - 0.0195(750)^2$$

$$Y = 10{,}976 \text{ kg/ha}$$

$$ET = \frac{10{,}976 + 3783}{25.65} = 575 \text{ mm}$$

$$Eff = \frac{575 \text{ mm}}{750 \text{ mm}} \times 100 = 76.7\%$$

Irrigation Systems Within a Hydrologic and Environmental Framework

As indicated in the first section of this chapter, irrigation systems locally alter the hydrologic cycle. Hydrology involves the study of the movement of water above, on, and below the earth's surface. The hydrologic cycle is depicted in Fig. 1-4. An irrigation project modifies the amount of water normally applied to the land, local evapotranspiration rates, seepage from unlined ditches and excessive irrigation into the subsoil, runoff into streams and rivers, and the amount of water held within groundwater aquifers. Different types of irrigation systems have an impact on different components of the hydrologic cycle. For example, a system which has water delivered from a distant surface water supply, such as a reservoir, and distributes the water through unlined canals to surface application systems, such as spreading basins, will tend to increase the volume of water stored in local groundwater aquifers. A system which perhaps uses sprinklers and extracts water from deep wells in excess of the amount normally recharged by precipitation or deep percolation of irrigation to groundwater on an annual basis will, over time, deplete the groundwater resources. This depletion may eventually make pumping for irrigation uneconomical. Different scenarios having various components of these two examples can be imagined.

In addition to shifts in the local water balance caused by irrigation systems, there are oftentimes changes in the distribution of salts and sediments within the irrigation project area. These changes can have effects far downstream of the project area. Long-term degradation of soils and natural waterways through excessive salinity build-up or sediment deposition can permanently take areas out of agricultural production. Such actions can cause disruption of farmers' livelihood and adverse impacts on local and perhaps regional economies.

The fact to bear in mind in the design of any system is that the local hydrologic cycle will be affected by irrigation. Seriously adverse effects can make the operation of an irrigation system economically unacceptable much sooner than the expected economic life used in system planning. The ultimate goal of all irrigation projects is in some way to improve the standard of living of the society affected by the project. For this long-term goal to be achieved, correct appraisal must be made of the expected changes in the hydrologic balance and environment resulting from system installation and operation along with procedures to avoid development of adverse impacts.

This book will concentrate on design of on-farm irrigation systems and leave for other publications the broad field of hydrologic and water resources analysis.

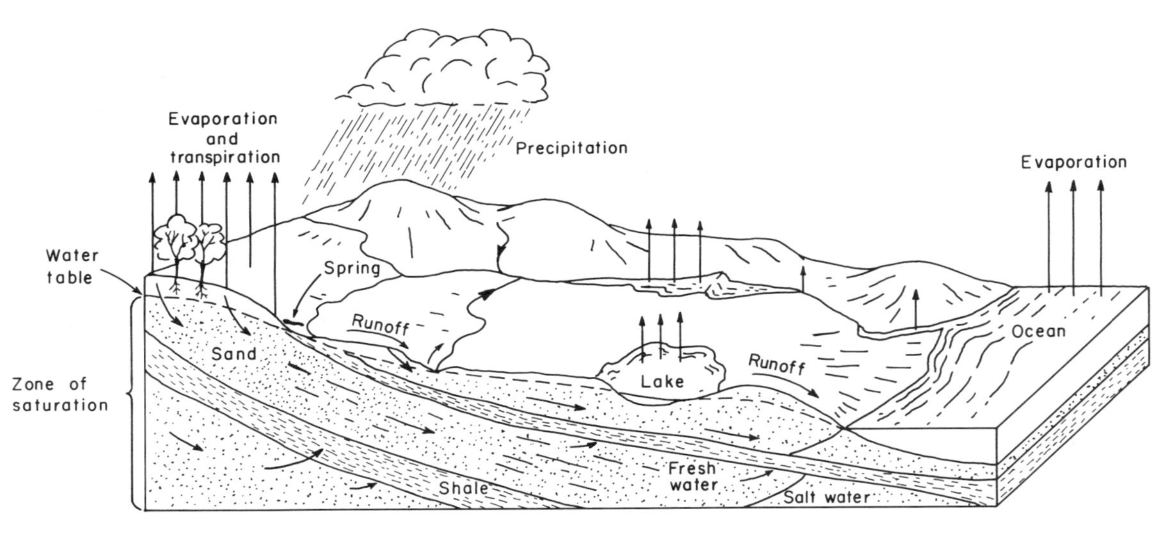

Figure 1-4 Components of the hydrologic cycle.

Nevertheless, the irrigation system design engineer must be aware of the fact that there are short- and long-term environmental impacts in the installation and operation of irrigation systems. To disregard these impacts is to invite development of a system which does not meet its expected economic benefit and which will degrade the reputation of the design engineer and the design firm.

1.2 Development and Distribution of Irrigation Systems

United States

The first irrigation systems in the United States were developed by American Indians, probably in the arid southwest where summer rainfalls were not adequate to produce a crop. Traces of canal distribution systems developed by Indians in Arizona are still visible today (Taylor and Ashcroft, 1972). In 1986 there were 24.0 million hectares (ha) of irrigated cropland in the United States (*Irrigation Journal,* 1986). Approximately 90 percent of this area was in the 14 western states of California, Texas, Nebraska, Idaho, Colorado, Kansas, Montana, Oregon, Washington, Wyoming, Nevada, Arizona, New Mexico, and Utah plus the three southern states of Arkansas, Florida, and Georgia. Detailed state-by-state surveys made by the *Irrigation Journal* since 1972 indicate that nationwide irrigated acreage increased continuously by a total of 24 percent in the 10 years from 1972 to 1982.

Worldwide

The origins of the first irrigation system are lost in the sediments of prerecorded history. It is known that many ancient civilizations had irrigation systems and some relied on them for a significant degree of their economic development. Civilizations in India, Egypt, China, and Iraq have practiced irrigation since prehistoric times (Taylor and Ashcroft, 1972). The annual flooding of the Nile River in Egypt, bringing water and nutrient-rich sediments to the surrounding fields, was the basis of an extensively developed agricultural society which continues today to use the Nile River as its main input to production. Figures 1-5 and 1-6 depict systems related to agricultural use of water resources from the days of the Roman Empire in North Africa and from the 18th century in Europe.

The Food and Agriculture Organization (FAO) of the United Nations estimated in 1977 that the total global irrigated area was 223 million ha and that this area would increase to approximately 273 million ha by 1990 (FAO, 1977). A comparison of cultivated land and the portion of that land irrigated by region and country is given in Table 1.1. The economic impact of irrigated development in contrast to non-irrigated agriculture is indicated by the fact that although irrigated agriculture represented only 13 percent of global arable land, the value of crop production from that land represented 34 percent of the world total (FAO, 1979).

Figure 1-5 Covered Roman aquaduct used to transport collected rainfall from highlands through arid region to Carthage adjacent to present day Tunis, capital of Tunisia.

Figure 1-6 Dam used to elevate flows of the Ebro River in Navarra region, Spain, to discharge into Imperial Canal originally constructed in 1772 and modified in 1780 and 1784. Building in background is former mill with canal entrance at right. Irrigation canal system is still fully operational.

Types of Systems

Irrigation systems can be broadly categorized into pressurized or non-pressurized systems. Non-pressurized systems are also termed gravity or surface systems. Due to the stimulus for low pressure systems brought on by increasing worldwide energy prices since the 1970s, there is sometimes an overlap between the actual pressure requirements of what were traditionally termed pressurized and non-pressurized systems.

Chronologically, non-pressurized systems in which water is flooded onto the soil surface predated pressurized systems by milleniums. Such surface application systems include contour levees, level basins (Fig. 1-7), and graded borders (Fig. 1-8). Other surface systems in which water is distributed through the field in channels are termed furrow and corrugation systems. Figure 1-9 indicates a furrow system. Chapter 6 deals with the design of surface systems.

TABLE 1-1 Major irrigated areas of the world. (Adapted from Jensen, 1980.)

	Agricultural Land		
	Area Cultivated	Area Irrigated	
Continent and Country	(1000 ha)	(1000 ha)	(percent)
Africa	214,000	8,930	4.2
Asia, excluding USSR	463,000	164,640	35.6
Australia and Oceania	47,000	1,700	3.6
Europe, excluding USSR	145,000	12,770	8.8
North and Central America	271,000	27,430	10.1
South America	84,000	6,660	7.9
USSR	233,000	11,500	4.9
TOTAL	1,457,000	233,630	16.0

Pressurized systems include all types of sprinkler systems and low-pressure nozzle systems. Typical sprinkler systems include solid set, hand-move, side-roll, center-pivot, and linear-move systems. Three of these systems are depicted in Figs. 1-10 through 1-12. Low pressure nozzles developed in the late 1970s and early 1980s are finding increased application, especially in center pivot and linear move systems. A high pressure system which utilizes a single nozzle is a big-gun system shown in Fig. 1-13. Sprinkler system design is the subject of Chapter 7.

Trickle, or drip, irrigation systems differ from those previously described in that they apply water specifically to the soil in the vicinity of the root zone and not to the field in general. Such systems usually require very low pressure and can be designed to more precisely deliver the estimated crop water requirement than other systems under normal operating conditions. Trickle systems may require a relatively high level of water filtration to ensure continued operations. A trickle irrigation system in an orchard is shown in Fig. 1-14. The design of trickle systems is discussed in Chapter 8.

Figure 1-7 Flooding of moderate size level basin systems used for cereal production in Nile River Valley, Egypt. Basin being irrigated is in foreground to right of supply ditch.

Figure 1-8 Graded border irrigation used for orchards in Aragon region, Spain. Inlet is at corner of rectangular field.

Figure 1-9 Graded furrow irrigation used in cotton production in San Juaquin Valley, California. Water source is submerged alfalfa valve in foreground supplied by buried mainline.

Figure 1-10 Hand-move sprinkler system used for vegetable production in Willamette Valley, Oregon.

Figure 1-11 Side-roll sprinkler system used for vegetable production in Willamette Valley, Oregon.

Figure 1-12 Center pivot system with low pressure nozzles and double end-gun used to irrigate cereals in eastern Oregon. A single pivot system typically irrigates approximately 68 ha in the western United States.

Figure 1-13 Big gun system with drag line used to irrigate pasture in Willamette Valley, Oregon.

Figure 1-14 Trickle irrigation system on young peach orchard in Willamette Valley, Oregon.

1.3 System Design Process

Statement of Problem

The design of any irrigation system may be divided into a number of steps which the design engineer follows consciously or intuitively. The first of these is the statement of the problem. What is the design supposed to accomplish physically? What is the goal that the physical system is to attain? Is the goal generally economic, or is it strategic or political? Finally, what measure will be used to determine whether or not the physical system meets the stated goals? Only when these questions have been answered is the design engineer prepared to proceed to the next step.

Data Collection

The most time-consuming task to the experienced system design engineer can easily be data collection. This is because irrigation systems operate in, and are expected to change, a natural environment. Proper system design requires information on physical and chemical properties of the soil, chemical properties of the water source, climatic parameters, expected crop response to irrigation, economic costs and benefits of all aspects of the irrigation system, as well as social impacts and constraints of the proposed system. No two systems are identical, particularly in terms of soil physical and chemical parameters, field topography, and water quality.

The time and energy required for data collection should not be underestimated. A false sense of security concerning data availability can arise from the solution of too many textbook problems in system design in which required input parameters are given. In solving the textbook problems, the reader should consider what data collection program was necessary to assemble the required input. Field measurements can quickly emphasize the difficulty in acquiring a good data base for system design. This is why most irrigation system design classes at universities in the United States are taught together with a field laboratory section. This practice is recommended to give the student an improved physical understanding of the meaning and significance of collecting field data.

System Choice

Once the physical, chemical, climatological, economic and social data are collected, it is necessary to proceed with the choice of system to accomplish the stated objective. Sometimes because of equipment availability, the operator's experience, or terrain characteristics, only a particular type of irrigation system will do the job. Other times, a number of systems are possible and alternative designs can be developed. The ultimate system choice may then be made considering expected economic return, the operator's experience, system maintenance requirements, and parts availability.

No single irrigation system is applicable or advantageous in all circumstances. Type of terrain and crop, power cost and availability, and sometimes water quality will dictate that one or more systems may be advantageous. An unbiased look by the design engineer at the realities of system cost, benefits, and constraints can balance the enthusiasm of equipment manufacturers for their particular product. Current availability of equipment for installation and long-term availability of replacement parts must also be considered.

Injecting Creativity

Creativity is the difference between the art and textbook solution of a design problem. A design by a creative engineer normally does the task better, more conveniently, and at less cost than a design done according to some strict format. There is no doubt that time in the field observing, making measurements, operating irrigation systems, and talking with operators who own and operate various types of systems enhances one's ability to prepare a more creative design.

Ultimately, creativity is not taught. It is formed by exposure, experience, and observation. It is developed from a mixture of technical expertise and experience in dealing with practical field problems. Although it is difficult to define creativity in design, it is an essential element.

Rigor and Reasonableness in the Design Solution

With the advent and application of computers and calculators in design, it is possible to produce an extremely precise, yet not necessarily accurate, design solution. Any solution to a design problem needs to be judged by two criteria: What is the expected accuracy of input data? and How reasonable is the final solution?

An example of the first criterion might be specification of system operating time per irrigation. If the estimated crop water requirement based on a combination of climatic, plant, and soil data is only accurate within plus or minus 10 percent, which is in fact better than can normally be expected (Burman et al., 1983; Warrick and Nielsen, 1980), there is no need to specify the system operating time to the nearest 15 minutes in a 12-hour application. The input data simply do not warrant that precise a solution. Engineers tend to pride themselves on accuracy in numerical problems. However, it must be recalled that in irrigation system design we are dealing with the interaction of natural systems which are by definition not deterministic. Such systems do not have a precisely predictable reaction to a given stimulus. Natural systems have a random or stochastic component in their response which precludes the possibility of precisely predicting the output.

Different factors required as input for design have a different expected accuracy. While it may be possible to predict friction loss in a new pipe with relative accuracy, the actual pumping level in a well which has a seasonal water level fluctuation is much more difficult to specify. Both of these data are necessary to compute the power requirements of a pump. The final solution is no more accurate than the least accurate input parameter.

The second criterion by which systems may be judged is reasonableness. The field experience of the design engineer and exposure in talking with operators enhances the ability of the engineer to specify a more reasonable design. An irrigation system design which requires the operator to move pipe at two o'clock in the morning is not feasible even though that may be the time at which the required amount of water is best applied. An alternative application system or operating schedule must be found.

It is this reasonableness of design which most clearly incorporates the social aspects of irrigated agriculture. The system operational constraints should be agreed upon between the engineer and operator before the design has progressed too far. The design may proceed along one path until a conflict is reached between two constraints—for example, intake rate of the soil and the operating schedule specified by the grower. At that point, it may be necessary to completely alter the proposed design, perhaps even going to a different type of irrigation system. Such design changes take time, but ultimately not as much time as trying to manage a poor design.

REFERENCES

BARRET, J. W. H., *Crop Yield Functions and the Allocation and Use of Irrigation Water*. Ph.D. Dissertation, Colorado State University, Fort Collins, Colorado, 1977.

BURMAN, R. D., R. H. CUENCA, AND A. WEISS, "Techniques for Estimating Irrigation Water Requirements," *Advances in Irrigation, Vol. 2,* ed. D. Hillel. New York: Academic Press, 1983.

CUENCA, R. H., *Transferable Simulation Model for Crop Soil Water Depletion*. Ph.D. Dissertation, University of California, Davis, California, 1978.

DOORENBOS, J. AND A. H. KASSAM, "Yield Response to Water," Irrigation and Drainage Paper no. 33. Rome: Food and Agriculture Organization, United Nations, 1979.

Food and Agriculture Organization, "Water for Agriculture," United Nations Water Conference, Mar del Plata, March, 1977.

Food and Agriculture Organization, "The On-farm Use of Water," Commission on Agriculture, Fifth Session, 1979.

Irrigation Journal, 1986 Irrigation Survey, vol. 37, no. 1, 1987, pp. 19–26.

JENSEN, M. E., ed., *Design and Operation of Farm Irrigation Systems*. St. Joseph, Michigan: American Society of Agricultural Engineers, Monograph Number 3, 1980.

STEWART, J. I. AND R. M. HAGAN, "Function to Predict Effects of Crop Water Deficits," *Journal of the Irrigation and Drainage Division,* American Society of Civil Engineers, vol. 99: IR4, 1973, pp. 421–439.

TAYLOR, S. A. AND G. L. ASHCROFT, *Physical Edaphology*. San Francisco, CA: W. H. Freeman and Co., 1972.

VAUX, JR., H. J., W. O. PRUITT, S. A. HATCHETT, AND F. DESOUZA, "Optimization of Water Use with Respect to Crop Production," Technical Completion Report, Agreement No. B-53395, California Department of Water Resources, June, 1981.

WARRICK, A. W. AND D. R. NIELSEN, "Spatial Variability of Soil Physical Properties in the Field," *Applications of Soil Physics,* ed. D. Hillel. New York: Academic Press, 1980.

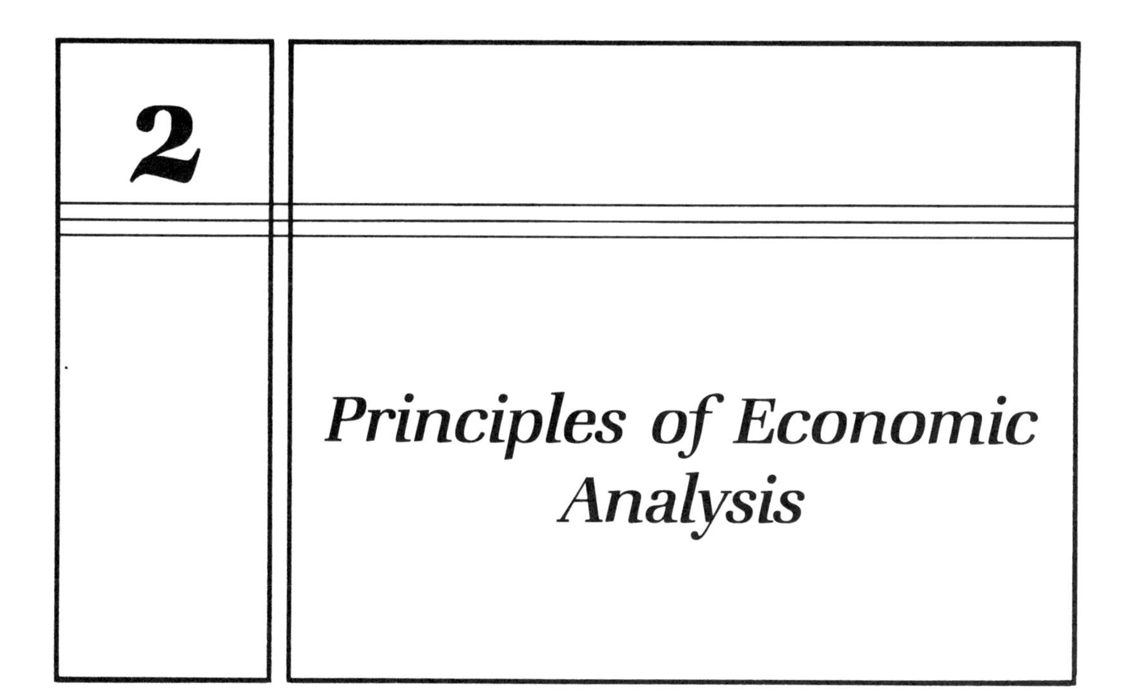

2.1 Economic Optimization

Scale of Analysis

This chapter deals with application of the principles of economic analysis to irrigation system design. In this book, it will be assumed that an irrigation system will be economically viable—that is, economic benefits derived from the system will be greater than the costs. This is the aim of the majority of irrigation systems in the quasi-free market system existing for agriculture in the United States. In fact, worldwide the general aim of irrigated agriculture is to produce benefits which exceed the costs. Governments may promote irrigated agriculture for other than economic reasons, such as less dependence on unstable foreign food and fiber supplies or to produce additional food sources to alleviate chronic hunger. But even under these conditions, the long-term goal is generally that the system be self-sustaining and not require continuous government subsidy.

In applying economic analysis to irrigation system design, at least two levels of analysis are possible. First is the farm scale at which a grower is expected to make decisions affecting his or her own operations. The irrigator's ability to finance capital investment and the cost of operations and management is assumed to be relatively restricted. It is not assumed that the operator can suffer a net economic loss due to an irrigation system for a long time. Within a few years, the system must start operating profitably to enable the operator to satisfy his or her creditors.

The second level of analysis is the regional or national scale. Regional or national governments typically have access to larger financial resources than growers for capital investment, operation, and management of irrigation systems. Governments can also generally afford to endure a longer time before an irrigation system has a net positive economic benefit.

The rules and formulas of economic analysis apply equally to irrigation systems at both the farm and regional or national scale, but the objectives may not be the same. The primary objective of the farm-scale system is to improve the net economic benefits to the farm. The regional- or national-scale system may more often have social objectives which are difficult, if not impossible, to quantify in an economic sense. These social objectives may range from improving the local standard of living to increasing production of crops of strategic importance—for example, certain oil seed crops used for special lubricants. Multiobjective economic analysis at the national or regional scale is covered in other references (Major and Lenton, 1979; Louch et al., 1981). This book will place emphasis on the design and economic analysis of on-farm irrigation systems. The mathematical fundamentals for economic analysis are the same at the various scales and are described in a later section.

Optimization at the Farm Scale

The question of economic evaluation at the farm scale is initially more straightforward than evaluation at the regional or national scale. At the farm scale, the design engineer must consider the total cost of the irrigation system and the total benefits. The costs and benefits may be considered a function of the level of irrigation. The level of irrigation in this case may refer to either the depth of water to be applied during the growing season or the extent of land area to be irrigated.

A typical plot of total costs and total benefits versus level of irrigation is indicated in Fig. 2-1. Looking first at the total cost curve, this curve initially has a steep slope. The steep part of the curve reflects the relatively high capital costs for such items as wells, pumping plants, and mainlines or head ditches for the distribution system. As the area of irrigated land increases, the slope of the total cost curve diminishes to reflect the relatively low cost of application devices such as additional graded borders or sprinkler lateral lines. Since the level of irrigation may also indicate increased depth of application, this decrease in slope of the total cost curve may simply reflect running the pumps for a longer period of time to apply more water.

The decrease in the slope of the total cost curve is commonly associated with economies of scale. Up to a point, some cost advantage can be realized by buying systems with larger capacity or in greater quantity. After this point, the slope of the total cost curve again increases. This may be because once the system reaches a certain capacity, more managers, labor, and equipment must be obtained to operate the system. It may no longer be feasible to operate one pump for a longer period of time—it may be necessary to buy a second pump. This latter portion of the total cost curve is said to reflect diseconomies of scale.

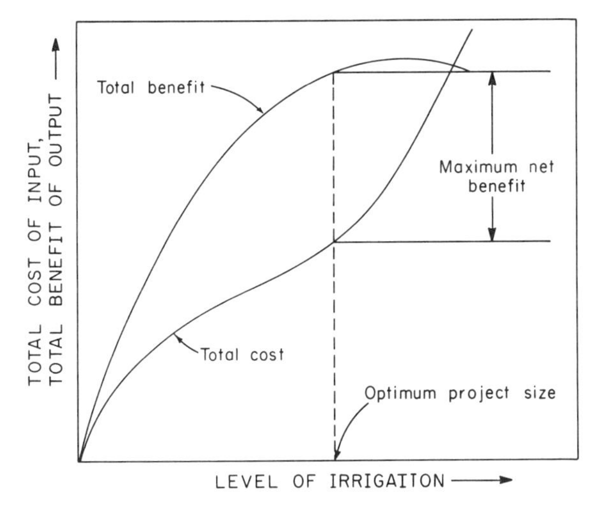

Figure 2-1 Plot of total cost and total benefit versus the level of irrigation.

The other function graphed in Fig. 2-1 is the total benefit curve. For the case of the on-farm irrigation system, the benefit is normally assumed directly related to the income the farmer derives from the marketable yield. The total benefit curve in Fig. 2-1 has been drawn to resemble the crop-water production function in Fig. 1-1. The initial slope is steep to reflect a sharp increase in yield with an incremental increase in depth of water applied. As a point of maximum total benefit is approached, in this case reflecting maximum yield, the slope decreases until it is zero. Beyond this point, the slope of the total benefit curve decreases, reflecting decreased yield due to over-irrigation.

The optimum project size, either in terms of area to be irrigated or depth of water to be applied to a given area, is at the point of maximum net benefit. Net benefit is defined as the difference between the total cost and total benefit curves. The procedure to determine optimum project size is indicated graphically in Fig. 2-2. This figure indicates plots of the marginal benefit and marginal cost curves versus level of irrigation for the same functions indicated in Fig. 2-1. The marginal benefit and marginal cost curves are the first derivatives of the functions in Fig. 2-1. The point at which the marginal curves are equal in Fig. 2-2 indicates the optimum project size. This point corresponds to the point of the maximum net benefit in Fig. 2-1.

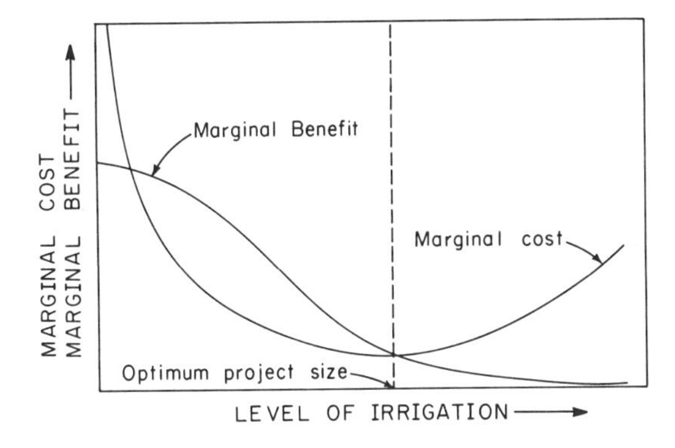

Figure 2-2 Locating optimum project size by plotting the marginal cost and the marginal benefit.

Example Problem 2-1

Total benefit and total cost functions are developed by a grower and design engineer working together. The functions are given as follows:

$$\text{Benefits:} \quad B = 8.2 + 16.4(I) - 0.4(I)^2$$

$$\text{Costs:} \quad C = -5.2 + 19.5(I) - 1.2(I)^2 + 0.03(I)^3$$

for

$$I \geq 3$$

where benefits and costs are given as dollars per hectare irrigated and I is the level of irrigation expressed as thousands of cubic meters of water allocated per hectare from the available water supply.

(a) Determine the optimum irrigation level assuming land and not water supply is the limiting resource.

(b) Compute the maximum net benefit.

Solution

(a) Compute marginal benefit and marginal cost and solve for the level of irrigation at which functions are equal.

$$\frac{dB}{dI} = 16.4 - 0.8(I)$$

$$\frac{dC}{dI} = 19.5 - 2.4(I) + 0.09(I)^2$$

Set

$$\frac{dB}{dI} = \frac{dC}{dI}$$

$$16.4 - 0.8(I) = 19.5 - 2.4(I) + 0.09(I)^2$$

$$0 = 3.1 - 1.6(I) + 0.09(I)^2$$

Solve the preceding equation using the quadratic formula:

$$I = \frac{1.6 \pm \sqrt{(-1.6)^2 - 4(0.09)(3.1)}}{2(0.09)}$$

$$I = 15.6 \times 10^3 \text{ m}^3/\text{ha}$$

which is the only solution for $I \geq 3$.

(b) At

$$I = 15.6 \times 10^3 \text{ m}^3/\text{ha}$$

$$B = 8.2 + 16.4(15.6) - 0.4(15.6)^2$$

$$B = \$167/\text{ha}$$

$$C = -5.2 + 19.5(15.6) - 1.2(15.6)^2 + 0.03(15.6)^3$$

$$C = \$121/\text{ha}$$

$$(\text{Net B})_{max} = B - C$$

$$= \$167/\text{ha} - \$121/\text{ha}$$

$$= \$46/\text{ha}$$

2.2 Mathematics of Economic Analysis

Cash Flow Diagrams

The purpose of economic analysis in the engineering sense is to be able to quantify and compare the costs and benefits of different economically and physically feasible design alternatives. To make this comparison, certain mathematical formulas are required to quantify the consequences of various design alternatives. This mathematical analysis is most conveniently carried out by using a cash flow diagram. The cash flow diagram graphically represents the economics of design alternatives to aid in system evaluation.

A sample of a cash flow diagram for an irrigation system is indicated in Fig. 2-3. In cash flow diagrams, time, usually in annual increments, increases horizontally to the right. Costs are drawn as vertical arrows down from the time line, and benefits as vertical arrows up from the time line. The cash flow diagram in Fig. 2-3 therefore represents the annual costs and benefits of an irrigation project or system.

Initially, large capital expenditures are required for project construction. These expenditures are required for two years in the sample diagram. After initial construction, benefits in terms of revenue from marketable yield begin to be realized and are shown increasing at a uniform rate for a total of five years. This uniform increase may be due to the irrigation system being extended to new lands, increasing production due to improved management, maturing of young orchards, or increased market value due to growing higher value crops which would not be possible without irrigation. During this time of increasing benefits, there is an annual cost of system operation and maintenance.

By the seventh year of the project as shown in Fig. 2-3, it is assumed that annual benefits due to crop production have stabilized. However, in the eighth year a

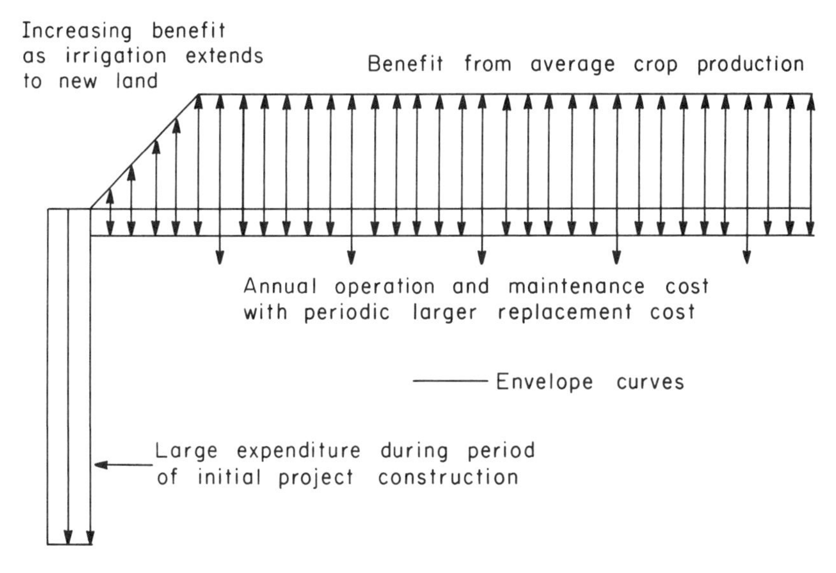

Figure 2-3 Cash flow diagram for hypothetical irrigation project.

replacement cost larger than the annual operation and maintenance cost for the previous five years is indicated. This larger cost may be required to replace worn equipment. The larger replacement cost is indicated every sixth year for the remainder of the project. The total period over which the analysis is made is termed the economic life of the project.

The question to be considered is: How can different design alternatives, all represented by their own cash flow diagrams, be compared so the one which most clearly meets project objectives can be selected? It is normally impossible to make this selection by visual inspection of the alternative cash flow diagrams. At this point, different economic factors must be applied to bring the alternatives to a comparable level. The remainder of this section discusses application of these economic factors.

Single Payment Factors

Single payment factors are used to compute the future worth of a present amount or the present amount required to have a given future value when invested at interest rate i for a period of N years. The present amount will be indicated by P and the future amount by F as shown in Fig. 2-4. There are two types of single payment factors.

Single payment compound-amount factor

Referring to Fig. 2-4, this factor is used to compute the future value (F) of a single payment made in the present (P) invested for N years at interest rate i. The objective is to calculate what is noted as the F/P factor. When this factor is multiplied by the present payment, P, the result is the future value F. The notation shown in Eq. (2-1) for the factor, interest rate, and number of years is standard and will be used throughout this book. The single payment compound-amount factor is given by

$$\left[\frac{F}{P}, i, N\right] = (1 + i)^N = \frac{F}{P} \tag{2-1}$$

where

$$i = \text{interest rate, fraction}$$
$$N = \text{number of years}$$

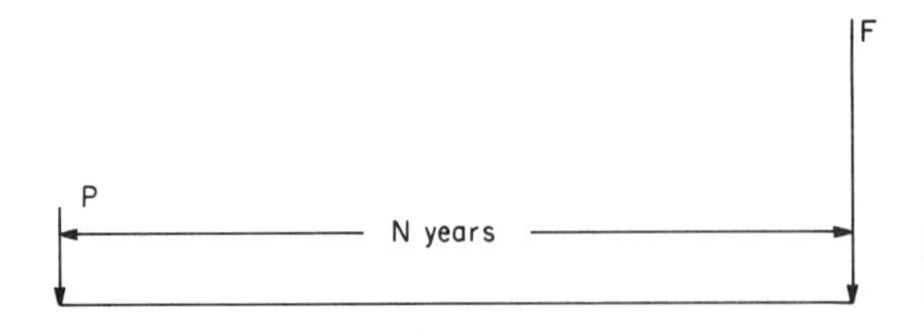

Figure 2-4 Single payment compound-amount and present-worth factors.

Single payment present-worth factor

This factor indicates the present amount (P) which must be invested at i percent interest to have future amount (F) after N years. Again, referring to Fig. 2-4, the single payment present worth factor is given by

$$\left[\frac{P}{F}, i, N\right] = \frac{1}{(1 + i)^N} = \frac{P}{F} \tag{2-2}$$

Uniform Annual Series Factors

Annual series factors are used to relate present or future costs or benefits to their equivalent value on an annual basis taking into account the expected interest rate and number of years involved in the economic analysis. Four types of uniform annual series factors will be demonstrated with respect to Fig. 2-5.

Sinking-fund factor

The sinking fund factor indicates the amount of annual investment (A) necessary to accumulate a future amount (F) when invested at i percent interest over a period of N years. The notation for this factor referring to Fig. 2-5 is given by

$$\left[\frac{A}{F}, i, N\right] = \frac{i}{(1 + i)^N - 1} = \frac{A}{F} \tag{2-3}$$

Capital-recovery factor

This factor indicates the annual amount (A) available for N years when present amount (P) is invested at i percent interest. It can also be thought of as indicating the equivalent present cost (P) of equal annual expenditures (A) over N years at i percent interest. Note that the capital-recovery factor can be determined by

$$\frac{A}{P} = \frac{A}{F}\frac{F}{P} \tag{2-4}$$

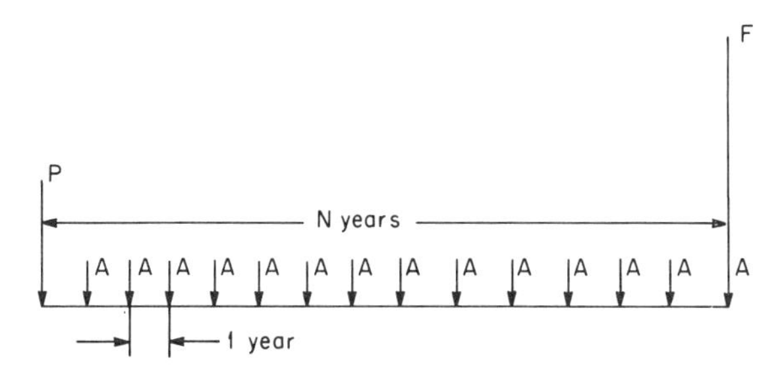

Figure 2-5 Uniform series sinking-fund, compound-amount, capital-recovery, and present-worth factors.

The ratio of A to P is therefore determined by combining Eqs. (2-1) and (2-3).

$$\left[\frac{A}{P}, i, N\right] = \frac{i(1 + i)^N}{(1 + i)^N - 1} = \frac{A}{P} \tag{2-5}$$

Series compound-amount factor

Referring to Fig. 2-5, this factor indicates the future amount (F) available after N years if annual investments (A) are made at i percent interest. The factor is given by

$$\left[\frac{F}{A}, i, N\right] = \frac{(1 + i)^N - 1}{i} = \frac{F}{A} \tag{2-6}$$

Equation (2-6) is the inverse of Eq. (2-3).

Series present-worth factor

This factor is used to compute the present amount (P) which must be invested at i percent interest to withdraw an annual amount (A) for N years. The series present-worth factor is specified as

$$\left[\frac{P}{A}, i, N\right] = \frac{(1 + i)^N - 1}{i(1 + i)^N} = \frac{P}{A} \tag{2-7}$$

As with the capital recovery factor, the series present-worth factor may be determined by combining Eqs. (2-2) and (2-6).

Uniform Gradient Series Factors

Uniform gradient series factors are applied to cash flows which are annual but are not equal. Cash flows requiring this type of analysis increase or decrease by a constant amount each year. Gradient series cash flows are demonstrated by Fig. 2-6 where G is the increasing annual increment or gradient.

Uniform gradient series compound-amount factor

The uniform gradient series compound-amount factor indicates the gradient amount (G) which must be invested annually at i percent interest to have accumulated amount F available N years in the future. This factor will be derived with reference to Fig. 2-6.

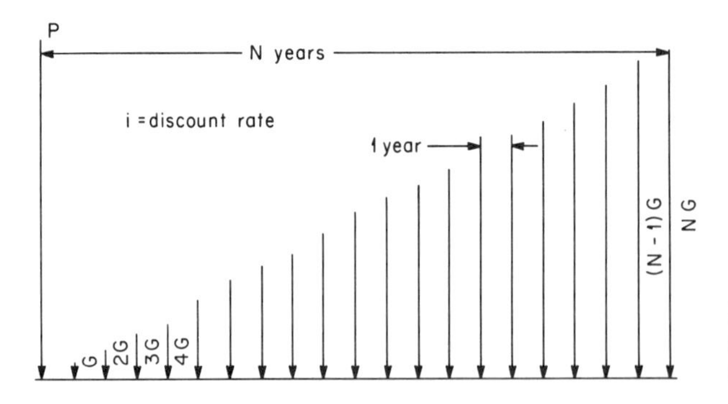

Figure 2-6 Representation of a gradient series. Case shown demonstrates gradient series present-worth factor.

Applying the single payment compound-amount factor individually to each annual investment and summing to obtain the accumulated amount after the last deposit, the following equation is arrived at (James and Lee, 1971):

$$F = G[N + (N - 1)(1 + i) + \ldots + 2(1 + i)^{N-2} + (1 + i)^{N-1}] \qquad (2\text{-}8)$$

Multiplying both sides of Eq. (2-8) by $1 + i$ gives

$$(1 + i)F = G[N(1 + i) + (N - 1)(1 + i)^2 + \ldots + 2(1 + i)^{N-1} + (1 + i)^N] \qquad (2\text{-}9)$$

Subtracting Eq. (2-8) from Eq. (2-9) term by term

$$iF = G[-N + (1 + i) + \ldots + (1 + i)^{N-1} + (1 + i)^N] \qquad (2\text{-}10)$$

Again multiplying both sides of Eq. (2-10) by $1 + i$ yields

$$(1 + i)iF = G[-N(1 + i) + (1 + i)^2 + \ldots + (1 + i)^N + (1 + i)^{N+1}] \qquad (2\text{-}11)$$

Subtracting Eq. (2-10) from Eq. (2-11) term by term gives this simplified equation:

$$i^2F = G[N - N(1 + i) - (1 + i) + (1 + i)^{N+1}] \qquad (2\text{-}12)$$

Upon rearranging, Eq. 2-12 yields the uniform gradient series compound amount factor as

$$\left[\frac{F}{G}, i, N\right] = \frac{(1 + i)^{N+1} - (1 + Ni + i)}{i^2} = \frac{F}{G} \qquad (2\text{-}13)$$

Uniform gradient series present-worth factor

This factor indicates the present amount (P) one must invest at i percent interest to withdraw the annual gradient amount G. It is determined by applying Eq. (2-2) to Eq. (2-13), such that

$$\frac{P}{G} = \frac{P}{F}\frac{F}{G} \qquad (2\text{-}14)$$

$$\left[\frac{P}{G}, i, N\right] = \frac{(1 + i)^{N+1} - (1 + Ni + i)}{i^2(1 + i)^N} = \frac{P}{G} \qquad (2\text{-}15)$$

Equivalent annual series

Conversion of gradient series to equivalent annual series is accomplished by combining Eq. (2-15) and (2-5). This is demonstrated by

$$\frac{A}{G} = \frac{A}{P}\frac{P}{G} \qquad (2\text{-}16)$$

$$\left[\frac{A}{G}, i, N\right] = \frac{i(1 + i)^N}{(1 + i)^N - 1}\,\frac{(1 + i)^{N+1} - (1 + Ni + i)}{i^2(1 + i)^N} = \frac{A}{G} \qquad (2\text{-}17)$$

Equation (2-17) is not conveniently simplified. At this point, the benefits of computerizing the economic analysis formulas can be appreciated. The Computer Application section of this chapter indicates a program which may be used for such analysis.

Uniformly decreasing gradient series

The gradient series discussed thus far has been uniformly increasing. Uniformly decreasing gradient series are managed by subtracting a uniformly increasing series from a uniform annual series. A cash flow diagram for this condition is indicated in Fig. 2-7. In the case illustrated, an increasing uniform gradient series indicated as series a exists for the first 10 years. A decreasing uniform gradient series exists for the last 10 years of analysis. This decreasing gradient series is simulated by subtracting a uniform gradient, indicated as series c, from the uniform annual series indicated by b. The result is the triangular cash flow diagram bounded by the envelope curve with all cash values indicated as positive.

Example Problems

The following example problems have been developed in support of this section on mathematics of economic analysis. The problems demonstrate application of the various factors defined and derived in this section. The drawing of cash flow diagrams is recommended to aid in understanding the solutions, especially for the more complex problems.

Example Problem 2-2

A 20 hp pump which costs $1600 has an economic life of 15 years and is assumed to have no salvage value. What is the equivalent annual cost for the pump if interest rates are 14 percent?

Solution

$$\left[\frac{A}{P}, 14\%, 15 \text{ yrs}\right] = \frac{0.14(1 + 0.14)^{15}}{(1 + 0.14)^{15} - 1}$$

$$\frac{A}{P} = 0.1628$$

$$A = 0.1628(P) = 0.1628(\$1600) = \$260.49/\text{year}$$

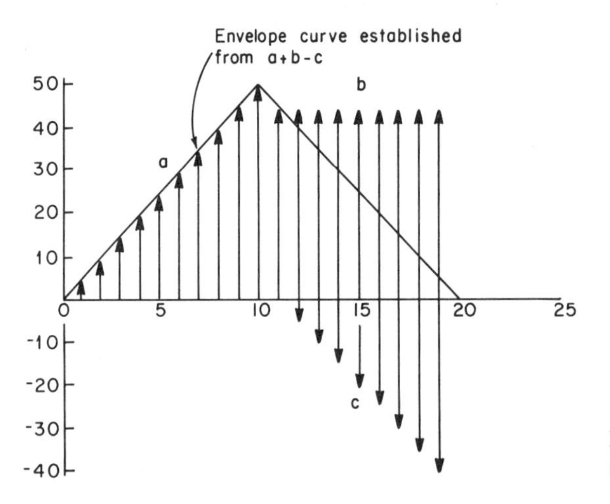

Figure 2-7 Cash flow diagram demonstrating decreasing gradient series.

Example Problem 2-3

The pump in example problem 2-2 is assumed to have a salvage value of 30 percent of the purchase price at the end of 15 years. What is the annual cost for this condition?

Solution

$$F = 0.30(\$1600) = \$480$$

$$A = \left[\frac{A}{P}, 14\%, 15 \text{ yrs}\right]\$1600 - \left[\frac{A}{F}, 14\%, 15 \text{ yrs}\right]\$480$$

$$A = 0.1628(\$1600) - \frac{0.14}{(1 + 0.14)^{15} - 1}(\$480)$$

$$A = \$260.49 - 0.0228(\$480) = \$249.54/\text{year}$$

Example Problem 2-4

\$3000 worth of 5-inch (127 mm) aluminum mainline is purchased along with the pump in example problem 2-3. If the mainline is assumed to have an economic life of 30 years and a salvage value of 20 percent of the purchase price, what is the annual cost of this system at 14 percent interest for the first 15 years?

Solution For mainline,

$$F = 0.20(\$3000) = \$600$$

$$A = \left[\frac{A}{P}, 14\%, 30 \text{ yrs}\right]\$3000 - \left[\frac{A}{F}, 14\%, 30 \text{ yrs}\right]\$600$$

$$A = 0.1428(\$3000) - 0.0028(\$600)$$

$$A = \$428.41 - \$1.68 = \$426.73/\text{year}$$

For the total system,

$$A = \$249.54 + \$426.73 = \$676.27/\text{year}$$

Example Problem 2-5

For the system in example problem 2-4 to be operational, a replacement pump must be purchased after 15 years. The replacement pump has a \$1900 purchase price in 15 years and a salvage value of 30 percent after 15 years' usage. What is the new annual cost for this system for the full 30 years?

Solution For analysis over a full 30 years, determine the present worth of the purchase price and salvage value for both pumps. Then use the capital-recovery factor to calculate the annual equivalent of the net present worth over 30 years.

Net present-worth cost of the first pump:

$$P_1 = \$1600 - \left[\frac{P}{F}, 14\%, 15 \text{ yrs}\right][0.30(\$1600)]$$

$$\left[\frac{P}{F}, 14\%, 15 \text{ yrs}\right] = \frac{1}{(1 + 0.14)^{15}} = 0.1401$$

$$P_1 = \$1600 - 0.1401(\$480) = \$1532.75$$

Net present-worth cost of the second pump:

$$F = 0.30(\$1900) = \$570$$

$$P_2 = \$1900\left[\frac{P}{F}, 14\%, 15 \text{ yrs}\right] - \$570\left[\frac{P}{F}, 14\%, 30 \text{ yrs}\right]$$

$$\left[\frac{P}{F}, 14\%, 30 \text{ yrs}\right] = \frac{1}{(1 + 0.14)^{30}} = 0.0196$$

$$P_2 = \$1900(0.1401) - \$570(0.0196) = \$255.02$$

Total net present-worth cost for pumps:

$$P_T = P_1 + P_2 = \$1532.72 + \$255.02 = \$1787.74$$

Equivalent annual cost of pumps over 30 years:

$$A = \$1787.74\left[\frac{A}{P}, 14\%, 30 \text{ yrs}\right]$$

$$\left[\frac{A}{P}, 14\%, 30 \text{ yrs}\right] = \frac{0.14(1 + 0.14)^{30}}{(1 + 0.14)^{30} - 1} = 0.1428$$

$$A = \$1787.74(0.1428) = \$255.29/\text{year}$$

Total annual costs for pumps and mainline over 30 years (using results from example problem 2-4):

$$A_T = \$255.29 + \$426.73 = \$682.02/\text{year}$$

2.3 Predicting Yield Response

Crop-Water Production Functions

It was observed in the first chapter that different amounts of irrigation water have different benefits in terms of crop production. This was shown in Fig. 1-1 in which the change in yield at one point on the production function, ΔY_1, was greater than at another point on the function, ΔY_2, for the same incremental increase in applied water, ΔAW. There is a cost associated with application of additional increments of irrigation water to a field. The normal objective is to insure that the benefit associated with the yield increase is greater than the cost.

Crop-water production functions of the type shown in Figs. 1-1 and 1-2 are useful tools to predict yield response to irrigation. These functions are normally developed in field experiments in which different amounts of irrigation are applied to a particular crop throughout the growing season. At the end of the season, account is made of the applied water, evapotranspiration, and resulting yield for each irrigation level. As previously indicated, the yield-evapotranspiration relationship is felt to be a crop characteristic and does not reflect the impact of the type of irrigation system as much as the applied water relationship. An experimental field used to develop a crop-water production function is depicted in Fig. 2-8.

There are difficulties in applying production functions which will be indicated in the following section. For the present, let us look at some typical crop-water production functions and their application in irrigation system design. Figures 2-9 through

Figure 2-8 Crop growth response to line source system used to evaluate yield versus applied water and evapotranspiration functions for cotton in San Juaquin Valley, California. Sprinkler line runs down right-center of photograph with diminishing irrigation levels and crop production shown to either side of line.

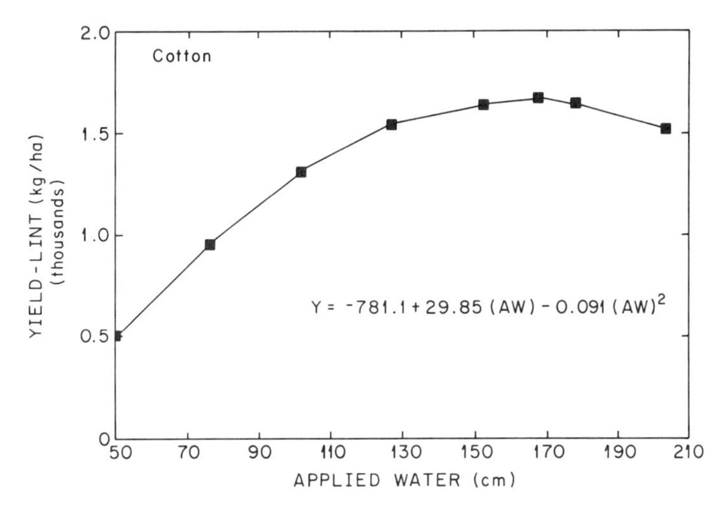

Figure 2-9 Yield versus applied water production function for cotton.

2-11 indicate example crop-water production functions expressed as a function of applied water for three crops. The shape of the production functions is typical, like that discussed for the function in Figs. 1-1. The main features to be noted are the maximum yield and decreasing slope of yield versus applied water as the maximum point is approached. The form of the regression equations given in the figure is the most direct in that yield is related to seasonal levels of applied water. Other forms of the

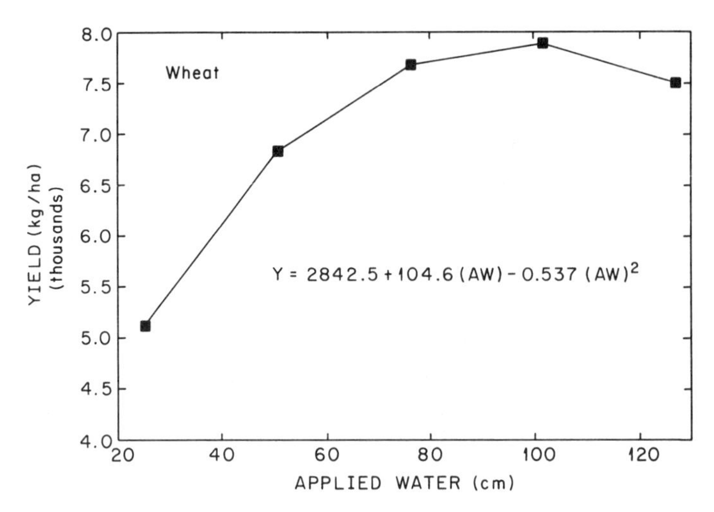

Figure 2-10 Yield versus applied water production function for wheat.

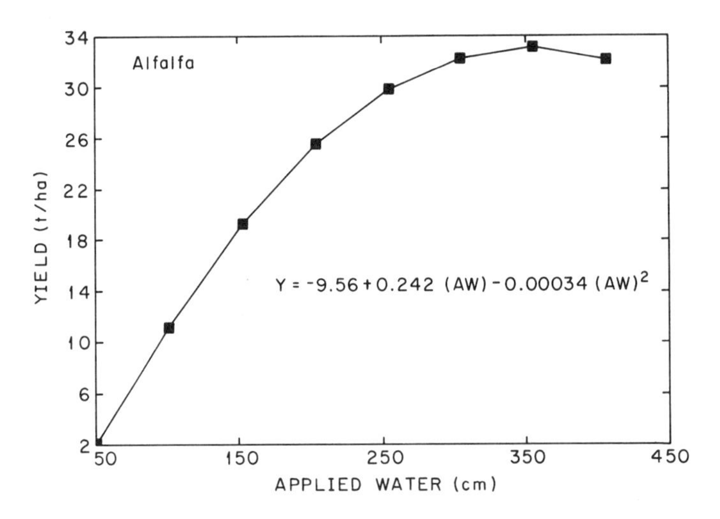

Figure 2-11 Yield versus applied water production function for alfalfa.

functions which account separately for levels of applied water or evapotranspiration for different growth periods have also been developed (Stewart et al., 1977). This book uses the simpler seasonal form of the functions due in part to the practical difficulty of obtaining reliable regression constants for different growth periods.

Difficulties in applying the functions are discussed in the next section. In spite of these difficulties, the functions can be applied as useful tools. This will be demonstrated through an example problem.

Example Problem 2-6

A grower has the opportunity to plant any of the three crops indicated in Figs. 2-9 through 2-11. The type of irrigation system and irrigation frequency the grower will use is the same as that used in the field experiment to develop the production functions. Assume other conditions such as climate, soil type, and nutrient level are also equivalent.

The grower has a limited water allocation for the growing season equivalent to 100 cm depth of water over the land to be irrigated. The seasonal water supply will be evenly distributed over the growing season corresponding to the method used to derive the production functions.

(a) Assume the price of water application and crop management is equal for all three crops. Which crop would you recommend to the grower to maximize economic return using the full seasonal water supply given the following selling prices for marketable yield:

Market Price

$$\text{Cotton:}\quad P_c = \$1.21/\text{kg}$$
$$\text{Wheat:}\quad P_w = \$0.13/\text{kg}$$
$$\text{Alfalfa:}\quad P_a = \$83/\text{t}$$

(where t is the abbreviation for metric ton equal to 1000 kg)

(b) What price would the second-ranked crop have to attain to surpass the value of the first-ranked crop given the same seasonal water supply?

(c) Given the indicated variable cost of crop production as a function of yield, which crop would have the highest overall economic return using the same seasonal water supply:

Variable Cost of Production per Hectare

$$\text{Cotton:}\quad C_c = \$620 + [\$0.33/(\text{kg/ha})][Y(\text{kg/ha})]$$
$$\text{Wheat:}\quad C_w = \$125 + [\$0.04/(\text{kg/ha})][Y(\text{kg/ha})]$$
$$\text{Alfalfa:}\quad C_a = \$185 + [\$22/(\text{t/ha})][Y(\text{t/ha})]$$

Solution

(a) Using full seasonal water supply of 100 cm, the following yields are determined from Figs. 2-9 through 2-11:

$$\text{Cotton:}\quad Y_c = 1300 \text{ kg/ha}$$
$$\text{Wheat:}\quad Y_w = 7900 \text{ kg/ha}$$
$$\text{Alfalfa:}\quad Y_a = 11.2 \text{ t/ha}$$

Compute return R for each crop:

$$R_i = P_i Y_i$$
$$\text{Cotton:}\quad R_c = \$1.21/\text{kg}(1300 \text{ kg/ha}) = \$1573/\text{ha}$$
$$\text{Wheat:}\quad R_w = \$0.13/\text{kg}(7900 \text{ kg/ha}) = \$1027/\text{ha}$$
$$\text{Alfalfa:}\quad R_a = \$83/\text{t}(11.2 \text{ t/ha}) = \$930/\text{ha}$$

All other factors being equal, the highest return at full seasonal water supply is for cotton.

(b) Compute the required price of wheat to exceed the return for cotton:

$$P_w \geq \frac{\$1573/\text{ha}}{\$1027/\text{ha}}(\$0.13/\text{kg})$$

$$P_w \geq \$0.20/\text{kg}$$

(c) Compute net return, NR_i, as R_i minus cost for yield levels indicated:

$$NR_i = R_i - C_i$$

Cotton: $NR_c = \$1573/\text{ha} - \{\$620 + [\$0.33/(\text{kg/ha})](1300\ \text{kg/ha})\}/\text{ha}$

$NR_c = \$524/\text{ha}$

Wheat: $NR_w = \$1027/\text{ha} - \{\$125 + [\$0.04/(\text{kg/ha})](7900\ \text{kg/ha})\}/\text{ha}$

$NR_w = \$586/\text{ha}$

Alfalfa: $NR_a = \$930/\text{ha} - \{\$185 + [\$22/(\text{t/ha})](11.2\ \text{t/ha})\}/\text{ha}$

$NR_a = \$499/\text{ha}$

Based on variable production costs and using full seasonal water supply, highest net return is for wheat.

An alternative and useful form of expressing the production function is in terms of the yield reduction ratio, YRR. This ratio is determined by applying the concept of yield deficit and evapotranspiration deficit defined by the following equations:

$$\text{ETDEF} = \text{ETMAX} - \text{ETACT} \tag{2-18}$$

where

ETDEF = evapotranspiration deficit, cm

ETMAX = maximum possible crop evapotranspiration, cm

ETACT = actual evapotranspiration, cm

and

$$\text{YLDEF} = \text{YLDMAX} - \text{YLDACT} \tag{2-19}$$

where the terms are analagous to the definitions given for evapotranspiration in Eq. (2-18). The yield and evapotranspiration deficits are normalized between zero and 1.0 by expressing them as ratios of maximum yield and maximum evapotranspiration. The yield reduction ratio is given by

$$\text{YRR} = \frac{\text{YLDEF}/\text{YLDMAX}}{\text{ETDEF}/\text{ETMAX}} \tag{2-20}$$

Figure 2-12 indicates the graphical presentation of the yield reduction ratios for two bean varieties and grain sorghum. It can be noted in the figure that the yield and evapotranspiration deficits go from zero to 1.0, with the coordinate pair (0, 0) representing conditions of maximum yield and maximum evapotranspiration. This presentation of the yield reduction ratio is a convenient means to visualize drought tolerance. For example, at any level of evapotranspiration deficit, the pink bean va-

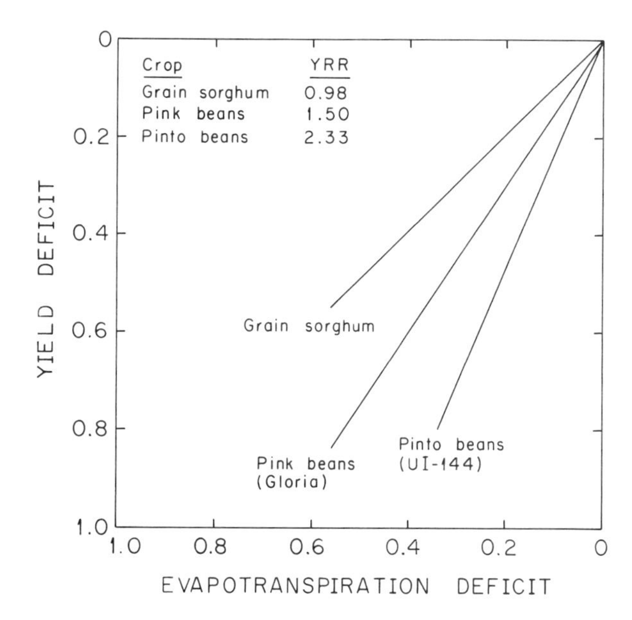

Figure 2-12 Yield reduction ratios for three sample crops. Yield and evapotranspiration deficits have been normalized by dividing by YLDMAX and ETMAX, respectively.

riety is seen to experience less yield reduction than the pinto bean. Therefore, if all other conditions are equal, the selection of pink beans would be recommended over pinto beans if water supplies inadequate to meet the crop water requirement are anticipated. Grain sorghum is clearly more drought resistant than either bean variety. However, the food value of grain sorghum and acceptability for local and export markets must also be considered before crop selection.

Scope of Problem

The problem facing engineers and others responsible for decisions regarding the allocation of water resources is how to correctly specify the irrigation water–crop production relationship. It is not difficult to imagine that many factors influence the shape of the crop-water production function. The following is a list of factors which most directly affect the shape of the production function for a given crop:

(a) Frequency of irrigation
(b) Timing of irrigation with respect to crop development
(c) Soil texture
(d) Soil fertility
(e) Water quality
(f) Type of irrigation system
(g) Uniformity of water application
(h) Plant density
(i) Climate

(j) Pest management
(k) Planting date
(l) Soil drainage

Assuming the influence of no more than the 12 factors listed, there are 2^{12} or 4096 permutations of the crop-water production function for a single crop. At this point the problem appears intractable.

Fortunately there are some approaches which make the problem more reasonable. The first is to eliminate those factors which can be assumed to have the least impact on the shape of the crop-water production functions. For example, most field experiments used to determine crop-water production functions are conducted using the same field conditions recommended by seed distributors and extension agents. As long as these typical conditions are to be maintained in a growers field, factors such as plant density and pest management can be assumed equivalent.

Some of the most serious questions in application of crop-water production functions are related to frequency and timing of irrigation. At this point it is helpful to look at results of field experiments used to develop crop-water production functions. Figure 2-13 indicates the yield response of maize to applied water in an experiment in which the water application was constrained during different growth periods. The season was divided into the major growth periods of vegetative, flowering (or tasseling in the case of maize), and maturation. A growth period in which irrigation was applied is designated by an I, and a period with no irrigation is designated with O. Thus treatment IOI indicates irrigation was applied only in the vegetative and maturation growth periods. Figure 2-13 demonstrates that the irrigation treatment with the highest yield response was that in which the irrigation deficit was distributed evenly over the growing season, treatment III. This indicates that if the maximum level required for irrigation cannot be delivered, the highest yield results for this crop will be attained by spreading the irrigation deficit evenly over the growing season.

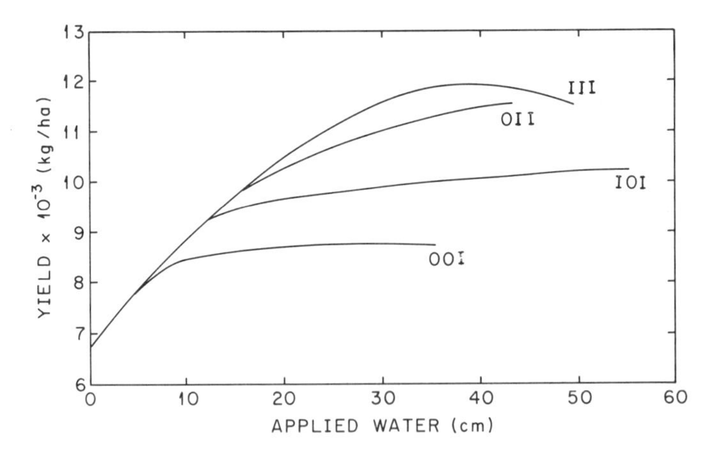

Figure 2-13 Examples of crop-water production functions for stressing of maize during different growth periods. (I = irrigated; 0 = unirrigated)

There are many crops for which a particular growth period may be more critical than others, particularly the flowering period. However a straightforward, logical explanation is available for the results shown in Fig. 2-13. If applied water is restricted from a plant during the vegetative period, smaller plants result with smaller total leaf area. These plants also tend to have a more developed root zone than fully irrigated plants because they have attempted to search as much of the soil profile as possible for moisture. As the flowering period of the season, critical for marketable yield, is reached, plants in treatment III have already become accustomed to reduced water amounts. The physical characteristics of smaller total leaf area and highly developed root zone can be called drought hardening. During the flowering period, the plants continue to be deprived of the full water requirement, but are able to resist this condition better than plants which were fully irrigated during the vegetative period. The plant is therefore under less stress and able to produce a higher final yield than the plant which was not drought hardened. The conclusion is that if an irrigation deficit is unavoidable for a crop which responds as that shown in Fig. 2-13, it is best to spread this deficit evenly over the growing season and let the natural reaction mechanisms of the plant aid in resisting the effects of reduced water amounts.

Example Problem 2-7

Projections indicate that a grower will have available a seasonal water supply of 25 cm to apply to a maize crop which responds as that shown in Fig. 2-13. Given a maximum yield of 11.9×10^3 kg/ha, compute the yield deficit as percent of maximum if

(a) the total irrigation deficit is applied during the tassling (flowering) period; and

(b) the deficit is distributed evenly over the total season.

Solution

(a) Referring to Fig. 2-13, treatment IOI, with applied water equal to 25 cm,

$$\text{YLDACT} = 9.8 \times 10^3 \text{ kg/ha}$$

$$\text{YLDEF} = \frac{\text{YLDMAX} - \text{YLDACT}}{\text{YLDMAX}}(100) = \frac{11.9 \times 10^3 - 9.8 \times 10^3}{11.9 \times 10^3}(100)$$

$$\text{YLDEF} = 17.6 \text{ percent}$$

(b) For treatment III with 25 cm applied water,

$$\text{YLDACT} = 11.2 \times 10^3 \text{ kg/ha}$$

$$\text{YLDEF} = \frac{11.9 \times 10^3 - 11.2 \times 10^3}{11.9 \times 10^3}(100) = 5.9 \text{ percent}$$

A second related problem of non-optimal timing of irrigation may be used to describe much of the scatter in crop-water production function results. This situation is indicated in Fig. 2-14 in which measured seasonal evapotranspiration amounts are plotted versus yield. There appears to be a maximum envelope to the data through which a line has been drawn. This envelope represents the optimum yield response to evapotranspiration produced by having the optimal timing of water application. Any points falling below the envelope curve are produced by timing of water appli-

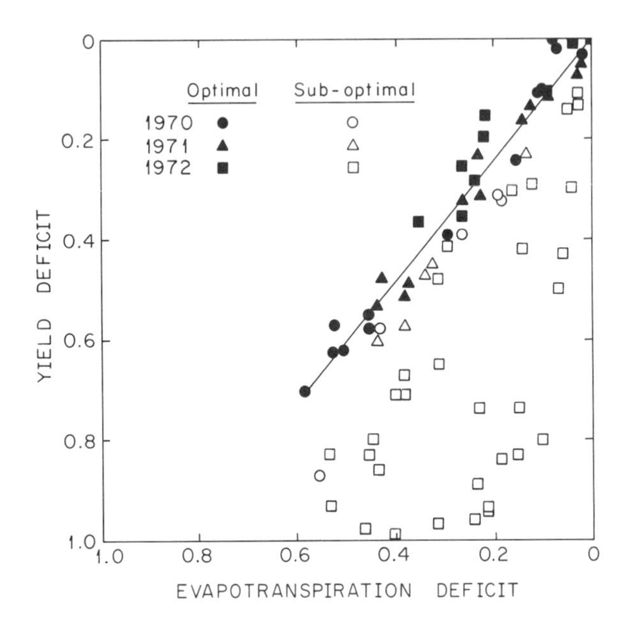

Figure 2-14 Plot of yield deficit versus evapotranspiration deficit indicating effects of sub-optimal timing of irrigation and upper envelope curve.

cation being less than optimal. Figure 2-13 indicates that for many crops this optimal timing is going to be realized by distributing the evapotranspiration deficit evenly over the growing season. The crop-water production function results often reported are those which exhibit optimal timing. Field results falling below those amounts may in part be described by non-optimal timing of water application.

The complete topic of application of crop-water production functions is beyond the scope of this book. It is a subject which has been studied by a Task Committee of the American Society of Civil Engineers (Cuenca, 1987). There are many factors affecting yield in the list previously presented which are difficult to quantify for a particular location but which may be approached using local information. For example, factors such as climate, soil texture, and water quality have significant impact on the maximum attainable yield. Yet these factors are often identical or similar within a given agricultural area. Local records for maximum crop growth may therefore be very useful in evaluating application of water production functions. This type of approach is demonstrated in the example problem.

Example Problem 2-8

The maximum measured yield for a particular variety of tomato at one location is 19.1 metric tons per hectare. This maximum yield corresponds to a maximum seasonal crop ET of 77.5 cm. The yield reduction ratio is equal to 1.27 for this tomato variety. Analysis of 10 years of production data at a different location reveals a local maximum yield of 13.8 t/ha for the same variety tomato. This lower maximum yield is felt to be due to a cooler growing season with less solar radiation—that is, climatic effects.

(a) Determine an estimate of the expected yield at the new location if a 15 percent ET deficit below the local maximum ET is imposed on the crop by limiting irrigation throughout the growing season.

(b) Compute an estimate of the seasonal ET with this level of deficit.

Solution

(a) Assume the model of a linear yield reduction ratio shown in Fig. 2-12 is applicable. Apply Eq. (2-20) using subscript 1 for the original site and subscript 2 for the new location.

$$\text{YRR} = \frac{\text{YLDEF/YLDMAX}}{\text{ETDEF/ETMAX}}$$

For the new location,

$$\begin{aligned}\text{ETDEF}_2/\text{ETMAX}_2 &= 0.15\\ \text{YLDEF}_2/\text{YLDMAX}_2 &= \text{YRR}(\text{ETDEF}_2/\text{ETMAX}_2)\\ &= 1.27(0.15)\\ &= 0.19\end{aligned}$$

Compute the estimated yield for this deficit:

$$\begin{aligned}\text{YLDEF}_2 &= 0.19(\text{YLDMAX}_2)\\ \text{YLDACT}_2 &= 13.8\ \text{t/ha} - 2.6\ \text{t/ha}\\ &= 11.2\ \text{t/ha}\end{aligned}$$

(b) Use the estimated yield at the new location to compute yield deficit based on maximum varietal yield:

$$\begin{aligned}\text{YLDEF} &= 19.1\ \text{t/ha} - 11.2\ \text{t/ha}\\ &= 7.9\ \text{t/ha}\end{aligned}$$

Compute deficit and actual ET based on maximum varietal ET:

$$\begin{aligned}\text{ETDEF/ETMAX} &= \frac{(7.9\ \text{t/ha})/(19.1\ \text{t/ha})}{\text{YRR}}\\ \text{ETDEF} &= \frac{0.41}{1.27}(77.5\ \text{cm})\\ &= 25.2\ \text{cm}\\ \text{ETACT}_2 &= 77.5\ \text{cm} - 25.2\ \text{cm}\\ &= 52.3\ \text{cm}\end{aligned}$$

2.4 Computer Program

The sample program which follows indicates application of single payment compound-amount and present-worth analysis. Specific commands in BASIC can be determined from any BASIC language manual. However, the following program demonstrates these important programming features:

(a) Liberal use of REM (remark) statements to divide the program and clearly indicate operations of different program sections.

(b) Use of variable names which are similar to the meaning of the variable to avoid confusion; for example, NYEAR for Number of Years and OPTID for Option Identification.
(c) Division of the program into convenient subroutines with specific tasks. This is especially helpful when editing large, complex programs in which an error can be tracked to a specific subroutine.
(d) Care has been taken to produce output which is complete and conveniently visualized by the user.

It is important to recognize that in an engineer-client relationship, the output of a program may be an important factor by which the client judges the competence of an engineering firm. An output table which is unclear, unlabeled and decipherable only by the person who wrote it is useless to anyone else and makes a very poor impression. In writing a program, the effort necessary to produce clear and thorough output is required only once, yet it may make a good impression on clients many times over.

2.5 Computer Problems

1. Modify the MAIN PROGRAM of the sample program as required and add subroutines to conduct annual series sinking-fund and capital recovery analysis.
2. Modify the MAIN PROGRAM as required and add subroutines to conduct annual series compound-amount and present-worth analysis.
3. Modify the MAIN PROGRAM as required and add subroutines to conduct gradient series compound-amount and present-worth analysis.

```
10 REM ########################################################################
20 REM #####                                                              ####
30 REM #####             ECONOMIC ANALYSIS - MAIN PROGRAM                 ####
40 REM #####                                                              ####
50 REM ########################################################################
60 REM
70 CLS
80 PRINT
90 PRINT
100 PRINT "SELECT ECONOMIC ANALYSIS OPTION: "
110 PRINT
120 PRINT "     1 = SINGLE PAYMENT COMPOUND-AMOUNT FACTOR (F/P)"
130 PRINT
140 PRINT "     2 = SINGLE PAYMENT PRESENT-WORTH FACTOR (P/F)"
150 PRINT
160 INPUT "ENTER NUMBER OF OPTION REQUIRED = ", OPTID
170 PRINT
180 INPUT "TIME HORIZON OF ANALYSIS (YEARS) = ", NYEAR
190 PRINT
200 INPUT "INTEREST RATE (PERCENT) = ", INTER
210 IF OPTID = 1 THEN GOSUB 1000
220 IF OPTID = 2 THEN GOSUB 2000
230 PRINT
240 INPUT "DO YOU WANT TO PERFORM ADDITIONAL ANALYSIS (Y/N) = ", ANAL$
250 IF ANAL$ = "Y" OR ANAL$ = "y" GOTO 70
260 END
1000 REM
```

```
1010 REM ######################################################################
1020 REM #####                                                             ####
1030 REM #####      SINGLE PAYMENT COMPOUND-AMOUNT FACTOR SUBROUTINE       ####
1040 REM #####                                                             ####
1050 REM ######################################################################
1060 PRINT
1070 INPUT "INPUT PRESENT PAYMENT (P) = ", P
1080 FACTOR = (1! + INTER/100!)^NYEAR
1090 F = FACTOR * P
1100 CLS
1110 PRINT
1120 PRINT "SINGLE PAYMENT COMPOUND-AMOUNT ANALYSIS"
1130 PRINT
1140 PRINT USING "     N = ### YEARS"; NYEAR
1150 PRINT
1160 PRINT USING "     I = ###.## PERCENT"; INTER
1170 PRINT
1180 PRINT USING "     P = ######"; P
1190 PRINT
1200 PRINT USING "     F/P FACTOR = ######.######"; FACTOR
1210 PRINT
1220 PRINT USING "     FUTURE VALUE (F) = ##########"; F
1230 RETURN
2000 REM
2010 REM ######################################################################
2020 REM #####                                                             ####
2030 REM #####      SINGLE PAYMENT PRESENT-WORTH FACTOR SUBROUTINE         ####
2040 REM #####                                                             ####
2050 REM ######################################################################
2060 PRINT
2070 INPUT "INPUT FUTURE AMOUNT (F) = ", F
2080 FACTOR = 1!/(1! + INTER/100!)^NYEAR
2090 P = FACTOR * F
2100 CLS
2110 PRINT
2120 PRINT "SINGLE PAYMENT PRESENT-WORTH ANALYSIS"
2130 PRINT
2140 PRINT USING "     N = ### YEARS"; NYEAR
2150 PRINT
2160 PRINT USING "     I = ###.## PERCENT"; INTER
2170 PRINT
2180 PRINT USING "     F = ######"; F
2190 PRINT
2200 PRINT USING "     P/F FACTOR = ######.######"; FACTOR
2210 PRINT
2220 PRINT USING "     PRESENT AMOUNT (P) = ##########"; P
2230 RETURN
```

REFERENCES

CUENCA, R. H., "Crop-Water Production Functions and System Design," Proceedings, American Society of Civil Engineers Irrigation and Drainage Division Specialty Conference, Portland, Oregon, 1987, pp. 271–278.

JAMES, L. D. and R. R. LEE, *Economics of Water Resources Planning*. New York: McGraw-Hill, 1971.

LOUCKS, D. P., J. R. STEDINGER, and D. A. HAITH, *Water Resource Systems Planning and Analysis*. Englewood Cliffs, NJ: Prentice-Hall, 1981.

MAJOR, D. C. and R. L. LENTON, *Applied Water Resource Systems Planning*. Englewood Cliffs, NJ: Prentice-Hall, 1979.

PROBLEMS

2-1. The yield versus applied water graph in Fig. 1-1 for SJ2 variety cotton can be described by the function

$$Y = 392.6 + 5.65(AW) - 0.0055(AW)^2$$

where

$$Y = \text{yield, kg/ha}$$
$$AW = \text{applied water, mm}$$

A grower is interested in growing the same variety of cotton on the same soil using an irrigation system with the same efficiency as that used to develop the function in Fig. 1-1. The grower may increase his or her applied water starting from zero by increments of 50 mm. Each 50 mm increment costs the grower an additional $5.00. The grower computes net benefits taking into account all costs except water to be $0.027/ha of cotton.

Start at zero applied water and using increments of 50 mm, compute the highest level of applied water you would recommend the grower use.

2-2. A grower and irrigation specialist working together determine that the grower's total benefits and total costs due to irrigation are defined by the following functions:

$$\text{Benefits:} \quad B = 5I - (I)^{1.5}$$
$$\text{Costs:} \quad C = 5I - 2(I)^2 + (I)^{2.21}$$

where

$$I = \text{level of irrigation}$$

(a) Compute the grower's benefits and costs for the following levels of irrigation:

$$I = 2, 5, 8, 11, 15$$

(b) What is the optimum level of irrigation for this grower's operations? (Can be solved using an iterative procedure such as Newton's method.)

(c) What is the maximum net benefit for this irrigated farm?

2-3. The data in the following tables are costs and yield data for beans and sugar beets. The water use is the equivalent depth of water diverted to the grower at the headworks of the irrigation system.

Costs of Production

	Sugar Beets	Beans
Fixed production costs ($/ha): includes land preparation, planting, seed, cultivating, labor and equipment, fertilizer and other chemicals	581	467
Variable costs:		
Water ($\$/m^3$)	0.0146	0.0146
Variable harvest costs ($/unit weight)	5.90/t	0.027/kg
Crop price ($/unit weight)	27.40/t	0.465/kg

Crop Yield

Sugar Beets		Beans	
Water Use (cm)	Yield (t/ha)	Water Use (cm)	Yield (kg/ha)
25.4	25.9	15.2	1202
30.5	28.1	20.3	1272
35.6	30.5	30.5	1403
40.6	31.4	40.6	1494
45.7	32.5	45.7	1530
50.8	33.5	50.8	1560
55.9	34.3	55.9	1586
61.0	34.8	61.0	1607
66.0	35.4	66.0	1626
71.1	35.8	71.1	1641
76.2	36.0	76.2	1654
81.3	36.2	81.3	1662
86.4	36.3	86.4	1672
91.4	36.4	91.4	1679
96.5	36.5		
101.6	36.5		
106.7	36.5		
111.8	36.5		
116.8	36.5		
121.9	36.5		

(a) Plot water use versus yield for each crop and draw a best fit curve through the data.

(b) Use the following formula to compute the profit for each value of water use. Tabulate the results.

$$P = CP(Y) - FPC - WC(WU) - VHC(Y)$$

where

P = profit, \$/ha
CP = crop price, \$/unit weight
Y = yield, unit weight/ha
FPC = fixed production costs, \$/ha
WC = water cost, \$/m^3
WU = water used, m^3/ha
VHC = variable harvest costs, \$/unit weight

(c) Plot the profit for each level of water use for both crops and draw a best fit curve through the data.

(d) Assume the grower's water supply is not limited.

(i) If the grower wants to plant all the land in beans, at what level should he or she irrigate?

(ii) If the grower wants to plant all the land in sugar beets, at what level should he or she irrigate?

(e) Market prices are variable and can change considerably from year to year. The grower faces a drought year and has a water allocation of only 40.6 cm. At the crop price used to draw the graphs, which crop should the grower plant for the greater profit? Using the equation given for profit, compute how much the crop price of the most favorable crop would have to decrease before you would recommend planting the other crop.

(f) If the water allocation is increased to 61.0 cm, which crop would you recommend the grower plant for greater profit? Compute how much the crop price of the least favorable crop would have to increase before you would recommend growing it.

2-4. Refer to Fig. P2-1 given indicating yield reduction ratios for different crops. You are hired as a consultant by the Agency for International Development to recommend cropping practices for the East African Highlands. This area has a relatively high probability of drought. The crops are to be grown for food consumption and not profit. The food choices acceptable to the local population are beans and corn.

(a) Given no further information, which crop would you recommend planting?

(b) What further information is necessary to make a more complete planning decision?

2-5. The slope of a total cost curve plotted versus level of irrigation typically changes at various levels of irrigation. Briefly describe (one sentence each) three reasons for this change in slope at different irrigation levels.

2-6. A grower has two choices to buy pumping equipment for an irrigation system expected to expand 15 years in the future:

(a) Buy a pump today for $3000 and buy a replacement pump in year 15 for $5500.

(b) Buy a pump today for $2400, buy a replacement pump in year 10 for $3200, and buy an additional pump in year 15 for $4100.

Assume operating costs are equal for option (a) or (b), that pump salvage value is negligible, that total project life is 30 years and that interest rate on capital is 11 percent. What option would you recommend to the farmer based on minimizing annual costs for the life of the project?

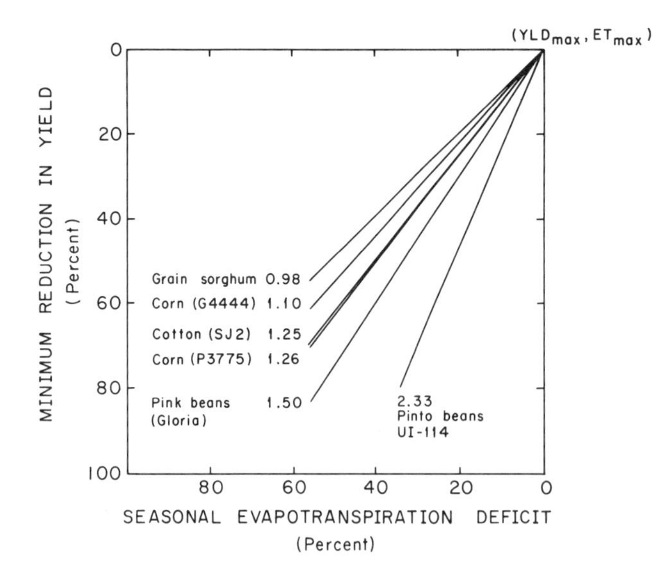

Figure P2-1 Minimum rates of yield reduction expected in response to optimally timed seasonal evapotranspiration deficits (relative values).

2-7. The total benefit function for a sasaras crop is given by

$$B = 200 + 610(I) - 61(I)^2$$

where

$$B = \text{benefit in dinars}$$
$$I = \text{depth of irrigation in todits}$$

The total cost of irrigation water is 200 dinars per todit. What depth of todits, to the nearest tenth of a todit, do you recommend the farmer apply to the crop and why?

2-8. A pumping system is planned which will last 30 years. This is to be accomplished by buying one pump today at $5600 and a replacement pump in year 20. If the interest rate is 10 percent and the annual cost of the complete system is $703/yr over the total 30 years, what is the cost of the replacement pump to the nearest dollar in year 20? Assume negligible salvage value for the pumps.

2-9. A water production function for corn expressed as yield (Y) versus applied water (AW) is given by the following equation:

$$Y = 393 + 5.6(AW) - 0.005(AW)^2$$

(a) Compute the magnitude of applied water at maximum yield.
(b) Compute the maximum yield.

2-10. A yield (Y in kg/ha) versus applied water (AW in mm) function is given by

$$Y = -8520 + 61.8(AW) - 0.0336(AW)^2$$

For the same crop, the yield versus evapotranspiration (ET in mm) relationship is given by

$$Y = -6525 + 46.0(ET)$$

If the irrigation efficiency is defined as the amount of ET for a given level of AW, compute the efficiency at an applied water level of 650 mm.

2-11 Two alternative pump and pipeline systems will deliver an equal water requirement and result in the same crop production. The two systems have different operating costs, initial costs, and salvage values at the end of 20 years. This information is given in the following table.

	DESIGN A		DESIGN B	
ITEM	Cost	Salvage Value, %	Cost	Salvage Value, %
Pump	$2500	50	$1900	40
Pipeline	$8500	40	$10800	30
Operations	$2200/yr	—	$1800/yr	—

Indicate which system you would recommend purchasing and why based on a constant 12 percent interest rate over an economic life of 20 years.

2-12. The maximum measured yield for a particular variety of tomato is 19.1 metric tons per hectare. This maximum yield corresponds to a maximum seasonal crop ET of 775 mm at a location where the maximum seasonal reference ET for a grass reference crop is 889 mm. The yield reduction ratio is equal to 0.865 for this tomato variety. For a new site, analysis of 10 years of climatic data indicates a mean seasonal reference ET for grass of 825 mm.

(a) Predict the mean maximum yield for this tomato variety in metric tons per ha at the new site.

(b) Predict the mean expected yield in metric tons per ha if a 15 percent ET deficit is imposed on the tomato crop by using deficit irrigation at the new site.

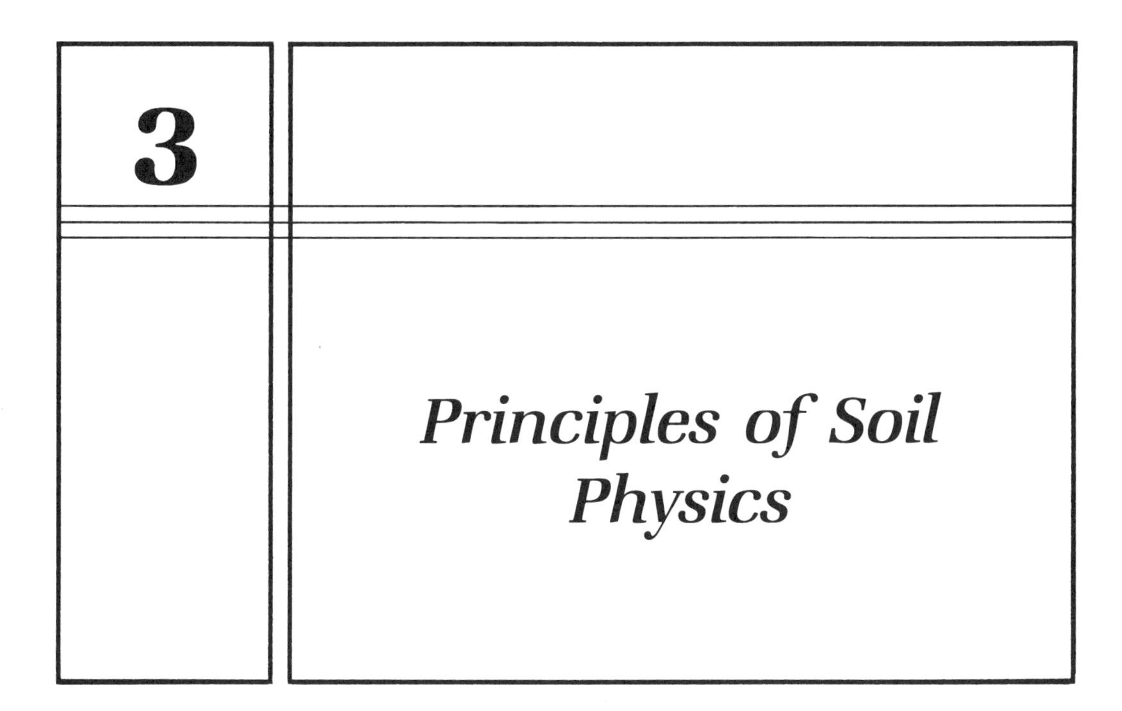

3.1 Descriptive Terminology

Soil Horizons

Some soil profiles are formed over the millenia by physical and chemical decomposition of the parent material. These soils are called weathered soils and tend to have gradual transitions between different soil layers. Such a condition is indicated on the left-hand side of Fig. 3-1. This type of soil is typically formed by the weathering of the parent material in the C horizon and by decomposition of organic material above the A horizon. Eluviation refers to the washing out of clay and other materials from the overlying A horizon. Illuviation refers to the accumulation of this material in the B horizon, which therefore differs in composition and structure from the A horizon.

Soils may also be formed by deposition of sediments which are transported by either wind or water. Water-borne sediments form what are termed alluvial soils. Soils formed by deposition of sediments tend to be stratified and have abrupt transitions between individual layers as indicated on the right-hand side of Fig. 3-1.

Whether formed by physical and chemical weathering or stratified due to deposition of sediments, soils do not tend to be as uniform over a field as most engineers would like to expect. Thicknesses of various layers, distribution of chemical constituents, and physical properties of the soil can vary substantially over a field, even

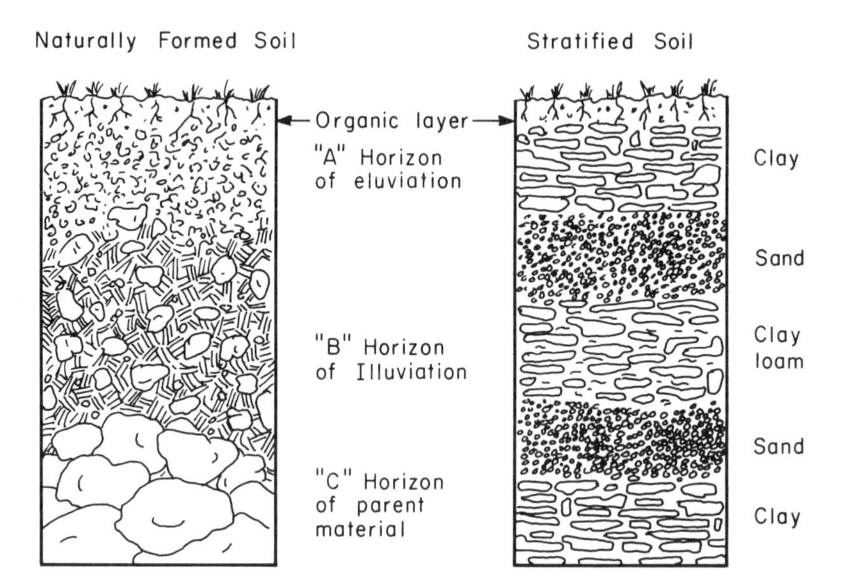

Figure 3-1 Naturally formed soil with gradual transitions between horizons and stratified soil with abrupt transitions. (Taken from Luthin, 1973.)

within the distance of a few meters. Distribution of thin lenses of various materials over a field can lead to large variation in properties. These physical and chemical realities of soils on a field scale make the question of proper soil sampling techniques a difficult one to answer. Currently, there are increased efforts on the part of soil scientists to develop criteria for adequate soil sampling. Results of these investigations will be extremely important to engineers involved in design of irrigation systems.

Soil Texture

The term soil texture refers to the size of particles which make up a soil sample. Particle size is represented by the diameter of a sphere which has equivalent volume to a soil particle. In practice, soils are graded by sifting them through a series of sieves in which each successive sieve has a smaller screen size. Such a set of sieves is indicated in Fig. 3-2. The amount of soil collected on each sieve is weighed and the percent of the total sample weight retained on each sieve is recorded. When the results are plotted, a grain size distribution curve such as shown in Fig. 3-3 results. If the particles tend to fall predominantly in one size category, resulting in a vertical line in Fig. 3-3, or if particles of a particular size are not found in the sample, resulting in a horizontal line in Fig. 3-3, the sample is said to be poorly graded. In contrast, if the soil tends to have a significant percentage of various particle sizes, it is said to be well graded.

There are numerous classification schemes for soil texture as a function of particle diameter. These schemes vary from country to country and sometimes between government agencies within the same country. An example of six such classification schemes is indicated in Fig. 3-4.

Figure 3-2 Agricultural soil sieve set.

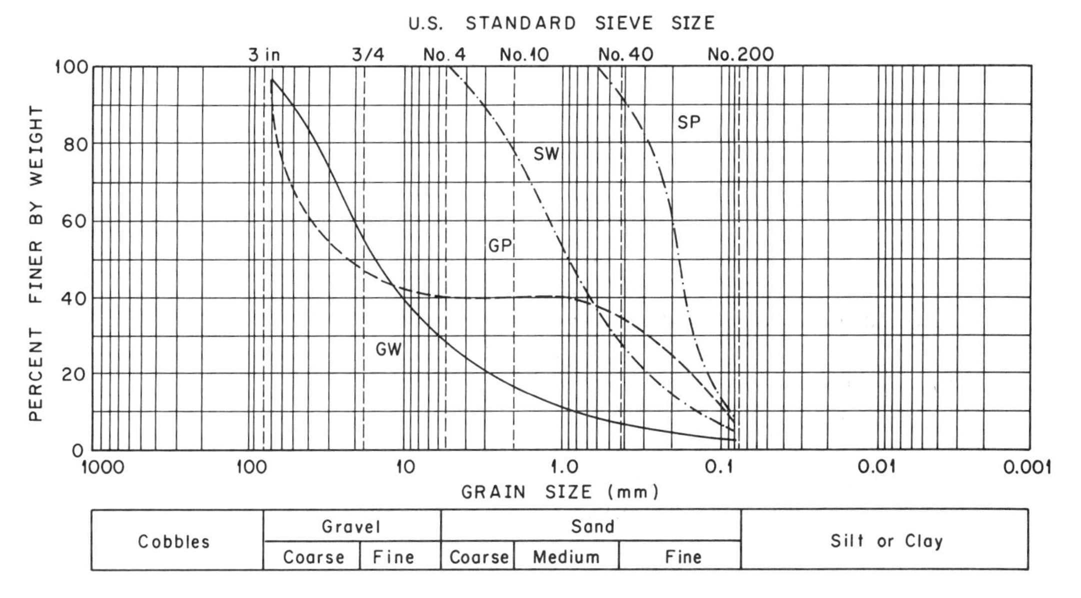

Figure 3-3 Typical grain size distribution for well-graded gravel (GW) and sand (SW) and poorly graded gravel (GP) and sand (SP). (Taken from Luthin, 1973.)

Soil texture denotes both a qualitative as well as a quantitative aspect of soil compounds. A soil with a high percent of gravel will be made up of harder particles than that with a high percent of silt. Note that the soil texture does not tell us information about the physical characteristics of the soil other than the equivalent diameter of its particle composition. Soil-water relationships which are important to irrigation system design require other physical tests than those used to determine texture. Information about texture, however, may give expected ranges of values for particular soil-water relationships.

A commonly used method of textural classification, and one which is used throughout this book, is that developed by the United States Department of Agriculture (USDA). It has been developed for agricultural soils and requires information on the amount by weight of sand, silt, and clay in the soil. The amount of sand may

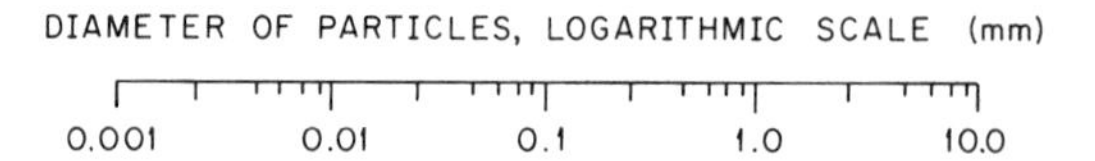

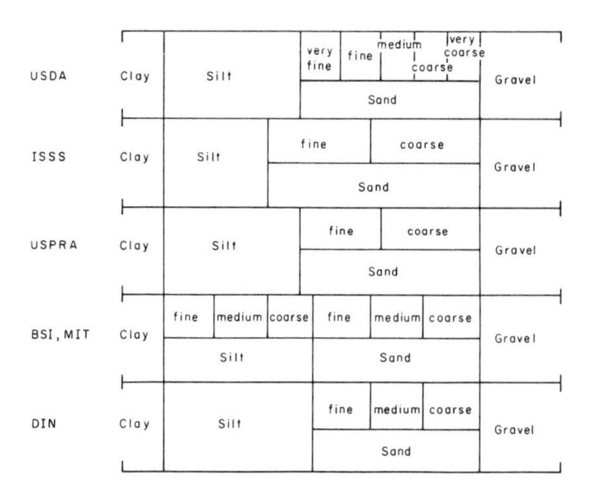

Figure 3-4 Different systems for distinguishing size classifications. [U.S. Department of Agriculture (USDA); International Soil Science Society (ISSS); U.S. Public Roads Administration (USPRA); German Standards (DIN); British Standards Institute (BSI); Massachusetts Institute of Technology (MIT).] (Taken from Hillel, 1980.)

be determined by measuring the quantity of sample retained on a No. 200 mesh (0.1 mm diameter) sieve. Separation of the silt and clay fractions may be accomplished by the pipette or hydrometer methods of analysis, both of which depend upon the physical laws of sedimentation. The relationship between the size of a spherical particle and its settling velocity is known as Stoke's law. It is applied as an approximation for size analysis to non-spherical soil particles.

Stoke's law is derived by setting the downward force due to gravity of a soil particle in suspension equal to the resisting drag force due to friction of the fluid at the terminal velocity of the particle. Setting the two forces equal and rearranging the equation, Stoke's law is given by

$$v_t = \frac{d^2 g}{18\eta}(\rho_s - \rho_f) \tag{3-1}$$

where

v_t = terminal velocity, m/s

d = particle diameter, m

g = acceleration of gravity, m/s^2

η = dynamic viscosity of fluid, Pa(s)

ρ_s = particle density, kg/m^3

ρ_f = fluid density, kg/m^3

Assuming the terminal velocity is reached instantly, the time t (seconds) required for a particle to fall through height h (meters) may be determined by rewriting Eq. (3-1) as:

$$t = \frac{18h\eta}{d^2 g(\rho_s - \rho_f)} \tag{3-2}$$

Equations (3-1) and (3-2) form the basis of particle size analysis by the pipette and hydrometer methods. The normal procedure involves measuring the density of the sustension at a given depth as a function of time using a hydrometer. Equation (3-2) is applied to solve for time t for particles of diameter corresponding to the clay and silt fractions to pass a given depth h at which the hydrometer is placed. Correction factors are required for temperature and initial suspension concentration.

Description of the steps for the complete procedure is beyond the scope of this book. For a description of the hydrometer method to separate the silt and clay fractions, the reader is referred to standard soil testing handbooks (Black et al., 1965).

Once the percent by weight of sand, silt, and clay has been determined, the USDA textural triangle can be used to classify soils into various textures which are important in agricultural applications. This textural triangle is shown in Fig. 3-5.

Soil Structure

Soil structure refers to the arrangement of soil particles *in situ*—that is, as the particles are found in the field. Regardless of texture, soil particles can have different structure depending upon whether a mass of particles is relatively open and porous to allow for rapid movement of air and water, or relatively dense with few interconnecting pores. As opposed to soil texture, soil structure is a qualitative descriptor of the soil condition (Hillel, 1980).

Soil structure affects the mechanical properties of a soil as well as the properties more commonly of interest in irrigation system design. A more densely packed soil mass will support a given load with less deflection than a loosely structured soil mass. Soil structure is one of the soil properties affected by the actions of man since compaction due to overland traffic and breaking up of soils from plowing affect the structure. Agronomists speak of optimum soil tilth as a desirable physical condition in which the soil is a loose, porous assemblage of aggregates which permits

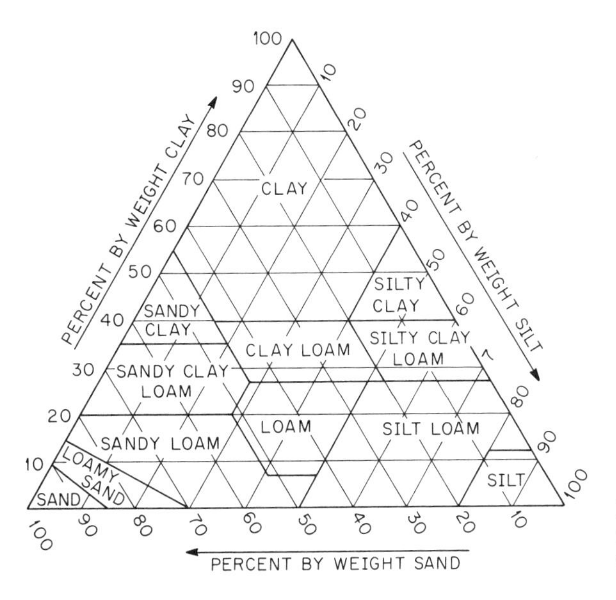

Figure 3-5 Textural triangle for determining soil texture given the sand, silt, and clay fractions as percent.

relatively rapid movement of air and water and unobstructed germination and root growth. The term friable is also used to indicate the desirable soil condition in which a soil can be easily broken up into smaller units in the palm of the hand.

The three broad categories of soil structure are single grained, massive and aggregated (Hillel, 1980). Single grained refers to a structureless soil in which there is no adhesion between individual particles. Desert or beach sand is a good example. Very large clumps of soil, on the order of tens of centimeters across, made up of chemically and/or physically bonded soil particles, are termed massive. Blocks of dry clay occurring under drought conditions are one example of this structure. In between these extremes is an aggregated soil structure made up of clods on the order of centimeters or fractions of a centimeter in diameter. These aggregates are sometimes also classified as peds. The aggregated structure is the most suitable for the majority of agricultural operations. Degree of aggregation is influenced not only by vehicular traffic over a field, but also by the physical, biological, and chemical relationships which occur as a function of root density, organic matter, and depth. Crops with vigorous root systems that provide a high degree of vegetative cover and do not require intensive mechanical cultivation during the growing season tend to promote soils of optimum tilth.

Classes of Soil Water Availability

Later in this chapter, we will discuss parameters which define the availability of soil-water to plants in a quantitative manner. Presently, we will concentrate on the description of categories of soil-water of general interest in a physical sense. The following definitions will refer to Fig. 3-6. Other terms indicated in Fig. 3-6 will be described later in this chapter.

Gravitational water is defined as that water which is rapidly drained from the soil profile by the force of gravity. The term rapid is relative and in soil-water studies normally refers to time periods of 24 to 48 hours. Capillary water is the water remaining after rapid drainage by gravity. This water may be removed by forces greater than gravity such as those exerted by plant roots. Hygroscopic water is water

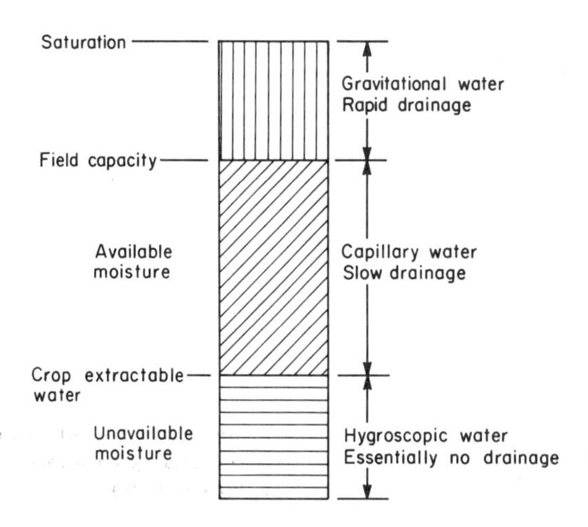

Figure 3-6 Classes of soil-water availability to plants and characteristics of drainage.

which adheres to soil particles which cannot generally be removed by forces found in nature. Hygroscopic water can be removed by oven drying a soil sample, but cannot be removed by plant roots.

3.2 Soil-Water Properties

Elemental Soil Volume

Many soil-water properties can be defined relative to the elemental soil volume depicted in Fig. 3-7. This is a volume of soil which has been separated into its air, water, and solid constituents. Relationships for mass are indicated on the right-hand side of the soil volume. The mass of air, M_a, is assumed negligible. The mass of water and solids are indicated as M_w and M_s, respectively. The total mass is shown as M_t.

The volume relationships are indicated on the left-hand side of the soil volume in Fig. 3-7. V_a is the volume of air which can be a significant percentage of the soil volume under field conditions. V_w represents the volume of water, and V_s the volume of solids. The volume of pores, designated by V_p, is the sum of the volume of air and water. The total volume is normally termed the bulk volume in agricultural applications and designated as V_b.

Soil-Water Properties Defined

Various soil-water properties may be defined in a number of different ways. The following listing indicates the definitions for soil-water properties which will be used throughout this book. The formulas refer to the elemental soil volume in Fig. 3-7.

Formulas for soil-water properties

Water content on mass basis, θ_m

$$\theta_m = \frac{\text{mass water}}{\text{mass dry soil}} = \frac{M_w}{M_s} \tag{3-3}$$

Volumetric water content, θ_v

$$\theta_v = \frac{\text{volume water}}{\text{bulk volume soil}} = \frac{V_w}{V_b} = \frac{V_w}{V_s + V_p} = \frac{V_w}{V_s + V_a + V_w} \tag{3-4}$$

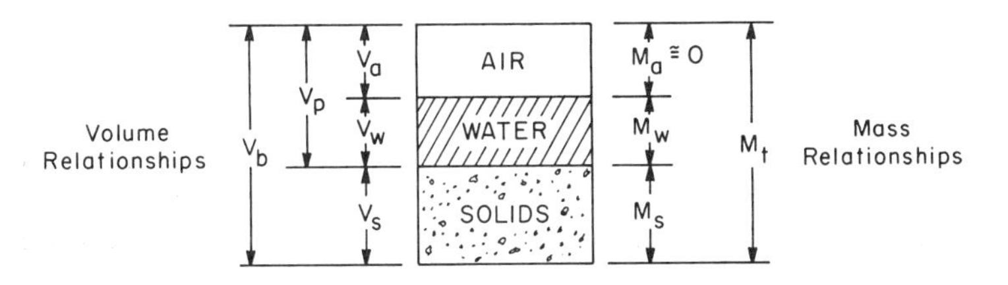

Figure 3-7 Schematic diagram of a soil block as a three-phase system.

Soil bulk density, ρ_b

$$\rho_b = \frac{\text{mass dry soil}}{\text{bulk volume soil}} = \frac{M_s}{V_b} = \frac{M_s}{V_s + V_a + V_w} \tag{3-5}$$

Soil porosity, N

$$N = \frac{\text{total pore volume}}{\text{bulk volume soil}} = \frac{V_p}{V_b} = \frac{V_a + V_w}{V_b} = \frac{V_a + V_w}{V_s + V_a + V_w} \tag{3-6}$$

Air filled porosity, N_a

$$N_a = \frac{\text{air filled pore volume}}{\text{bulk volume soil}} = \frac{V_a}{V_b} \tag{3-7}$$

Saturated water content (volume basis), θ_{vs}

$$\theta_{vs} = \frac{\text{volume of water when saturated}}{\text{bulk volume soil}} = \frac{V_p}{V_b} \tag{3-8}$$

Saturated water content (mass basis), θ_{ms}

$$\theta_{ms} = \frac{\text{mass of water when saturated}}{\text{mass dry soil}} = \frac{\rho_w V_p}{M_s} \tag{3-9}$$

where

$$\rho_w = \text{density of water}$$

Different parameters may be derived using the values defined under soil-water properties. As an example, the bulk density may be derived as a function of the soil porosity and density of the soil particles. Equation (3-6) for soil porosity may be rewritten as

$$N = \frac{V_p}{V_b} = \frac{V_b - V_s}{V_b} \tag{3-10}$$

Defining density as mass divided by volume, Eq. (3-10) may be rewritten

$$N = \frac{M_s/\rho_b - M_s/\rho_s}{M_s/\rho_b} = 1 - \frac{\rho_b}{\rho_s} \tag{3-11}$$

where

$$\rho_s = \text{particle density}$$

Equation (3-11) may be rewritten to describe bulk density as a function of porosity and particle density:

$$\rho_b = (1 - N)\rho_s \tag{3-12}$$

A common value for ρ_s in agricultural soils is approximately 2.65 g/cm^3, the particle density of silica sand, granite, and quartz rock, which are common parent materials. A common value of ρ_b for agricultural soils is in the range of 1.3 g/cm^3, indicating that porosities of agricultural soils are typically in the neighborhood of 0.5 or 50 percent air plus water.

Another relationship which may be derived is that between water content on a mass basis, which is relatively easy to measure, and water content on a volume basis, which is more useful in irrigation system design. Rewriting Eq. (3-4) as

$$\theta_v = \frac{V_w}{V_b} = \frac{M_w/\rho_w}{M_s/\rho_b} \tag{3-13}$$

Substituting the water content on a mass basis from Eq. (3-3) into Eq. (3-13)

$$\theta_v = \theta_m\left[\frac{\rho_b}{\rho_w}\right] \tag{3-14}$$

The quantity in brackets in Eq. (3-14), ρ_b/ρ_w, is termed the apparent specific gravity of the soil.

Procedures for field measurement of the soil-water properties indicated in this section are discussed in standard soil testing handbooks (Black et al., 1965). The following example problem demonstrates application of field data to determine soil-water properties.

Example Problem 3-1

A 100.0 cm^3 soil sample is taken in the field which weighs 174 g at the time of sampling. The oven dry weight of the sample is 155 g. Assume $\rho_w = 1.00\ g/cm^3$ and $\rho_s = 2.65\ g/cm^3$. Compute θ_m, θ_v, ρ_b, porosity N, and air filled porosity, N_a.

Solution

$$\theta_m = \frac{M_w}{M_s} = \frac{174\ g - 155\ g}{155\ g} = \frac{19\ g}{155\ g} = 0.123 = 12.3\%$$

$$\theta_v = \frac{V_w}{V_b} = \frac{19\ g/(1.00\ g/cm^3)}{100\ cm^3} = 0.190 = 19.0\%$$

$$\rho_b = \frac{155\ g}{100\ cm^3} = 1.55\ g/cm^3$$

$$N = \frac{V_p}{V_b} = \frac{V_w + V_a}{V_b}$$

$$V_b = V_w + V_a + V_s$$

$$V_s = \frac{M_s}{\rho_s} = \frac{155\ g}{2.65\ g/cm^3} = 58.5\ cm^3$$

$$V_a = V_b - V_w - V_s = 100.0 - 19.0 - 58.5 = 22.5\ cm^3$$

$$N = \frac{19.0\ cm^3 + 22.5\ cm^3}{100.0\ cm^3} = 0.415 = 41.5\%$$

$$N_a = \frac{V_a}{V_b} = \frac{22.5\ cm^3}{100.0\ cm^3} = 0.225 = 22.5\%$$

Relationship between pressure, head, and tension

It is useful when discussing soil-water properties and water movement within the soil matrix to differentiate between pressure, head, and tension. Pressure, p, is defined as force per unit area. Considering System International (SI) units with force in newtons (N) and pressure in pascals (Pa),

$$p = \text{Pa} = \text{N/m}^2 = [\text{kg(m)/s}^2]/\text{m}^2 \tag{3-15}$$

The head at any point in a soil-water system is directly proportional to pressure and inversely proportional to the fluid density, ρ_f. The proportionality constant is 1 divided by the acceleration of gravity, g. This is shown by

$$h = p/[\rho_f(g)] = [\text{kg(m/s}^2)]/[\text{kg/m}^3)(\text{m/s}^2)] = \text{m} \tag{3-16}$$

In many studies of the soil-water system, water within the soil matrix is at considerably less than atmospheric pressure—that is, the soil is unsaturated. The pressure or head measured is therefore negative. Instead of referring to a negative pressure or head, the water at the point of measurement is said to be subject to a positive tension. The tension is equal in magnitude but opposite in sign to the negative pressure or head.

Concepts of the forces acting in a soil-water system are described more fully later in this chapter. For the present discussion, we can consider that under unsaturated conditions the soil matrix exerts a tension on the water within the soil-water system. This tension is directly proportional to the capillary tension which is inversely proportional to the diameter of the interconnected pores. As the field becomes drier, more and more force is required to remove additional water from the soil matrix. Therefore the soil matrix must be exerting more and more tension on the water within the system. This tension may be measured in various equivalent units. Table 3-1 may be used for conversions between different measures of soil-water tension. Other conversions for SI and English units are found in Appendix A.

Field Capacity

Field capacity is defined as the soil-moisture content attained in an originally thoroughly wet field—that is, at or near saturation, after the rate of drainage by gravity has markedly decreased. As can be seen in Fig. 3-6, field capacity corresponds to the soil moisture content after gravitational water has been drained from the soil. The rate of drainage depends on the particular soil, but field capacity is normally assumed to occur 24 to 48 hours after thorough wetting by irrigation or rainfall. This is demonstrated in Fig. 3-8.

Field capacity is a function of many parameters in addition to soil texture and structure (Taylor and Ashcroft, 1972). Temperature and previous soil-moisture his-

TABLE 3-1 Equivalent values for measurement of soil-water tension, pressure, or potential.

Bar	Atmosphere (atm)	Pascal (Pa)	m of water at 4°C
1	0.9869	100,000	10.1981
1.013	1	101,325	10.3322
0.00001	0.000009869	1	0.00010198
0.000009806	0.000009677	0.98058	1

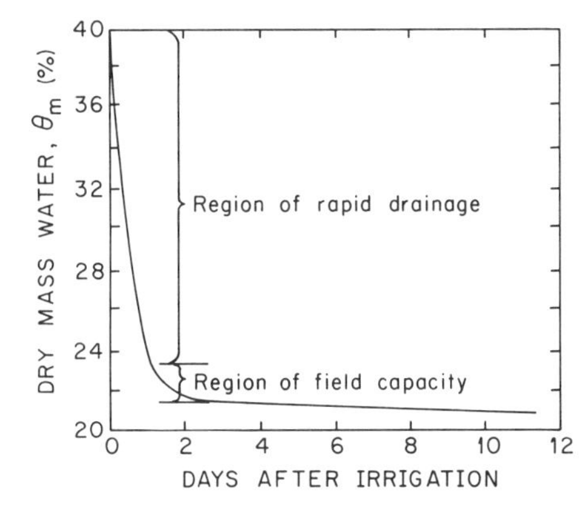

Figure 3-8 Decrease in water content as a function of time following an irrigation.

tory affect field capacity. In practical applications, field capacity is assumed constant over a growing season and is expressed in terms of soil-moisture tension—that is, the tension with which soil particles attract the surrounding water molecules. Field capacity is said to exist when water within the soil matrix is under a tension of from approximately 1.0 m of water for sandy soils to 3.4 m of water for fine, silty, or clay soils (approximately equivalent to 1/10 to 1/3 bar).

Although field capacity is normally used to indicate the upper range of soil moisture available to plants, it should be noted that this results in a slightly conservative estimate of available water. This is because gravitational water is available for plant use during the 24 to 48 hours that it drains through the soil profile.

Crop Extractable Water

The other end of the soil-moisture spectrum important to irrigation system design has been traditionally termed the permanent wilting point. This parameter is defined as the soil-moisture content at which plants cannot recover overnight from excessive drying during the day. The soil-moisture content at this point has generally been determined in greenhouse experiments to correspond to the range of 102 m to 204 m of water tension (10 to 20 bars), with 153 m (15 bars) accepted as a median value.

As defined, permanent wilting point is also a function of parameters other than soil texture and structure (Taylor and Ashcroft, 1972). It is a function of temperature, size of layers in the soil profile, atmospheric evaporative demand, and distribution of the root system. Most important, permanent wilting point is not just a function of the soil but also of the plant variety since different varieties of plants are able to effectively extract water to different tensions.

Recognizing the importance of the plant in defining the lower limit of available water, and realizing that crops in field situations do not respond the same way as individual pots in greenhouse experiments, another term has been chosen to improve on the permanent wilting point concept. This term is crop extractable water which is defined as the soil-moisture content which exists when plants in a field condition cannot recover overnight from excessive drying during the day. This term emphasizes that the plant variety is of vital importance in describing the lower limit of available moisture as well as the soil texture and structure in the field. Normally, plants undergoing wilting in field conditions do not reach as high a tension as crops in greenhouse experiments. (This may be one reason why in practical applications the quantity of readily available water, defined as 40 to 75 percent of the available water between field capacity and permanent wilting point, has been used for irrigation system design.)

The problem for the design engineer is that to date much less information is available for soil moisture content at the limit of crop extractable water measured under field conditions than for permanent wilting point measured in greenhouse experiments. Use of permanent wilting point in system design could result in overestimation of the amount of available water if the 153 m (15 bar) median value is accepted. It should be noted that this overestimation may be very small due to the flatness of the soil-water characteristics curve as it approaches the range of permanent wilting. (Refer to Fig. 3-9).

While the concept of crop extractable water for individual crops is an important one, the practicalities of readily available data require an intermediate solution. For this reason, the lower limit of available water in this book is assumed to occur at 102 m of water tension (10 bars) unless otherwise specified.

Table 3-2 indicates representative ranges of soil-water properties for six classes of soil texture. These values are only approximate and should be replaced by field measured quantities for any type of sophisticated design. Nevertheless, the values indicated can often be of significant importance in rough approximations and preliminary discussions with clients.

Example Problem 3-2

At full development, a tomato crop is measured in an unrestricted soil profile to have an active root zone of 1.5 m. The maximum equivalent crop evapotranspiration at the midpoint of the growing season is 9 mm/day. Assume that each irrigation fills the soil profile up to field capacity.

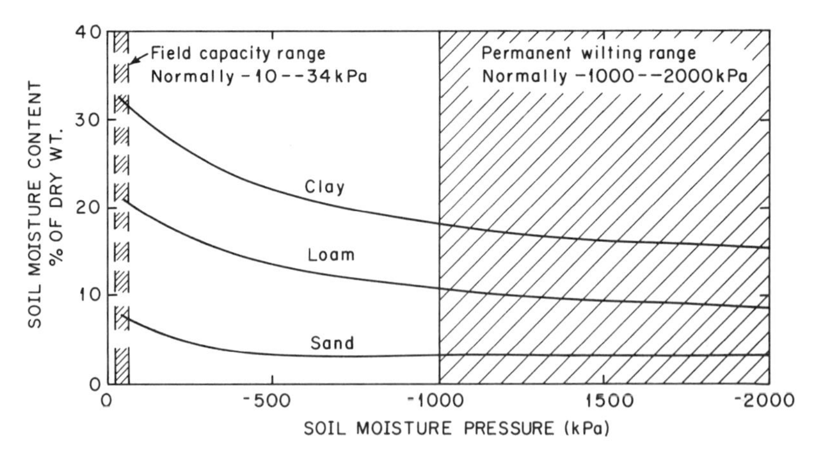

Figure 3-9 Typical soil-water characteristic curves for different soil types and ranges of field capacity and permanent wilting point.

(a) For a sandy loam soil, how many days are allowable between irrigations if 40 percent depletion of available water is allowed?—if 60 percent depletion is allowed?

(b) Compute the same values if the soil is a silty clay.

Solution Compute the total available moisture (TAM) as a function of the difference in volumetric moisture content between field capacity (FC) and the limit of crop extractable water (CEW).

(a) Using Table 3-2 for a sandy loam,

$$\text{TAM} = \text{FC} - \text{CEW}$$

$$\text{TAM} = 21\% - 9\% = 12\% = 120 \text{ mm/m}$$

Compute available moisture (AM) at 40 percent depletion:

$$\text{AM}_{40} = (0.40)120\,\frac{\text{mm}}{\text{m}}(1.50 \text{ m}) = 72.0 \text{ mm}$$

Compute frequency of irrigation, I_{fr}:

$$I_{fr} = \frac{72.0 \text{ mm}}{9 \text{ mm/day}} = 8 \text{ days}$$

Compute available moisture at 60 percent depletion:

$$\text{AM}_{60} = \frac{(0.60)}{(0.40)}(72.0 \text{ mm}) = 108 \text{ mm}$$

$$I_{fr} = \frac{108 \text{ mm}}{9 \text{ mm/day}} = 12 \text{ day}$$

(b) For a silty clay soil,

$$\text{TAM} = 40\% - 20\% = 20\% = 200 \text{ mm/m}$$

At 40 percent depletion,

$$\text{AM}_{40} = (0.40)200\,\frac{\text{mm}}{\text{m}}(1.50) = 120 \text{ mm}$$

$$I_{fr} = \frac{120.0 \text{ mm}}{9 \text{ mm/day}} = 13 \text{ days}$$

At 60 percent depletion,

$$I_{fr} = \frac{(0.60)}{(0.40)}13 \text{ days} = 20 \text{ days}$$

3.3 Concept of Soil-Water Potential

Definition and Representation of Potential

The water potential of a soil-water system refers to the ability of the system to do work measured relative to some reference state. The reference state applied is pure free water at the same temperature as water in the soil-water system and at atmospheric pressure.

TABLE 3-2 Representative physical properties of soils.

Soil Texture	Infiltration Rate[a] (mm/h)	Porosity (percent)	Specific Gravity	Field Capacity (percent)	Crop Extractable Water (percent)	Total Available Moisture	
						Volume Basis (percent)	Depth Basis (mm/m)
Sand	50 (25–250)	38 (32–42)	1.65 (1.55–1.80)	15 (10–20)	7 (3–10)	8 (6–10)	80 (60–100)
Sandy loam	25 (13–76)	43 (40–47)	1.50 (1.40–1.60)	21 (15–27)	9 (6–12)	12 (9–15)	120 (90–150)
Loam	13 (8–20)	47 (43–49)	1.40 (1.35–1.50)	31 (25–36)	14 (11–17)	17 (14–20)	170 (140–200)
Clay loam	8 (2.5–15)	49 (47–51)	1.35 (1.30–1.40)	36 (31–42)	18 (15–20)	19 (16–22)	190 (160–220)
Silty clay	2.5 (0.3–5)	51 (49–53)	1.30 (1.30–1.40)	40 (35–45)	20 (17–22)	20 (18–23)	200 (180–230)
Clay	0.5 (0.1–1)	53 (51–55)	1.25 (1.20–1.30)	44 (39–49)	21 (19–24)	23 (20–25)	230 (200–250)

Note: Normal ranges are shown in parenthesis.
[a]Intake rates vary greatly with soil structure and stability, even beyond the normal ranges shown.

Considering a macroscopic viewpoint, the difference in potential between two soil-water systems is the work that one system is capable of doing with reference to work that can be done by water in the second system. This difference is the driving force which causes movement of water in a soil-water system. As such, it is a key element in our discussions of infiltration and hydraulic conductivity which follow later in this chapter. In the traditional analysis of soil-water systems, water is always considered to flow from a point of higher to a point of lower potential.

There has been some discussion regarding the proper measurement and interpretation of soil-water potential (Corey and Klute, 1983). This discussion centers around the fact that in a soil-water system water is said to flow from a higher to lower potential and there is no flow if the total potential is equal everywhere in the system. Corey and Klute (1983) contend that soil moisture movement is made up of a combination of bulk flow plus diffusion of chemical water. They demonstrate their case by considering flow of solutes and chemical water across a membrane in a system for which the total potentials on both sides of the membrane are equal. If the membrane is porous only to water, flow will occur in spite of equal total potentials due to diffusion of chemical water. This reasoning is particularly appropriate to agricultural systems since root systems can certainly be considered as a membrane.

Although the discussion of Corey and Klute (1983) appears to be a valid one, it is not clear at this time the relative magnitude of error in neglecting the diffusion flux nor the recommended procedure for measuring the diffusion flux or its gradient. For the macroscopic measurements normally of interest to the design engineer, the error is probably negligible. In this book the traditional method of measuring total potential and assuming that water flows from a higher to lower potential is accepted.

Types of Potential

Soil-water potential is typically divided into different components. In this book, gravitational, pressure, and osmotic potential are considered. Gravitational potential is the potential due to position of a point relative to some datum and will be designated by z. The position of the datum in soil-water movement problems is normally situated such that z is positive. The position of the datum is arbitrary in calculating the contribution of gravitational potential to total potential as long as the convention is followed that the direction above the datum corresponds to a positive gravitational potential and the direction below the datum corresponds to a negative potential. This convention is demonstrated in example problem 3-3.

In this book, pressure potential, p, is considered as being both positive and negative. If the water at the point in the soil at which potential is being measured is at or above atmospheric potential, the pressure potential is positive. If the water at the point at which potential is being measured is at less than 1 atmosphere pressure, the pressure potential is negative. When the point of measurement is in an unsaturated state and at less than atmospheric pressure, the negative pressure potential is caused by the attraction of the soil matrix structure for the water. This is the condition under which capillary tension is said to be exerted.

The terminology in this book differs from traditional textbooks on soil-water properties in which pressure is considered zero or positive if water at the point of measurement is at or above atmospheric pressure and is termed matric potential if water at the point of measurement is at less than atmospheric pressure. Using such traditional terminology, at any given point in the soil either the pressure potential is positive and the matric potential is zero or the matric potential is negative and the pressure potential is zero. This type of terminology in which the name of the potential changes whether it is negative or positive is confusing to readers and will not be used.

The third type of potential to be considered is osmotic potential, p_{os}. Osmotic potential is the potential due to the concentration of salts in the soil-water solution. The impact of the osmotic potential in an agronomic situation can be determined by considering the activity of a root system in a saline solution. Under normal midday growing conditions, both the roots and the dissolved salts in a soil-water system will be exerting an attractive force on the water. Increasing the salt concentration in the soil-water solution will cause an increase in the attractive forces due to the salts, which is called the osmotic potential. From the perspective of the root membrane, this increase in salt concentration will cause an increase in negative pressure forces drawing water away from the roots and towards the salts. Osmotic potential therefore becomes increasingly negative with increasing concentration of salts in the soil-water solution. This increase in osmotic potential corresponds to an increase in the difficulty roots will have in extracting water from the soil-water solution.

The total potential, T, is the sum of the individual potentials. This is indicated by the following equation:

$$T = z + p + p_{os} \tag{3-17}$$

Moisture movement from one point to another in soil-water systems is governed by the difference in total potential for the general case. In the absence of roots and when the osmotic potential is everywhere equal or negligible, moisture movement is governed by the difference in hydraulic potential, h, which is equal to the sum of the gravitational and pressure potentials:

$$h = z + p \tag{3-18}$$

The potential of a soil-water system may be expressed in a number of different ways. Specific potential is defined as energy per unit mass—that is, joules/kg. A convenient way to express the potential of a soil-water system is weight potential which is given as

$$\text{weight potential} = \frac{\text{specific potential}}{\text{gravitational constant}} \tag{3-19}$$

In SI units, weight potential is given as

$$\text{weight potential} = \frac{\text{joules/kg}}{\text{m/s}^2} = \frac{\text{newton (m/kg)}}{\text{m/s}^2} = \text{m} \tag{3-20}$$

Therefore, weight potential is expressed as a head (m).

An alternative way to describe potential is a volumetric potential. Volumetric potential is given as

$$\text{volumetric potential} = \text{specific potential(density)} \tag{3-21}$$

Again applying SI units:

$$\text{volumetric potential} = (\text{joules/kg})(\text{kg/m}^3) \tag{3-22}$$

$$= \frac{\text{newton}}{\text{m}^2} = \text{Pa} \tag{3-23}$$

Volumetric potential is therefore expressed as a pressure (Pa).

Measurement of Soil-Water Potential

Soil-water potential in the field is commonly measured by installation of tensiometers or gypsum blocks. Water-filled tensiometers can only be applied in the limited range of 0 to 8 m tension. Beyond this range, the water column is broken by the vacuum created and the readings are no longer valid.

Gypsum blocks are able to give valid readings over a much larger range of tensions. The blocks actually read an electrical resistance which is correlated to tension by means of a calibration procedure. Thus the gypsum block method is an indirect procedure for measuring tension which requires an additional calibration curve which contributes additional uncertainty to the measurement.

The range of sensitivity of the blocks is a function of the ratio of gypsum to other materials used in block construction. Blocks can be manufactured to read in the range of 0 to 150 m tension, but the sensitivity and resulting accuracy of the block readings will not be the same over the entire range. For this reason, blocks are normally installed to work either in the more moist ranges of soil-water tension, from 0 to about 50 m, or in the dryer ranges of from about 10 m to 150 m. Gypsum blocks are typically not as responsive nor as accurate as tensiometers in the 0 to 8.0 m range and are best used in fine-textured soils which experience higher tensions in normal irrigated field conditions than sandy soils. Gypsum blocks are also sensitive to soil salinity and should be independently calibrated for saline conditions.

A description of installation and measurement of tension using tensiometers and gypsum blocks is given in standard soil testing handbooks (Black et al., 1965). Photographs of tensiometers and gypsum blocks are shown in Figs. 3-10 and 3-11.

Figure 3-10 Tensiometer installed adjacent to trickle emitter line to monitor soil-moisture status.

Figure 3-11 Installation of white gypsum block using Oakfield soil probe. Wires to block are passed through opening at end of probe as block is pushed into bottom of previously prepared hole.

Example Problem 3-3

As shown in the accompanying diagram, a field is to be monitored for potential using a tensiometer connected to a mercury manometer. The top of the manometer is open to the atmosphere.

(a) Assuming osmotic potential is negligible, compute the individual potentials of point C.
(b) Determine if point C is above or below the water table.

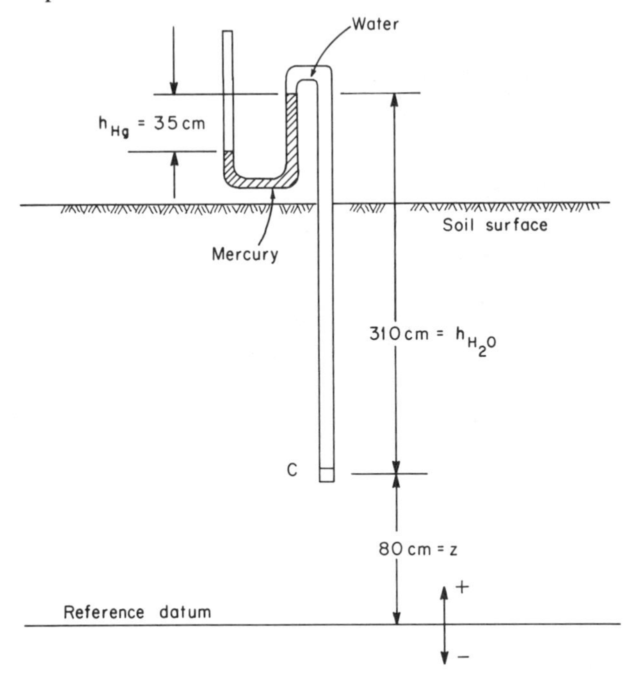

Solution

(a) Gravitational potential:

$$z = +80 \text{ cm}$$

Pressure potential: Considering that the specific gravity of mercury equals 13.6,

$$p = -h_{Hg}(13.6) + h_{H2O}$$
$$p = -(35 \text{ cm})(13.6) + 310 \text{ cm}$$
$$p = -166 \text{ cm}$$

Total potential at point C:

$$T = z + p + p_{os}$$
$$T = +80 - 166 = -86 \text{ cm}$$

(b) Is point C above or below the water table?

Point C is above the water table since the pressure potential at C is less than atmospheric.

The total potential at some other point could now be computed with respect to the same reference datum to determine the direction of water movement—that is, from higher to lower potential.

Soil-Water Characteristic Curves

The relationship between soil-water content and soil-water potential is termed the soil-water characteristic curve. Representative characteristic curves for seven soil types are indicated in Fig. 3-12. Knowledge of the soil-water characteristic curve can be important in the design and management of irrigation systems. Such curves are necessary in the application of tensiometer or gypsum block readings in tension to convert to the equivalent soil-moisture content. The characteristic curve was previously demonstrated in Fig. 3-9 to give an indication of the water available for plant growth for different soil types. As can be noted in Fig. 3-12, soil types made up of fine particles, such as clays, have a higher moisture content at the same tension as soils with coarser particles, such as sands. As shown in Fig. 3-9, finer soils have a higher water holding capacity—that is, more moisture available for plant growth—than coarser soils.

3.4 Infiltration

Definition

Infiltration is defined as the process by which water passes through the soil surface and enters the subsoil, generally the root zone for applications in irrigation. The rate at which infiltration can be maintained in a particular soil is an extremely important parameter in the design of irrigation systems. The type of irrigation system which may be applied at a given site is often governed by the infiltration characteristics of the soil. The infiltration rate also usually plays a key role in the management and operating schedule of an irrigation system.

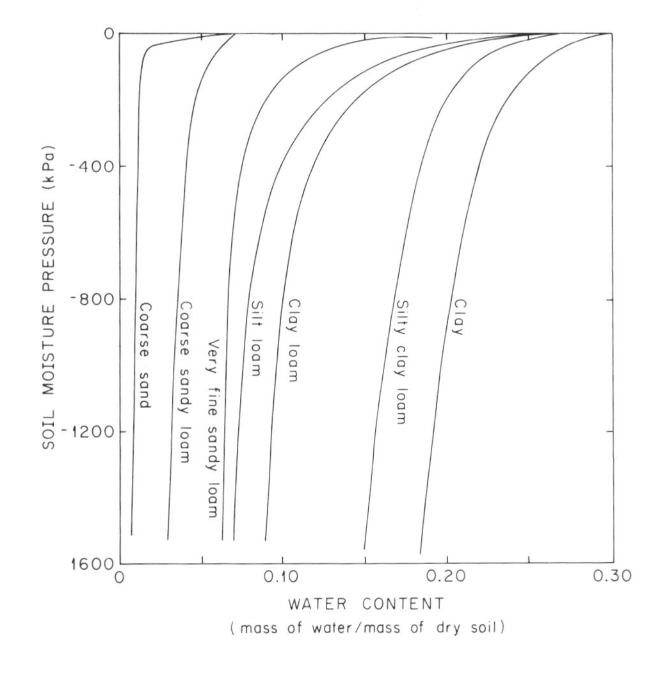

Figure 3-12 Soil-water characteristic curves for several soil types.

Quantification of Infiltration

For the following discussion, it will be convenient to define the variables listed:

i = depth of infiltration, cm

i_0 = depth of infiltration at $t = t_0 = 0$

t = time, min

i_1 = depth of infiltration at $t = t_1$

I = rate of infiltration, cm/min

Figure 3-13 indicates the variation in cumulative infiltration versus time for different representative soil types. This figure demonstrates characteristic infiltration curves which start out with a steep slope which diminishes with time. Using the terminology described in this section, we say that for the curves shown in Fig. 3-13, di/dt is constantly decreasing.

Figure 3-14 indicates other characterizations of infiltration which are useful in irrigation system design. Those values are defined as the following:

$$\text{instantaneous infiltration rate} = \left.\frac{di}{dt}\right|_t = I_t \tag{3-24}$$

$$\text{accumulated depth of infiltration at time } t = i_t$$

$$\text{average infiltration rate} = \frac{i_t - i_0}{t - t_0} \tag{3-25}$$

As demonstrated in Fig. 3-14, the instantaneous and average infiltration rates normally decrease with time in agricultural soils.

The rate of infiltration is also a function of the initial moisture content of the soil. This is demonstrated graphically in Fig. 3-15 in which the infiltration rate is

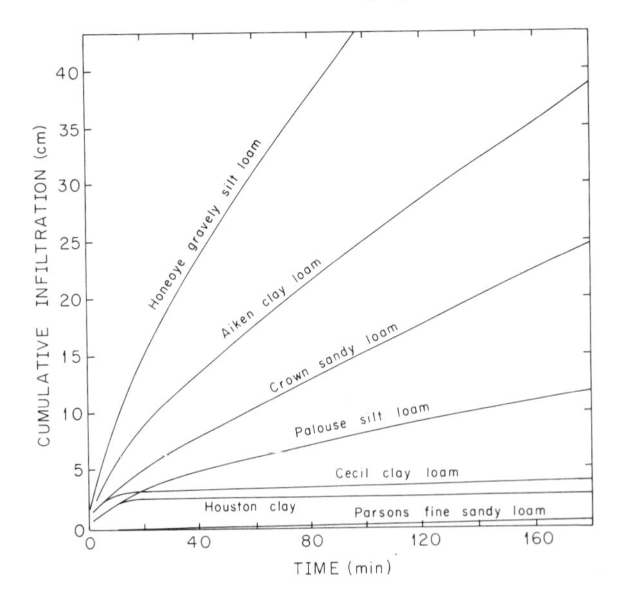

Figure 3-13 Infiltration curves for different soil types. (Adapted from Taylor and Ashcroft, 1972.)

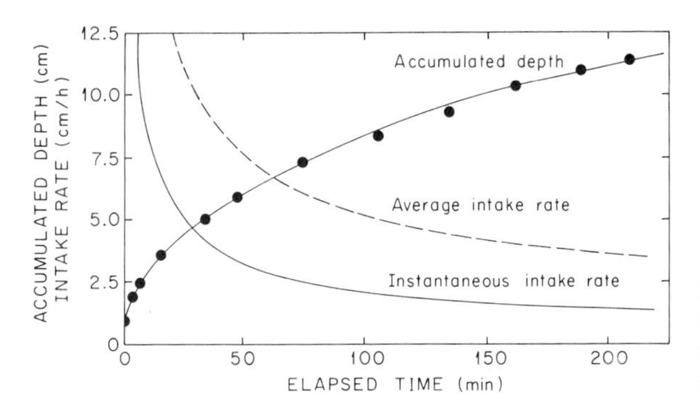

Figure 3-14 Typical patterns for average and instantaneous intake rates and cumulative depth of infiltration. (Adapted from Hansen et al., 1980.)

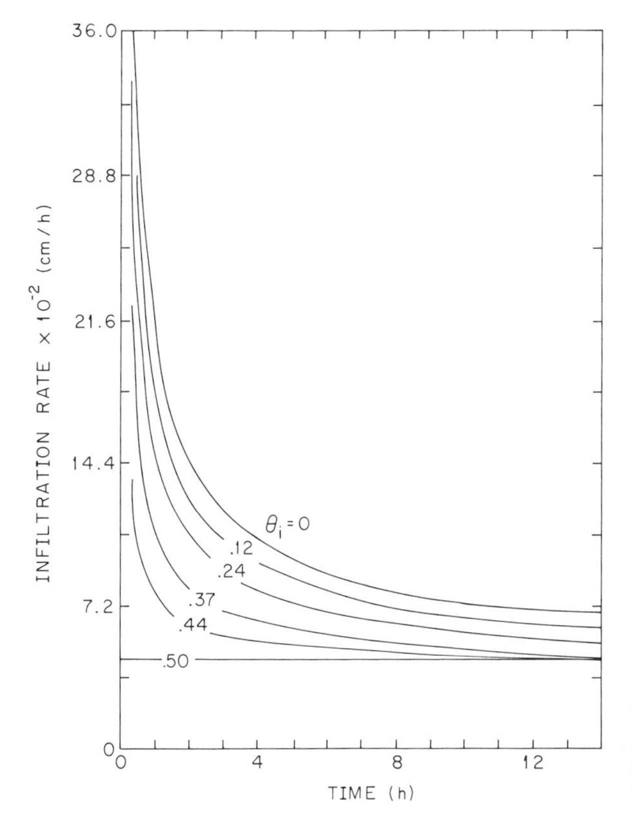

Figure 3-15 Different infiltration rate curves caused by variation in initial water content for Yolo light clay. (Taken from Taylor and Ashcroft, 1972.)

plotted versus time for six different initial moisture concentrations of the same soil. The figure demonstrates that the magnitude of the infiltration rate increases with decreasing initial moisture content.

Infiltration Equations

Numerous equations have been developed to represent the infiltration phenomena. Most of these equations are empirical in nature and have been developed to match observed data sets. Systems of equations have also been applied to produce numerical simulation models of the infiltration phenomenon. Many of these numerical simulation models have empirical terms within them and may therefore be termed quasi-numerical models (Cuenca and Corey, 1981). These numerical simulation models are not generally used in irrigation system design at present, but will find increased application in the future with additional computerization of design procedures. Based on present conditions, this book emphasizes application of empirical equations which have proven useful in system design.

Kostiakov equation

An early equation to quantify infiltration was that developed by Kostiakov (1932). It is described by the following:

$$i = c(t)^{\alpha} \tag{3-26}$$

where

i = depth of infiltration, cm

t = time of infiltration, min

c and α = empirical constants

The Kostiakov equation has been found to fit field measured infiltration data especially over relatively short periods—that is, in the range of a few hours. The fact that this method fits data well over the time of a few hours makes it particularly adaptable to irrigation system design.

Philip equation

A slightly more complex equation, but one which may be derived from the same field data set as the Kostiakov equation, is that developed by Philip (1957). Philip's equation can be derived from theoretical analysis of one-dimensional vertical infiltration into a uniform soil. It is given as:

$$i = S_p(t)^{0.5} + A_p(t) \tag{3-27}$$

where

i = depth of infiltration, cm

t = time of infiltration, min

S_p = sorptivity constant, cm/(min)$^{0.5}$

A_p = conductivity constant, cm/min

The sorptivity term, S_p, is dominant during the early stages of infiltration and the conductivity term, A_p, dominant during the late stages of infiltration. The conductivity constant approximates the saturated hydraulic conductivity for the soil type being analyzed. Philip's equation has been shown to produce relatively accurate results over moderate periods of time with improving accuracy for periods longer than a few hours.

It should be noted that both the Kostiakov and Philip equations are normally given with different constants for different soil types. In reality, infiltration rates change not only with soil type but also with soil cover, soil temperature, previous soil-moisture history and other conditions (Taylor and Ashcroft, 1972). In practical application to system design, these refinements are normally ignored and the constants are assumed to be a function of soil type only. This will lead to some decrease in accuracy in the system design. As in all design decisions of this type, it is assumed that the resulting decrease in accuracy will not significantly compromise the overall design. This is generally a valid assumption in the case of infiltration since, in reality, soil-water properties like infiltration vary stochastically over a field with a significant degree of random fluctuation (Warrick and Nielsen, 1980).

Soil Conservation Service equation

The United States Soil Conservation Service (SCS) has made a large number of field trials to measure and categorize infiltration rates. The SCS has used a slightly modified form of the Kostiakov equation to represent infiltration. Application of this method has been aided by use of the intake family concept. Figure 3-16 indicates cumulative depth of infiltration versus time of infiltration in minutes. By locating the results of a field test on the graph for the soil in question, the nearest intake family to represent the particular soil type can be chosen. The governing equation for infiltration using the SCS method is given by

$$i = a(t)^b + c \tag{3-28}$$

in which i and t are as previously defined except that i may represent the depth of infiltration in inches or centimeters, and a and b are given as a function of intake family. a, which varies depending on whether i is determined in inches or centimeters, and b are listed for different intake families in Table 3-3. c is equal to 0.275 for i in inches and equal to 0.6985 for i in centimeters.

The number of the SCS intake family in Table 3-3 approximates the level of the long-term intake rate in inches per hour. The long-term intake rate is defined as the point on a Kostiakov type infiltration curve at which the infiltration rate decreases by 5 percent within a one-hour period. This may be expressed as

$$(d/dt)(di/dt) = d^2i/dt^2 = (0.05/60 \text{ min})di/dt \tag{3-29}$$

The time at which the long-term intake rate occurs can be calculated using either of the following two equations:

$$IF/60 = di/dt = a(b)(t_L)^{(b-1)} \tag{3-30}$$

or

$$(0.05\ IF)/(60)^2 = d^2i/dt^2 = a(b)(b-1)(t_L)^{(b-2)} \tag{3-31}$$

where

IF = number of the SCS intake family

t_L = time to long-term infiltration rate, min

Solving either of the two preceding equations for t_L approximates the time at which the long-term intake rate occurs. The intake rate in inches per hour at time t_L is equal to IF, the intake family number.

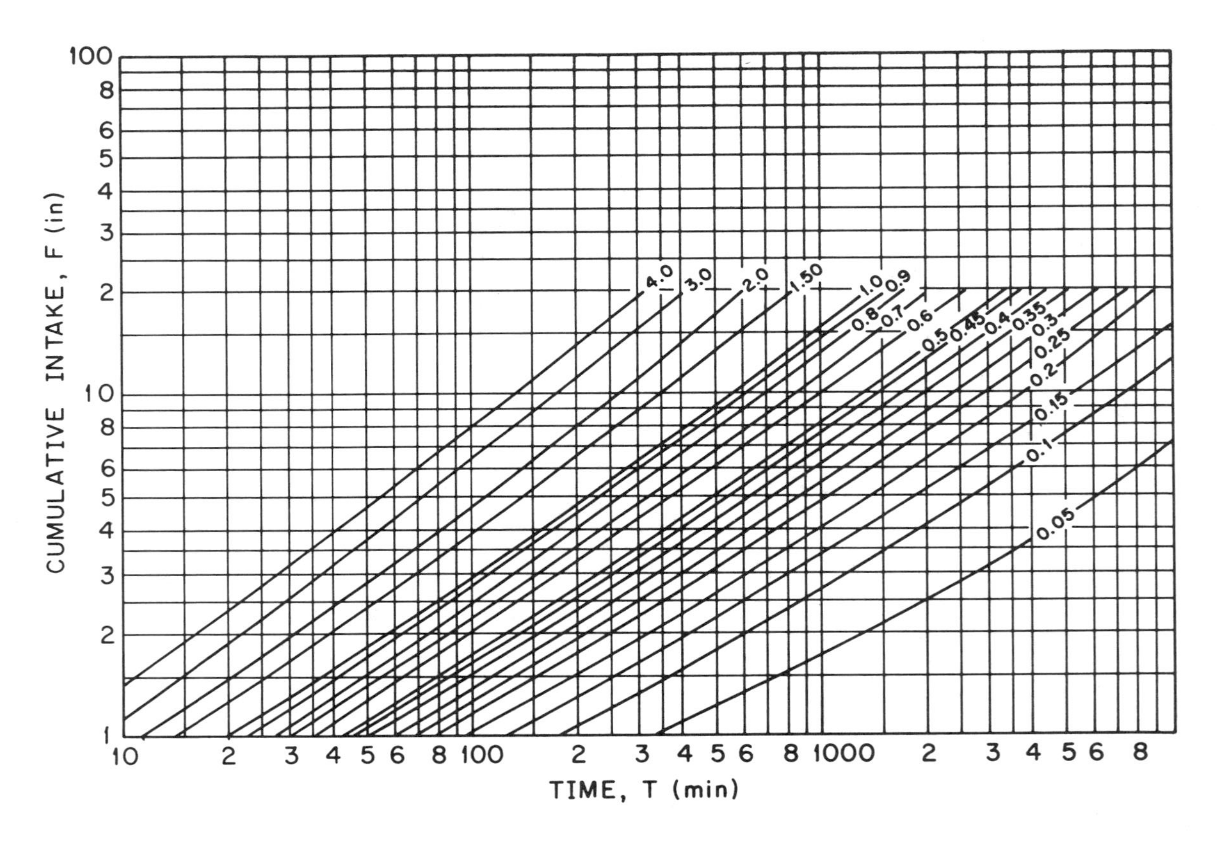

Figure 3-16 Intake families for the SCS method of infiltration.

TABLE 3-3 Parameters for calculation of accumulated infiltration using the SCS intake family concept.

Intake Family	a (cm)	a (in)	b
0.05	0.0533	0.0210	0.6180
0.10	0.0620	0.0244	0.6610
0.15	0.0701	0.0276	0.6834
0.20	0.0771	0.0306	0.6988
0.25	0.0853	0.0336	0.7107
0.30	0.0925	0.0364	0.7204
0.35	0.0996	0.0392	0.7285
0.40	0.1064	0.0419	0.7356
0.45	0.1130	0.0445	0.7419
0.50	0.1196	0.0471	0.7475
0.60	0.1321	0.0520	0.7572
0.70	0.1443	0.0568	0.7656
0.80	0.1560	0.0614	0.7728
0.90	0.1674	0.0659	0.7792
1.00	0.1786	0.0703	0.785
1.50	0.2283	0.0899	0.799
2.00	0.2753	0.1084	0.808
3.00	0.3650	0.1437	0.816
4.00	0.4445	0.1750	0.823

Example Problem 3-4

Compute the time to the long-term infiltration rate in minutes and the cumulative depth of infiltration in cm at this time for the 0.5 and 1.5 SCS intake families.

Solution For the 0.5 intake family the long-term infiltration rate is 0.5 in/hr or 1.3 cm/hr. From Table 3-3 for depth of infiltration in inches, a = 0.0471 and b = 0.7475. Rearranging Eq. (3-30) and solving for t_L,

$$t_L = [(IF/60)1/(ab)]^{[1/(b-1)]}$$

$$t_L = \{(0.5/60)1/[(0.0471)(0.7475)]\}^{[1/(0.7475-1)]}$$

$$t_L = (0.2367)^{-3.9604} = 301 \text{ min}$$

Cumulative depth of infiltration in cm at time t_L using a and c for cm from Table 3-3 and Eq. (3-28),

$$i_{301} = 0.1196(301)^{0.7475} + 0.6985$$

$$i_{301} = 9.2 \text{ cm}$$

For the 1.5 intake family the long-term infiltration rate is 1.5 in/hr or 3.8 cm/hr. From Table 3-3 for depth of infiltration in inches, a = 0.0899 and b = 0.799. Solving for t_L,

$$t_L = \{(1.5/60)1/[(0.0866)(0.799)]\}^{[1/(0.799-1)]}$$

$$t_L = (0.3480)^{-4.9751} = 191 \text{ min}$$

Cumulative depth of infiltration in cm using Eq. (3-28),

$$i_{191} = 0.2283(191)^{0.799} + 0.6985$$

$$i_{191} = 15.9 \text{ cm}$$

The relationship between the value of the intake family and the long-term infiltration rate is very useful in developing general design parameters for surface or sprinkler systems. It should be carefully noted that the coefficients listed in Table 3-3 assume that the soil moisture has been depleted by about 50 percent of the difference between field capacity and crop extractable water before infiltration is initiated. Any significant difference in the antecedent moisture content of the soil from this value will substantially alter the shape and magnitude of the infiltration curve. Irrigation systems are typically designed and operated such that irrigation begins when on the order of 50 percent of the available moisture has been used by the plants. The SCS intake family concept is therefore a reasonable one for application to typical irrigation systems.

Measurement of Infiltration

Infiltration rates are measured in a number of different ways depending upon whether the irrigation system is being designed to serve a border or furrow type surface system or a sprinkler system. The rate of infiltration must be measured in different ways since the effect of conditions at the soil surface which significantly change infiltration rates varies depending on the type of irrigation system. In any case, the best measurement of infiltration is one made in the field for which a design is being developed. Infiltration measurements made in a laboratory are extremely questionable for design purposes since the correlation with field conditions can rarely be guaranteed.

Infiltration rates for border type, and sometimes furrow type, irrigation systems are commonly made with a single-ring or double-ring type infiltrometer. Using either type of instrument, the depth of water infiltrated as a function of time is measured to allow for solution of the parameters in either Kostiakov or Philip type infiltration equations. Common measurement times are 15 and 160 minutes. Fig. 3-17 depicts a sectional view of a double-ring infiltrometer. The double-ring infiltrometer has been proposed as a more representative method of measuring infiltration under steady-state conditions than the single-ring. This is due to the fact that the influence

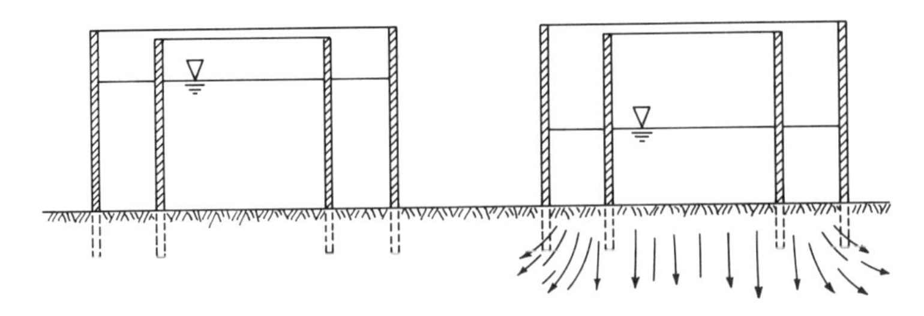

Figure 3-17 Section view of double-ring infiltrometer. View at left is at t = 0 and view at right is after infiltration has proceeded for some time.

of boundary conditions is minimized because infiltration between the inner and outer concentric rings causes near vertical infiltration in the inner ring. The depth of infiltration is measured in the inner ring. It is important that water levels within the two rings are maintained at equal depths to promote vertical infiltration from the inner ring.

Example Problem 3-5

The following infiltration data are taken from a double-ring infiltrometer test:

$$t = 60 \text{ min} \qquad i = 3.5 \text{ cm}$$
$$t = 180 \text{ min} \qquad i = 7.6 \text{ cm}$$

Compute the coefficients for the Kostiakov equation for the soil tested. Compute the predicted cumulative depth of infiltration in cm and the infiltration rate in cm/hr at a time of 5 hours.

Solution Rewriting the Kostiakov Eq. (3-26) using logarithms,

$$\log i = \log c + \alpha \log t$$

Using data from the first measurement period,

$$\log(3.5 \text{ cm}) = \log c + \alpha \log(60 \text{ min})$$
$$0.5441 = \log c + \alpha(1.7782)$$
$$\log c = 0.5441 - \alpha(1.7782)$$

Using data from the second measurement period,

$$\log(7.6 \text{ cm}) = \log c + \alpha \log(180 \text{ min})$$

Substituting the expression for log c from the first time period into the foregoing equation,

$$0.8808 = 0.5441 - \alpha(1.7782) + \alpha(2.2553)$$
$$\alpha = 0.3367/0.4771 = 0.7058$$

Applying this result in the equation for log c,

$$\log c = 0.5541 - 0.7058(1.7782) = -0.7109$$
$$c = 0.1946$$

Computing the cumulative depth of infiltration at t = 5 hrs = 300 min using Eq. (3-26),

$$i_{300} = 0.1946(300 \text{ min})^{0.7058}$$
$$i_{300} = 10.9 \text{ cm}$$

Compute the rate of infiltration at t = 300 min by taking the derivative of Eq. (3-26),

$$di/dt = \alpha[c(t)^{(\alpha-1)}]$$
$$di/dt = 0.7058[0.1946(300 \text{ min})^{(0.7058-1)}]$$
$$di/dt = 0.0256 \text{ cm/min} = 1.54 \text{ cm/hr}$$

Note that the preceding example problem was worked out for the simple case of having two data points for the solution of the two unknowns of the Kostiakov equation. Normally in infiltration tests, one has data from a number of time periods.

The logarithmic form of the Kostiakov equation just shown is linear. Therefore linear regression analysis may be applied to fit a straight line to the log of the depth of infiltration versus the log of the time. The slope of the resulting line is the coefficient and the intercept is the c value. Thus the Kostiakov coefficients can be solved for directly and the standard error of the estimate calculated using linear regression analysis.

The data points given in the foregoing example problem may also be used with the SCS intake families shown in Fig. 3-16. To use this figure the data must be collected at a soil moisture content close to 50 percent depletion between field capacity and crop extractable water. A straight line drawn between the two data points from the problem plotted on Fig. 3-16 approximates the 0.6 intake family line. The constants for this family from Table 3-3 could therefore be used to solve for the cumulative depth of infiltration and infiltration rate at $t = 300$ min.

Infiltration rates for furrow systems are sometimes measured using ring infiltrometers and other times using small Parshall flumes or other discharge measuring devices inserted directly into the furrow at two points separated by a known distance. The difference in discharge between the two devices is the amount of water infiltrated over the length between the measurement sites. Furrow spacing must be taken into account in such tests so the infiltration rate can be expressed as volume per unit area which can be converted to an equivalent depth. The equivalent unit area is equal to the distance between the flumes times the spacing between the furrows. The distance between the flumes must be large enough so errors of flow rate measured with the flumes, which are on the order of ± 10 percent, do not overly influence the estimated amount of infiltration.

Infiltration rates for sprinkler systems are normally measured using a single sprinkler nozzle which is directed to discharge along a particular line in the field along which catch-cans are placed at a fixed interval. This type of system works on the principle that, at proper operating pressures, the amount of water applied to the surface decreases with radial distance from the sprinkler nozzle. At the time of interest, if no water is visible on the surface at a particular point in the field, then it is assumed that the application rate at that point is less than the infiltration rate. Observations are then made along the line of catch-cans moving closer to the sprinkler nozzle. If ponding of water is noticeable on the surface, then the application rate is assumed to be greater than the infiltration rate. By observing field surface conditions and measuring the equivalent depth of water caught in the line of catch-cans, it is possible to determine the proper design application and infiltration rate for the field in question.

3.5 HYDRAULIC CONDUCTIVITY

Concept of Hydraulic Conductivity

Hydraulic conductivity is the proportionality constant between the potential for flow and velocity of soil water movement through the soil. The potential for flow is defined by the hydraulic gradient which is equivalent to the change in total potential measured over some distance. For the general case, the hydraulic gradient measured between two points 1 and 2 may be expressed as

$$\frac{dT}{ds} = \frac{T_1 - T_2}{s_1 - s_2} \tag{3-32}$$

where

ds = distance measured between points 1 and 2

When the magnitude of the osmotic potential between two points is negligible, the total potential is equal to the sum of the gravitational plus pressure potentials. Equation (3-32) can therefore be rewritten in terms of the hydraulic potential, h, and hydraulic gradient, dh/ds:

$$T = z + p = h \tag{3-33}$$

$$\frac{dh}{ds} = \frac{h_1 - h_2}{s_1 - s_2} = \frac{(z_1 + p_1) - (z_2 + p_2)}{s_1 - s_2} \tag{3-34}$$

If v is the equivalent velocity of flow of soil water across the entire cross-sectional area—that is, not just the pore area—then v can be defined as a function of the hydraulic conductivity and hydraulic gradient. This relationship is given by

$$v = -K\frac{dh}{ds} \tag{3-35}$$

where

K = hydraulic conductivity

The hydraulic conductivity, K, is therefore a proportionality factor with the same units as velocity. The minus sign in Eq. (3-35) is used to determine the direction of flow with respect to the reference datum and is required since flow is from the point of higher to lower potential.

Note that the velocity term in Eq. (3-35) is the apparent velocity across the entire cross-sectional area which is made up of soil and pore space. This velocity term is also called the Darcian velocity after Henry Darcy who developed the concept expressed by Eq. (3-35) which is referred to as Darcy's law (Darcy, 1856). Water moves only through the pore space and the actual fluid velocity, termed the pore velocity, v_p, is given by the following equation under conditions of a saturated soil:

$$v_p = \frac{v}{N} \tag{3-36}$$

where

N = soil porosity

In most applications in irrigation system design, we are concerned with the Darcian rather than the pore velocity. The pore velocity is an important parameter in the analysis of chemical transport by convection and diffusion in the soil-water system.

Hydraulic conductivity is a function of a number of parameters including texture, temperature, and porosity. It is most notably a function of soil-water content. In irrigation system design we are typically interested in the value of conductivity under saturated soil conditions. The hydraulic conductivity should actually be expressed as a function of θ_v, the volumetric soil-water content. Darcy's law may therefore be rewritten as

$$v = -K(\theta_v)\frac{dh}{ds} \tag{3-37}$$

for a soil-water system and as

$$v = -K(\theta_v)\frac{dT}{ds} \tag{3-38}$$

for a soil-water plant system in which osmotic potential may arise across the root membrane. Variation of hydraulic conductivity as a function of soil-water pressure, which may be related to soil-water content by the soil-water characteristic curve, is demonstrated in Fig. 3-18 for four different soil types. These curves indicate a decrease in hydraulic conductivity with decreasing soil moisture content—that is, increasingly negative soil-water pressure. Relationships in this figure help explain the concept of field capacity in that K decreases significantly in the range of -1 m to -3 m potential, which defines the range of rapid drainage. Drainage will continue to levels less than -10 m (1 bar) but at much reduced rates.

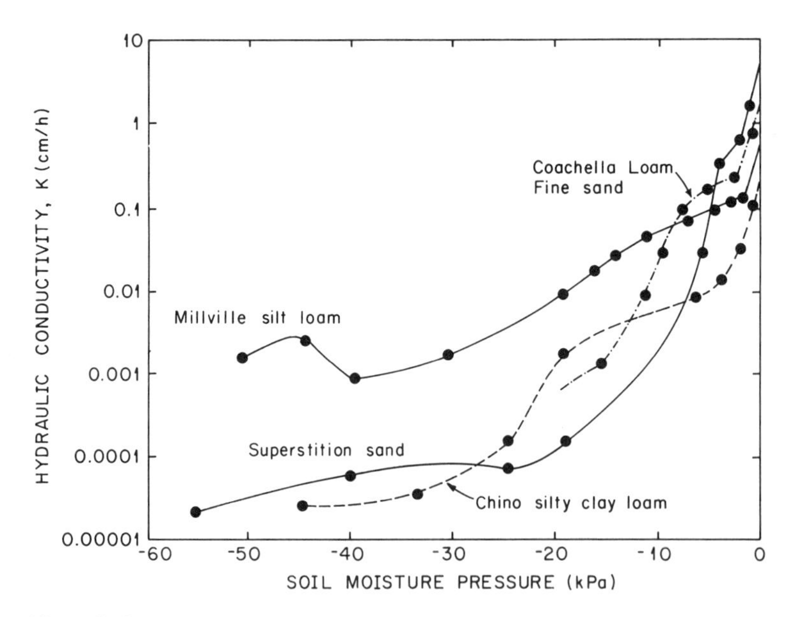

Figure 3-18 Variation of hydraulic conductivity with pressure potential for different soils.

Measurement of Hydraulic Conductivity

Hydraulic conductivity is typically measured in the laboratory using undisturbed soil cores or in the field under conditions of saturation using a shallow hole. An example of an apparatus for measuring saturated conductivity in the laboratory is indicated in Fig. 3-19. Figure 3-20 indicates application of the auger hole method to measure saturated conductivity in the field. The method of measuring hydraulic conductivity by the auger hole method is described in various references (Jensen, 1980; Smedema and Rycroft, 1983; Black et al., 1965).

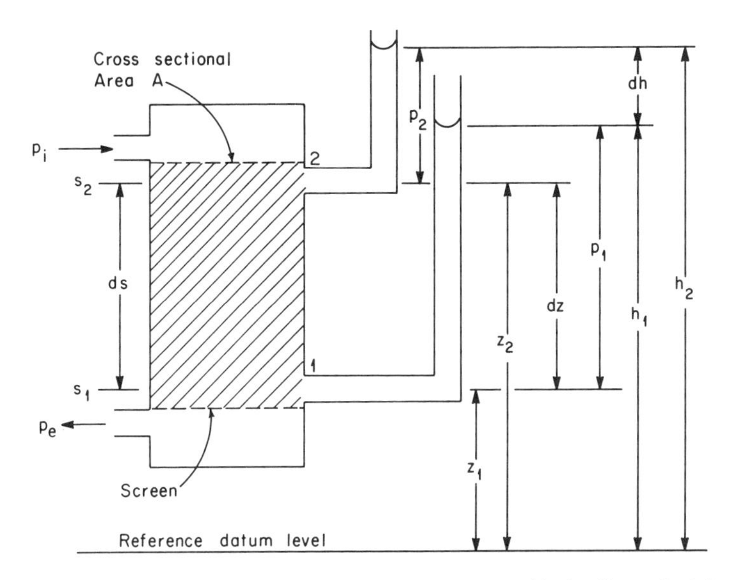

Figure 3-19 Schematic of apparatus for measuring saturated hydraulic conductivity.

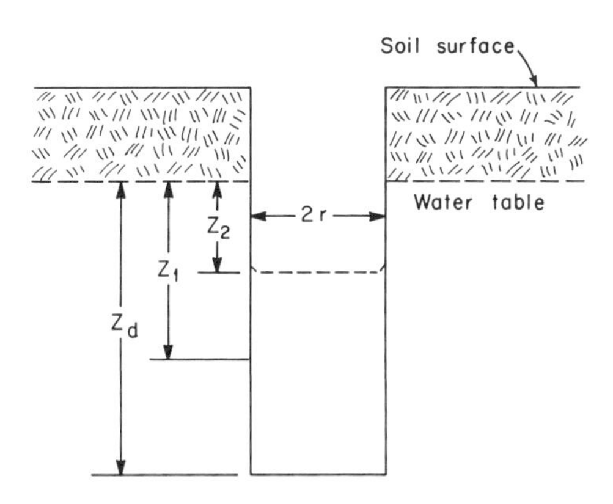

Figure 3-20 Auger hole method for determination of *in situ* hydraulic conductivity.

Example Problem 3-6

Referring to Fig. 3-19 (not drawn to scale), assume the following values:

$$z_1 = 10 \text{ cm}$$
$$z_2 = 40 \text{ cm}$$
$$p_1 = 17 \text{ cm}$$
$$p_2 = 8 \text{ cm}$$

(a) Assuming steady state, what is the hydraulic gradient for flow between points 1 and 2?

Solution

$$h_1 = p_1 + z_1 = 17 \text{ cm} + 10 \text{ cm} = 27 \text{ cm}$$
$$h_2 = 8 \text{ cm} + 40 \text{ cm} = 48 \text{ cm}$$
$$ds = s_1 - s_2 = 10 \text{ cm} - 40 \text{ cm} = -30 \text{ cm}$$
$$\text{hydraulic gradient} = \frac{dh}{ds} = \frac{27 \text{ cm} - 48 \text{ cm}}{-30 \text{ cm}} = 0.700 \text{ cm/cm}$$

(b) What is the direction of flow?

Solution From higher to lower potential. Since $h_2 > h_1$, flow is from point 2 to point 1, or downward in the cylinder.

(c) Assume cross-sectional area $A = 300 \text{ cm}^2$ and steady-state $Q = 75 \text{ cm}^3/\text{hr}$. Compute the hydraulic conductivity in cm/hr.

Solution Since flow is downward with respect to reference datum, we may assign a negative sign to velocity in computing K.

$$-K = v\frac{ds}{dh} = -0.250 \text{ cm/h}\left[\frac{1}{0.700 \text{ cm/cm}}\right]$$
$$K = 0.357 \text{ cm/h}$$

As with other soil-water properties, values of saturated hydraulic conductivity vary over a wide range under field conditions. At present, there are no generally accepted guidelines specifying the number of samples required to have a representative measure of hydraulic conductivity on a field scale. Research results indicate that something on the order of 1300 measurements would have to be made in a 10-hectare field to accurately measure saturated hydraulic conductivity to within 10 percent of the mean value (Warrick and Nielsen, 1980). This type of sampling program is impractical to carry out for the design of irrigation systems. Those working in engineering must therefore continue to rely on previous experience gained in field measurement of conductivity until reasonable guidelines which address the question of cost effectiveness of each subsequent measurement for hydraulic conductivity are developed by soil scientists.

REFERENCES

BLACK, C. A., D. D. EVANS, J. L. WHITE, L. E. ENSMINGER, and F. E. CLARK, *Methods of Soil Analysis,* Part 1-Physical and Mineralogical Properties, Including Statistics of Measurement and Sampling. Madison, Wisconsin: American Society of Agronomy, 1965.

COREY, A. T. and A. KLUTE, "Application of the Potential Concept to Soil Water Equilibrium and Transport," U.S. Department of Agriculture, Rocky Mountain Area Agricultural Research Service in Cooperation with Agricultural Experiment Station, Colorado State University, 1983.

CUENCA, R. H. and P. D. COREY, "Revised Parametric Simulation Model for Soil Water Depletion by Crops," Proceedings of the 1981 Summer Computer Simulation Conference, Washington, D.C., 1981, pp. 282–285.

DARCY, H., *Les Fontaines Publiques de la Ville de Dijon*. Paris: Victor Dalmont, 1856, pp. 570, 590–594.

HANSEN, V. E., O. W. ISRAELSEN, and G. E. STRINGHAM, *Irrigation Principles and Practices,* 4th edition. New York: John Wiley and Sons, 1980.

HILLEL, D., *Fundamentals of Soil Physics*. New York: Academic Press, 1980.

JENSEN, M. E., ed., *Design and Operation of Farm Irrigation Systems*. St. Joseph, Michigan: American Society of Agricultural Engineers, Monograph Number 3, 1980.

LUTHIN, J. N., *Drainage Engineering*. Huntington, New York: Robert E. Krieger Publishing Co., 1973.

SMEDEMA, L. K. and D. W. RYCROFT, *Land Drainage*. Ithaca, New York: Cornell University Press, 1983.

TAYLOR, S. A. and G. L. ASHCROFT, *Physical Edaphology*. San Francisco, CA: W. H. Freeman and Co., 1972.

WARRICK, A. W. and D. R. NIELSEN, "Spatial Variability of Soil Physical Properties in the Field," in *Applications of Soil Physics,* ed., D. Hillel. New York: Academic Press, 1980, pp. 319–344.

PROBLEMS

3-1. Happy Sam irrigates his 25 ha field once a week. Because he is uncertain of his crop water requirement and in an effort to thoroughly irrigate the field, he loses an areal average of 2 cm to deep percolation with each irrigation. His pumping plant requires 75 kW of power and the average application rate of his irrigation system is 1 cm/hr. Assume water costs are 55 cents per 100 m^3 and energy costs 5 cents per kWh. How much is Sam paying for this deep percolation over a 12-week growing season?

3-2. Define the following terms using one sentence for each:

(a) Gravitational water
(b) Capillary water
(c) Hygroscopic water
(d) Field capacity
(e) Crop extractable water
(f) Available water

3-3. Given a saturated soil sample with mass equal to 283 g and an oven dry mass of 202 g, find the water content on a mass and volume basis, the soil bulk density, and the porosity. Assume density of water equals 1.00 g/cm^3 and density of soil particles equals 2.65 g/cm^3.

3-4. How much water (m^3) must be added to a field of area 3 ha to increase the volumetric water content of the top 40 cm from 16 to 28 percent? Assume all water added to the field stays in the top 40 cm.

3-5. A soil sample with a wet weight of 300 g has a 28.0 percent water content on a mass basis. Its saturated water content is 36.1 percent on a mass basis. Assume density of water equals 1.00 g/cm^3 and density of soil particles equals 2.65 g/cm^3.

(a) Find the mass of water, soil porosity, and air filled porosity of the sample at 28.0 percent water content.

(b) If the sample at 28.0 percent water content is representative of the top 45 cm of a 0.3 ha plot, how much water, in m^3 and as an average areal depth in cm, must drain from the top 45 cm to increase the air filled porosity to 25.0 percent?

3-6. A soil sample taken in the field has a wet weight of 373 g. The sample is dried at 105–110°C to a constant weight of 295 g. The apparent specific gravity of the soil is 1.4 and the density of the soil particles is 2.58 g/cm^3. Find the water content on a mass and volume basis, volumetric water content as percent, soil porosity, and the aeration porosity.

3-7. A tensiometer open to the atmosphere with a porous ceramic cup at the end is placed in a field as shown in the figure.

(a) Let h_{Hg} = 10.0 cm. Is point A above or below the water table? What is the total potential?

(b) Let h_{Hg} = 1.5 cm. Is point A above or below the water table? What is the total potential?

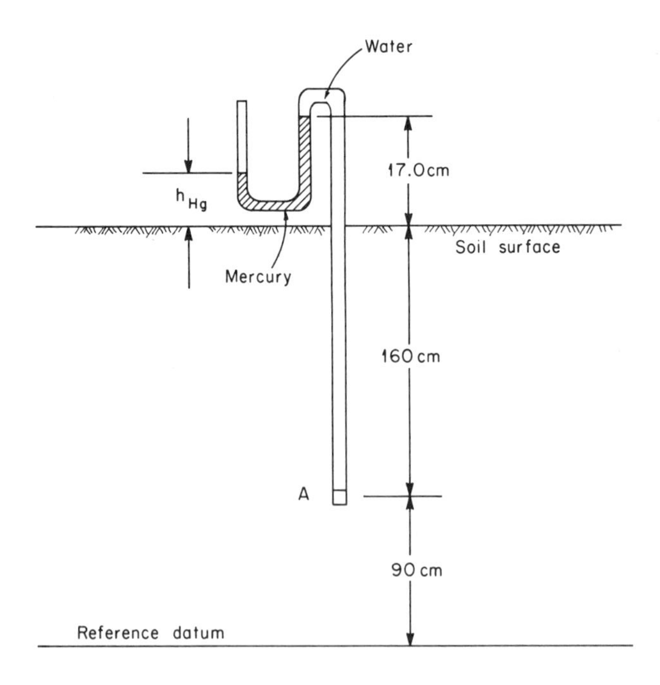

3-8. A soil which is homogeneous in the horizontal and vertical directions is cropped in tomatoes. Three resistance blocks calibrated for tension are placed in the soil and tension readings are made as shown in the figure.

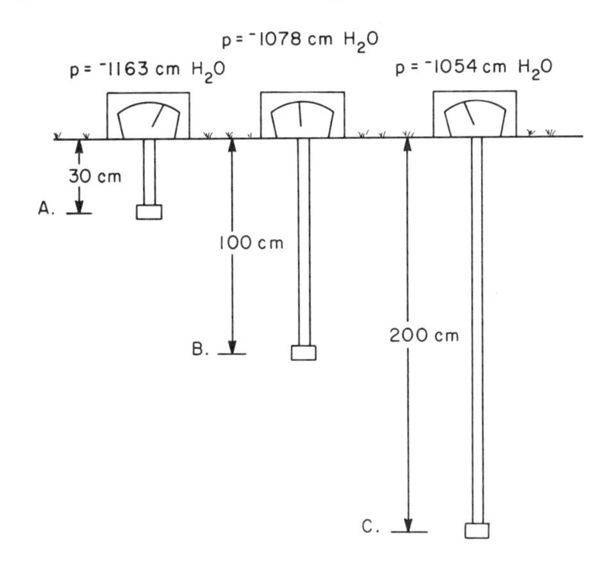

Is the soil water flowing up or down between points A and B; between points B and C? How could you logically explain your results to a grower who asked you to monitor his or her field?

3-9. A soil-water characteristic curve is approximated by the function

$$\theta_v = 0.35 - [(\log(t) - 2.0)(0.10)]$$

where

θ_v = volumetric water content, fraction

t = soil water tension, cm H_2O

Assume that at the time of harvest a wheat crop grown in this soil will uniformly deplete 85 percent of the available water in the soil profile to a depth of 2.5 m. Assume that field capacity is at a tension of 180 cm and the limit of crop extractable water is at a tension of 12,000 cm. What volume of water in cubic meters will be required to bring a 45-hectare field to field capacity following harvest of the wheat? How many acre-feet is this?

3-10. An infiltration test is made on a silt loam soil. The following data points are determined from the graph of the field data.

t = 15 min.   i = 2.5 cm

t = 160 min.   i = 9.2 cm

Apply Philip's equation with time in minutes and solve for S_p and A_p. Using the computed values of S_p and A_p, calculate the total infiltration after 5 hours.

3-11. The following data are obtained from a ring infiltrometer test on a clay loam soil.

Time, t min	Infiltration, i in
1	0.52
15	3.15
30	4.80
60	7.20
150	14.05
200	15.30

(a) Plot the data on log-log graph paper with the time axis as the abscissa. Approximate the best fit straight line through the data.

(b) Use the logarithmic form of the Kostiakov equation,

$$\log i = \log c + \alpha \log t$$

(i) Determine c from the intercept of the plotted line with the ordinate (infiltration) axis (i.e., at $\log t = 0$)

(ii) Determine α as the slope of the plotted line. Note that α in the preceding equation is a function of log i and log t (and not i and t which are plotted).

(iii) Using the computed values of c and α, determine the infiltration after 6 hours.

3-12. Two wells penetrate an inclined groundwater aquifer as shown in the sketch. The hydraulic conductivity of the soil has been determined by previous tests to be 7.4 cm/hr. A contaminant is accidentally spilled into well A. Given the conditions shown and assuming the most direct solution (i.e., no miscible displacement), how long a time in years will it be before the contaminant is noticed in well B if the aquifer porosity is 37 percent?

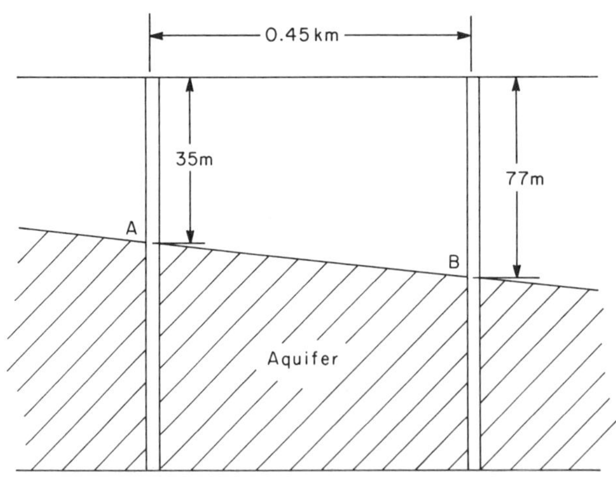

3-13. An experimental soil column to measure unsaturated hydraulic conductivity is set up in a cylinder as shown in the sketch. The conductivity is assumed to correspond to the average pressure potential. When the system comes to steady state, $Q = 119\ cm^3/hr$.

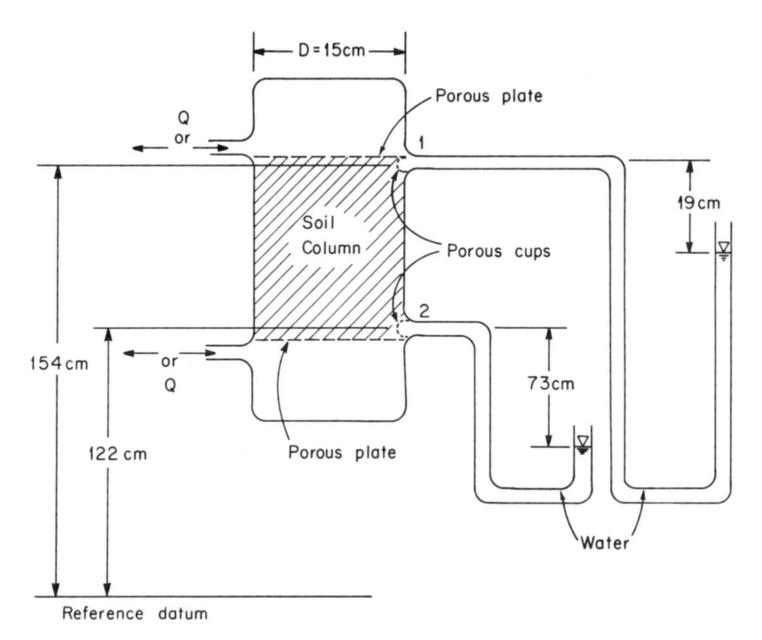

What is the direction of flow through the sample (up or down), the average pressure potential (cm), and the value of the hydraulic conductivity (cm/hr)?

3-14. A similar column to that in the previous problem is set up except mercury is used in the manometer to be able to record higher tensions. (See sketch.)

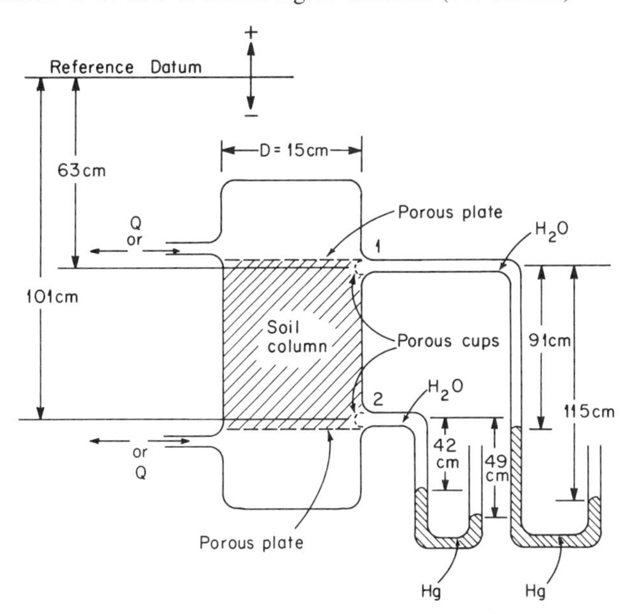

When the system comes to steady state, Q = 53 cm^3/hr. What is the direction of flow through the soil column, the average pressure potential (cm), and the hydraulic conductivity (cm/hr) of the soil sample?

3-15. A dry sandy soil is underlain by a dry clay loam soil in a shallow ditch. Water is applied to the sandy soil.

For conditions before the wetting front reaches the clay loam,

(a) Describe the relative movement of water vertically and laterally away from the ditch.

(b) Describe the relative hydraulic conductivity in the transmission zone, at the wetting front, and ahead of the wetting front.

After the wetting front penetrates the clay loam,

(c) Describe the relative rate and direction of movement of soil water in the sandy soil and in the clay loam.

(d) Describe the relative hydraulic conductivity in the sandy soil, in the transmission zone of the clay loam soil and ahead of the wetting front in the clay loam.

After the water is shut off and the sandy soil has some time to drain,

(e) Describe the relative hydraulic conductivity in the sandy and clay loam soils.

(f) Describe the relative rate of soil water movement in the sandy and clay loam soils.

COMPUTER PROBLEM

The sketch for Problem 3-13 indicates a soil column apparatus used to determine values of hydraulic conductivity in the laboratory by application of Darcy's law in the following form:

$$Q = K(A)\frac{dT}{ds}$$

where

Q = volumetric flow rate, L^3/T

K = hydraulic conductivity, L/T

A = cross-sectional flow area, L^2

s = distance between points of measurement of total potential, L

T = total potential, L

$= z + p$

where

z = gravitational potential above an arbitrary datum, L

p = pressure potential measured by manometers, L

Write an interactive computer program to evaluate the hydraulic conductivity which allows for the following variable inputs:

(a) Q
(b) D
(c) z_1
(d) p_1
(e) z_2
(f) p_2

The program should output the following:

(a) Listing of input variables
(b) Computed average total potential
(c) Computed value of the hydraulic conductivity

Run a test case of the program for the conditions shown in Fig. P3-5 with Q = 119 cm^3/hr.

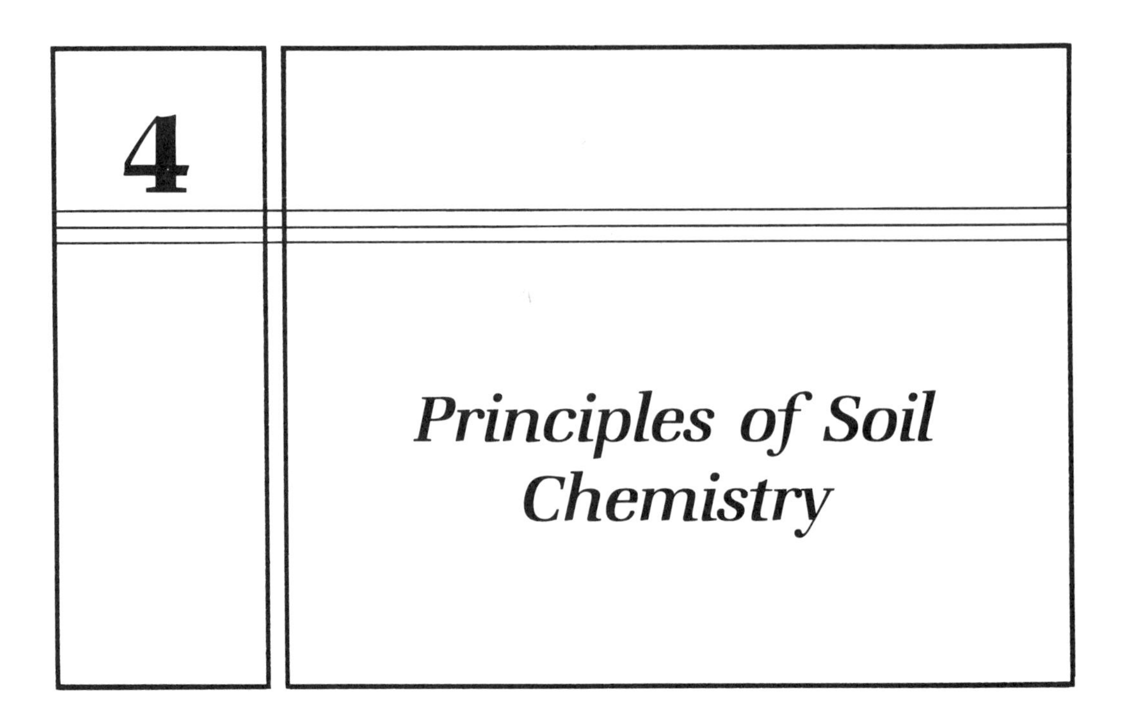

4.1 Introduction

Concepts of soil chemistry and the reactions of soil and water compounds are important to the irrigation engineer because of the ultimate impact on crop production. An example of this impact is demonstrated in Fig. 4-1. Concerns involving soil chemistry and water quality interaction can be divided into two major categories. The first of these is the accumulation of salts in the soil profile and their effect on crop growth. The second is the impact of specific chemical elements in the water or the soil on crop growth or even crop survival.

Problems of salinity or toxic element build-up can in extreme cases cause abandonment of potential or ongoing irrigation projects. An example of a project abandoned due to salinity effects is shown in Fig. 4-2. At the very least, if potential salt or toxic element problems exist, adjustments must be made in the design procedure so the effects will not make crop production totally infeasible. These adjustments will affect the design crop water requirement and often require design of drainage facilities in the irrigated area.

Problems of salinity affect agricultural production throughout the world. Many areas with long histories of irrigated agriculture have serious salinity problems due to the build-up of salts over the centuries. One survey indicates that out of 756 million hectares cultivated worldwide, 50 million hectares have decreased production

Figure 4-1 Damage to cereal production caused by saline conditions in Aragon Province, Spain.

Figure 4-2 Previously irrigated fields abandoned due to salt build-up in Aragon Province, Spain. Vegetation shown is an extremely salt-tolerant species of no commercial value.

potential due to salinity (Shalhevet and Kamburov, 1976). Another report indicates that salinity is a threat to agricultural productivity on approximately half of the 20 million hectares of irrigated land in the western United States and that crop production is already limited by salinity on approximately 25 percent of this land

(Wadleigh, 1968). Specific chemical elements can also cause severe crop stress or crop destruction if present at high enough concentrations. Chloride and boron are the most common of these types of compounds.

This chapter will begin with a review of basic concepts in chemistry required to understand and analyze problems involving chemical interactions in soil-water systems. Specific concepts related to chemical analysis of soil and water resources for agricultural production will be presented next. The impacts of different concentrations of salinity and specific compounds on crop production will be indicated. The final section will discuss design procedures and adjustments necessary to manage soil chemical concentrations.

4.2 Fundamentals of Chemistry

The purpose of this section is to review fundamental principles in chemistry required for analysis of soil-water systems. These principles will serve as the basis of the analyses discussed in later sections of this chapter.

Periodic Table of the Elements

The Periodic Table gives information about the atomic number, atomic weight, and oxidation states of different elements. This information is required to compute both the concentration of different elements and the potential for reactions. We begin by looking at the amount of material contained in a mole of an element or compound and the procedure to compute the molecular weight.

One mole contains 6.02×10^{23} atoms of an element or 6.02×10^{23} molecules of a compound, where 6.02×10^{23} is Avogadro's number. The gram molecular weight or simply the molecular weight is given by

$$MW = \sum_{i=1}^{N} x_i \tag{4-1}$$

where

x_i = atomic weight of element, g

N = number of elements in compound

The molecular weight is equivalent to 1 mole of a substance. Atomic weights from the Periodic Table of the Elements are given in Table 4-1 and listed in Table 4-2.

Example Problem 4-1

(a) Calculate the molecular weight of $CaCO_3$.
Solution

$$\begin{aligned} CaCO_3 &= 40.03\text{ g} + 12.01\text{ g} + 3(16.0\text{ g}) \\ &= 100.04\text{ g} \end{aligned}$$

(b) Calculate the molecular weight of H_2SO_4.

Solution

$$H_2SO_4 = 2(1.008 \text{ g}) + 32.06 \text{ g} + 4(16.00 \text{ g})$$
$$= 98.08 \text{ g}$$

A molar solution consists of 1 molecular weight (i.e., 1 mole) of a substance dissolved in enough water to make a total of 1 liter of solution. The concentration of such a solution is termed 1 molar (1 M).

Example Problem 4-2

Compute the molar concentration of a solution containing 10 g/L of

(a) NaOH
(b) Na_2SO_4

Solution First compute the molecular weight and then the concentration.

(a)
$$NaOH = 22.94 \text{ g} + 16.00 \text{ g} + 1.008 \text{ g}$$
$$= 39.948 \text{ g}$$

Denoting concentrations by square brackets,

$$[NaOH] = (10 \text{ g/L})/(39.948 \text{ g/mole}) = 0.25 \text{ M}$$

(b)
$$Na_2SO_4 = 2(22.94 \text{ g}) + 32.06 \text{ g} + 4(16.00 \text{ g})$$
$$= 141.94 \text{ g}$$
$$[Na_2SO_4] = (10 \text{ g/L})/(141.94 \text{ g/mole}) = 0.0705 \text{ M}$$

Valency and Oxidation-Reduction Reactions

Atoms consist of neutrons, protons (+), and electrons (−). The number of electrons are arranged in orderly rings surrounding the nucleus made up of neutrons and protons. The number of electrons is equal to the number of protons.

In a chemical reaction, the metal or metal-like element loses electrons to gain or approach a stable condition with no electrons in its outer ring. This atom therefore becomes a positively charged ion. The nonmetal element takes electrons from the metal to bring its outer ring to a stable condition with eight electrons. This simple type of reaction is illustrated in Fig. 4-3 showing the combination of sodium and chloride.

Some elements, such as chlorine, may take on different valences depending on the reaction. Chlorine may take on the valences of 1^+, 2^+, 3^+, 4^+, 5^+, and 7^+. Three of these valence states are indicated in Fig. 4-4.

Note that the valences involved in a reaction always cancel such that the final molecule has no charge or zero valence, meaning that it has an equal number of protons and electrons. For the reactions in Fig. 4-4,

(a)
$$2Cl^+ + O^{2-} \longrightarrow Cl_2O \tag{4-2}$$
(b)
$$Cl^{4+} + 2O^{2-} \longrightarrow ClO_2 \tag{4-3}$$
(c)
$$2Cl^{7+} + 7O^{2-} \longrightarrow Cl_2O_7 \tag{4-4}$$

TABLE 4-1 Periodic table of the elements indicating for each element from top to bottom the atomic number, atomic weight, chemical symbol, and oxidation states found in soils and plants. (Adapted from Bohn et al., 1979.)

PERIOD	GROUP IA	IIA	IIIB	IVB	VB	VIB	VIIB	VIII	VIII	VIII	IB	IIB	IIIA	IVA	VA	VIA	VIIA	
1	1 1.008 H 1																	2 4.003 He 0
2	3 6.94 Li 1	4 9.01 Be 2											5 10.81 B 3	6 12.01 C ±4,2,0	7 14.01 N 5,±3,0	8 16.00 O −2,0	9 19.00 F −1	10 20.18 Ne 0
3	11 22.94 Na 1	12 24.31 Mg 2											13 26.98 Al 3	14 28.09 Si 4	15 30.97 P 5	16 32.06 S 6,−2,0	17 35.43 Cl −1	18 39.95 Ar 0
4	19 39.10 K 1	20 40.03 Ca 2	21 44.96 Sc 3	22 47.90 Ti 4	23 50.94 V 5,4	24 52.00 Cr 6,3	25 54.94 Mn 4,3,2	26 55.85 Fe 3,2	27 58.93 Co 2	28 58.71 Ni 2	29 63.54 Cu 2	30 65.37 Zn 2	31 69.72 Ga 3	32 72.59 Ge 4	33 74.92 As 3,5	34 78.96 Se 6,4,0	35 79.91 Br −1	36 83.80 Kr 0
5	37 85.47 Rb 1	38 87.62 Sr 2	39 88.91 Y 3	40 91.22 Zr 4	41 92.91 Nb 5,3	42 95.94 Mo 6,3	43 (99) Tc 7	44 101.0 Ru 4	45 102.9 Rh 3	46 106.4 Pd 2,0	47 107.9 Ag 1,0	48 112.4 Cd 2	49 114.8 In 3	50 118.7 Sn 4,2	51 121.8 Sb 3,5	52 127.6 Te 4	53 126.9 I −1	54 131.3 Xe 0
6	55 132.9 Cs 1	56 137.3 Ba 2	57 138.9 La 3	72 178.5 Hf 4	73 181.0 Ta 5	74 183.8 W 6	75 186.2 Re 7	76 190.2 Os 4	77 192.2 Ir 4	78 195.1 Pt 4,0	79 197.0 Au 3,0	80 200.6 Hg 2,1,0	81 204.4 Tl 1	82 207.2 Pb 4,2	83 209.0 Bi 3	84 (210) Po 2	85 (210) At −1	86 (222) Rn 0
7	87 (223) Fr 1	88 (226) Ra 2	89 (227) Ac 3															

PERIOD														
6	58 140.1 Ce 3,4	59 140.9 Pr 3,4	60 144.2 Nd 3	61 (147) Pm 3	62 150.4 Sm 3	63 152.0 Eu 3	64 157.2 Gd 3	65 158.9 Tb 3	66 162.5 Dy 3	67 164.9 Ho 3	68 167.3 Er 3	69 168.9 Tm 3	70 173.0 Yb 3	71 175.0 Lu 3
7	90 232 Th 4	91 (231) Pa 5	92 238 U 6	93 (237) Np 5	94 (242) Pu 4	95 (243) Am 3	96 (247) Cm 3							

TABLE 4-2 Table of atomic numbers, atomic weights (based on Carbon—12), and symbol of chemical elements. Value of brackets is mass number of longest lived or best-known isotope.

	Symbol	Atomic Number	Atomic Weight		Symbol	Atomic Number	Atomic Weight
Actinium	Ac	89	227	Mercury	Hg	80	200.59
Aluminum	Al	13	26.9815	Molybdenum	Mo	42	95.94
Americium	Am	95	[243]	Neodymium	Nd	60	114.24
Antimony	Sb	51	121.75	Neon	Ne	10	20.183
Argon	Ar	18	39.948	Neptunium	Np	93	[237]
Arsenic	As	33	74.9216	Nickel	Ni	28	58.71
Astatine	At	85	[210]	Niobium	Nb	41	92.906
Barium	Ba	56	137.34	Nitrogen	N	7	14.0067
Berkelium	Bk	97	[249]	Nobelium	No	102	[253]
Beryllium	Be	4	9.0122	Osmium	Os	76	190.2
Bismuth	Bi	83	208.980	Oxygen	O	8	15.9994
Boron	B	5	10.811	Palladium	Pd	46	106.4
Bromine	Br	35	79.909	Phosphorus	P	15	30.9738
Cadmium	Cd	48	112.40	Platinum	Pt	78	195.09
Calcium	Ca	20	40.08	Plutonium	Pu	94	[242]
Californium	Cf	98	[251]	Polonium	Po	84	210
Carbon	C	6	12.01115	Potassium	K	19	39.102
Cerium	Ce	58	140.12	Praseodymium	Pr	59	140.907
Cesium	Cs	55	132.905	Promethium	Pm	61	[145]
Chlorine	Cl	17	35.453	Protactinium	Pa	91	[231]
Chromium	Cr	24	51.996	Radium	Ra	88	[226]
Cobalt	Co	27	58.9332	Radon	Rn	86	[222]
Copper	Cu	29	63.54	Rhenium	Re	75	186.2
Curium	Cm	96	[247]	Rhodium	Rh	45	102.905
Dysprosium	Dy	66	162.50	Rubidium	Rb	37	85.47
Einsteinium	Es	99	[254]	Ruthenium	Ru	44	101.07
Erbium	Er	68	167.26	Samarium	Sm	62	150.35
Europium	Eu	63	151.96	Scandium	Sc	21	44.956
Fermium	Fm	100	[253]	Selenium	Se	34	78.96
Fluorine	F	9	18.9984	Silicon	Si	14	28.086
Francium	Fr	87	[223]	Silver	Ag	47	107.870
Gadolinium	Gd	64	157.25	Sodium	Na	11	22.9898
Gallium	Ga	31	69.72	Strontium	Sr	38	87.62
Germanium	Ge	32	72.59	Sulfur	S	16	32.064
Gold	Au	79	196.967	Tantalum	Ta	73	180.948
Hafnium	Hf	72	178.49	Technetium	Tc	43	[99]
Helium	He	2	4.0026	Tellurium	Te	52	127.60
Holmium	Ho	67	164.930	Terbium	Tb	65	158.924
Hydrogen	H	1	1.00797	Thallium	Tl	81	204.37
Indium	In	49	114.82	Thorium	Th	90	232.038
Iodine	I	53	126.9044	Thulium	Tm	69	168.934
Iridium	Ir	77	192.2	Tin	Sn	50	118.69
Iron	Fe	26	55.847	Titanium	Ti	22	47.90
Krypton	Kr	36	83.80	Tungsten	W	74	183.85
Lanthanum	La	57	138.91	Uranium	U	92	238.03

TABLE 4-2 Table of atomic numbers, atomic weights (based on Carbon—12), and symbol of chemical elements. Value of brackets is mass number of longest lived or best-known isotope. (continued)

	Symbol	Atomic Number	Atomic Weight		Symbol	Atomic Number	Atomic Weight
Lawrencium	Lw	103	[257]	Vanadium	V	23	50.942
Lead	Pb	82	207.19	Xenon	Xe	54	131.30
Lithium	Li	3	6.939	Ytterbium	Yb	70	173.04
Lutetium	Lu	71	174.97	Yttrium	Y	39	88.905
Magnesium	Mg	12	24.312	Zinc	Zn	30	65.37
Manganese	Mn	25	54.9380	Zirconium	Zr	40	91.22
Mendelevium	Md	101	[256]				

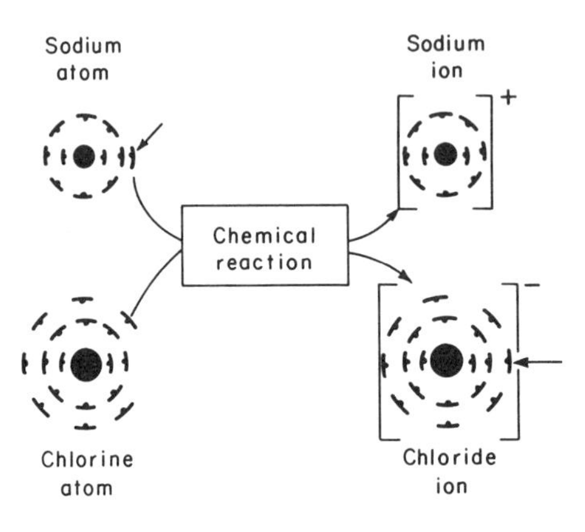

Figure 4-3 Reaction in which an electron is transferred to produce a sodium ion and chloride ion.

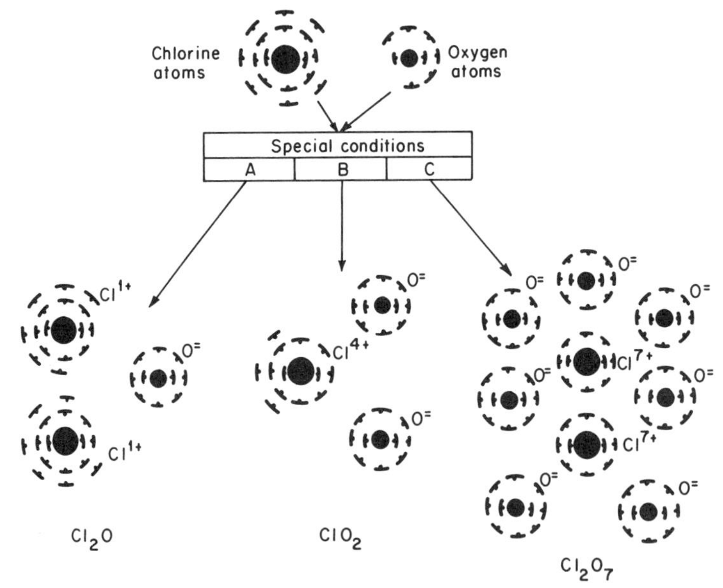

Figure 4-4 Graphical representation of three of the potential six valence states of chlorine. (Adapted from Sawyer and McCarty, 1967.)

Sulfur, nitrogen, and the halogens are nonmetals capable of exhibiting a wide range of valences due to their ability to take on electrons to complete the outer shell of eight electrons or to give up one or more electrons to reach a stable configuration. Manganese, chromium, copper, and iron are metals that yield one or more electrons to obtain different valence states.

An *oxidizing agent* is any substance that can add electrons. Examples are

$$O^2, Cl^0, Fe^{3+}, Cr^{6+}, Mn^{4+}, Mn^{7+}, N^{5+}, N^{3+}, S^0, S^{4+}, S^{6+}$$

A *reducing agent* is any substance that can give up electrons. For example,

$$H^0, Fe^0, Mg^0, Fe^{2+}, Cr^{2+}, Mn^{4+}, N^{3+}, Cl^-, S^0, S^{2-}, S^{4+}$$

Note that Mn^{4+}, N^{3+}, S^0, and S^{4+} can serve as both oxidizing and reducing agents. In any oxidation-reduction reaction, the number of electrons lost by the reducing agent equals the number of electrons gained by the oxidizing agent. Table 4-1 indicates the oxidation or valence states found in soils and plants for all chemical elements. The following equations are examples of simple oxidation-reduction reactions:

(a) $$H_2{}^0 + Cl_2{}^0 \longrightarrow 2H^+Cl^- \qquad (4\text{-}5)$$

(b) $$4Fe^0 + 3O_2{}^0 \longrightarrow 2Fe_2{}^{3+}O_3{}^{2-} \qquad (4\text{-}6)$$

(c) $$Mg^0 + H_2{}^+SO_4{}^{2-} \longrightarrow Mg^{2+}SO_4{}^{2-} + H_2{}^0 \qquad (4\text{-}7)$$

(d) $$2Fe^{2+} + Cl_2{}^0 \longrightarrow 2Fe^{3+} + 2Cl^- \qquad (4\text{-}8)$$

(Note: Other references should be referred to for procedures to balance complex oxidation-reduction reactions—e.g., Sawyer and McCarty, 1967.)

Equivalent Weight and Normality

The *equivalent weight* equals the weight of a compound that reacts with one mole of hydrogen ion (H^+) or hydroxide ion (OH^-). For practical purposes, the equivalent weight equals the molecular weight divided by the valence of the ion. This is demonstrated in the following Example Problem.

Example Problem 4-3

Compute the equivalent weight of the compounds given.

(a) $BaCl_2 \rightarrow Ba^{2+} + 2Cl^-$

Solution

$$\text{equivalent weight} = [137.34\text{ g} + 2(35.45\text{ g})]/2 = 104.1\text{ g}$$

(b) $Al_2(SO_4)_3 \rightarrow 2Al^{3+} + 3(SO_4{}^{2-})$

Solution

$$\text{equivalent weight} = \{2(26.98\text{ g}) + 3[32.06\text{ g} + 4(16.00\text{ g})]\}/6$$
$$= 57.02\text{ g}$$

The *normality* of a solution is the number of equivalents per liter. Thus a 1 normal (1 N) solution is prepared by diluting 1 equivalent weight of a compound to 1 liter.

Example Problem 4-4

Compute the number of grams of $AgNO_3$ required to prepare 500 ml of 0.1 N solution for a precipitation reaction.

Solution From Table 4-1, the valence of Ag is 1+. Therefore, the number of grams is equal to

$$x = 0.1(500\text{ ml}/1000\text{ ml})[107.87\text{ g} + 14.01\text{ g} + 3(16.00\text{ g})]/1$$
$$x = 8.49\text{ g}$$

4.3 Soil Chemical Properties

Classification of Salt-Affected Soils

Salt-affected soils are classified according to the electrical conductivity of a saturation extract (EC_e) measured at 25°C and the sodium absorption ratio. Soils may be classified as normal, saline, sodic, or saline-sodic as indicated in Table 4-3. Soil reclamation and treatment for salinity depends on this classification.

The EC_e is measured using a modified Wheatstone bridge which is commercially available for analysis of this type. The SAR is calculated using the following equation:

$$SAR = [Na^+]/[(Ca^{2+} + Mg^{2+})/2]^{0.5} \quad (4\text{-}9)$$

where

SAR = sodium absorption ratio, $(mmol/L)^{1/2}$
Na^+ = sodium ion concentration, meq/L
Ca^{2+} = calcium ion concentration, meq/L
Mg^{2+} = magnesium ion concentration, meq/L

TABLE 4-3 Classification of salt-affected soils based on analysis of saturation extracts. (Adapted from James et al., 1982.)

Criteria	Normal	Saline	Sodic	Saline-Sodic
EC_e (mmhos/cm)	<4	>4	<4	>4
SAR	<13	<13	>13	>13

The following example problem demonstrates application of Eq. 4-9 and Table 4-3.

Example Problem 4-5

Classify a soil sample which has been analyzed giving the following results:

(a) EC_e = 2.53 mmhos/cm (b) Ca^{2+} = 1.41 meq/L
(c) Mg^{2+} = 1.01 meq/L (d) Na^+ = 21.5 meq/L

Solution Compute the SAR using Eq. 4-9.

$$SAR = [21.5/\text{meq/L}]/[(1.41\ \text{meq/L} + 1.01\ \text{meq/L})/2]^{0.5}$$

$$SAR = 19.55\ (\text{mmol/L})^{1/2}$$

Referring to Table 4-3, the soil is sodic.

Cation Exchange Capacity and Exchangeable Sodium Ratio

The relationship between SAR, the cation exchange capacity (CEC) and the exchangeable sodium ratio (ESR) is given by the following equations:

$$kg'SAR = NaX/(CEC - NaX) \tag{4-10}$$

where

kg' = modified Gapon selectivity coefficient for relationship between $[Ca^{2+}]$ and $[Na^+]$

NaX = exchangeable sodium, meq/100 g

CEC = cation exchange capacity, meq/100 g

The exchangeable sodium ratio is given by

$$ESR = NaX/(CEC - NaX) \tag{4-11}$$

Combining Eqs. 4-10 and 4-11,

$$ESR = kg'(SAR) \tag{4-12}$$

An empirical relationship has been derived between SAR and ESR based on statistical analysis of saturation extract samples. This relationship is given as

$$ESR = 0.01475(SAR) - 0.0126 \tag{4-13}$$

where the slope of the ESR function is equivalent to kg' in Eq. 4-12:

$$kg' = 0.01475\ (\text{mmol/L})^{-1/2} \tag{4-14}$$

A relationship between the SAR and exchangeable sodium percentage (ESP) is developed by accepting the conclusion that the Gapon equation, which is the basis for SAR, is of practical significance at the field level as long as the exchangeable sodium to calcium ratio does not exceed approximately 1.5 to 2.0. The ESP is given by

$$ESP = [NaX/CEC]100 \tag{4-15}$$

Substituting this expression into Eq. 4-11 and rearranging, we have

$$\text{ESP}/(100 - \text{ESP}) = \text{kg}'(\text{SAR}) = 0.01475(\text{SAR}) \qquad (4\text{-}16)$$

where kg′ is taken from Eq. 4-13 for the ESR.

Effect of Salt Concentration on SAR

If the concentration of total salts in solution changes, the SAR varies according to the following expression:

$$\text{SAR}_{\text{final}} = [\Delta \text{ concentration}]^{1/2}\text{SAR}_{\text{initial}} \qquad (4\text{-}17)$$

where

Δ concentration = ratio of final concentration of total salts to initial concentration

The SAR increases as the square root of an increase in the total salt concentration of the soil solution. The relationship expressed in Eq. 4-17 may be useful when combined with other approximations for the soil water system. For example, it is sometimes assumed that the water content on a mass basis at saturation is 2 times the field capacity for a medium to fine textured soil. If we label the analysis for SAR made on the saturation extract as $\text{SAR}_{\text{initial}}$, then the SAR at field capacity is given by

$$\text{SAR}_{\text{final}} = \sqrt{2}\ \text{SAR}_{\text{initial}} \qquad (4\text{-}18)$$

Relationships between water content at saturation and other soil moisture levels are given in the following section.

Saturation Percentage

The saturation percentage is defined as the soil-water content on a mass basis for a saturated sample. Relationships between the saturation percentage and other soil-moisture content levels as a function of soil texture are given approximately by following:

Medium and fine textured soils

$$\text{SP} = 2(\theta_m)_{\text{FC}}(100) \qquad (4\text{-}19)$$

$$(\theta_m)_{\text{FC}} = 2(\theta_m)_{\text{CEW}} \qquad (4\text{-}20)$$

or

$$\text{SP} = 4(\theta_m)_{\text{CEW}}(100) \qquad (4\text{-}21)$$

where

SP = saturation percentage, percent

$(\theta_m)_{\text{FC}}$ = soil moisture content on mass basis at field capacity, g/g

$(\theta_m)_{\text{CEW}}$ = soil moisture content on mass basis at limit of crop extractable water, g/g

Coarse textured soils

$$SP = 3(\theta_m)_{FC}(100) \quad (4\text{-}22)$$

$$(\theta_m)_{FC} = 2(\theta_m)_{CEW} \quad (4\text{-}23)$$

or

$$SP = 6(\theta_m)_{CEW}(100) \quad (4\text{-}24)$$

Practical Relationships

The following general relationships are useful in some practical applications. They should be applied with caution because they are not rigorously derived and may be affected by complex ionic relationships at high levels of salinity.

(a) Osmotic potential, ψ_{os}, and EC (James et al., 1982)

$$\psi_{os} = -36EC \quad (4\text{-}25)$$

where

ψ_{os} = osmotic potential, kPa

EC = electrical conductivity at 25°C, mmhos/cm

(b) Salinity concentration, C_s, of a solution in the units shown and EC (James et al., 1982)

$$C_s\ (mg/l) = 640EC \quad (4\text{-}26)$$

$$C_s\ (meq/l) = 10EC \quad (4\text{-}27)$$

(c) EC of soil water (sw), irrigation water (iw) and saturation extract (Ayers and Wescot, 1976)

$$3EC_{iw} = EC_{sw} \quad (4\text{-}28)$$

$$2EC_e = EC_{sw} \quad (4\text{-}29)$$

or

$$EC_e = (3/2)(EC_{iw}) \quad (4\text{-}30)$$

4.4 Impacts of Soil and Water Chemical Concentrations on Yields

Crop Sensitivity to Chloride, Sodium, and Boron

Crop sensitivity to different concentrations of chloride measured in a saturation extract is given in Table 4-4. Beyond the chloride concentrations indicated in the table, chloride is considered toxic to the crop.

Crop tolerance to sodium concentration as indicated by the exchangeable sodium percentage (ESP) of the soil is indicated in Table 4-5. Note that crops can be severely affected by toxic levels of chloride and sodium which are not at sufficiently

TABLE 4-4 Hazardous chloride levels in saturation extracts for various fruit varieties and rootstocks. (Adapted from Bernstein, 1967.)

Variety or Rootstock	Chloride (meq/l) Saturation Extract
Citrus rootstocks	
Rungpur lime, Cleopatra mandarin	25
Rough lemon, Tangelo. Sour orange	15
Sweet orange, Citrange	10
Stone fruits rootstocks	
Mariana	25
Lovel, Shalil	10
Yunnan	7
Avocado rootstocks	
West Indian	8
Mexican	5
Grape varieties	
Thompson seedless, Perlette	25
Cardinal, Black Rose	10
Strawberry	5-8

TABLE 4-5 Tolerance of various crops to exchangeable sodium percentage. (Adapted from Pearson, 1960.)

Tolerance to ESP (range at which affected)	Crop	Growth Responsible under Field Conditions
Extremely sensitive (ESP = 2–10)	Deciduous fruits Nuts Citrus Avocado	Sodium toxicity symptoms even at low ESP values
Sensitive (ESP = 10–20)	Beans	Stunted growth at low ESP values even though the physical condition of the soil may be good
Moderately tolerant (ESP = 20–40)	Clover Oats Tall fescue Rice Dallisgrass	Stunted growth due to both nutritional factors and adverse soil conditions
Tolerant (ESP = 40–60)	Wheat Cotton Alfalfa Barley Tomatoes Beets	Stunted growth usually due to adverse physical conditions of soil
Most tolerant (ESP > 60)	Crested and Fairway wheatgrass Tall wheatgrass Rhodes grass	Stunted growth usually due to adverse physical conditions of soil

high levels to cause soil salinity or sodicity problems. This is especially true under sprinkler irrigation in which chloride or sodium may become concentrated on the leaf surface following evaporation of irrigation water.

Crop sensitivity to boron concentrations in irrigation water is indicated in Table 4-6.

Salinity Effects on Yield

The effects on yield of different salinity levels in soils as measured by the EC of a saturation extract are indicated in Table 4-7 for fruit, vegetable, field, and forage crops. The salinity effects are indicated as the expected percent of yield reduction over different ranges of the EC_e. Note that even within a particular type of crop category, the yield impacts can vary widely for the same level of EC_e. Note also that forage crops are most resistant to salinity, followed by field crops, vegetable crops, and fruit crops which are generally the most sensitive. The level of yield decrease expected can affect the decision of what crop to grow. Quantifying the impact on yield may also be necessary to compute the potential economic benefits from salt leaching practices or installation of land drainage systems.

TABLE 4-6 Limits of boron in irrigation water for different degrees of boron tolerance. Concentrations indicated in parts per million (ppm) of boron which is essentially equivalent to mg/L. Concentration range goes from top to bottom of column. (Adapted from Wilcox, 1960.)

Tolerant (4.0 to 2.0 ppm)	Semitolerant (2.0 to 1.0 ppm)	Sensitive (1.0 to 0.3 ppm)
Athel (Tamarix aphylla)	Sunflower (native)	Pecan
Asparagus	Potato	Walnut (black or Persian or English)
Palm (Phoenix canariensis)	Cotton (Acala and Pima)	Jerusalem artichoke
Date palm (P. dactylifera)	Tomato	Navy bean
Sugarbeet	Sweetpea	American elm
Mangel	Radish	Plum
Garden beet	Field pea	Pear
Alfalfa	Ragged-robin rose	Apple
Gladiolus	Olive	Grape (Sultanina and Malaga)
Broadbean	Barley	Kadota fig
Onion	Wheat	Persimmon
Turnip	Corn	Cherry
Cabbage	Milo	Peach
Lettuce	Oat	Apricot
Carrot	Zinnia	Thornless blackberry
	Pumpkin	Orange
	Bell pepper	Avocado
	Sweet potato	Grapefruit
	Lima bean	Lemon

TABLE 4-7 Salt tolerance levels for different crops. (Adapted from Ayers and Westcot, 1976.)

	Yield Potential								
	100%		90%		75%		50%		Maximum
Crop	EC_e	EC_{iw}	EC_e	EC_{iw}	EC_e	EC_{iw}	EC_e	EC_{iw}	EC_e
Field Crops									
Barley[a]	8.0	5.3	10.0	6.7	13.0	8.7	18.0	12.0	28
Beans (field)	1.0	0.7	1.5	1.0	2.3	1.5	3.6	2.4	7
Broad beans	1.6	1.1	2.6	1.8	4.2	2.0	6.8	4.5	12
Corn	1.7	1.1	2.5	1.7	3.8	2.5	5.9	3.9	10
Cotton	7.7	5.1	9.6	6.4	13.0	8.4	17.0	12.0	27
Cowpeas	1.3	0.9	2.0	1.3	3.1	2.1	4.9	3.2	9
Flax	1.7	1.1	2.5	1.7	3.8	2.5	5.9	3.9	10
Groundnut	3.2	2.1	3.5	2.4	4.1	2.7	4.9	3.3	7
Rice (paddy)	3.0	2.0	3.8	2.6	5.1	3.4	7.2	4.8	12
Safflower	5.3	3.5	6.2	4.1	7.6	5.0	9.9	6.6	15
Sesbania	2.3	1.5	3.7	2.5	5.9	3.9	9.4	6.3	17
Sorghum	4.0	2.7	5.1	3.4	7.2	4.8	11.0	7.2	18
Soybean	5.0	3.3	5.5	3.7	6.2	4.2	7.5	5.0	10
Sugarbeet	7.0	4.7	8.7	5.8	11.0	7.5	15.0	10.0	24
Wheat[a]	6.0	4.0	7.4	4.9	9.5	6.4	13.0	8.7	20
Vegetable Crops									
Beans	1.0	0.7	1.5	1.0	2.3	1.5	3.6	2.4	7
Beets[b]	4.0	2.7	5.1	3.4	6.8	4.5	9.6	6.4	15
Broccoli	2.8	1.9	3.9	2.6	5.5	3.7	8.2	5.5	14
Cabbage	1.8	1.2	2.8	1.9	4.4	2.9	7.0	4.6	12
Cantaloupe	2.2	1.5	3.6	2.4	5.7	3.8	9.1	6.1	16
Carrot	1.0	0.7	1.7	1.1	2.8	1.9	4.6	3.1	8
Cucumber	2.5	1.7	3.3	2.2	4.4	2.9	6.3	4.2	10
Lettuce	1.3	0.9	2.1	1.4	3.2	2.1	5.2	3.4	9
Onion	1.2	0.8	1.8	1.2	2.8	1.8	4.3	2.9	8
Pepper	1.5	1.0	2.2	1.5	3.3	2.2	5.1	3.4	9
Potato	1.7	1.1	2.5	1.7	3.8	2.5	5.9	3.9	10
Radish	1.2	0.8	2.0	1.3	3.1	2.1	5.0	3.4	9
Spinach	2.0	1.3	3.3	2.2	5.3	3.5	8.6	5.7	15
Sweet corn	1.7	1.1	2.5	1.7	3.8	2.5	5.9	3.9	10
Sweet potato	1.5	1.0	2.4	1.6	3.8	2.5	6.0	4.0	11
Tomato	2.5	1.7	3.5	2.3	5.0	3.4	7.6	5.0	13

4.5 Management of Soil Chemical Concentrations

Irrigation Water Quality

All waters contain various concentrations and different species of salt. The ions generally analyzed to determine the suitability of water for irrigation include Ca^{2+}, Mg^{2+}, Na^+, K^+, SO_4^{2-}, Cl^-, HCO_3^-, and CO_3^{2-}. Total salinity can be measured by the EC of the irrigation water denoted as EC_{iw}. Boron is often included in analysis of

TABLE 4-7 Salt tolerance levels for different crops. (Adapted from Ayers and Westcot, 1976.) (continued)

	Yield Potential								
	100%		90%		75%		50%		Maximum
Crop	EC_e	EC_{iw}	EC_e	EC_{iw}	EC_c	EC_{iw}	EC_e	EC_{iw}	EC_e
Forage Crops									
Alfalfa	2.0	1.3	3.4	2.2	5.4	3.6	8.8	5.9	16
Barley hay[a]	6.0	4.0	7.4	4.9	9.5	6.3	13.0	8.7	20
Bermuda grass	6.9	4.6	8.5	5.7	10.8	7.2	14.7	9.8	23
Clover, berseem	1.5	1.0	3.2	2.1	5.9	3.9	10.3	6.8	19
Corn (forage)	1.8	1.2	3.2	2.1	5.2	3.5	8.6	5.7	16
Harding grass	4.6	3.1	5.9	3.9	7.9	5.3	11.1	7.4	18
Orchard grass	1.5	1.0	3.1	2.1	5.5	3.7	9.6	6.4	18
Perennial rye	5.6	3.7	6.9	4.6	8.9	5.9	12.2	8.1	19
Soudan grass	2.8	1.9	5.1	3.4	8.6	5.7	14.4	9.6	26
Tall fescue	3.9	2.6	5.8	3.9	8.6	5.7	13.3	8.9	23
Tall wheat grass	7.5	5.0	9.9	6.6	13.3	9.0	19.4	13.0	32
Trefoil, big	2.3	1.5	2.8	1.9	3.6	2.4	4.9	3.3	8
Trefoil, small	5.0	3.3	6.0	4.0	7.5	5.0	10.0	6.7	15
Wheat grass	7.5	5.0	9.0	6.0	11.0	7.4	15.0	9.8	22
Fruit Crops									
Almond	1.5	1.0	2.0	1.4	2.8	1.9	4.1	2.7	7
Apple, pear	1.7	1.0	2.3	1.6	3.3	2.2	4.8	3.2	8
Apricot	1.6	1.1	2.0	1.3	2.6	1.8	3.7	2.5	6
Avocado	1.3	0.9	1.8	1.2	2.5	1.7	3.7	2.4	6
Date Palm	4.0	2.7	6.8	4.5	10.9	7.3	17.9	12.0	32
Fig, olive, pomegranate	2.7	1.8	3.8	2.6	5.5	3.7	8.4	5.6	14
Grape	1.5	1.0	2.5	1.7	4.1	2.7	6.7	4.5	12
Grapefruit	1.8	1.2	2.4	1.6	3.4	2.2	4.9	3.3	8
Lemon	1.7	1.1	2.3	1.6	3.3	2.2	4.8	3.2	8
Orange	1.7	1.1	2.3	1.6	3.2	2.2	4.8	3.2	8
Peach	1.7	1.1	2.2	1.4	2.9	1.9	4.1	2.7	7
Plum	1.5	1.0	2.1	1.4	2.9	1.9	4.3	2.8	7
Strawberry	1.0	0.7	1.3	0.9	1.8	1.2	2.5	1.7	4
Walnut	1.7	1.1	2.3	1.6	3.3	2.2	4.8	3.2	8

[a]During germination and seedling stage EC_e should not exceed 4 or 5 mmhos/cm. Data may not apply to new semi-dwarf varieties of wheat.

[b]During germination EC_e should not exceed 3 mmhos/cm.

the suitability of water for irrigation as is pH and the calculated SAR. Results of analyses are reported in meq/l or mg/l which are related by

$$(\text{mg/l})/\text{equivalent wt} = \text{meq/l} \tag{4-31}$$

Total salt concentration can be expressed as total dissolved solids (TDS) in mg/l. Because of the lower concentration of salt in irrigation water, EC in μmhos/cm is often used as the measure of salinity for irrigation water compared to EC in mmhos/cm for soil water. Table 4-8 indicates representative chemical compositions for various irrigation waters from which an understanding for the magnitudes and ranges of reasonable values can be developed.

The effect of water quality on the soil and plant growth is related to the chemical and physical properties of the soil, crop salt tolerance, climatic regime of the area, and method, frequency, and amount of irrigation water applied. Due to these diverse properties and their complex relationships, no universal, comprehensive irrigation water classification scheme is currently accepted.

Irrigation Water Quality Criteria

Attempts have been made at general classification schemes for irrigation water quality based on salinity (total salt concentration), sodicity (sodium) hazard, and toxicity of certain chemical constituents to crop growth. Detrimental effects from salinity are normally associated with high osmotic potential which reduces plant evapotranspiration and therefore reduces yield. This effect was quantified in Table 4-8. The sodicity hazard is related to the detrimental effect of ESP on the soil structure and permeability as well as the direct toxic effects on sodium sensitive plants. The toxic effect on plants was specified in Table 4-5. The effects of irrigation water quality on permeability are given in Table 4-9 for the broad ranges of none, moderate, and severe. An effort by the U.S. Salinity Laboratory to classify irrigation water in terms of both sodicity and salinity hazard is indicated in Fig. 4-5.

A general classification scheme based on analysis of the irrigation water alone, such as that indicated in Fig. 4-5, has generally been found unsatisfactory because it did not apply well to certain plants and certain soils. This has caused the development of a site-specific approach for salinity and sodium control which considers local soil, plant, and climatic conditions. The following sections deal with specific criteria to develop management schemes related to salinity and sodicity hazards.

Salinity Control—Site Specific Approach

Development of a salinity hazard in an originally nonsaline soil depends on the initial salinity of the irrigation water and the amount of applied water which filters through the root zone in excess of evapotranspiration demands. This latter amount may be used to push accumulated salts past the root zone where they are no longer directly detrimental to crop production. The excess water is removed from the field by a drainage system which may incorporate the types of open drain shown in Fig. 4-6.

The fraction of water which infiltrates into the soil that passes through the root zone is termed the leaching fraction and is expressed as

$$LF = D_d/D_i = (D_i - ET_c)/D_i \tag{4-32}$$

where

LF = leaching fraction

D_d = depth of drainage water, cm

TABLE 4-8 Composition of representative irrigation waters and average river waters of the world. (Adapted from Rhoades and Bernstein, 1971.)

Water	EC (μmhos/cm at 25°C)	Total Concentration (mg/l)	Total Concentration (meq/l)	B (mg/l)	Ca	Mg	Na	K	Alkalinity	SO_4	Cl	NO_3	SAR
San Joaquin, Biola, CA	60	46	0.52	0.10	0.23	0.08	0.19	0.02	0.36	0.06	0.09	0.02	0.5
Rogue, Grants Pass, OR	82	74	0.82	0.03	0.39	0.23	0.20	—	0.79	0.04	0.06	0.01	0.4
Sacramento, Knights Landing, CA	148	99	1.42	0.10	0.65	0.41	0.34	0.02	1.10	0.14	0.17	0.01	0.5
Snake, Heise, ID	371	218	3.85	0.05	2.34	0.98	0.48	0.05	2.69	0.90	0.31	0.02	0.4
Missouri, Nebraska City, NE	694	456	7.26	0.13	2.89	1.49	2.74	0.14	3.05	3.75	0.51	0.01	1.8
Colorado, Lees Ferry, AZ	864	547	8.80	0.11	3.84	1.72	3.13	0.11	2.72	4.37	1.72	0.02	1.9
Arkansas, Ralston, OK	1260	724	11.75	—	2.74	0.94	8.07	—	1.77	1.98	7.95	0.05	5.9
Salt, below Stewart Mt. Dam, AZ	1520	832	14.04	0.27	2.94	1.15	9.79	0.16	2.80	1.23	10.01	0.01	6.8
Sevier, Lynndyl, UT	1950	1190	20.37	0.31	3.76	5.80	10.66	0.15	5.44	6.23	8.46	0.14	4.9
Pecos, Shumla, TX	3160	2000	31.48	0.22	7.18	5.51	18.79	—	2.54	8.59	20.57	0.03	7.4
Gila, Gellespie Dam, AZ	8160	5260	87.02	2.60	14.87	10.94	60.90	0.31	3.18	27.90	59.96	0.34	17.0
Average Rivers													
North America	220	142	1.89	—	1.05	0.41	0.39	0.04	1.11	0.42	0.23	0.02	0.5
Europe	270	182	2.28	—	1.55	0.46	0.23	0.04	1.56	0.50	0.19	0.06	0.2
Australia	95	59	0.58	—	0.19	0.22	0.13	0.04	0.52	0.50	0.28	trace	0.3
World	190	120	1.42	—	0.75	0.34	0.27	0.06	0.96	0.23	0.22	0.02	0.4

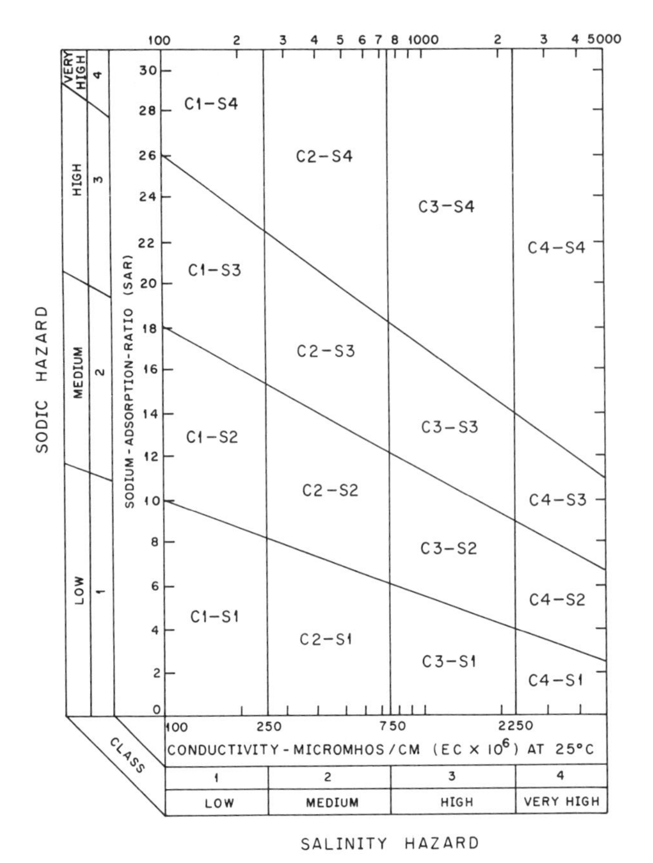

Figure 4-5 Diagram for classification of irrigation water based on total salts (EC) and relative sodium content (SAR). (Adapted from Richards, 1954.)

TABLE 4-9 Effect of irrigation water quality on permeability. (Adapted from Doorenbos and Pruitt, 1977.)

Parameter	Effect: None	Moderate	Severe
EC_{iw} (mmhos/cm)	>0.5	0.5–0.2	<0.2
SAR_{adj}			
Montmorillonite	<6	6–9	>9
Illite	<8	8–16	>16
Kaolinite	<16	16–24	>24

$$D_i = \text{depth of water infiltrated, cm}$$

$$ET_c = \text{crop evapotranspiration, cm}$$

The depth of infiltrated irrigation water can be expressed as the product of the average infiltration rate, I_{avg} (cm/d), or rate of water application, d_{ap} (cm/d),

Figure 4-6 Deep field drainage ditch used to remove leaching water for salinity management in Aragon Province, Spain.

whichever is the lesser, multiplied by the time of irrigation, t_i (d). The depth of water infiltrated can be written as

$$D_i = I_{avg}(t_i) \tag{4-33a}$$

or

$$D_i = d_{ap}(t_i) \tag{4-33b}$$

whichever is smaller. The LF equation can be written by substituting Eq. (4-33) into (4-32) as

$$LF = 1 - ET_c/[I_{avg}(t_i)] \tag{4-34a}$$

or

$$LF = 1 - ET_c/[d_{ap}(t_i)] \tag{4-34b}$$

In the foregoing equations, I_{avg}, d_{ap}, t_i, and ET_c all have to be considered over the same period of time, whether it is daily, weekly, monthly, or seasonally. t_i refers to the time of irrigation during the same time interval.

The previous equations pertain to field conditions in which drainage is not limiting—that is, fields in which the rate of drainage below the root zone is equal to or greater than the rate of infiltration. If the drainage rate is limiting, the LF must be modified to account for this fact. If this modification is not made, the result could very well be a rising water table situation. If the water table intersects the root zone, root development will be stunted where the soil becomes saturated, and salts may be brought up into instead of washed out of the root zone. When the rate of internal drainage is limiting, the depth of drainage is given by

$$D_d = I_d t_f \tag{4-35}$$

where

I_d = average internal drainage rate, cm/d

t_f = irrigation frequency, d

The preceding equation assumes the soil profile is continuously draining between irrigations. When drainage is limiting and the LF is controlled to avoid a rising water table, the leaching fraction is calculated as

$$\begin{aligned} LF &= (I_d t_f)/(ET_c + I_d t_f) \qquad (4\text{-}36) \\ &= I_d/(ET_c/t_f + I_d) \end{aligned}$$

which is equivalent to the depth of drainage divided by the depth of irrigation, as expressed in Eq. (4-32). Crop evapotranspiration is evaluated over the same time interval as t_f. The upper limit for t_f is determined using standard irrigation scheduling principles which consider available water in the root zone. Procedures to perform irrigation scheduling are described in the Soil Physics chapter.

Calculation of the leaching fraction using parameters related to a particular crop management situation can be made using the foregoing equations. In addition, the permissible EC of the irrigation water, EC_{iw}, may be determined if it is assumed that *no net change in the amount of soluble salt occurs either by precipitation, dissolution of soil minerals, or uptake by crops*. Based on this assumption, which is critical, the leaching fraction can be defined as

$$LF = D_d/D_i = EC_{iw}/EC_d \tag{4-37}$$

The permissible EC of the irrigation water can therefore be calculated by

$$EC_{iw} = LF(EC_d) \tag{4-38}$$

EC_d in Eqs. (4-37) and (4-38) is the EC of the drainage water. For salinity management decisions, it is assumed that it is equal to the EC_e of the soil sample. This assumption allows for application of allowable EC_e values in relation to percent yield reduction—that is, from Table 4-7—to compute the LF or EC_{iw}. The assumption that EC_d equals EC_e is somewhat conservative for salinity management in that EC_d occurs at the soil-water potential of field capacity and EC_e occurs at a potential of zero by definition.

It should be reemphasized that natural or indigenous salts that may exist in the soil profile are not taken into account in application of Eqs. (4-37) and (4-38), nor are possible chemical reactions in the profile. The following example problem demonstrates application of the equations in this section to compute the leaching fraction.

Example Problem 4-6

The following information is given for crop production, irrigation procedure, and water quality.

$$EC_{iw} = 3.2 \text{ mmhos/cm (saline)}$$

Crop = tomatoes

Allowable yield reduction = 25 percent

Soil infiltration rate = 1.3 cm/h (loam)

$$ET_c = 8 \text{ mm/d}$$

Irrigation frequency = 10 days

Calculate the required time of irrigation in hours per irrigation interval if soil drainage is not limiting.

Solution Applying Eq. (4-38),

$$EC_{iw} = LF(EC_d)$$

Assume that the EC_d is equal to the EC_e for application of Table 4-7. From Table 4-7 for tomatoes with 75 percent yield, EC_d = 5.0 mmhos/cm. Applying Eq. (4-37),

$$LF = EC_{iw}/EC_d = (3.2 \text{ mmhos/cm})/(5.0 \text{ mmhos/cm})$$
$$= 0.64$$

Applying Eq. (4-34a) and calculating required input parameters,

$$LF = 1 - ET_c/(I_{avg}t_i)$$
$$ET_c = 8 \text{ mm/d}(10 \text{ d}) = 80 \text{ mm} = 8.0 \text{ cm}$$
$$ET_c/(I_{avg}t_i) = 1 - LF = 1 - 0.64 = 0.36$$
$$I_{avg}t_i = ET_c/0.36 = 8.0 \text{ cm}/0.36 = 22.2 \text{ cm}$$
$$t_i = 22.2 \text{ cm}/(1.3 \text{ cm/h}) = 17.1 \text{ h}$$

Therefore, 17.1 hours of irrigation are required for each 10-day irrigation cycle.

Use of Eq. (4-38) to compute the leaching fraction has been brought into question because its application assumes that the entire root zone is at a uniform concentration and that this concentration is constant with time. In fact, between irrigations the root zone dries out exhibiting generally wetter conditions toward the bottom and drier conditions toward the top. The roots seek out moisture from the wetter depth layers which have less salt concentration.

Ayers and Westcot (1976) worked out a formula for leaching fraction that takes these conditions into account. Implicit in the formula is the assumption that the average EC in the root zone is equal to the sum of EC_d and EC_{iw} divided by 2 and that the EC at field capacity equals 2.5 times EC_e. Based on these relations, the following expression is derived for the leaching fraction

$$LF = \frac{EC_{iw}}{5(EC_e) - EC_{iw}} \tag{4-39}$$

Use of Eq. (4-39) is felt to result in values of the leaching fraction which better reflect actual conditions in the root zone. The difference in the resulting leaching fraction is demonstrated in the following example problem, which is based on Example Problem 4-6.

Example Problem 4-7

Rework Example Problem 4-6 using Eq. (4-39).

Solution Applying Eq. (4-39) with the same data given in the previous example problem including an EC_e of 5.0 mmhos/cm from Table 4-7

$$LF = \frac{3.2 \text{ mmhos/cm}}{5(5.0 \text{ mmhos/cm}) - 3.2 \text{ mmhos/cm}} = 0.15$$

This result is approximately four times less than the result in Example Problem 4-6. Continuing the analysis as shown in the previous example problem

$$\frac{ET_c}{I_{avg}t_i} = 1 - 0.15 = 0.85$$

$$I_{avg}t_i = \frac{ET_c}{0.85} = \frac{8.0 \text{ cm}}{0.85} = 9.4 \text{ cm}$$

$$t_i = \frac{9.4 \text{ cm}}{1.3 \text{ cm/h}} = 7.2 \text{ h}$$

The reduced time of irrigation for each 10-day irrigation cycle computed using Eq. (4-39) is therefore 7.2 hours.

Sodicity Control-Site Specific Approach

Sodicity hazard is related to the harmful effects of sodium concentration on soil permeability as indicated in Table 4-9 and on the toxicity of sodium concentrations on plants indicated in Table 4-5. Sodium concentration is expressed in terms of the amount of exchangeable sodium compared to other cations in the soil-water complex. This is quantified by the exchangeable sodium percentage, ESP, or sodium absorption ratio, SAR.

Under field conditions, the SAR of the irrigation water, SAR_{iw}, does not ordinarily equal the SAR of the soil water, SAR_{sw}, due to the following conditions:

(a) Evapotranspiration increases the constituent concentration of the soil water relative to that of the irrigation water. As indicated in Eq. (4-17),

$$SAR_{final} = [\Delta \text{ concentration}]^{1/2} SAR_{initial}$$

when considering salt concentration alone.

(b) Salt concentration in the soil varies with time between irrigations and depth in the profile. Salt concentration usually increases with depth and elapsed time following an irrigation.

(c) Soil-water composition is affected both by salt precipitation and soil mineral weathering during irrigation.

The SAR_{sw} rather than SAR_{iw} reflects the ESP of the soil and the sodium hazard. Methods have been developed to account for the aforementioned effects. Two empirical equations have been developed; (a) one to estimate ESP of the surface soil from the SAR_{iw}—the result is termed the adjusted SAR, SAR_{adj}; (b) another to estimate ESP at the bottom of the root zone—the result is termed the drainage water SAR, SAR_{dw}. They are given by the following:

$$SAR_{adj} = SAR_{iw}[1 + (8.4 - pH_c)] \tag{4-40}$$

$$SAR_{dw} = SAR_{iw}[Y^{(1+2LF)}/\sqrt{LF}][1 + (8.4 - pH_c)] \tag{4-41}$$

where

$$Y = \text{soil mineral weathering coefficient}$$

A generally accepted value for Y is 0.7. The $(8.4 - pH_c)$ term indicates the tendency of the applied water to precipitate or dissolve $CaCO_3$. When $(8.4 - pH_c) > 0$, $CaCO_3$ precipitates in the soil when water is applied. When $(8.4 - pH_c) < 0$, the irrigation water dissolves $CaCO_3$ available in the soil. pH_c is calculated by

$$pH_c = (pK_2 - pK_s) - p[(Ca^{2+} + Mg^{2+})/2] + p(HCO_3^- + CO_3^{2-}) \tag{4-42}$$

where

pK_2 = negative logarithm (base 10) of second dissociation constant of H_2CO_3

pK_s = negative logarithm (base 10) of solubility-product constant of $CaCO_3$

All concentrations in the preceding equation are in eq/l. Both pK_2 and pK_s must be corrected for the ionic strength of the water. For applications in this text, Table 4-10 will be applied to evaluate $pK_2 - pK_s$ as a function of total cation concentration—that is, the sum of the concentration of positively charged ions.

The last term in Eq. (4-42) is the negative logarithm of the equivalent concentration of titratable base and is closely related to the alkalinity (i.e., $HCO_3^- + 2CO_3^{2-} + OH^- + H^+$) of the water. The foregoing equation assumes both Ca^{2+} and Mg^{2+} precipitate as carbonates, but there is doubt that $MgCO_3$ precipitates under ordinary soil-water conditions.

Using the computed SAR_{adj} and SAR_{dw} values, the minimum (surface) and maximum (root zone) ESP values can be calculated based on the SAR_{iw} and the sodium hazard evaluated. A maximum SAR_{dw} can be selected to avoid a sodicity hazard and Eq. (4-41) applied to compute the LF for exchangeable sodium control. This result is termed the exchangeable sodium control leaching requirement, LR_{SAR}. The procedure to compute this leaching requirement is indicated in the following example problem.

TABLE 4-10 Values of $pK_2 - pK_s$ as a function of total cation concentration. (Adapted from Taylor and Ashcroft, 1972, based on personal communication with L.V. Wilcox, Soil Scientist, U.S. Salinity Lab, retired.)

Sum of Cations (meq/L)	0	0.1	0.2	0.3	0.4	0.5	0.6	0.7	0.8
0	2.03	—	—	—	—	2.11	2.11	2.12	2.12
1	2.13	2.13	2.14	2.14	2.14	2.14	2.15	2.15	2.15
2	2.16	2.16	2.16	2.16	2.17	2.17	2.17	2.17	2.18
3	2.18	2.18	2.18	2.19	2.19	2.19	2.19	2.19	2.19
4	2.20	2.20	2.20	2.20	2.20	2.21	2.21	2.21	2.21
5	2.21	2.22	2.22	2.22	2.22	2.22	2.22	2.22	2.23
6	2.23	2.23	2.23	2.23	2.23	2.23	2.24	2.24	2.24
7	2.24	2.24	2.24	2.24	2.25	2.25	2.25	2.25	2.25
8	2.25	2.25	2.25	2.26	2.26	2.26	2.26	2.26	2.26
9	2.26	2.26	2.27	2.27	2.27	2.27	2.27	2.27	2.27
Sum of Cations (meq/L)	**0**	**1**	**2**	**3**	**4**	**5**	**6**	**7**	**8**
0	2.03	2.13	2.16	2.18	2.20	2.21	2.23	2.24	2.25
10	2.27	2.28	2.29	2.30	2.31	2.32	2.32	2.33	2.34
20	2.35	2.35	2.36	2.37	2.37	2.38	2.38	2.39	2.39
30	2.40	2.40	2.41	2.41	2.42	2.42	2.42	2.43	2.43
40	2.44	2.44	2.45	2.45	2.45	2.46	2.46	2.46	2.47
50	2.47	2.48	2.48	2.48	2.48	2.49	2.49	2.49	2.49
60	2.50	2.50	2.50	2.51	2.51	2.51	2.51	2.52	2.52
70	2.52	2.53	2.53	2.53	2.53	2.53	2.54	2.54	2.54
80	2.54	—	—	—	—	—	—	—	—

Example Problem 4-8

Analysis of well water from Kasserine, Tunisia yields the following results:

$$SAR_{iw} = 2.59\ (\text{mmol/L})^{1/2}$$
$$[Ca^{2+}] = 6.58\ \text{meq/L}$$
$$[Mg^{2+}] = 8.48\ \text{meq/L}$$
$$[HCO_3^-] = 4.39\ \text{meq/L}$$
$$\sum[\text{cations}] = 22.3\ \text{meq/L}$$

Assume this water is to be evaluated for irrigation on citrus. Compute the required leaching requirement for sodium control.

Solution Based on Table 4-5, choose a maximum ESP_{dw} level of 5 percent as a reasonable value. Applying Eq. (4-16),

$$ESP_{dw}/(100 - ESP_{dw}) = kg'SAR_{dw} = 0.01475(SAR_{dw})$$
$$5/(100 - 5) = 0.01475(SAR_{dw})$$
$$SAR_{dw} = (5/95)(1/0.01475) = 3.57(\text{mmol/L})^{1/2}$$

This value is the target SAR_{dw}. Using Table 4-10 with Σ [cations] equal to 22.3 meq/L,

$$(pK_2 - pK_s) = 2.36$$

Calculate other factors in Eq. (4-42) for pH_c, recognizing that concentrations are in eq/L.

$$p\{([Ca^{2+}] + [Mg^{2+}])/2\} = p[(0.00658 \text{ eq/L} + 0.00848 \text{ eq/L})/2]$$
$$= p(0.00753) = 2.12$$
$$p\{[HCO_3^-] + [CO_3^{2-}]\} = p(0.00439 \text{ eq/L})$$
$$= 2.36$$

Evaluating pH_c,

$$pH_c = 2.36 + 2.12 + 2.36$$
$$pH_c = 6.84$$

Substituting the preceding values into Eq. (4-41) for SAR_{dw},

$$SAR_{dw} = SAR_{iw}[Y^{(1+2LF)}/\sqrt{LF}](1 + 8.4 - pH_c)$$
$$3.57 = 2.59[0.7^{(1+2LF)}/\sqrt{LF}](1 + 8.4 - 6.84)$$
$$0.538 = 0.7^{(1+2LF)}/\sqrt{LF}$$

which can be rearranged as

$$0 = 0.7^{(1+2LF)} - 0.538(LF)^{0.5}$$

The foregoing nonlinear equation may be solved for using the numerical technique of finding a root called the secant method. Expressing the equation as a function of LF which is set equal to zero,

$$f(LF) = 0 = 0.7^{(1+2LF)} - 0.538(LF)^{0.5}$$

The secant method converges on the root by an iterative scheme given as

$$LF_{n+1} = LF_n - \{f(LF_n)/[f(LF_n) - f(LF_{n-1})]\}(LF_n - LF_{n-1})$$

where

$$n = \text{number of iterations}$$

and LF_{n-1} and LF_n are of opposite sign. When f(LF) is close enough to zero using a given convergent criteria, the value of LF_n will be the required root. For this problem, the convergence criteria will be set equal to ± 0.001. Set up the solution in tabulated form with LF_n and $f(LF_n)$ and use an initial estimate of 2 for LF.

LF_n	$f(LF_n)$
2	−0.5932
1	−0.1953
0.5	+0.1094
0.6795	−0.0126
0.6610	−0.0008

The solution converges on a value of LF equal to,

$$LF_{SAR} = 0.66$$

This indicates that a leaching fraction of 0.66 or 66 percent would be required for sodium control given the chemical composition of the water source and type of crop. This value should be compared to the LF computed using Eq. (4-39) and the larger value used in design and management of the irrigation system.

REFERENCES

AYERS, R. S. and D. W. WESTCOT, "Water Quality for Agriculture," Food and Agriculture Organization (FAO) of the United Nations, Irrigation and Drainage Paper No. 29, Rome, Italy, 1976.

BERNSTEIN, L., "Quantitative Assessment of Irrigation Water Quality," American Society of Testing and Materials, Special Technical Publication No. 416, 1967.

BOHN, H., B. MCNEAL and G. O'CONNOR, *Soil Chemistry*. New York: John Wiley and Sons, New York, 1979.

DOORENBOS, J. and W. O. PRUITT, "Guidelines for Predicting Crop Water Requirements," Food and Agriculture Organization (FAO) of the United Nations, Irrigation and Drainage Paper No. 24, Revised. Rome, Italy, 1977.

JAMES, D. W., R. J. HANKS and J. H. JURINAK, *Modern Irrigated Soils*. New York: John Wiley and Sons, 1982.

PEARSON, G. A., "Tolerance of Crops to Exchangeable Sodium," U.S. Dept. of Agriculture, Information Bulletin No. 216, 1960.

RHOADES, J. D. and L. BERNSTEIN, "Chemical, Physical and Biological Characteristics of Irrigation and Soil Water," in *Water and Water Pollution Handbook, Vol. 1*, ed. L. L. Ciaccio. New York: Marcel Decker, Inc., 1971.

RICHARDS, L. A., ed., "Diagnosis and Improvement of Saline and Alkali Soils," U.S. Dept. of Agriculture, Agricultural Handbook No. 60, 1954.

SAWYER, C. N. and P. L. MCCARTY, *Chemistry for Sanitary Engineers*, 2nd edition. New York: McGraw-Hill Book Co., 1967.

SHALHEVET, J. and J. KAMBUROV, "Irrigation and Salinity—A World-Wide Survey," International Commission on Irrigation and Drainage, New Delhi, India, 1976.

TAYLOR, S. A. and G. L. ASHCROFT, *Physical Edaphology*. San Francisco, CA: W. H. Freeman and Co., 1972.

WADLEIGH, C. H., "Wastes in Relation to Agriculture and Forestry," U.S. Dept. of Agriculture, Miscellaneous Publication No. 1065, 1968.

WILCOX, L. V., "Boron Injury to Plants," U.S. Dept. of Agriculture, Information Bulletin No. 211, 1960.

PROBLEMS

4-1. Describe the procedure to make up 1 liter of a 0.3 N solution of $CaCO_3$ in acid.

4-2. Calculate the ESP of water draining from the bottom of the root zone in a fine-textured soil if the following chemical concentrations are measured from a soil water sample taken at ⅓ bar tension:

$$[Ca^{2+}] = 120 \text{ mg/L} \qquad [HCO_3^-] = 286 \text{ mg/L}$$
$$[Mg^{2+}] = 81 \text{ mg/L} \qquad [Na^+] = 326 \text{ mg/L}$$
$$[SO_4^{2-}] = 379 \text{ mg/L}$$

4-3. An irrigation water with an EC of 1.8 mmhos/cm is to be applied to production of lettuce. Seasonal evapotranspiration is 65 cm, seasonal time of irrigation is to be 165 hrs, and the water application rate, which is less than the average infiltration rate, is 0.9 cm/hr. Estimate the expected yield if the maximum yield is 450 kg/ha.

4-4. Chemical analysis indicates the following constituent concentration in an irrigation water source. If the irrigation project is to be managed with a leaching fraction of 0.25, estimate the SAR of the drainage water in this project.

$$pK_2 - pK_s = 2.23 \qquad p(HCO_3^- + CO_3^{2-}) = 2.50$$
$$p\left[\frac{Ca^{2+} + Mg^{2+}}{2}\right] = 3.19 \qquad SAR_{iw} = 4.6\ (\text{mmol/L})^{1/2}$$

4-5. An irrigation water source with an EC of 2.1 mmhos/cm is to be used to irrigate tomatoes in a clay-loam soil. The peak period crop water requirement for 7 days is 80.0 mm. The drainage rate of water past the lower boundary of the root zone is 8 mm/d. The target crop yield is 90 percent of maximum. Can this yield be attained assuming uniform distribution of salts and leaching water in the root zone, performing the analysis over the 7-day peak period, and allowing for no rise in the water table? If not, estimate the yield which can be attained as a percent of the maximum possible.

4-6. The measured soil water potential for a medium-textured soil is −33 kPa. A sample is taken and the soil water extracted using a pressure-plate apparatus. Chemical analyses indicate that the soil water contains 70 meq/L of salt. Using approximate relationships, estimate

(a) The solute potential of the soil water at −33 kPa

(b) The EC_e at 25°C in mmhos/cm

(c) The EC of the soil water when the soil water potential equals −1500 kPa

4-7. A saturation extract of fine-textured soil was analyzed. The analyses were as follows: Ca = 8.5 mmol/L, Mg = 3.0 mmol/L, and Na = 30 mmol/L. Calculate

(a) EC_e

(b) The SAR of the extract

(c) The SAR at −33 kPa soil water potential

(d) The ESP at −33 kPa soil water potential if $kg' = 0.015\ (\text{mmol/L})^{-1/2}$

4-8. The following chemical analyses results are for a water sample taken from a well near Kasserine in central Tunisia. No other ions were found in significant concentrations except those listed.

$$Ca^{++} = 132 \text{ mg/L}$$
$$Mg^{++} = 103 \text{ mg/L}$$
$$Na^+ = 163 \text{ mg/L}$$
$$K^+ = 5 \text{ mg/L}$$
$$SO_4^{2-} = 480 \text{ mg/L}$$
$$Cl^- = 366 \text{ mg/L}$$
$$HCO_3^- = 268 \text{ mg/L}$$

The bicarbonate ion concentration (HCO_3^-) was determined by titration. The water is to be used to irrigate a medium-textured sandy loam soil.

(a) It is decided that to promote the economic viability of irrigation, crops should not be grown which will exhibit more than a 10 percent reduction below maximum yield due to salinity effects. Which crops under the categories of fruits, vegetables, field crops, and forage crops would you recommend as possibilities for this project area?

(b) Assume water is a free good for this project—that is, there is no cost to the farmer for a volume of water nor for application of the water. Due to the possibility of supplying the European import market, assume the market value of fruits and vegetables is twice that of field crops and three times that of forage crops. How would you modify your recommendations and possibly the yield criteria from part (a)?

(c) What additional considerations would have to be made if water was not a free good—that is, if there was a cost to the farmer association with the volume and/or the application of water?

(d) Assess the chloride hazard of this water.

(e) Assess the sodium hazard of this water.

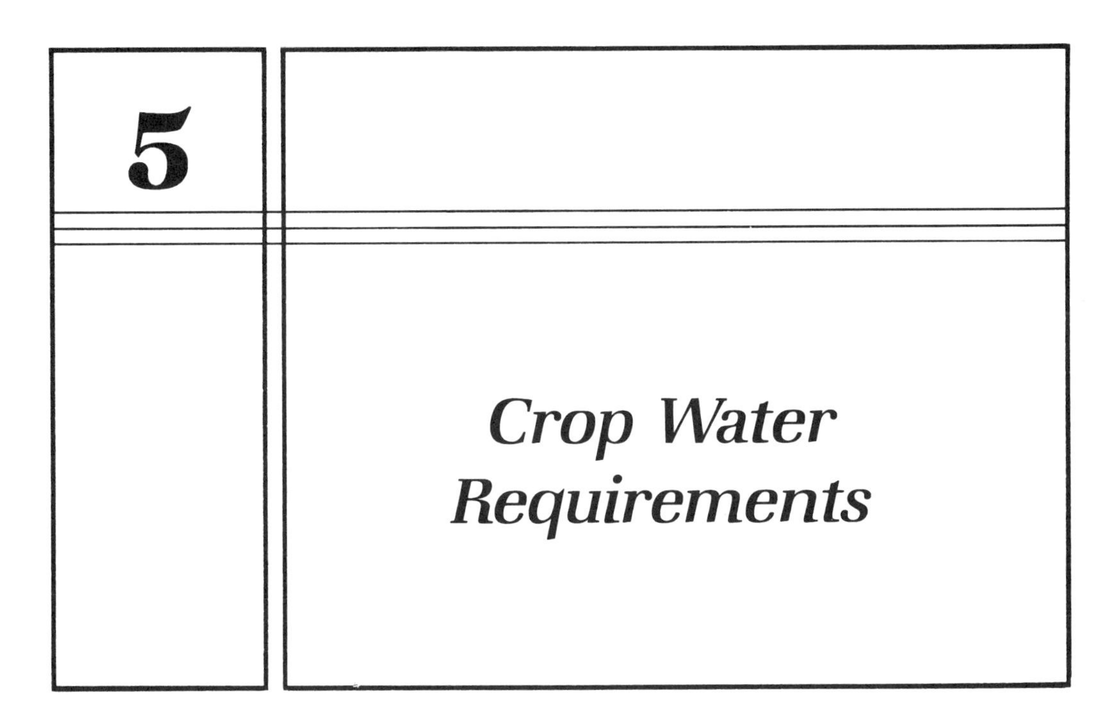

5.1 Introductory Comments

Irrigation systems are designed and constructed to meet deficiencies in crop water requirements caused by shortages in precipitation or soil-moisture storage capacities. This is the reason for their existence. In spite of this fact, it is astounding how little effort is sometimes put into estimation of crop water requirements for design and management of irrigation systems involving capital costs in the range of tens or hundreds of thousands of dollars. The percent of system capital costs spent on improved estimates of crop water requirements may be negligible compared to that spent on proper pipeline sizing or other hydraulic considerations, even though the crop water requirement is the driving force for the entire system. It is not unreasonable to expect that an improved crop water requirement estimate may make a substantial change in system size specifications and profitability. There are numerous examples of systems in the United States and other countries which have proven to be economically unprofitable and have eventually failed due to improper estimates of crop water requirements.

The explanation for this secondary importance typically applied to crop water requirements is probably twofold. First of all, engineers are normally not well versed in, and in fact can oftentimes be uncomfortable with, principles of agronomy and crop production. Training in this area is rarely given much priority in an engineering curriculum and oftentimes even the terminology appears forbidding. Secondly, there are a multitude of methods used to estimate crop water requirements and this plethora of methods is often confusing to engineers. This confusion is gradually being rectified by the leading work done by specialized committees or consultants for the American Society of Civil Engineers (Jensen, 1974), American Society of Agricultural Engineers (Jensen, 1980), and Food and Agricultural Organization of the United Nations (Doorenbos and Pruitt, 1977).

This chapter will not provide extensive training in agronomic principles or crop production. In fact, the discussion on development and application of water production functions in chapter 2 pointed out the difficulty in applying yield versus water use functions in economic analysis for system design over wide variations in climate or soil types. This chapter will present adequate terminology and review of crop water requirement estimating methods to give the engineer a sound basis for making such estimates and applying the results to develop system design parameters. The estimating methods are presented to allow for various levels of sophistication depending upon the database available. Examples are worked out for individual methods to improve understanding of application. The methods chosen do not represent one researcher's or one consulting engineer's favorite method, but represent methods which have been calibrated and applied over a wide range of climates, worldwide if possible. In general, these methods represent those recommended by a consensus of experienced consultants working with American engineering societies and the United Nations as just referenced. In all cases, they represent methods for which considerable experience and therefore improved understanding has been gained by consulting engineers.

5.2 Definition of Terms

Energy Balance

In crop water requirement estimates, the energy balance refers to components of energy which act on the earth's surface on a daily basis. This energy is available to evaporate water from plant, soil, and water surfaces. The first component considered is net radiation, R_n, given by the following equation:

$$R_n = (1 - \alpha)R_s + I\downarrow - I\uparrow \tag{5-1}$$

where

R_n = net radiation

α = albedo (reflectance)

R_s = incoming solar radiation (shortwave) at surface of the earth

$I\downarrow$ = infrared energy (longwave) reflected from atmosphere to surface of the earth

$I\uparrow$ = infrared energy from surface of the earth to the atmosphere

Components of the radiation balance from the preceding equation are demonstrated in Fig. 5-1. Note that in accordance with Wien's displacement law relating temperature with wavelength, radiation emanating from the sun (solar) is always shortwave and radiation emanating from the earth (infrared) is always longwave (Rosenberg et al., 1983). Values for the albedo term in Eq. (5-1) over different surfaces are given in Table 5-1. A recommended value of 0.25 for field crops is applied in a later example.

With net radiation defined by Eq. (5-1), we are prepared to define the energy balance on the earth's surface by the following:

$$R_n = G + E + H \tag{5-2}$$

where

G = soil heat flux (+ if soil is warming)

E = evaporation (+ if evaporation is occurring)

H = sensible heat (+ if air is warming)

The net amount of radiative energy at the earth's surface (R_n) must be transformed to or from the components on the right-hand side of Eq. (5-2). A study of the right-

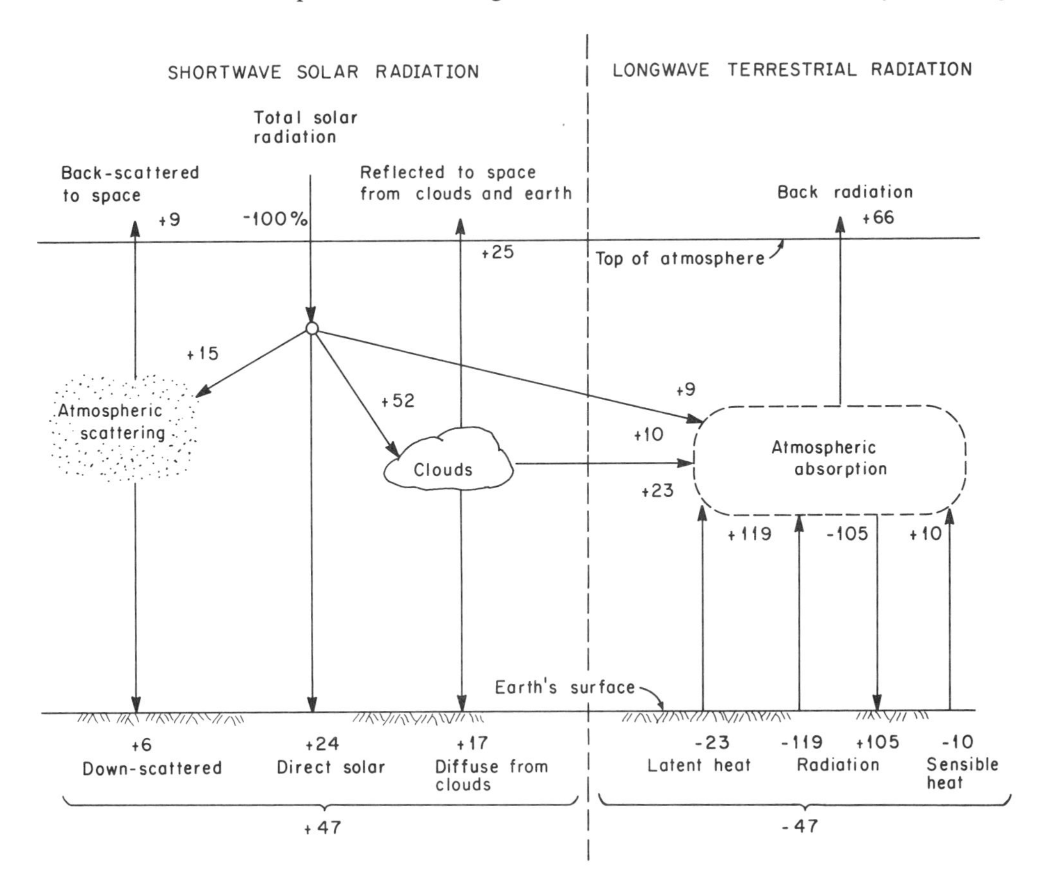

Figure 5-1 Mean vertical heat balance in the northern hemisphere.

TABLE 5-1 Shortwave albedo of natural surfaces. (Adapted from Rosenberg et al., 1983.)

Surface	Shortwave Albedo	Surface	Shortwave Albedo
Fresh snow	0.80–0.95	Bare dark soil	0.16–0.17
Old snow	0.42–0.70	Dry clay soil	0.20–0.35
Lake ice, clear	0.10	Peat soil	0.05–0.15
Lake ice with snow	0.46	Savannah	0.22
Sea surface, calm	0.07–0.08	Most field crops	0.18–0.30
Sea surface, windy	0.12–0.14	Deciduous forest	0.15–0.20
Niger River water, clear	0.06	Coniferous forest	0.10–0.15
Niger River water, dirty	0.12	Mangrove swamp	0.12
Dry sandy soil	0.25–0.45	Vineyard	0.18–0.19

hand side components can provide insight into the relationship of plant evaporation to energy availability. Components of the energy balance on a daily basis over different surfaces are indicated in Fig. 5-2. This figure can be inspected to determine the effects of different types of surfaces on the energy balance. On the dry lake bed at El Mirage, California, evaporation is negligible and the net radiation is basically partitioned into warming the air and soil. Over the sudangrass field in Tempe, Arizona, energy is extracted from the dry air mass passing over the field causing a decrease in air temperature (negative value for H) and causing evaporation to exceed net radiation. In the alfalfa-grass field in Hancock, Wisconsin, net radiation is more typically proportioned into E, H, and G than in the other two more extreme cases.

5.3 Temperature Based Estimating Methods

Soil Conservation Service Modified Blaney-Criddle

Temperature is the most commonly measured meteorological parameter around the world. Virtually all meteorological stations in the world measure maximum and minimum temperature on a daily basis with inexpensive yet accurate equipment. (See Figs. 5-3 and 5-4.) It is not surprising therefore that one of the most basic crop water requirement estimating methods is temperature based. This is the original Blaney-Criddle method (Blaney and Criddle, 1950) which is described by the following equations:

$$f = \frac{pT}{100} \tag{5-3}$$

where

f = monthly consumptive use factor

p = percentage of total annual daylight hours during the month

T = mean monthly temperature, °F

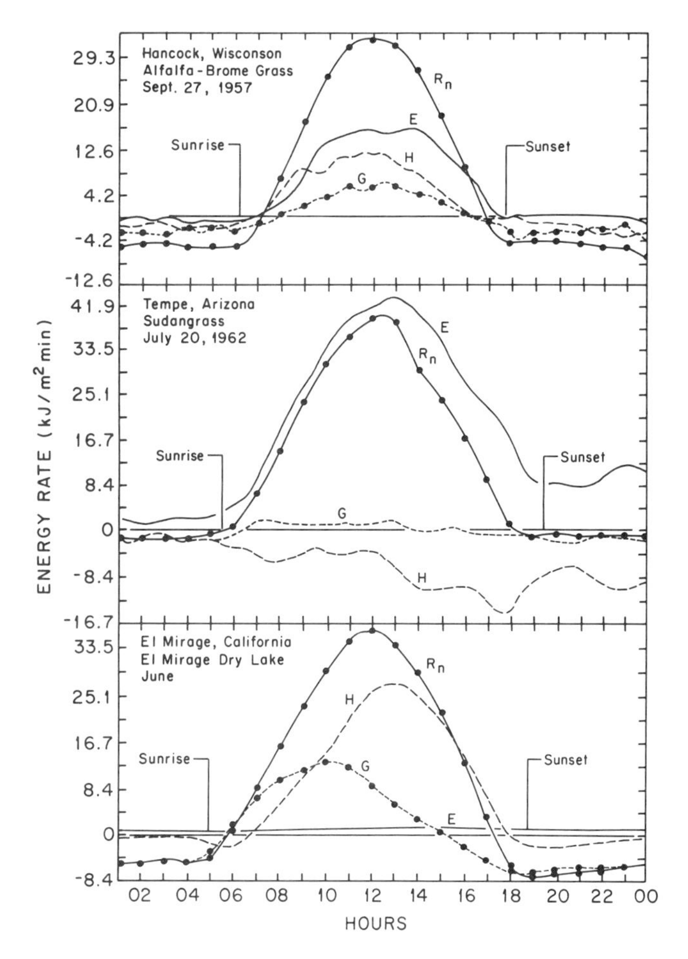

Figure 5-2 Average diurnal variation of components of surface energy balance over different surfaces. (Adapted from Sellers, 1965.)

Values for p are tabulated as a function of month of year and latitude and given in Table 5-2. The monthly consumptive use factor, f_i, is used in the following equation for evapotranspiration:

$$ET = \sum_{i=1}^{n} k_i f_i = KF \tag{5-4}$$

where

ET = seasonal crop water requirement for full yield, inches

n = number of months in growing season

k_i = monthly Blaney-Criddle crop coefficient

K = seasonal Blaney-Criddle crop coefficient

F = $\sum f_i$ over individual months of growing season

Figure 5-3 Weather station site with standard louvered wooden shelter raised on a wooden stand in Nile River Valley, Egypt.

Figure 5-4 Interior of weather station shelter showing the basic thermometers required for minimum and maximum temperature at the top. Shelter also contains strip-chart recording devices for temperature and relative humidity at the sides and naturally ventilated wet and dry bulb thermometers and Piche evaporimeter.

Note that crop water requirement estimating methods are normally developed for the maximum evapotranspiration at full yield. *This convention will be applied in this book unless otherwise specifically stated.*

TABLE 5-2 Percent of annual daylight hours per month from latitude 46° south to 60° north.

Latitude 0° North	Jan.	Feb.	Mar.	Apr.	May	June	July	Aug.	Sept.	Oct.	Nov.	Dec.
0	8.50	7.66	8.49	8.21	8.50	8.22	8.50	8.49	8.21	8.50	8.22	8.50
5	8.32	7.57	8.47	8.29	8.65	8.41	8.67	8.60	8.23	8.42	8.07	8.30
10	8.13	7.47	8.45	8.37	8.81	8.60	8.86	8.71	8.25	8.34	7.91	8.10
15	7.94	7.36	8.43	8.44	8.98	8.80	9.05	8.83	8.28	8.26	7.75	7.88
20	7.74	7.25	8.41	8.52	9.15	9.00	9.25	8.96	8.30	8.18	7.58	7.66
25	7.53	7.14	8.39	8.61	9.33	9.23	9.45	9.09	8.32	8.09	7.40	7.42
30	7.30	7.03	8.38	8.72	9.53	9.49	9.67	9.22	8.33	7.99	7.19	7.15
32	7.20	6.97	8.37	8.76	9.62	9.59	9.77	9.27	8.34	7.95	7.11	7.05
34	7.10	6.91	8.36	8.80	9.72	9.70	9.88	9.33	8.36	7.90	7.02	6.92
36	6.99	6.85	8.35	8.85	9.82	9.82	9.99	9.40	8.37	7.85	6.92	6.79
38	6.87	6.79	8.34	8.90	9.92	9.95	10.10	9.47	8.38	7.80	6.82	6.66
40	6.76	6.72	8.33	8.95	10.02	10.08	10.22	9.54	8.39	7.75	6.72	6.52
42	6.63	6.65	8.31	9.00	10.14	10.22	10.35	9.62	8.40	7.69	6.62	6.37
44	6.49	6.58	8.30	9.06	10.26	10.38	10.49	9.70	8.41	7.63	6.49	6.21
46	6.34	6.50	8.29	9.12	10.39	10.54	10.64	9.79	8.42	7.57	6.36	6.04
48	6.17	6.41	8.27	9.18	10.53	10.71	10.80	9.89	8.44	7.51	6.23	5.86
50	5.98	6.30	8.24	9.24	10.68	10.91	10.99	10.11	8.46	7.45	6.10	5.65
52	5.77	6.19	8.21	9.29	10.85	11.13	11.20	10.12	8.49	7.39	5.93	5.43
54	5.55	6.08	8.18	9.36	11.03	11.38	11.43	10.26	8.51	7.30	5.74	5.18
56	5.30	5.95	8.15	9.45	11.22	11.67	11.69	10.40	8.53	7.21	5.54	4.89
58	5.01	5.81	8.12	9.55	11.46	12.00	11.98	10.55	8.55	7.10	5.31	4.56
60	4.67	5.65	8.08	9.65	11.74	12.39	12.31	10.70	8.57	6.98	5.04	4.22
South												
0	8.50	7.66	8.49	8.21	8.50	8.22	8.50	8.49	8.21	8.50	8.22	8.50
5	8.68	7.76	8.51	8.15	8.34	8.05	8.33	8.38	8.19	8.56	8.37	8.68
10	8.86	7.87	8.53	8.09	8.18	7.86	8.14	8.27	8.17	8.62	8.53	8.88
15	9.05	7.98	8.55	8.02	8.02	7.65	7.95	8.15	8.15	8.68	8.70	9.10
20	9.24	8.09	8.57	7.94	7.85	7.43	7.76	8.03	8.13	8.76	8.87	9.33
25	9.46	8.21	8.60	7.84	7.66	7.20	7.54	7.90	8.11	8.86	9.04	9.58
30	9.70	8.33	8.62	7.73	7.45	6.96	7.31	7.76	8.07	8.97	9.24	9.85
32	9.81	8.39	8.63	7.69	7.36	6.85	7.21	7.70	8.06	9.01	9.33	9.96
34	9.92	8.45	8.64	7.64	7.27	6.74	7.10	7.63	8.05	9.06	9.42	10.08
36	10.03	8.51	8.65	7.59	7.18	6.62	6.99	7.56	8.04	9.11	9.51	10.21
38	10.15	8.57	8.66	7.54	7.08	6.50	6.87	7.49	8.03	9.16	9.61	10.34
40	10.27	8.63	8.67	7.49	6.97	6.37	6.76	7.41	8.02	9.21	9.71	10.49
42	10.40	8.70	8.68	7.44	6.85	6.23	6.64	7.33	8.01	9.26	9.82	10.64
44	10.54	8.78	8.69	7.38	6.73	6.08	6.51	7.25	7.99	9.31	9.94	10.80
46	10.69	8.86	8.70	7.32	6.61	5.92	6.37	7.16	7.96	9.37	10.07	10.97

The Soil Conservation Service (SCS) attempted to improve the accuracy of the original Blaney-Criddle method. Measured and estimated crop water use were compared and a linear regression equation using temperature was derived to improve the estimate of crop water use. This procedure did not really bring any new meteorological variables into the estimating method, but employed the larger database available at the time to improve the crop water use estimate (Soil Conservation Service, 1970). The SCS modified Blaney-Criddle is given by the following equation:

$$ET = \sum_{i=1}^{n} k_{ti}k_i f_i = K \sum_{i=1}^{n} k_{ti} f_i \qquad (5\text{-}5)$$

where

$$ET = \text{seasonal crop water requirement, inches, and}$$

$$k_{ti} = 0.0173T_i - 0.314 \qquad (5\text{-}6)$$

where

$$T_i = \text{mean temperature for month i, °F}$$

Table 5-3 indicates SCS Blaney-Criddle seasonal crop coefficients, K, as well as crop coefficients for the month of maximum crop water requirements during the growing season for various crops. These coefficients should not be used for any method other than the SCS Blaney-Criddle.

Although the SCS modification of the Blaney-Criddle method did not bring additional meteorological variables into the procedure, it did improve the estimate for some locations. Fig. 5-5 indicates crop water use estimates made using the SCS modified Blaney-Criddle method with lysimeter-measured values of evapotranspiration for three different climatic conditions. A lysimeter is a large container filled with soil, plants, and water. Typical sizes for the more accurate lysimeters are in the range of 3 to 30 m^2 surface area and 1 to 2 m in depth. (Howell et al., 1985; Grebet, 1982; Mukammal et al., 1971; Pruitt and Angus, 1960) The lysimeters from which data were taken for Fig. 5-5 were of the weighing type in which crop water use may be measured accurately on a daily or even hourly basis. The lysimeter from Davis, California is pictured in Figs. 5-6 and 5-7. Drainage type lysimeters, which measure crop water use by applying a water balance method, are useful for measurements made on a minimum interval of approximately 5 to 7 days.

Part (c) of Fig. 5-5 indicates lysimeter measurements compared to a Blaney-Criddle estimate for a low elevation, relatively humid site. Part (b) is representative of a low elevation, semi-arid site, and part (a) is for a high elevation, arid location. It can be noted that the SCS Blaney-Criddle method does very well at the humid location. At the semi-arid low elevation site, the method underpredicts actual crop water use for the entire season and underpredicts the peak demand by about 19 percent. At the higher elevation, arid site, the performance of this method is even less satisfactory with an underestimation of the peak demand of about 29 percent.

The main cause of this underestimation is probably that the SCS Blaney-Criddle method only includes locally measured average temperature data and evenly weights nighttime and daytime temperatures. At high elevation arid sites, there is an appreciable difference between day and night temperatures but plant growth and water usage are affected predominantly by the higher daytime temperature coupled with high rates of solar radiation.

TABLE 5-3 SCS Blaney-Criddle seasonal consumptive-use crop coefficients (K) for irrigated crops.

Crop	Length of Normal Growing Season or Period[a]	Consumptive Use Coefficient K[b]	Maximum Monthly k[c]
Alfalfa	Between frosts	0.80 to 0.90	0.95 to 1.25
Bananas	Full year	0.80 to 1.00	—
Beans	3 months	0.60 to 0.70	0.75 to 0.85
Cocoa	Full year	0.70 to 0.80	—
Coffee	Full year	0.70 to 0.80	—
Corn (Maize)	4 months	0.75 to 0.85	0.80 to 1.20
Cotton	7 months	0.60 to 0.70	0.75 to 1.10
Dates	Full year	0.65 to 0.80	—
Flax	7 to 8 months	0.70 to 0.80	—
Grains, small	3 months	0.75 to 0.85	0.85 to 1.00
Grain, sorghums	4 to 5 months	0.70 to 0.80	0.85 to 1.10
Oilseeds	3 to 5 months	0.65 to 0.75	—
Orchard crops:			
Avocado	Full year	0.50 to 0.55	—
Grapefruit	Full year	0.55 to 0.65	—
Orange and lemon	Full year	0.45 to 0.55	0.65 to 0.75[d]
Walnuts	Between frosts	0.60 to 0.70	—
Deciduous	Between frosts	0.60 to 0.70	0.70 to 0.95
Pasture crops:			
Grass	Between frosts	0.75 to 0.85	0.85 to 1.15
Ladino whiteclover	Between frosts	0.80 to 0.85	—
Potatoes	3 to 5 months	0.65 to 0.75	0.85 to 1.00
Rice	3 to 5 months	1.00 to 1.10	1.10 to 1.30
Soybeans	140 days	0.65 to 0.70	—
Sugar beet	6 months	0.65 to 0.75	0.85 to 1.00
Sugarcane	Full year	0.80 to 0.90	—
Tobacco	4 months	0.70 to 0.80	—
Tomatoes	4 months	0.65 to 0.70	—
Truck crops, small	2 to 4 months	0.60 to 0.70	—
Vineyard	5 to 7 months	0.50 to 0.60	—

[a]Length of season depends largely on variety and time of year when the crop is grown. Annual crops grown during the winter period may take much longer than if grown in the summertime.

[b]The lower values of K for use in the Blaney-Criddle formula are for the more humid areas, and the higher values are for the more arid climates.

[c]Dependent upon mean monthly temperature and crop growth stage.

[d]Given by Criddle as "Citrus orchard."

Source: From "Irrigation Water Requirements," Technical Release No. 21, Soil Conservation Service, U.S.D.A., September, 1970.

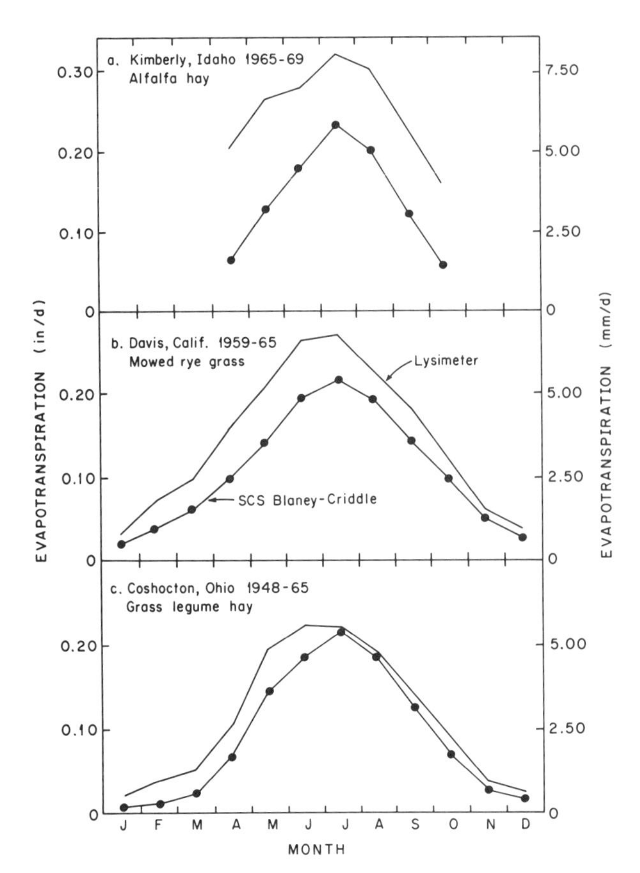

Figure 5-5 Comparison of crop water use at three sites measured by lysimeter with estimates using SCS modified Blaney-Criddle method. (Adapted from Jensen, 1974.)

Figure 5-6 Sensitive weighing lysimeter at Davis, California being prepared for planting of tomato crop. Lysimeter is 6.08 m in diameter and 0.97 m in depth.

Figure 5-7 Same weighing lysimeter at Davis, California as Fig. 5-6 showing tomato crop development in lysimeter to be virtually identical and representative of growth in surrounding field.

Food and Agriculture Organization Modified Blaney-Criddle

Recognizing the problems indicated in Fig. 5-5 with the SCS modified Blaney-Criddle method and the inherent strength of a method which relied predominantly on temperature data, work was conducted under the auspices of the Food and Agricultural Organization (FAO) of the United Nations in Rome to modify the Blaney-Criddle method. This work was conducted by Pruitt using meteorological and lysimeter data available to the United Nations from 13 sites in various climates around the world. Based on statistical analysis of the data, it was decided to bring into the method additional effects of local meteorological conditions by using either specific data or general ranges of relative humidity, sunshine hours, and wind speed. This resulted in the following equation (Doorenbos and Pruitt, 1977):

$$ET_r = a + b[p(0.46T + 8.13)] \tag{5-7}$$

where

ET_r = reference crop ET for grass, mm/d

p = percent of annual sunshine during month on a daily basis

T = mean temperature, °C

a, b = climatic calibration coefficients

Values for p tabulated as a function of month of year and latitude are given in Table 5-4. (These are equivalent to the values given in Table 5-2 but given on a daily rather than monthly basis.)

a and b are climatic calibration coefficients arrived at by performing step-wise regression analysis on meteorological and lysimetric data from 13 sites worldwide. The objective of the regression analysis was to include those climatic variables which gave the greatest degree of improvement to the ET estimate when temperature, as indicated in Eq. (5-7), is the main climatic parameter. a and b are a function of minimum relative humidity, RH_{min}, ratio of actual to maximum possible sunshine hours, n/N, and daytime wind velocity, U_{day}. Relative humidity may be determined using measurements of wet and dry bulb temperature. Actual sunshine hours may be measured using the sunshine recorder shown in Fig. 5-8. Figure 5-9 indicates an anemometer which may be used to compute daytime windspeed. The corrected value of ET_r from Eq. (5-7) may be solved for graphically, by computerized look-up table, or by a regression equation fit through the look-up table.

At first sight, the inclusion of RH_{min}, n/N, and U_{day} in the calibration factors would seem to require a good deal of additional data found at a limited number of locations worldwide. However, approximate values for the additional climatic data in ranges of low, medium, or high can be conveniently applied to arrive at a calibrated estimate of ET_r using graphical procedures. This is demonstrated by Fig. 5-10 which indicates nine boxes with columns representing low, medium, and high categories of RH_{min} and rows representing low, medium, and high categories of n/N.

TABLE 5-4 Mean daily percentage (p) of annual daytime hours for different latitudes.

North	Jan	Feb	Mar	Apr	May	June	July	Aug	Sept	Oct	Nov	Dec
Latitude South[1]	July	Aug	Sept	Oct	Nov	Dec	Jan	Feb	Mar	Apr	May	June
60°	.15	.20	.26	.32	.38	.41	.40	.34	.28	.22	.17	.13
58	.16	.21	.26	.32	.37	.40	.39	.34	.28	.23	.18	.15
56	.17	.21	.26	.32	.36	.39	.38	.33	.28	.23	.18	.16
54	.18	.22	.26	.31	.36	.38	.37	.33	.28	.23	.19	.17
52	.19	.22	.27	.31	.35	.37	.36	.33	.28	.24	.20	.17
50	.19	.23	.27	.31	.34	.36	.35	.32	.28	.24	.20	.18
48	.20	.23	.27	.31	.34	.36	.35	.32	.28	.24	.21	.19
46	.20	.23	.27	.30	.34	.35	.34	.32	.28	.24	.21	.20
44	.21	.24	.27	.30	.33	.35	.34	.31	.28	.25	.22	.20
42	.21	.24	.27	.30	.33	.34	.33	.31	.28	.25	.22	.21
40	.22	.24	.27	.30	.32	.34	.33	.31	.28	.25	.22	.21
35	.23	.25	.27	.29	.31	.32	.32	.30	.28	.25	.23	.22
30	.24	.25	.27	.29	.31	.32	.31	.30	.28	.26	.24	.23
25	.24	.26	.27	.29	.30	.31	.31	.29	.28	.26	.25	.24
20	.25	.26	.27	.28	.29	.30	.30	.29	.28	.26	.25	.25
15	.26	.26	.27	.28	.29	.29	.29	.28	.28	.27	.26	.25
10	.26	.27	.27	.28	.28	.29	.29	.28	.28	.27	.26	.26
5	.27	.27	.27	.28	.28	.28	.28	.28	.28	.27	.27	.27
0	.27	.27	.27	.27	.27	.27	.27	.27	.27	.27	.27	.27

[1]Southern latitudes: apply 6-month difference as shown.

Figure 5-8 Campbell-Stokes sunshine recorder. Number of hours of sunshine is measured by length of charred line on paper strip placed behind glass ball. The ball focuses sun's rays on the paper as with a magnifying glass.

Figure 5-9 Totalizing cup anemometer to measure wind run placed on stand at height of 2 m at meteorological station in Nile River Valley, Egypt. Evaporation pan to the right of anemometer is explained in the next section.

The example indicated in the figure is for Cairo in July with high n/N and medium RH_{min}. Locating the proper box, the uncorrected f factor given by the portion of Eq. 5-7 in brackets is entered in the bottom of the box and projected vertically upward to the correct daytime wind velocity line. For the example shown, the uncorrected f factor is 6.6 mm/d. The proper line in this case is for moderate daytime wind velocities indicated by the number 2. From the wind line, one projects hori-

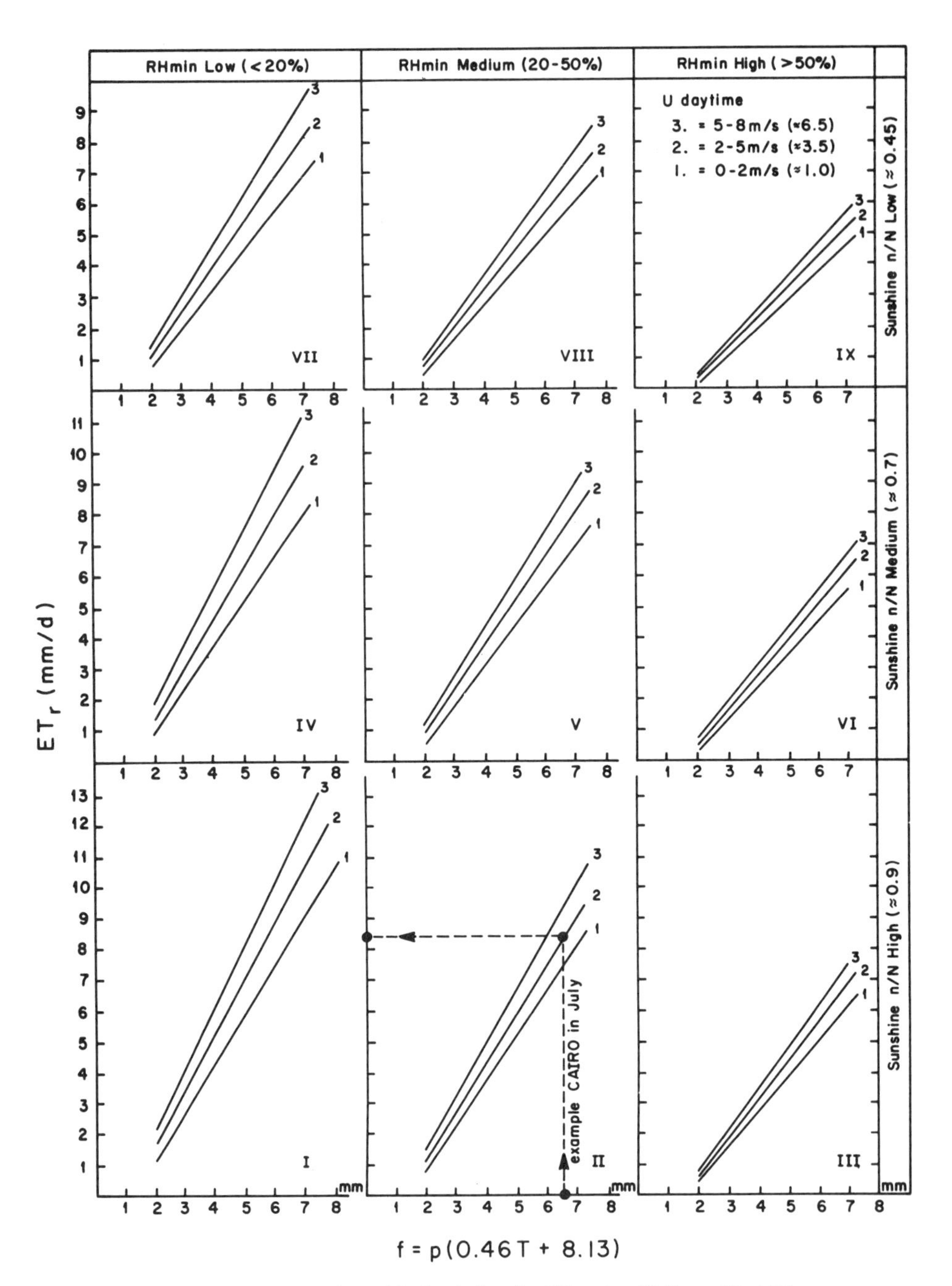

Figure 5-10 Example of graphical solution for ET_r using FAO modified Blaney-Criddle method. (From Doorenbos and Pruitt, 1977.)

zontally to the left for the calibrated ET_r indicated as 8.5 mm/d. This represents a correction of 29 percent due to the FAO modification of the Blaney-Criddle method, which is clearly significant.

The FAO modifications are based on passing regression lines through recorded data. The FAO report gives a regression equation for the a coefficient in Eq. (5-7) and a computerized look-up table for the b coefficient (Doorenbos and Pruitt, 1977). The a coefficient is given by

$$a = 0.0043(RH_{min}) - n/N - 1.41 \tag{5-8}$$

where

RH_{min} = minimum relative humidity, percent

n/N = ratio of actual to maximum sunshine hours, fraction

The tabulated values for the b coefficient as a function of RH_{min}, U_{day}, and n/N are given in Table 5-5.

Frevert et al. (1983) published a regression equation involving independent and interaction variables based on step-wise regression analysis of data in Table 5-5. The tabulated values are the actual data base and will always be more accurate than the regression equation because the equation does not have a perfect fit for every coefficient in the table. However, such an equation may be convenient for a computer program or with a programmable calculator. The equation of Frevert et al. (1983) was further simplified by Cuenca and Jensen (1988) with no appreciable loss in accuracy resulting in an equation for the b term of Eq. (5-7) of the following form:

$$\begin{aligned} b = 0.82 &- 0.0041(RH_{min}) + 1.07(n/N) + 0.066(U_{day}) \\ &- 0.006(RH_{min})(n/N) - 0.0006(RH_{min})(U_{day}) \end{aligned} \tag{5-9}$$

where

U_{day} = daytime wind speed at 2 m height, m/s

Units for RH_{min} and n/N are as indicated for Eq. (5-8). Note that daytime hours for the FAO methods are arbitrarily defined from 0700 to 1900 hours. If wind speed is measured at a height other than 2 m, it is modified to the equivalent 2 m wind speed by application of the log-wind law in the following form:

$$U_{2m} = U_z \left[\frac{2.0}{z}\right]^{0.2} \tag{5-10}$$

where

U_{2m} = equivalent wind speed at 2 m

U_z = wind speed measured at height z

z = height of measurement, m

TABLE 5-5 Values of b as a function of RH-min, U-day, and n/N.

	RH-min (percent)						U-day
n/N	0	20	40	60	80	100	(m/s)
0.0	0.84	0.80	0.74	0.64	0.52	0.38	0
0.2	1.03	0.95	0.87	0.76	0.63	0.48	
0.4	1.22	1.10	1.01	0.88	0.74	0.57	
0.6	1.38	1.24	1.13	0.99	0.85	0.66	
0.8	1.54	1.37	1.25	1.09	0.94	0.75	
1.0	1.68	1.50	1.36	1.18	1.04	0.84	
0.0	0.97	0.90	0.81	0.68	0.54	0.40	2
0.2	1.19	1.08	0.96	0.84	0.66	0.50	
0.4	1.41	1.26	1.11	0.97	0.77	0.60	
0.6	1.60	1.42	1.25	1.09	0.89	0.70	
0.8	1.79	1.59	1.39	1.21	1.01	0.79	
1.0	1.98	1.74	1.52	1.31	1.11	0.89	
0.0	1.08	0.98	0.87	0.72	0.56	0.42	4
0.2	1.33	1.18	1.03	0.87	0.69	0.52	
0.4	1.56	1.38	1.19	1.02	0.82	0.62	
0.6	1.78	1.56	1.34	1.15	0.94	0.73	
0.8	2.00	1.74	1.50	1.28	1.05	0.83	
1.0	2.19	1.90	1.64	1.39	1.16	0.92	
0.0	1.18	1.06	0.92	0.74	0.58	0.43	6
0.2	1.44	1.27	1.10	0.91	0.72	0.54	
0.4	1.70	1.48	1.27	1.06	0.85	0.64	
0.6	1.94	1.67	1.44	1.21	0.97	0.75	
0.8	2.18	1.86	1.59	1.34	1.09	0.85	
1.0	2.39	2.03	1.74	1.46	1.20	0.95	
0.0	1.26	1.11	0.96	0.76	0.60	0.44	8
0.2	1.52	1.34	1.14	0.93	0.74	0.55	
0.4	1.79	1.56	1.32	1.10	0.87	0.66	
0.6	2.05	1.76	1.49	1.25	1.00	0.77	
0.8	2.30	1.96	1.66	1.39	1.12	0.87	
1.0	2.54	2.14	1.82	1.52	1.24	0.98	
0.0	1.29	1.15	0.98	0.78	0.61	0.45	10
0.2	1.58	1.38	1.17	0.96	0.75	0.56	
0.4	1.86	1.61	1.36	1.13	0.89	0.68	
0.6	2.13	1.83	1.54	1.28	1.03	0.79	
0.8	2.39	2.03	1.71	1.43	1.15	0.89	
1.0	2.63	2.22	1.86	1.56	1.27	1.00	

Numerous publications have indicated the FAO modification of the Blaney-Criddle method to be a surprisingly accurate estimating method even when compared with more complex methods such as the Penman combination method described later in this chapter (Cuenca et al., 1981; Burman et al., 1983; Allen and Brockway, 1983). In fact, the unusual accuracy of this method compared to lysimetric measurements even on a daily basis has surprised some research engineers. The originators of this method recommend application on a monthly basis for the best accuracy based on input data used for the original calibration (Doorenbos and Pruitt, 1977). However, Allen and Brockway (1983) and Allen and Pruitt (1986) have shown excellent results with the FAO Blaney-Criddle applied on a daily basis.

Example Problem 5-1 (Adapted from Doorenbos and Pruitt, 1977)

Compute the monthly reference evapotranspiration for grass using the FAO modified Blaney-Criddle method given the following data for July for Cairo, Egypt.

Given: Latitude = 30°N

Altitude = 95 m

Monthly data:

$$T_{max} = \sum T_{max}(\text{daily})/31 = 35°\text{C}$$

$$T_{min} = \sum T_{min}(\text{daily})/31 = 22°\text{C}$$

$$T_{mean} = \sum T_{mean}(\text{daily})/31$$

$$\text{or} = (T_{max} + T_{min})/2 = 28.5°\text{C}$$

Solution

$$p = 0.31 (\text{Table 5-4})$$

$$f = p(0.46T_{mean} + 8.13) = 6.6 \text{ mm/d}$$

Using general climatic data (Muller, 1982),

$$RH_{min} = 20 \text{ to } 50\% \quad \text{Medium}$$

$$n/N = 0.8 \text{ to } 0.9 \quad \text{High}$$

$$U_{day} = 2 \text{ to } 5 \text{ m/s} \quad \text{Moderate}$$

Using Fig. 5-10 for graphical calibration,

$$ET_r = 8.5 \text{ mm/d}$$

On a monthly basis,

$$ET_r = 8.5\frac{\text{mm}}{\text{d}}(31 \text{ days}) = 264 \text{ mm}$$

This result may be compared with the solution in equation form using median values for the secondary variables. Using Eq. (5-9) with $RH_{min} = 35\%$, $n/N = 0.85$ and $U_{day} = 3.5$ m/s,

$$b = 1.57$$

and using Eq. (5-8),

$$a = 2.11$$

$$ET_r = -2.11 + 1.57(6.6) = 8.3 \text{ mm/d} = 256 \text{ mm/month}$$

5.4 Pan Evaporation

Pan Description and Installation

Evaporation pans can be effectively used for estimates of crop water use over moderate periods of time—that is, approximately 10 days—if proper adjustment is made to account for pan siting.

At least 9 types of evaporation pans are used around the world. The most common type is the class A pan of the U.S. Weather Service. The dimensions of this pan are as follows:

diameter = 121 cm (46.5 in)
depth = 25.4 cm (10.0 in)
water depth = 5 to 7.5 cm (2 to 3 in) below pan rim

The standard installation for this pan is mounted on a wooden open frame platform with its bottom 15 cm (6 in) above ground level and soil built up to within 5 cm (2 in) of the bottom of the pan. Figure 5-11 indicates the installation of a number of pans for comparative tests in the East African Highlands.

Pan Coefficient

The relationship between ET_r and E_{pan} is given by the following equation:

$$ET_r = K_p E_{pan} \tag{5-11}$$

Figure 5-11 Evaporation pans being tested for response to size, screening, and interior paint at the Kenya Agricultural Research Institute at Muguga. All pans are located within a well-watered grass environment.

where

ET_r = evapotranspiration of grass reference crop, mm/d

K_p = pan coefficient.

= f(type of pan, pan siting, RH_{mean}, wind run, fetch distance)

E_{pan} = pan evaporation, mm/d

In a pan installation, fetch refers to the distance upwind for the prevailing daytime wind conditions. There are two cases of pan siting which vary depending on the ground surface in the upwind condition. As shown in Fig. 5-12, for case A there is a green crop of variable distance upwind of the pan. For case B there is a dry surface of variable distance in the upwind direction. For all cases, wind direction refers to prevailing daytime wind. The wind run is assumed to be measured at 2 m height or converted to the equivalent 2 m wind run using Eq. 5-10.

Tabulated values for the pan coefficient for a class A pan as a function of pan environment, mean relative humidity, and 24-hour wind run are given in Table 5-6 (Jensen, 1974; Doorenbos and Pruitt, 1977). As can be noted from the table, incorrect accounting for pan environment and local climate can cause errors in estimates of crop water use on the order of plus or minus 40 percent.

As for the FAO Blaney-Criddle method, Frevert et al. (1983) developed a regression equation to fit part of Table 5-6. This relationship was only derived for case A pan sites, which are the most common. The original equation was simplified by Cuenca and Jensen (1988) and the result is given by the following:

$$\begin{aligned} K_p = {} & 0.475 - 0.24 \times 10^{-3}(U_{2m}) + 0.00516(RH_{mean}) + 0.00118(d) \\ & - 0.16 \times 10^{-4}(RH_{mean})^2 - 0.101 \times 10^{-5}(d)^2 \\ & - 0.8 \times 10^{-8}(RH_{mean})^2(U_{2m}) - 1.0 \times 10^{-8}(RH_{mean})^2(d) \end{aligned} \tag{5-12}$$

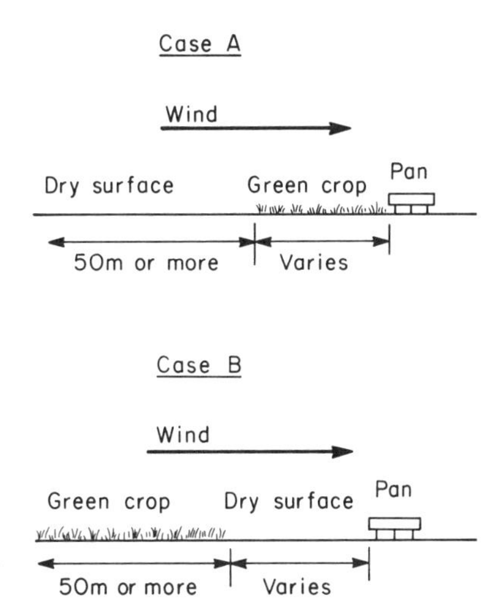

Figure 5-12 Classification of evaporation pan environment.

TABLE 5-6 Pan coefficient (K_p) for class A pan for different groundcover and levels of mean relative humidity and 24-hour wind.

Class A pan		Case A: Pan placed in short green cropped area				Case B[1] Pan placed in dry fallow area		
RHmean %		low <40	medium 40–70	high >70		low <40	medium 40–70	high >70
Wind km/d	Windward side distance of green crop m				Windward side distance of dry fallow m			
Light <175	1	.55	.65	.75	1	.7	.8	.85
	10	.65	.75	.85	10	.6	.7	.8
	100	.7	.8	.85	100	.55	.65	.75
	1 000	.75	.85	.85	1 000	.5	.6	.7
Moderate 175–425	1	.5	.6	.65	1	.65	.75	.8
	10	.6	.7	.75	10	.55	.65	.7
	100	.65	.75	.8	100	.5	.6	.65
	1 000	.7	.8	.8	1 000	.45	.55	.6
Strong 425–700	1	.45	.5	.6	1	.6	.65	.7
	10	.55	.6	.65	10	.5	.55	.65
	100	.6	.65	.7	100	.45	.5	.6
	1 000	.65	.7	.75	1 000	.4	.45	.55
Very strong >700	1	.4	.45	.5	1	.5	.6	.65
	10	.45	.55	.6	10	.45	.5	.55
	100	.5	.6	.65	100	.4	.45	.5
	1 000	.55	.6	.65	1 000	.35	.4	.45

[1]For extensive areas of bare-fallow soils and no agricultural development, reduce K_p by 20% under hot, windy conditions; by 5–10% for moderate wind, temperature, and humidity conditions.

where

$$U_{2m} = \text{wind run at 2 m height, km/d}$$

$$RH_{mean} = \text{mean relative humidity, percent}$$

$$d = \text{fetch distance of green crop, m}$$

d must be limited to less than or equal to 1000 m in Eq. 5-12.

As was discussed concerning the regression equation for the Blaney-Criddle coefficient, the data in Table 5-6 represent the database through which Eq. (5-12) was fit. The regression equation is only an approximation of the tabulated values.

Example Problem 5-2

Compute the reference evapotranspiration for grass for the first 10-day period in August, 1979 using the following data from Moro, Oregon.

Location—Agricultural Experiment Station, Moro, Oregon
Siting—Case B, fetch = 300 m
10-day period—mean values 01 to 10 Aug. 1979

$$RH_{max} = 87.9\%$$
$$RH_{min} = 47.4\%$$
$$RH_{mean} = (87.9 + 47.4)/2 = 68\%$$
$$U_z = 117 \text{ mi/d}$$

where

$$z = 0.46 \text{ m}$$

Converting to equivalent wind run at 2 m height,

$$U_{2m} = U_z \left[\frac{2.0 \text{ m}}{0.46 \text{ m}}\right]^{0.2} \left(1.609 \frac{\text{km}}{\text{mi}}\right)$$
$$U_{2m} = 253 \text{ km/d}$$
$$E_{pan} = 0.45 \frac{\text{in}}{\text{d}} \left(25.4 \frac{\text{mm}}{\text{in}}\right) = 11.4 \text{ mm/d}$$

Solution Apply pan coefficient method from Jensen (1974) or Doorenbos and Pruitt (1977).

$$ET_r(\text{grass}) = K_p E_{pan}$$

From Table 5-6, RH_{mean} = Medium, Case B, U_{2m} = Moderate
Use linear interpolation for fetch condition:

$$f = 100 \text{ m} \qquad K_p = 0.6$$
$$f = 1000 \text{ m} \qquad K_p = 0.55$$

For f = 300 m,

$$K_{p300} = 0.6 - \frac{(300 \text{ m} - 100 \text{ m})}{1000 \text{ m} - 100 \text{ m})}(0.6 - 0.55)$$
$$K_{p300} = 0.59$$
$$ET_r = 0.59(11.4 \text{ mm/d})$$
$$= 6.7 \text{ mm/d}$$

Total ET_r for first 10 days in August:

$$ET_r = 6.7 \frac{\text{mm}}{\text{d}} (10 \text{ days}) = 67 \text{ mm}$$

5.5 Combination Method

Penman Method

The term combination method is normally used to indicate methods that apply a radiation balance plus aerodynamic approach to estimating crop evapotranspiration. The best known combination method is that developed by Perman (1948). This method estimates evapotranspiration using the radiation balance to indicate part of the amount of energy available for evaporation and an aerodynamic term to quantify the influence of advection conditions over the crop canopy in removing water vapor from the field. A definition equation for the Penman method may be given as:

$$ET = f(\text{radiation balance} + \text{aerodynamic term}) \tag{5-13}$$

Particular Forms of the Penman Method

There are a number of representations of the Penman method, most of which vary in the representation of the aerodynamic term (Cuenca and Nicholson, 1982). The basic or original form of the Penman equation (Penman, 1948; Penman, 1952) with units converted to SI is given by

$$ET_r = \frac{\Delta}{\Delta + \gamma}(R_n - G) + \frac{\gamma}{\Delta + \gamma} f(u)\,\Delta e \tag{5-14}$$

where

ET_r = reference evapotranspiration for grass, mm/d

$\frac{\Delta}{\Delta + \gamma}$ = weighting function for elevation and temperature

Δ = slope of saturation vapor pressure versus temperature curve at T_{mean}

γ = psychrometric constant (relationship between vapor pressure deficit and wet bulb depression)

R_n = net radiation, equivalent mm/d

G = soil heat flux (+ if soil is warming), equivalent mm/d

$$\frac{\gamma}{\Delta + \gamma} = 1 - \frac{\Delta}{\Delta + \gamma} \tag{5-15}$$

$$f(u) = \text{wind function} = m(w_1 + w_2 U_{2m}) \tag{5-16}$$

where

m = conversion constant

w_1, w_2 = empirical constants

U_{2m} = wind run at 2 m height, km/d

Δe = vapor pressure deficit, mb
$= e_s - e_a$

e_s = saturation vapor pressure at T_{air}, mb

e_a = mean actual vapor pressure of air, mb

The ET_r parameter in Eq. (5-14) was originally (Penman, 1948), and for decades thereafter, referred to as potential evapotranspiration. However, it is clear by careful reading of Penman's paper (1948) and realization that the original wind function was calibrated against measured evapotranspiration for grass that the result is reference evapotranspiration as defined in this text with grass as the reference crop.

FAO Modified Penman Method

A particular form of the Penman equation which will be stressed in this book is the FAO modified Penman method developed by Pruitt (Doorenbos and Pruitt, 1975; Doorenbos and Pruitt, 1977). Like the FAO Blaney-Criddle method previously discussed, this method was developed by comparing lysimeter-measured evapotranspiration from locations worldwide with Penman type estimates using meteorological data collected at the same sites. Statistical analysis was performed to determine which additional meteorological parameters would reduce the error between measured and estimated evapotranspiration. The FAO form of the Penman equation is given by

$$ET_r = c\left[\frac{\Delta}{\Delta + \gamma}R_n + \frac{\gamma}{\Delta + \gamma}f(u)\,\Delta e\right] \tag{5-17}$$

where

c = calibration coefficient based on meteorological data

$$c = f(RH_{max}, U_{day}/U_{night}, R_s, U_{2day}) \tag{5-18}$$

where

U_{day}/U_{night} = day to night wind run ratio (daytime arbitrarily chosen as 0700–1900 hrs)

R_s = solar radiation, mm/d

U_{2day} = daytime wind speed at 2 m height, m/s

The c term in Eq. (5-17) is the FAO calibration coefficient determined by statistical analysis of meteorological parameters and lysimeter data. The soil heat flux term, G, does not appear in the FAO form of the Penman equation since it was not included when lysimetric and meteorological data were analyzed to determine the calibration constant.

Different forms of the Penman equation have different wind functions and different methods of computing the vapor pressure deficit. In applying any form of the Penman equation, one must be certain that the method of computing the vapor pressure deficit (Δe) is in agreement with the wind function being applied or appreciable errors can result (Cuenca and Nicholson, 1982). This is because different forms of the wind function have been empirically solved for with everything else in Eq. (5-14), including reference evapotranspiration, being measured. Since there are different ways of calculating the vapor pressure deficit, one must be certain that the method used for calculation agrees with the method used to develop the wind function to be applied.

The complete expression for the FAO modified Penman method is given by

$$ET_r = c\left[\frac{\Delta}{\Delta + \gamma}R_n + \frac{\gamma}{\Delta + \gamma}(0.27)(1.0 + 0.01U_{2m})(e_s - e_a)\right] \tag{5-19}$$

In the originally published form of the FAO Penman (Doorenbos and Pruitt, 1975), only the term in brackets in Eq. (5-19) was given. A graphical adjustment procedure was given for seven climatic conditions indicated as being unusual. The typical conditions for which the unadjusted equation was felt to be applicable included a daytime average wind speed of approximately twice the nighttime wind speed and maximum relative humidities equal to or greater than 60 percent. However, Allen (1987) has found the unadjusted FAO Penman method (Doorenbos and Pruitt, 1975) to be more accurate than the adjusted FAO Penman method (Doorenbos and Pruitt, 1977) for some locations. In general, it is recommended that the adjusted FAO Penman method with the c factor be applied because it has been calibrated for a wide range of climates at 10 locations worldwide. But if measured evapotranspiration data indicate an overprediction using the adjusted method, as found by Allen (1987), the unadjusted FAO Penman method could be applied.

Tabulated values for c given as a function of the variables indicated in Eq. (5-18) are shown in Table 5-7. Multiple linear interpolation and extrapolation may be used to determine c from Table 5-7. This can be conveniently done using a computerized version of Table 5-7 (Doorenbos and Pruitt, 1977). An approximate value of c factor, normally accurate to within much less than ±5 percent for all but the most extreme weather conditions in Table 5-7, is given by the regression relationship in Eq. (5-20). This relationship was originally derived by Frevert et al. (1982) using step-wise regression analysis on Table 5-7 and later simplified by Cuenca and Jensen (1988):

$$\begin{aligned} c = {} & 0.68 + 0.0028(RH_{max}) + 0.018(R_s) - 0.068(U_{2day}) \\ & + 0.013(U_{day}/U_{night}) + 0.0097(U_{2day})(U_{day}/U_{night}) \\ & + 0.43 \times 10^{-4}(RH_{max})(R_s)(U_{2day}) \end{aligned} \tag{5-20}$$

where

RH_{max} = maximum relative humidity, percent

R_s = solar radiation, mm/d

U_{day} = mean daytime wind velocity at 2 m height, m/s

U_{day}/U_{night} = day to night wind run ratio (daytime arbitrarily chosen as 0700–1900)

The c values in Table 5-7 are more accurate than the results of Eq. (5-20) which only approximate the table. However, the equation is conveniently programmable.

TABLE 5-7 Adjustment factor (c) in FAO modified Penman equation.

	RHmax = 30%				RHmax = 60%				RHmax = 90%			
Rs mm/d	3	6	9	12	3	6	9	12	3	6	9	12
Uday m/s	Uday/Unight = 4.0											
0	.86	.90	1.00	1.00	.96	.98	1.05	1.05	1.02	1.06	1.10	1.10
3	.79	.84	.92	.97	.92	1.00	1.11	1.19	.99	1.10	1.27	1.32
6	.68	.77	.87	.93	.85	.96	1.11	1.19	.94	1.10	1.26	1.33
9	.55	.65	.78	.90	.76	.88	1.02	1.14	.88	1.01	1.16	1.27
	Uday/Unight = 3.0											
0	.86	.90	1.00	1.00	.96	.98	1.05	1.05	1.02	1.06	1.10	1.10
3	.76	.81	.88	.94	.87	.96	1.06	1.12	.94	1.04	1.18	1.28
6	.61	.68	.81	.88	.77	.88	1.02	1.10	.86	1.01	1.15	1.22
9	.46	.56	.72	.82	.67	.79	.88	1.05	.78	.92	1.06	1.18
	Uday/Unight = 2.0											
0	.86	.90	1.00	1.00	.96	.98	1.05	1.05	1.02	1.06	1.10	1.10
3	.69	.76	.85	.92	.83	.91	.99	1.05	.89	.98	1.10	1.14
6	.53	.61	.74	.84	.70	.80	.94	1.02	.79	.92	1.05	1.12
9	.37	.48	.65	.76	.59	.70	.84	.95	.71	.81	.96	1.06
	Uday/Unight = 1.0											
0	.86	.90	1.00	1.00	.96	.98	1.05	1.05	1.02	1.06	1.10	1.10
3	.64	.71	.82	.89	.78	.86	.94	.99	.85	.92	1.01	1.05
6	.43	.53	.68	.79	.62	.70	.84	.93	.72	.82	.95	1.00
9	.27	.41	.59	.70	.50	.60	.75	.87	.62	.72	.87	.96

Wright Penman Method

Another form of the Penman method which was developed in the arid climate of the western United States was that published by Wright (1982). It was calibrated using lysimeter measured evapotranspiration for alfalfa at the U.S. Department of Agriculture Snake River Conservation Research Center in Kimberly, Idaho. This is a relatively high elevation (1195 m), arid location representative of much of the intermountain plateau in the western United States. This method is interesting to inspect for a number of reasons including use of alfalfa as the reference crop, method of computing net radiation, and method of computing the wind function and vapor pressure deficit.

The Wright (1982) form of the Penman equation is given by

$$ET_r = \frac{\Delta}{\Delta + \gamma}(R_n - G) + \frac{\gamma}{\Delta + \gamma}\frac{(15.36)}{0.1(L)}(a_w + b_w U_{2m})(e_s - e_a) \qquad (5\text{-}21)$$

where

ET_r = reference evapotranspiration for alfalfa, mm/d

L = latent heat of vaporization, cal/g

and other terms are as previously defined. The 15.36 and 0.1(L) terms on the wind function are required for converting the results to mm/d.

Wright (1982) originally expressed the a_w and b_w terms of the wind function as fifth degree polynomials which were a function of the Julian day of the year for the period of 01 April (Day 91) to 31 October (Day 304). These expressions were later simplified to a form applicable for the complete year (Wright, 1987, personal communication) given by

$$a_w = 0.4 + 1.4 \exp\{-[(D - 173)/58]^2\} \tag{5-22}$$

$$b_w = 0.007 + 0.004 \exp\{-[(D - 243)/80]^2\} \tag{5-23}$$

where

$$D = \text{Julian day of the year}$$

Coefficients of the type given in Eqs. (5-22) and (5-23) essentially indicate the influence of varying day length throughout the year on the effects of the daily wind run. The Wright Penman method also uses a function described later which varies with day of year for albedo.

Because of the increased data requirements for the Penman method, it is generally assumed to be applicable for crop water use estimates on a daily basis (Doorenbos and Pruitt, 1977). There are conflicting indications in the literature as to whether or not, if adequately detailed data are available, the Penman method could be applied on less than a daily basis—for example, hourly (Pruitt and Doorenbos, 1977; Mahrt and Ek, 1984). For applications in irrigation system design and irrigation scheduling, use of the Penman method on a daily basis is usually more than adequate.

Due to the data requirements, the Penman method is the most sophisticated and complex method normally used to evaluate crop water use for irrigation system design or management. Certain physical constants and meteorological parameters, such as net radiation and the vapor pressure deficit, must be computed to use this method. Procedures to calculate these parameters are presented followed by example problems which indicate application of the FAO modified Penman method (Doorenbos and Pruitt, 1977) and the Wright Penman method (Wright, 1982).

Calculation of Input Parameters

The expressions given in Eqs. (5-24), (5-26), (5-28), (5-29), and (5-30) are used to calculate input parameters required in the Penman method. They are numerical or empirical approximations for certain variables and are referenced in Burman et al. (1982) and Jensen (1974), among others. The expressions indicated are sufficiently accurate for application with the Penman method.

Slope of saturation pressure, Δ(mb/°C)

$$\Delta = 2.00(0.00738T_{mean} + 0.8072)^7 - 0.00116 \tag{5-24}$$

where

$$T_{mean} = \text{mean air temperature over period of interest, °C}$$

Psychrometric constant, γ(mb/°C)

The psychrometric constant is the relationship between the vapor pressure deficit and the wet bulb depression. This relationship can be expressed as

$$(e_s - e_a) = \gamma(T_{dry} - T_{wet}) \tag{5-25}$$

The expression for the psychrometric constant is

$$\gamma = \frac{c_p P}{L\varepsilon} \tag{5-26}$$

where

c_p = specific heat at constant pressure of dry air

c_p = 1.0035 kJ/kg°C

P = atmospheric pressure, mb

L = latent heat of vaporization, kJ/kg

ε = ratio of mass of water vapor to mass of dry air

ε = 0.62198

For our purposes, Eq. (5-26) can be simplified to

$$\gamma = 1.6134\frac{P}{L} \tag{5-27}$$

For the U.S. Standard Atmosphere, the relationship between atmospheric pressure and elevation is given as

$$P = 1013 - 0.1055(E) \tag{5-28}$$

where

E = elevation, m

The general expression for Eq. (5-28) which may be applied to any standard atmosphere is

$$P = P_0\{[T_0 - \delta(E - E_0)]/T_0\}^{g/(\delta R)} \tag{5-29}$$

where

P_0 = standard sea level atmospheric pressure, mb

T_0 = standard sea level temperature, K

δ = standard lapse rate, K/m

R = universal gas constant for air, 287 J/kgK

g = acceleration of gravity, 9.81 m/s

E_0 = base elevation, m

E = elevation, m

For the U.S. Standard Atmosphere, the lapse rate is given as 6.5 K/km, standard sea level pressure is 1013 mb, and standard sea level temperature is 288K. Under conditions found at equatorial or polar regions, different standards may be required (Burman et al., 1983; Burman et al., 1987). Variation of atmospheric pressure due to passage of frontal systems may be ignored without loss of accuracy for evapotranspiration estimates. Barometric pressure is therefore not an important data requirement for estimating evapotranspiration.

The latent heat of vaporization can be expressed as a function of wet or dry bulb temperature. It is often assumed to be a constant at a standard temperature of 20°C for evapotranspiration estimates without significant loss of accuracy. The recommended expression for latent heat of vaporization as a function of dry bulb temperature is

$$L = 2500.78 - 2.3601T_{air} \qquad (5\text{-}30)$$

with L in kJ/kg and T_{air} in °C (Burman et al., 1987). Values of L tabulated as a function of temperature are given in Appendix Table B-1.

Saturation vapor pressure, e_s (mb)

$$e_s = 33.8639[(0.00738T_{mean} + 0.8072)^8 - 0.000019\left|1.8T_{mean} + 48\right| + 0.001316] \qquad (5\text{-}31)$$

with T_{mean} in °C. Values of e_s as a function of temperature are given in Table 5-8.

TABLE 5-8 Saturation vapor pressure over water.

	Tenth of °C									
	.0	.1	.2	.3	.4	.5	.6	.7	.8	.9
°C	mb	mb	mb	mb	mb	mb	mb	mb	mb	mb
0	6.12	6.15	6.21	6.25	6.30	6.34	6.39	6.44	6.48	6.53
1	6.58	6.62	6.67	6.72	6.77	6.82	6.87	6.92	6.97	7.02
2	7.07	7.12	7.17	7.22	7.27	7.32	7.38	7.43	7.48	7.54
3	7.59	7.54	7.70	7.75	7.81	7.85	7.92	7.97	8.03	8.09
4	8.14	8.20	8.26	8.32	8.38	8.44	8.49	8.55	8.61	8.67
5	8.74	8.80	8.85	8.92	8.98	9.04	9.11	9.17	9.23	9.30
6	9.36	9.43	9.49	9.56	9.63	9.69	9.76	9.83	9.89	9.96
7	10.0	10.1	10.2	10.2	10.3	10.4	10.5	10.5	10.6	10.7
8	10.7	10.8	10.9	11.0	11.0	11.1	11.2	11.3	11.3	11.4
9	11.5	11.6	11.6	11.7	11.8	11.9	12.0	12.0	12.1	12.2
10	12.3	12.4	12.5	12.5	12.5	12.7	12.8	12.9	13.0	13.1
11	13.1	13.2	13.3	13.4	13.5	13.6	13.7	13.8	13.9	13.9
12	14.0	14.1	14.2	14.3	14.4	14.5	14.6	14.7	14.8	14.9
13	15.0	15.1	15.2	15.3	15.4	15.5	15.6	15.7	15.8	15.9
14	16.0	16.1	16.2	16.3	16.4	16.5	16.6	16.7	16.8	17.0
15	17.1	17.2	17.3	17.4	17.5	17.6	17.7	17.8	18.0	18.1
16	18.2	18.3	18.4	18.5	18.7	18.8	18.9	19.0	19.1	19.3
17	19.4	19.5	19.6	19.8	19.9	20.0	20.1	20.3	20.4	20.5
18	20.5	20.8	20.9	21.0	21.2	21.3	21.4	21.5	21.7	21.8
19	22.0	22.1	22.2	22.4	22.5	22.7	22.8	23.0	23.1	23.2

TABLE 5-8 Saturation vapor pressure over water. (continued)

	Tenth of °C									
	.0	.1	.2	.3	.4	.5	.6	.7	.8	.9
°C	mb	mb	mb	mb	mb	mb	mb	mb	mb	mb
20	23.4	23.5	23.7	23.8	24.0	24.1	24.3	24.4	24.6	24.7
21	24.9	25.0	25.2	25.3	25.5	25.6	25.8	26.0	26.1	26.3
22	26.4	26.5	26.8	26.9	27.1	27.2	27.4	27.6	27.7	27.9
23	28.1	28.3	28.4	28.5	28.8	28.9	29.1	29.3	29.5	29.6
24	29.8	30.0	30.2	30.4	30.6	30.7	30.9	31.1	31.3	31.5
25	31.7	31.9	32.0	32.2	32.4	32.5	32.8	33.0	33.2	33.4
26	33.5	33.8	34.0	34.2	34.4	34.5	34.8	35.0	35.2	35.4
27	35.6	35.8	36.1	36.3	36.5	36.7	36.9	37.1	37.3	37.5
28	37.8	38.0	38.2	38.4	38.7	38.9	39.1	39.3	39.5	39.8
29	40.0	40.3	40.5	40.7	41.0	41.2	41.4	41.7	41.9	42.2
30	42.4	42.7	42.9	43.1	43.4	43.6	43.9	44.1	44.4	44.7
31	44.9	45.2	45.4	45.7	45.9	46.2	46.5	46.7	47.0	47.3
32	47.5	47.8	48.1	48.3	48.6	48.9	49.2	49.4	49.7	50.0
33	50.3	50.5	50.9	51.1	51.4	51.7	52.0	52.3	52.6	52.9
34	53.2	53.5	53.8	54.1	54.4	54.7	55.0	55.3	55.6	55.9
35	56.2	56.5	56.9	57.2	57.5	57.8	58.1	58.4	58.8	59.1
36	59.4	59.7	60.1	60.4	60.7	61.1	61.4	61.7	62.1	62.4
37	62.5	63.1	63.5	63.8	64.2	64.5	64.9	65.2	65.6	65.9
38	66.3	66.6	67.0	67.4	67.7	68.1	68.5	68.8	69.2	69.6
39	70.0	70.3	70.7	71.1	71.5	71.9	72.3	72.7	73.0	73.4
40	73.8	74.2	74.6	75.0	75.4	75.8	76.2	76.6	77.1	77.5
41	77.9	78.3	78.7	79.1	79.5	80.0	80.4	80.8	81.3	81.7
42	82.1	82.6	83.0	83.4	83.9	84.3	84.8	85.2	85.7	86.1
43	86.6	87.0	87.5	87.9	88.4	88.9	89.3	89.6	90.3	90.7
44	91.2	91.7	92.2	92.7	93.1	93.6	94.1	94.6	95.1	95.5
45	96.1	96.6	97.1	97.6	98.1	98.5	99.1	99.6	100.1	100.7
46	101.2	101.7	102.2	102.8	103.3	103.8	104.4	104.9	105.4	106.0
47	106.5	107.1	107.6	108.2	108.7	109.3	109.8	110.4	111.0	111.5
48	112.1	112.7	113.2	113.8	114.4	115.0	115.6	116.1	116.7	117.3
49	117.9	118.5	119.1	119.7	120.3	120.9	121.5	122.1	122.8	123.4
50	124.0	124.6	125.3	125.9	126.5	127.1	127.8	128.4	129.1	129.7

Net radiation, R_n (mm/d)

$$R_n = (1 - \alpha)R_s - R_b \quad (5\text{-}32)$$

where

α = albedo

R_s = incoming short-wave (solar) radiation

R_b = net outgoing long-wave (terrestrial) radiation

For the FAO modified Penman method over a cropped surface,

$$\alpha = 0.25 \quad (5\text{-}33)$$

For the Wright Penman method, α is given as a function of time of year to account for sun angle effects. For generally clear days when R_s is greater than 70 percent of clear-sky solar radiation, R_{so},

$$\alpha = 0.29 + 0.06 \sin[30(M + 0.0333N + 2.25)] \tag{5-34}$$

where

$$M = \text{number of month (Jan} = 1)$$

$$N = \text{day of month}$$

and the sine function is for degrees. Equation (5-34) was specifically calibrated for Kimberly, Idaho for dates between 01 April and 31 October. For cloudy days in which $(R_s/R_{so}) < 0.7$,

$$\alpha = 0.30 \tag{5-35}$$

In temperate latitudes such as Kimberly which have low sun angles during much of the winter season, Eq. (5-34) is probably valid from September to May. For tropical latitudes in which the sun angle remains high, the albedo estimated by Eq. (5-34) from September to May is potentially too large and it is recommended that Eq. (5-33) be used for conditions over cropped surfaces.

If R_s is measured, it can be applied directly in Eq. (5-32). Note that net radiation must be expressed as equivalent mm/d for input into either Eq. (5-19) or (5-21). For radiation expressed as energy per unit area (J/m^2), this conversion is made by dividing the radiation by the latent heat of vaporization (kJ/kg) and density of water (kg/m^3) and converting the resulting depth to mm. This conversion will be demonstrated in the example problems.

If the number of sunshine hours, n, is measured, as with the Campbell-Stokes sunshine recorder shown in Fig. 5-8, solar radiation can be estimated using the linear equation of Fritz and MacDonald (1949):

$$R_s = \left(0.35 + 0.61\frac{n}{N}\right)R_{so} \tag{5-36}$$

where

$$N = \text{maximum possible sunshine hours}$$

Table 5-9 indicates values of N as a function of latitude and month of year. Table 5-10 gives values of R_{so} as a function of the same parameters. Note that the values given are for the middle of the month and linear or other smoother interpolation techniques should be used on Tables 5-9 and 5-10 for other dates. Other expressions can be found for R_s as a function of extraterrestrial radiation, R_A (i.e., radiation above the earth's atmosphere) (Doorenbos and Pruitt, 1977; Jensen, 1974).

Net outgoing long-wave radiation is computed by

$$R_b = \left[a\frac{R_s}{R_{so}} + b\right]R_{bo} \tag{5-37}$$

TABLE 5-9 Mean daily duration of maximum possible sunshine hours (N) for different months and latitudes.

Northern Lats	Jan	Feb	Mar	Apr	May	June	July	Aug	Sept	Oct	Nov	Dec
Southern Lats	July	Aug	Sept	Oct	Nov	Dec	Jan	Feb	Mar	Apr	May	June
50	8.5	10.1	11.8	13.8	15.4	16.3	15.9	14.5	12.7	10.8	9.1	8.1
48	8.8	10.2	11.8	13.6	15.2	16.0	15.6	14.3	12.6	10.9	9.3	8.3
46	9.1	10.4	11.9	13.5	14.9	15.7	15.4	14.2	12.6	10.9	9.5	8.7
44	9.3	10.5	11.9	13.4	14.7	15.4	15.2	14.0	12.6	11.0	9.7	8.9
42	9.4	10.6	11.9	13.4	14.6	15.2	14.9	13.9	12.6	11.1	9.8	9.1
40	9.6	10.7	11.9	13.3	14.4	15.0	14.7	13.7	12.5	11.2	10.0	9.3
35	10.1	11.0	11.9	13.1	14.0	14.5	14.3	13.5	12.4	11.3	10.3	9.8
30	10.4	11.1	12.0	12.9	13.6	14.0	13.9	13.2	12.4	11.5	10.6	10.2
25	10.7	11.3	12.0	12.7	13.3	13.7	13.5	13.0	12.3	11.6	10.9	10.6
20	11.0	11.5	12.0	12.6	13.1	13.3	13.2	12.8	12.3	11.7	11.2	10.9
15	11.3	11.6	12.0	12.5	12.8	13.0	12.9	12.6	12.2	11.8	11.4	11.2
10	11.6	11.8	12.0	12.3	12.6	12.7	12.6	12.4	12.1	11.8	11.6	11.5
5	11.8	11.9	12.0	12.2	12.3	12.4	12.3	12.3	12.1	12.0	11.9	11.8
0	12.1	12.1	12.1	12.1	12.1	12.1	12.1	12.1	12.1	12.1	12.1	12.1

where

$$a, b = \text{empirical constants from Table 5-11}$$

$$R_{bo} = \text{net outgoing clear sky long-wave radiation, kJ/m}^2\text{(d)}$$

$$R_{bo} = \varepsilon\sigma(T^4_{max} + T^4_{min})/2 \tag{5-38}$$

where

$$\varepsilon = \text{emissivity of the surface}$$

$$\sigma = \text{Stephan-Boltzmann constant}$$

$$\sigma = 4.8995 \times 10^{-3}\ \text{J/m}^2\text{(d)K}^4$$

and temperatures are given in degrees Kelvin (K = °C + 273). The emissivity is computed as a function of the saturation vapor pressure at the mean dew point temperature. If dew point temperature is not measured but the maximum relative humidity is close to 100 percent, the minimum temperature can be assumed equal to the dew point temperature without loss of accuracy. The dew point temperature changes very little on any given day unless there is sudden frontal movement. The emissivity can therefore be computed as

$$\varepsilon = a_1 + b_1[e_{s-dp}]^{1/2} \tag{5-39}$$

where

$$a_1, b_1 = \text{empirical constants from Table 5-11}$$

$$e_{s-dp} = \text{saturation vapor pressure at dew point, mb}$$

TABLE 5-10 Total daily clear sky solar radiation at surface of the earth (cal/cm^2).

Latitude	Jan.	Feb.	Mar.	Apr.	May	June	July	Aug.	Sept.	Oct.	Nov.	Dec.
						Month						
60° N	58	152	319	533	671	763	690	539	377	197	87	35
55	100	219	377	558	690	780	706	577	430	252	133	74
50	155	290	429	617	716	790	729	616	480	313	193	126
45	216	365	477	650	729	797	748	648	527	371	260	190
40	284	432	529	677	742	800	755	674	567	426	323	248
35	345	496	568	700	742	800	761	697	603	474	380	313
30	403	549	600	713	742	793	755	703	637	519	437	371
25	455	595	629	720	742	780	745	703	660	561	486	423
20	500	634	652	720	726	760	729	697	680	597	537	474
15	545	673	671	713	706	733	706	684	697	623	580	519
10	584	701	681	707	684	700	681	665	707	648	617	565
5	623	722	690	700	652	663	645	645	710	665	650	606
0	652	740	694	680	623	627	616	623	707	684	680	619
5° S	648	758	690	663	590	587	577	590	693	690	727	677
10	710	772	681	640	571	543	526	558	680	690	727	710
15	729	779	665	610	516	497	497	519	657	687	747	739
20	748	779	645	573	474	447	445	481	630	677	753	761
25	761	779	626	533	419	400	406	439	600	665	767	777
30	771	772	600	497	384	353	358	390	567	648	767	793
35	774	754	568	453	335	300	310	342	530	629	767	806
40	774	729	529	407	281	243	261	290	477	603	760	813
45	774	704	490	357	229	183	203	235	447	571	747	813
50	761	669	445	307	174	127	148	177	400	535	727	806
55	748	630	397	250	123	77	97	123	343	497	707	794
60	729	588	348	187	77	33	52	74	283	455	700	787

If the dew point temperature cannot be determined, the emissivity can be calculated as

$$\varepsilon = -0.02 + 0.261 \exp[-7.77 \times 10^{-4}(T_{mean})^2] \qquad (5\text{-}40)$$

with temperature in °C (Burman et al., 1983).

If dew point temperature is measured, the following expression can be used to relate relative humidity with air temperature and dew point temperature:

$$RH = 100\left[\frac{112 - 0.2T_{air} + T_{dp}}{112 + 0.9T_{air}}\right]^8 \qquad (5\text{-}41)$$

where

RH = relative humidity, percent

T_{air} = drybulb temperature, °C

T_{dp} = dew point temperature, °C

Equation (5-41) approximates relative humidity to within 0.6 percent over the temperature range of −25 to +45°C. To compute RH_{max}, T_{min} is substituted into Eq. (5-41) for T_{air}, and T_{max} is substituted to compute RH_{min}.

TABLE 5-11 Net radiation coefficients. (Adapted from Jensen, 1974.)

Region	Experimental Coefficients for Net Radiation Eqs. 5-37 and 5-39			
	a	b	a_1	b_1
Davis, California	1.35	−0.35	0.35	−0.046
Southern Idaho	1.22	−0.18	0.325	−0.044
England	N.A.[a]	N.A.	0.47	−0.065
	N.A.	N.A.	0.44	−0.080
Australia	N.A.	N.A.	0.35	−0.042
General	1.2	−0.2	0.39	−0.05
	1.0	0	—	—

[a]N.A., not available

If relative humidity is measured along with air temperature, Eq. (5-41) can be rearranged to solve for dew point temperature as follows,

$$T_{dp} = \left[\frac{RH}{100}\right]^{1/8}(112 + 0.9T_{air}) - 112 + 0.1T_{air} \tag{5-42}$$

where the units are as indicated for Eq. (5-41).

For the FAO modified Penman, Doorenbos and Pruitt (1977) recommend computing net outgoing longwave radiation as a function of measured sunshine hours using the following expression:

$$R_b = \left(0.9\frac{n}{N} + 0.1\right)[0.34 - 0.044(e_{s-dp})^{1/2}]\sigma(T_{mean})^4 \tag{5-43}$$

with temperature in degrees Kelvin.

For the a_1 term in emissivity Eq. (5-39), Wright (1982) uses a term which changes with time of year to account for changes in the earth's emissivity caused by day length and upper atmospheric conditions. It is given by

$$a_1 = 0.26 + 0.1\exp\{-[0.0154(30M + N - 207)]^2\} \tag{5-44}$$

Wright (1982) uses −0.044 for b_1 in Eq. (5-39), agreeing with the value given in Table 5-11 for southern Idaho.

Vapor pressure deficit, $e_s - e_a$ (mb)

For the FAO modified Penman method, one of three methods is recommended for computing the vapor pressure deficit, depending on the data available (Doorenbos and Pruitt, 1977). All three methods result in answers of similar magnitude. For the Wright Penman method, a different method is recommended, the result of which will vary significantly from the FAO methods. It is important to apply the method to compute the vapor pressure deficit recommended for a particular form of the Penman method since that was the method applied when the empirical wind function was calibrated. Using a different method for the vapor pressure deficit than that used to calibrate the wind function could result in appreciable errors (Cuenca and Nicholson, 1982).

The following three methods are recommended for computing the e_s and e_a vapor pressure deficit terms for use in the FAO modified Penman:

(a) Given:

$$T_{max} \qquad T_{min}$$
$$RH_{max} \qquad RH_{min}$$

Compute e_s using Eq. (5-31) or Table 5-8 with T_{mean}. Compute e_a using,

$$e_a = e_s(RH_{mean}/100) \tag{5-45}$$

with the relative humidity given in percent.

(b) Given:

$$T_{max} \qquad T_{min}$$
$$T_{dry} \qquad T_{wet}$$

Compute the actual vapor pressure using the psychrometric constant and difference in temperature between a dry bulb thermometer and a thermometer with a saturated wick surrounding the bulb using Eq. (5-25) rearranged as

$$e_a = e_{s-wet} - \gamma(T_{dry} - T_{wet}) \tag{5-46}$$

where

e_{s-wet} = saturated vapor pressure computed using Eq. (5-31) with T_{wet} for the temperature

Compute e_s using Eq. (5-31) with T_{mean} as shown.

(c) Given:

$$T_{max} \qquad T_{min}$$
$$T_{dewpoint}$$

Compute e_s using Eq. (5-31) with T_{mean}.
Compute e_a using Eq. (5-31) substituting $T_{dewpoint}$ into the equation.

For the Wright Penman, the vapor pressure deficit is computed using T_{max}, T_{min}, and $T_{dewpoint}$. Recall that T_{min} is approximately equal to $T_{dewpoint}$ if the maximum relative humidity is close to 100 percent. For Wright's method,

$$e_s = \frac{e_{s-Tmax} + e_{s-Tmin}}{2} \tag{5-47}$$

and

$$e_a = e_{s-dp} \tag{5-48}$$

Wright's procedure to compute the vapor pressure deficit must be applied in the Wright Penman since it was the method applied when the wind function was calibrated. The combination of Eqs. (5-47) and (5-48) probably gives a better estimate of the daily mean vapor pressure deficit in arid climates than those methods recommended for the FAO modified Penman (Jensen, 1974).

The Wright Penman method also requires calculation of the soil heat flux, G. The magnitude of the daily soil heat flux over 10- to 30-day periods or longer is small enough that it can normally be neglected for practical estimates of evapotranspiration (Jensen, et al., 1988). Nevertheless, Wright (1982) indicates the following expression to compute the daily soil heat flux:

$$G = (T_{mean} - T_p)c_s \tag{5-49}$$

where

G = soil heat flux, equivalent mm/d

T_{mean} = mean daily air temperature, °C

T_p = mean air temperature for preceding three days, °C

c_s = specific heat coefficient of the soil, mm/d(°C)

The specific heat coefficient is an empirical parameter. Wright (1982) determined this parameter to be equal to 0.15 mm/d(°C) for Kimberly conditions. Due to the typically small value of the soil heat flux parameter, the same value for c_s can probably be assumed for other locations without significant loss of accuracy.

Another expression applicable to calculation of the soil heat flux over long time periods is obtained by assuming that the soil temperature to a depth of 2 m changes approximately with average air temperature and that the volumetric heat capacity of the soil is equal to 0.5 cal/cm^3 (°C) (Jensen et al., 1988). The soil heat flux then becomes

$$G = \frac{(T_{i-1} - T_{i+1})}{\Delta t} = \frac{(T_{i+1} - T_{i-1})}{\Delta t} \tag{5-50}$$

where

G = average soil heat flux, cal/cm^2 (d)

T = mean air temperature for period i, °C

Δt = time between midpoints of the two periods, d

Equation (5-50) is applicable for calculating the soil heat flux for periods longer than 14 days. Additional expressions for the soil heat flux are found in Jensen et al. (1988).

Brutsaert (1982) presents an expression for the soil heat flux as a function of net radiation which can be conveniently applied in some situations. The expression is

$$G = c_r R_n \tag{5-51}$$

where

c_r = empirical constant

For bare soil conditions, Brutsaert (1982) recommends a value of 0.3 for c_r based on review of experimental results of other researchers. For vegetative cover with R_n measured at the top of the crop canopy, Brutsaert (1982) indicates that the value of G is often negligible and can be disregarded in practical applications. This is equivalent to setting c_r to zero in Eq. (5-51). For engineering applications of the Penman equation, the net radiation balance can be considered at the top of the canopy. Therefore c_r can be set to zero under full cover conditions for application of the Penman equation on a daily basis without significant loss of accuracy.

Example Problem 5-3

Compute the reference evapotranspiration for grass using the FAO modified Penman method for Moro, Oregon given the following data:

Latitude = 45° 29′	Elevation = 570 m
Date: 01 May	
$T_{max} = 21°C$	$T_{min} = 9.4°C$
$RH_{max} = 98\%$	$RH_{min} = 48\%$
$U_z = 189$ mi/d	$z = 0.46$ m
$U_{day}/U_{night} = 4$	
$R_s = 512$ cal/cm^2/d	

Solution Compute basic meteorological variables and convert to required units. (Note: Many of the conversion factors applied are listed in Appendix Table A-3.)

$$T_{mean} = (21 + 9.4)/2 = 15.2°C$$

$$RH_{mean} = (98 + 48)/2 = 73\%$$

$$U_2 = 189 \text{ mi/d} \left(1.609344 \frac{\text{km}}{\text{mi}}\right) = 304 \text{ km/d}$$

$$U_{2m} = 304 \text{ km/d} \left[\frac{2.0 \text{ m}}{0.46 \text{ m}}\right]^{0.2} = 408 \text{ km/d}$$

The day/night wind ratio indicates that 4 units of wind occurred during the day for each 1 unit which occurred at night. Daytime wind run and wind speed are therefore computed as

$$U_{2m-day} = \frac{4}{5}(408 \text{ km/d}) = 326 \text{ km/12 h}$$

$$U_{2day} = \frac{326 \text{ km}}{12 \text{ hr}} \left[\frac{1 \text{ hr}}{3600 \text{ s}}\right] \frac{1000 \text{ m}}{1 \text{ km}} = 7.56 \text{ m/s}$$

Calculate the slope of saturated vapor pressure curve using Eq. (5-24) and psychrometric constant using Eq. (5-27) to compute weighting factors.

$$\Delta = 2.00[0.00738(15.2°C) + 0.8072]^7 - 0.00116$$

$$\Delta = 1.109 \text{ mb/°C}$$

Use Eqs. (5-28) and (5-30) to compute P and L required for Eq. (5-27). Assuming U.S. Standard Atmosphere,

$$P = 1013 - 0.1055(570 \text{ m})$$

$$P = 953 \text{ mb}$$

$$L = 2500.78 - 2.3601(15.2°\text{C})$$

$$L = 2464.9 \text{ kJ/kg}$$

$$\gamma = 1.6134(953/2464.9)$$

$$\gamma = 0.624 \text{ mb/°C}$$

$$\frac{\Delta}{\Delta + \gamma} = 0.640$$

$$\frac{\gamma}{\Delta + \gamma} = 1 - 0.640 = 0.360$$

Convert radiation data to equivalent mm/d and compute net radiation.

$$R_s = \frac{512 \text{ cal/cm}^2\text{/d}}{L\rho_w}\left[41{,}868\frac{\text{J/m}^2}{\text{cal/cm}^2}\right]$$

$$R_s = \frac{21{,}436.416 \text{ kJ/m}^2\text{/d}(1000 \text{ mm/m})}{2464.91 \text{ kJ/kg}(1000 \text{ kg/m}^3)}$$

$$R_s = 8.7 \text{ mm/d}$$

Note that the conversion factor for radiation with the computed value of latent heat of vaporization is equivalent to 0.01699 mm/(cal/cm^2).

Compute emissivity using Eq. (5-39). Apply empirical constants from Table 5-11 for southern Idaho which is geographically close and climatologically similar to Moro, Oregon. Assume that T_{min} is equal to $T_{dewpoint}$ since RH_{max} is close to 100 percent. Compute saturated vapor pressure using Eq. (5-31).

$$e_{s-dp} = 33.8639\{[0.00738(9.4°\text{C}) + 0.8072]^8 - 0.000019|1.8(9.4°\text{C}) + 48| + 0.001316\}$$

$$e_{s-dp} = 11.8 \text{ mb}$$

$$\varepsilon = 0.325 - 0.044(11.8 \text{ mb})^{1/2}$$

$$\varepsilon = 0.174$$

Compute net outgoing clear-sky long-wave radiation using Eq. (5-38),

$$R_{bo} = 0.174(4.8995 \times 10^{-3} \text{ J/m}^2\text{(d)K}^4)[(21 + 273)^4 + (9.4 + 273)^4]/2$$

$$R_{bo} = 5{,}895.642 \text{ kJ/m}^2\text{(d)}$$

Converting to equivalent mm/d,

$$R_{bo} = \frac{5{,}895.642 \text{ kJ/m}^2\text{(d)}(1000 \text{ mm/m})}{2464.91 \text{ kJ/kg}(1000 \text{ kg/m}^3)}$$

$$R_{bo} = 2.4 \text{ mm/d}$$

Compute net outgoing long-wave radiation using Eq. (5-37) and constants for southern Idaho from Table 5-11. Interpolate R_{so} from Table 5-10 for 01 May and convert to equivalent mm/d.

$$R_{so} = (650 + 729)/2 = 689.5 \text{ cal/cm}^2\text{/d}$$

$$R_{so} = 689.5 \text{ cal/cm}^2\text{/d}\left[0.01699\frac{\text{mm}}{\text{cal/cm}^2}\right]$$

$$R_{so} = 11.7 \text{ mm/d}$$

$$R_b = \left[1.22\frac{8.7}{11.7} - 0.18\right]2.4 \text{ mm/d}$$

$$R_b = 1.75 \text{ mm/d}$$

Compute net radiation using Eq. (5-32) with albedo equal to 0.25,

$$R_n = (1 - 0.25)8.7 \text{ mm/d} - 1.75 \text{ mm/d}$$

$$R_n = 4.8 \text{ mm/d}$$

Compute vapor pressure deficit using Eq. (5-31) (or Table 5-8) with T_{mean} and Eq. (5-45) with RH_{mean}.

$$e_s = 33.8639\{[0.00738(15.2°\text{C}) + 0.8072]^8 - 0.000019|1.8(15.2°\text{C}) + 48| + 0.001316\}$$

$$e_s = 17.3 \text{ mb}$$

$$e_a = 17.3 \text{ mb}(73\%/100)$$

$$e_a = 12.6 \text{ mb}$$

Calculate FAO c factor using linear interpolation for daytime wind speed on Table 5-7. Enter the table with $U_{day}/U_{night} = 4$, $RH_{max} \cong 90\%$ and $R_s \cong 9$ mm/d.

$$U_{2day} = 6 \text{ m/s} \qquad c = 1.26$$

$$U_{2day} = 9 \text{ m/s} \qquad c = 1.16$$

Performing linear interpolation for $U_{2day} \cong 8$ m/s,

$$c = 1.26 - \frac{8 \text{ m/s} - 6 \text{ m/s}}{9 \text{ m/s} - 6 \text{ m/s}}(1.26 - 1.16)$$

$$c = 1.19$$

[This result can be compared with the solution of applying Eq. (5-20).

$$c = 0.68 + 0.0028(98\%) + 0.018(8.7 \text{ mm/d}) - 0.068(7.56 \text{ m/s}) + 0.013(4) + 0.0097(7.56 \text{ m/s})(4) + 0.43 \times 10^{-4}(98\%)(8.7 \text{ mm/d})(7.56 \text{ m/s})$$

$$c = 1.22$$

The results of interpolating on Table 5-7 will be applied since it is more accurate than the regression equation. In reality, the difference between the two results—less than 3 percent—is insignificant in evapotranspiration estimates.]

Substitute the required values into Eq. (5-19) to solve for the reference evapotranspiration for grass.

$$ET_r = 1.19\{0.640(4.8 \text{ mm/d}) + 0.360(0.27)$$
$$[1.0 + 0.01(408 \text{ km/d})](17.3 \text{ mb} - 12.6 \text{ mb})\}$$
$$ET_r = 6.4 \text{ mm/d}$$

The calibration factor, c, for the FAO modified Penman method can be seen to adjust the result without modification upward by 19 percent.

Example Problem 5-4

For the same data for Moro, Oregon given in example problem 5-3, compute the reference evapotranspiration for alfalfa using the Wright Penman method. Additional datum required is T_p, mean air temperature for the preceding three days,

$$T_p = 13.6°C$$

Solution Many of the parameters solved for in example problem 5-3 can be applied to the Wright Penman method. The following indicates computation of only those parameters which must be recalculated for the Wright Penman. Convert latent heat of vaporization to cal/g.

$$L = 2464.9 \text{ kJ/kg}\left[\frac{1 \text{ cal}}{4.1868 \text{ J}}\right]$$
$$L = 588.7 \text{ cal/g}$$

Compute albedo using Eq. (5-34) or (5-35) after calculating ratio R_s/R_{so}.

$$\frac{R_s}{R_{so}} = \frac{8.7 \text{ mm/d}}{11.7 \text{ mm/d}} = 0.74 > 0.70$$

Therefore apply Eq. (5-34).

$$M = 5$$
$$N = 1$$
$$\alpha = 0.29 + 0.06 \sin\{30[5 + 0.0333(1) + 2.25]\}$$
$$\alpha = 0.25$$

Compute emissivity using Eq. (5-39) after calculating a_1 using Eq. (5-44). Apply b_1 from Table 5-11 for southern Idaho.

$$a_1 = 0.26 + 0.1 \exp(-\{0.0154[30(5) + 1 - 207]\}^2)$$
$$a_1 = 0.31$$
$$\varepsilon = 0.31 - 0.044(11.8 \text{ mb})^{1/2}$$
$$\varepsilon = 0.159$$

Compute net outgoing clear-sky longwave radiation using Eq. (5-38) and same conversion as applied in example problem 5-3,

$$R_{bo} = 0.159(4.8995 \times 10^{-3} \text{ J/m}^2\text{(d)K}^4)[(294)^4 + (282.4)^4]/2$$
$$R_{bo} = 5{,}387.397 \text{ kJ/m}^2\text{(d)}$$
$$R_{bo} = 2.2 \text{ mm/d}$$

Compute net outgoing longwave radiation as in example problem 5-3 but using R_{bo} just calculated:

$$R_b = \left[1.22\left[\frac{8.7}{11.7}\right] - 0.18\right]2.2 \text{ mm/d}$$

$$R_b = 1.60 \text{ mm/d}$$

Compute net radiation using Eq. (5-32):

$$R_n = (1 - 0.25)8.7 \text{ mm/d} - 1.60 \text{ mm/d}$$

$$R_n = 4.9 \text{ mm/d}$$

Compute vapor pressure deficit using Eqs. (5-47) and (5-48) together with Eq. (5-31) or Table 5-8:

$$e_{s-Tmax} = 24.9 \text{ mb}$$

$$e_{s-Tmin} = 11.8 \text{ mb}$$

$$e_s = (24.9 \text{ mb} + 11.8 \text{ mb})/2$$

$$e_s = 18.4 \text{ mb}$$

$$e_a = e_{s-dp} = 11.8 \text{ mb}$$

Compute wind function coefficients using Eqs. (5-22) and (5-23) for day of year equal to 121:

$$a_w = 0.4 + 1.4 \exp\{-[(121 - 173)/58]^2\}$$

$$a_w = 1.03$$

$$b_w = 0.007 + 0.004 \exp\{-[(121 - 243)/80]^2\}$$

$$b_w = 0.0074$$

Compute soil heat flux using Eq. (5-49). Assume the same empirical specific heat coefficient for soil as for Kimberly.

$$G = (15.2 - 13.6)\,0.15 \text{ mm/d(°C)}$$

$$G = 0.2 \text{ mm/d}$$

Substitute the required values into Eq. (5-21) to solve for reference evapotranspiration for alfalfa.

$$ET_r = 0.640(4.9 \text{ mm/d} - 0.2 \text{ mm/d}) + 0.360\left[\frac{(15.36)}{0.1(588.7 \text{ cal/g})}\right]$$

$$[1.03 + 0.0074(408 \text{ km/d})]\,(18.4 \text{ mb} - 11.8 \text{ mb})$$

$$ET_r = 5.5 \text{ mm/d}$$

5.6 Crop Coefficient Curves

FAO Crop Coefficient Method

The FAO methods, all modifications of the Penman equation, and many other evapotranspiration estimating methods result in an evapotranspiration estimate for a reference surface of water or reference crop of grass or alfalfa. To determine water use for a crop other than the reference, crop coefficients must be applied. This is demonstrated by the following equation:

$$ET_c = K_c ET_r \tag{5-52}$$

where

ET_c = crop evapotranspiration

K_c = crop coefficient

ET_r = reference evapotranspiration

Note that in Eq. (5-52), the K_c values have to be with respect to the same crop as the ET_r values used in the equation. That is, if the ET_r values in Eq. (5-52) are for grass, then the K_c values also have to be with respect to grass. Some methods, such as the SCS modification of the Blaney-Criddle method, actually have no reference crop. In this case, the method must be applied with crop coefficient curves especially developed for the method (Soil Conservation Service, 1970). (Seasonal and peak month coefficients for this method were given in Table 5-3.) The SCS crop curves should not be applied with any other ET estimating method because they are unrelated.

Development and application of the FAO method for determining crop coefficients will be described in this section. The FAO method divides the crop coefficient curve into four linear line segments which approximate the curve. The first step is to divide the growing season into the following four growth periods:

(a) Initial period: Time of planting to time of 10 percent ground cover.

(b) Crop development period: From end of initial period to time of effective full cover—that is, 70 to 80 percent ground cover.

(c) Mid-season period: From end of crop development period to start of plant maturity as indicated by leaf discoloration (e.g., beans, maize), leaves falling off (e.g., cotton), or leaves curling and discoloring (e.g., tomatoes, potatoes).

(d) Late season period: From end of mid-season period to time of full maturity or harvest.

Time of growth periods can often be determined by talking with growers or extension agents in the area of interest. When no local information, which would be superior, is available, the information for growth periods from Table 5-12 may be applied. This table indicates total season and crop growth period information for a wide range of crops and for different climatic regions.

After the season has been divided into the four growth periods, coefficients must be determined corresponding to the initial, mid- and late season periods to allow plotting of the total seasonal K_c function. The initial period crop coefficient, K_{ci}, is related to the evaporation from basically a bare soil. This coefficient is a function of soil wetness and the reference evapotranspiration rate during the initial period. This function is demonstrated graphically in Fig. 5-13.

TABLE 5-12 Length of growing season and crop development stages of selected field crops. (From Doorenbos and Pruitt, 1977.)

Crop	Description
Artichokes	Perennial, replanted every 4–7 years; Coastal California planting in April 40/40/250/30 and (360)[1]; subsequent crops with crop growth cutback to ground level in late spring each year at end of harvest or 20/40/220/30 and (310).
Barley	Also wheat and oats; varies widely with variety; wheat Central India November planting 15/25/50/30 and (120); early spring sowing, semi-arid, 35°–45° latitudes; and November planting Rep. of Korea 20/25/60/30 and (135); wheat sown in July in East African highlands at 2500 m altitude and Rep. of Korea 15/30/65/40 and (150).
Beans (green)	February and March planting California desert and Mediterranean 20/30/30/10 and (90); August–September planting California desert, Egypt, Coastal Lebanon 15/25/25/10 and (75).
Beans (dry) Pulses	Continental climates late spring planting 20/30/40/20 and (110); June planting Central California and West Pakistan 15/25/35/20 and (95); longer season varieties 15/25/50/20 and (110).
Beets (table)	Spring planting Mediterranean 15/25/20/10 and (70); early spring planting Mediterranean climates and pre-cool season in desert climates 25/30/25/10 and (90).
Carrots	Warm season of semi-arid to arid climates 20/30/30/20 and (100); for cool season up to 20/30/80/20 and (150); early spring planting Mediterranean 25/35/40/20 and (120); up to 30/40/60/20 and (150) for late winter planting.
Castorbeans	Semi-arid and arid climates, spring planting 25/40/65/50 and (180).
Celery	Pre-cool season planting semi-arid 25/40/95/20 and (180); cool season 30/55/105/20 and (210); humid Mediterranean mid-season 25/40/45/15 and (125).
Corn (maize) (sweet)	Philippines, early March planting (late dry season) 20/20/30/10 and (80); late spring planting Mediterranean 20/25/25/10 and (80); late cool season planting desert climates 20/30/30/10 and (90); early cool season planting desert climates 20/30/50/10 and (110).
Corn (maize) (grains)	Spring planting East African highlands 30/50/60/40 and (180); late cool season planting, warm desert climates 25/40/45/30 and (140); June planting sub-humid Nigeria, early October India 20/35/40/30 and (125); early April planting Southern Spain 30/40/50/30 and (150).
Cotton	March planting Egypt, April–May planting Pakistan, September planting South Arabia 30/50/60/55 and (195); spring planting, machine harvested Texas 30/50/55/45 and (180).
Crucifers	Wide range in length of season due to varietal differences; spring planting Mediterranean and continental climates 20/30/20/10 and (80); late winter planting Mediterranean 25/35/25/10 and (95); autumn planting Coastal Mediterranean 30/35/90/40 and (195).
Cucumber	June planting Egypt, August–October California desert 20/30/40/15 and (105); spring planting semi-arid and cool season arid climates, low desert 25/35/50/20 and (130).
Eggplant	Warm winter desert climates 30/40/40/20 and (130); late spring–early summer planting Mediterranean 30/45/40/25 and (140).
Flax	Spring planting cold winter climates 25/35/50/40 and (150); pre-cool season planting Arizona low desert 30/40/100/50 and (220).
Grain, small	Spring planting Mediterranean 20/30/60/40 and (150); October–November planting warm winter climates; Pakistan and low deserts 25/35/65/40 and (165).
Lentil	Spring planting in cold winter climates 20/30/60/40 and (150); pre-cool season planting warm winter climates 25/35/70/40 and (170).
Lettuce	Spring planting Mediterranean climates 20/30/15/10 and (75) and late winter planting 30/40/25/10 and (105); early cool season low desert climates from 25/35/30/10 and (100); late cool season planting, low deserts 35/50/45/10 and (140).

[1]40/40/250/30 and (360) stand respectively for initial crop development, mid-season and late season crop development stages in days, and (360) for total growing period from planting to harvest in days.

TABLE 5-12 Length of growing season and crop development stages of selected field crops. (From Doorenbos and Pruitt, 1977.) (continued)

Melons	Late spring planting Mediterranean climates 25/35/40/20 and (120); mid-winter planting in low desert climates 30/45/65/20 and (160).
Millet	June planting Pakistan 15/25/40/25 and (105); central plains U.S.A. spring planting 20/30/55/35 and (140).
Oats	See Barley.
Onion (dry)	Spring planting Mediterranean climates 15/25/70/40 and (150); pre-warm winter planting semi-arid and arid desert climates 20/35/110/45 and (210).
(green)	Respectively 25/30/10/5 and (70) and 20/45/20/10 and (95).
Peanuts (groundnuts)	Dry season planting West Africa 25/35/45/25 and (130); late spring planting Coastal plains of Lebanon and Israel 35/45/35/25 and (140).
Peas	Cool maritime climates early summer planting 15/25/35/15 and (90); Mediterranean early spring and warm winter desert climates planting 20/25/35/15 and (95); late winter Mediterranean planting 25/30/30/15 and (100).
Peppers	Fresh—Mediterranean early spring and continental early summer planting 30/35/40/20 and (125); cool coastal continental climates mid-spring planting 25/35/40/20 and (120); pre-warm winter planting desert climates 30/40/110/30 and (210).
Potato (Irish)	Full planting warm winter desert climates 25/30/30/20 and (105); late winter planting arid and semi-arid climates and late spring–early summer planting continental climate 25/30/45/30 and (130); early–mid-spring planting central Europe 30/35/50/30 and (145); slow emergence may increase length of initial period by 15 days during cold spring.
Radishes	Mediterranean early spring and continental summer planting 5/10/15/5 and (35); coastal Mediterranean late winter and warm winter desert climates planting 10/10/15/5 and (40).
Safflower	Central California early–mid-spring planting 20/35/45/25 and (125) and late winter planting 25/35/55/30 and (145); warm winter desert climates 35/55/60/40 and (190).
Sorghum	Warm season desert climates 20/30/40/30 and (120); mid-June planting Pakistan, May in mid-West U.S.A. and Mediterranean 20/35/40/30 and (125); early spring planting warm arid climates 20/35/45/30 and (130).
Soybeans	May planting Central U.S.A. 20/35/60/25 and (140); May–June planting California desert 20/30/60/25 and (135); Philippines late December planting, early dry season—dry: 15/15/40/15 and (85); vegetables 15/15/30/— and (60); early–mid-June planting in Japan 20/25/75/30 and (150).
Spinach	Spring planting Mediterranean 20/20/15/5 and (60); September–October and late winter planting Mediterranean 20/20/25/5 and (70); warm winter desert climates 20/30/40/10 and (100).
Squash (winter) pumpkin	Late winter planting Mediterranean and warm winter desert climates 20/30/30/15 and (95); August planting California desert 20/35/30/25 and (110); early June planting maritime Europe 25/35/35/25 and (120).
Squash (zucchini) crookneck	Spring planting Mediterranean 25/35/25/15 and (100+); early summer Mediterranean and maritime Europe 20/30/25/15 and (90+); winter planting warm desert 25/35/25/15 and (100).
Sugarbeet	Coastal Lebanon, mid-November planting 45/75/80/30 and (230); early summer planting 25/35/50/50 and (160); early spring planting Uruguay 30/45/60/45 and (180); late winter planting warm winter desert 35/60/70/40 and (205).
Sunflower	Spring planting Mediterranean 25/35/45/25 and (130); early summer planting California desert 20/35/45/25 and (125).
Tomato	Warm winter desert climates 30/40/40/25 and (135); and late autumn 35/45/70/30 and (180); spring planting Mediterranean climates 30/40/45/30 and (145).
Wheat	See Barley.

[1]40/40/250/30 and (360) stand respectively for initial crop development, mid-season and late season crop development stages in days, and (360) for total growing period from planting to harvest in days.

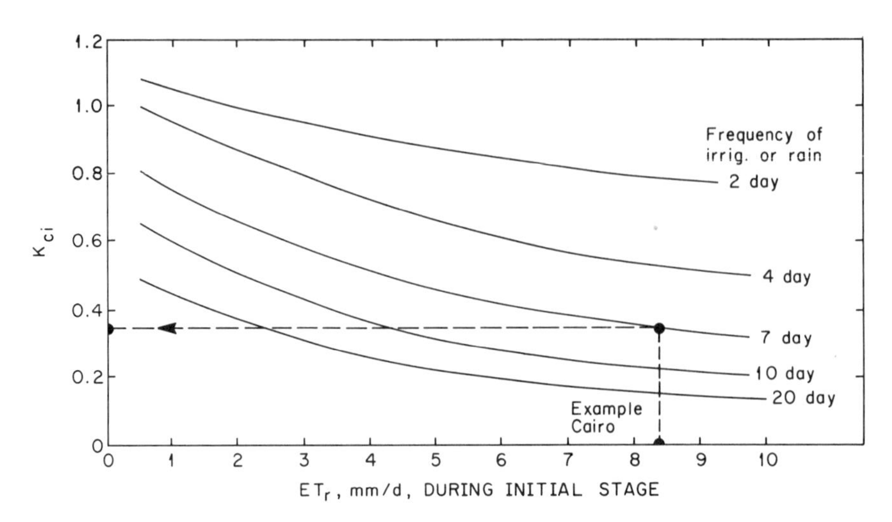

Figure 5-13 Average K_c for initial stage as function of average ET_r during initial stage and frequency of irrigation or significant rainfall. (Adapted from Doorenbos and Pruitt, 1975.)

A regression equation has also been developed to allow for the convenient computerization of Fig. 5-13 and calculation of K_{ci} (El Kayal, 1983; Ryan and Cuenca, 1984). It divides Fig. 5-13 into two parts and is given by

For $I_f < 4$ days,

$$K_{ci} = (1.286 - 0.27 \ln I_f) \exp[(-0.01 - 0.042 \ln I_f)ET_{ri}] \qquad (5\text{-}53)$$

For $I_f \geq 4$ days,

$$K_{ci} = 2(I_f)^{-0.49} \exp[(-0.02 - 0.04 \ln I_f)ET_{ri}] \qquad (5\text{-}54)$$

where

I_f = normal springtime interval between irrigations or significant rainfall, days

K_{ci} = initial stage crop coefficient

ET_{ri} = average initial period reference evapotranspiration, mm/d

ln = natural logarithm

For the FAO method, the mid- and late season coefficients have been developed as a function of ranges of minimum relative humidity and daytime wind speed. These coefficients are listed for a wide variety of field and vegetable crops in Table 5-13. The minimum relative humidity and daytime wind speed to be applied are the averages over the specific growth period—that is, mid-season or late season.

The first step in the graphical application of the FAO crop coefficient method is to plot the length of the growth periods. This is demonstrated along the bottom of Fig. 5-14 which contains data corresponding to the results for example problem 5-5 which follows. A horizontal line is drawn through the value of K_{ci} for the complete initial period. The mid-season coefficient is plotted and a horizontal line drawn

TABLE 5-13 FAO crop coefficient (K_c) for field and vegetable crops for different stages of crop growth and prevailing climatic conditions. (From Doorenbos and Pruitt, 1977.)

	Humidity		RHmin	>70%	RHmin	<20%
Crop	Wind m/s		0–5	5–8	0–5	5–8
	Crop stage					
All field crops	initial	1	Use Fig. 5-13			
"	crop dev.	2	by interpolation			
Artichokes (perennial-clean cultivated)	mid-season	3	.95	.95	1.0	1.05
	at harvest or maturity	4	.9	.9	.95	1.0
Barley		3	1.05	1.1	1.15	1.2
		4	.25	.25	.2	.2
Beans (green)		3	.95	.95	1.0	1.05
		4	.85	.85	.9	.9
Beans (dry)		3	1.05	1.1	1.15	1.2
Pulses		4	.3	.3	.25	.25
Beets (table)		3	1.0	1.0	1.05	1.1
		4	.9	.9	.95	1.0
Carrots		3	1.0	1.05	1.1	1.15
		4	.7	.75	.8	.85
Castorbeans		3	1.05	1.1	1.15	1.2
		4	.5	.5	.5	.5
Celery		3	1.0	1.05	1.1	1.15
		4	.9	.95	1.0	1.05
Corn (sweet)		3	1.05	1.1	1.15	1.2
(maize)		4	.95	1.0	1.05	1.1
Corn (grain)		3	1.05	1.1	1.15	1.2
(maize)		4	.55	.55	.6	.6
Cotton		3	1.05	1.15	1.2	1.25
		4	.65	.65	.65	.7
Crucifers (cabbage)		3	.95	1.0	1.05	1.1
cauliflower, broccoli, Brussels sprout)		4	.80	.85	.9	.95
Cucumber		3	.9	.9	.95	1.0
Fresh market		4	.7	.7	.75	.8
Machine harvest		4	.85	.85	.95	1.0
Eggplant		3	.95	1.0	1.05	1.1
(aubergine)		4	.8	.85	.85	.9
Flax		3	1.0	1.05	1.1	1.15
		4	.25	.25	.2	.2
Grain		3	1.05	1.1	1.15	1.2
		4	.3	.3	.25	.25
Lentil		3	1.05	1.1	1.15	1.2
		4	.3	.3	.25	.25
Lettuce		3	.95	.95	1.0	1.05
		4	.9	.9	.9	1.0
Melons		3	.95	.95	1.0	1.05
		4	.65	.65	.75	.75
Millet		3	1.0	1.05	1.1	1.15
		4	.3	.3	.25	.25

TABLE 5-13 FAO crop coefficient (K_c) for field and vegetable crops for different stages of crop growth and prevailing climatic conditions. (From Doorenbos and Pruitt, 1977.) (continued)

Crop	Humidity		RHmin >70%		RHmin <20%	
	Wind m/s		0–5	5–8	0–5	5–8
Oats		3	1.05	1.1	1.15	1.2
		4	.25	.25	.2	.2
Onion (dry)		3	.95	.95	1.05	1.1
		4	.75	.75	.8	.85
(green)		3	.95	.95	1.0	1.05
		4	.95	.95	1.0	1.05
Peanuts	mid-season	3	.95	1.0	1.05	1.1
(groundnuts)	harvest/maturity	4	.55	.55	.6	.6
Peas		3	1.05	1.1	1.15	1.2
		4	.95	1.0	1.05	1.1
Peppers (fresh)		3	.95	1.0	1.05	1.1
		4	.8	.85	.85	.9
Potato		3	1.05	1.1	1.15	1.2
		4	.7	.7	.75	.75
Radishes		3	.8	.8	.85	.9
		4	.75	.75	.8	.85
Safflower		3	1.05	1.1	1.15	1.2
		4	.25	.25	.2	.2
Sorghum		3	1.0	1.05	1.1	1.15
		4	.5	.5	.55	.55
Soybeans		3	1.0	1.05	1.1	1.15
		4	.45	.45	.45	.45
Spinach		3	.95	.95	1.0	1.05
		4	.9	.9	.95	1.0
Squash		3	.9	.9	.95	1.0
		4	.7	.7	.75	.8
Sugarbeet		3	1.05	1.1	1.15	1.2
		4	.9	.95	1.0	1.0
	no irrigation last month	4	.6	.6	.6	.6
Sunflower		3	1.05	1.1	1.15	1.2
		4	.4	.4	.35	.35
Tomato		3	1.05	1.1	1.2	1.25
		4	.6	.6	.65	.65
Wheat		3	1.05	1.1	1.15	1.2
		4	.25	.25	.2	.2

Note—Many cool season crops cannot grow in dry, hot climates. Values of k_c are given for latter conditions since they may occur occasionally, and result in the need for higher k_c values, especially for tall rough crops.

through this value for all of the mid-season period. The late season coefficient is plotted at the end of the growing season. Two straight lines are drawn connecting the end of the initial period to the beginning of the mid-season and the end of the mid-season to the late season coefficient.

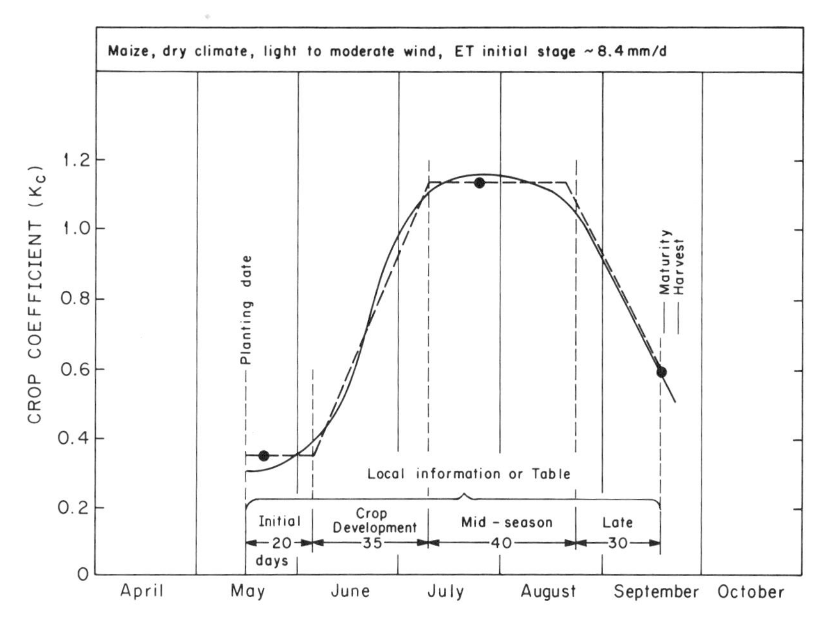

Figure 5-14 Example of crop coefficient curve. Solid line represents actual K_c curve. Dashed line represents linear approximation of K_c curve using FAO method. (Adapted from Doorenbos and Pruitt, 1977.)

The four connected straight-line segments make up the FAO K_c curve. These segments approximate the actual K_c curve which is represented in Fig. 5-14 by the smooth curve. It should be recognized that even though the reference evapotranspiration may be computed on a daily basis, the FAO K_c values come from coefficients which have average values over different growth periods. There is therefore some loss in accuracy and precision in going from an evapotranspiration computed for a reference crop over a short period to a crop evapotranspiration, even if it is computed over the same time interval. Application of the Wright K_c coefficient, described in the next section, indicates certain adjustments for short periods—for example, for days following an irrigation. In any case, going from a reference evapotranspiration to crop evapotranspiration using Eq. (5-52) adds one degree of uncertainty to the evapotranspiration estimate.

The determination and application of the FAO crop coefficients is demonstrated in the following example problems.

Example Problem 5-5

Determine the K_c curve for maize for grain for Cairo, Egypt. The planting date is 15 May and the length of growth periods are as given.

Growth periods: Based on local information

Initial period: 20 days

Crop development period: 35 days

Mid-season period: 40 days

Late season period: 30 days

Solution The growth periods are plotted as shown along the bottom of Fig. 5-14. The initial period ET_r is based on analysis of climatic data and given as

$$ET_{ri} = 8.4 \text{ mm/d}$$

The frequency of irrigation or rainfall is given as 7 days. The frequency of wetting—that is, wetting of the soil surface during the initial period—and initial period ET_r are used with Fig. 5-13 to determine the K_{ci}. As demonstrated in the figure,

$$K_{ci} = 0.35$$

The regression equations Eq. (5-53) or (5-54) can also be used to approximate the curves in Fig. 5-13. Using Eq. (5-54) with the data given,

$$K_{ci} = 0.34$$

For the example shown, a horizontal line is drawn through 0.35 for the entire initial period as indicated in Fig. 5-14. It is now necessary to determine $K_{c\ mid}$ and $K_{c\ late}$. This is done by applying Table 5-13.

For Cairo during mid- and late season periods:

RH_{min} is low

wind is light to moderate

From the values of K_c in Table 5-13 for maize,

$$K_{c\ mid} = 1.15$$
$$K_{c\ late} = 0.6$$

A horizontal line is drawn through 1.15 for the entire mid-season period. The end of the initial period line is joined to the start of the mid-season period line with a straight line. This is the K_c line for the crop development period. The $K_{c\ late}$ coefficient is plotted at the end of growing season. The end of the mid-season period line is joined to this point by a straight line. This is the K_c line for the late season period. The four connected straight-line segments in Fig. 5-14 make up the FAO K_c curve.

The crop coefficient curve must now be applied to the calculated values of reference evapotranspiration. It is important to solve for weighted values of K_c corresponding to those time periods for which ET_r has been calculated. The calculations must be carried out recognizing the date of change of the slope of the K_c curve and the date of change of the ET_r rate. The following example problem indicates the correct procedure. It is a continuation of the data presented in example problem 5-5.

Example Problem 5-6

Apply the K_c values from example problem 5-5 with mean monthly ET_r values and growth period data given.

Location: Cairo, Egypt
Crop: Maize for grain

Month	ET_r mm/d
May	8.4
June	9.6
July	8.3
August	7.4
September	6.0

Crop Growth Period	Number of Days	Dates
Initial	20	15 May–03 June
Crop Development	35	04 June–08 July
Mid-season	40	09 July–17 August
Late season	30	18 Aug–16 September

Solution

(a) Compute mean monthly K_c

From the plot of K_c, all of May is in the initial period. Therefore

$$K_{cMay} = 0.35$$

K_c for June: 30 days total

3 days initial period

27 days crop development period

Total crop development period = 35 days

K_c at start of crop development period = 0.35

K_c at end of crop development period = 1.15

K_c at end of June equals

$$K_{c30\ June} = 0.35 + \frac{27\ \text{days}}{35\ \text{days}}(1.15 - 0.35)$$

$$K_{c30\ June} = 0.97$$

Weighted average K_{cJune} equals

$$K_{cJune} = \frac{3 \text{ days}}{30 \text{ days}}(0.35) + \frac{27 \text{ days}}{30 \text{ days}} \frac{(0.35 + 0.97)}{2}$$

$$K_{cJune} = 0.63$$

K_c for July: 31 days total

8 days crop development period

23 days mid-season period

Weighted average K_{cJuly} equals

$$K_{cJuly} = \frac{8 \text{ days}}{31 \text{ days}} \frac{(0.97 + 1.15)}{2} + \frac{23 \text{ days}}{31 \text{ days}}(1.15)$$

$$K_{cJuly} = 1.13$$

K_c for August: 31 days total

17 days mid-season period

14 days late season period

Total late season period = 30 days

K_c at start of late season period = 1.15

K_c at end of late season period = 0.60

K_c at end of August equals

$$K_{c31\ Aug} = 1.15 - \frac{14 \text{ days}}{30 \text{ days}}(1.15 - 0.60)$$

$$K_{c31\ Aug} = 0.89$$

Weighted average $K_{c_{Aug}}$ equals

$$K_{cAug} = \frac{17 \text{ days}}{31 \text{ days}}(1.15) + \frac{14 \text{ days}}{31 \text{ days}} \frac{(1.15 + 0.89)}{2}$$

$$K_{cAug} = 1.09$$

K_c for September: 16 days late season

Weighted average $K_{c_{Sept}}$ equals

$$K_{cSept} = \frac{0.89 + 0.60}{2} = 0.75$$

(b) Calculate monthly and seasonal ET_c as indicated in the following table.

Month	ET_r mm/d	K_c	ET_c mm/d	Number of Crop Days	ET_c mm
May	8.4	0.35	2.9	17	49
June	9.6	0.63	6.0	30	180
July	8.3	1.13	9.4	31	291
August	7.4	1.09	8.1	31	251
Sept	6.0	0.75	4.5	16	72
Total				125	843

The crop coefficient method demonstrated in this section can be applied to a wide variety of crops, including all of those indicated in Table 5-13. Other crops for which the growing season cannot be conveniently divided into growth periods as indicated in Fig. 5-14 must be handled using other procedures. Alfalfa and grass for hay or pasture are crops for which the water requirement abruptly decreases several times during the growing season following cutting. Table 5-14 indicates applicable K_c values for these crops based on a grass reference evapotranspiration. In this table, $K_{c\ mean}$ refers to the mean value between cuttings, $K_{c\ low}$ to the value just after cutting, and $K_{c\ peak}$ to the value just before a harvest. The values indicated in Table 5-14 are for dry soil conditions. For wet soil conditions, these values may be increased up to 30 percent.

For other crops such as bananas, citrus, fruit and nut trees, grapes, and rice, specific tables have been developed for K_c values which can be applied to a grass-based reference ET_r. These tables are given in the FAO publication by Doorenbos and Pruitt (1977).

Wright Crop Coefficient Method

The reference crop for the Wright form of the Penman equation, Eq. (5-21), is alfalfa. Application of this form of the Penman method to compute crop water requirements using Eq. (5-52) therefore requires use of crop coefficients related to alfalfa as

TABLE 5-14 K_c values for alfalfa, clover, grass-legumes, and pasture. (Adapted from Doorenbos and Pruitt, 1977.)

Climatic Conditions		Alfalfa	Grass for Hay	Clover, Grass-legumes	Pasture
Humid	mean	0.85	0.8	1.0	0.95
Light to	peak	1.05	1.05	1.05	1.05
moderate wind	low	0.5	0.6	0.55	0.55
Dry	mean	0.95	0.9	1.05	1.0
Light to	peak	1.15	1.1	1.15	1.1
moderate wind	low	0.4	0.55	0.55	0.5
Strong wind	mean	1.05	1.0	1.1	1.05
	peak	1.25	1.15	1.2	1.15
	low	0.3	0.5	0.55	0.5

opposed to grass which is the basis of the FAO coefficients. Note that in principle the FAO coefficients could be related to alfalfa by applying the alfalfa coefficients given in Table 5-14 for the FAO grass reference. Either the crop coefficient or reference crop could be converted between grass and alfalfa using these coefficients. This is given in equation form as

$$K_{c-alf} = K_{c-FAO}/(K_{c-alf/grass}) \tag{5-55}$$

where

K_{c-alf} = crop coefficient based on alfalfa reference crop
K_{c-FAO} = FAO crop coefficient (grass reference)
$K_{c-alf/grass}$ = FAO coefficient for alfalfa given in Table 5-14 for peak conditions

Alternatively, the reference crop evapotranspiration computed using the Wright Penman may be converted to grass using the same FAO crop coefficient:

$$ET_{r-grass} = ET_{r-alf}/(K_{c-alf/grass}) \tag{5-56}$$

where

$ET_{r-grass}$ = reference evapotranspiration for grass
ET_{r-alf} = reference evapotranspiration for alfalfa

In principle, the results of Eq. (5-55) could be applied with the reference evapotranspiration (for alfalfa) computed using the Wright Penman method, or the results of Eq. (5-56) could be used with the FAO crop coefficients.

However, neither method indicated by Eqs. (5-55) or (5-56) is totally satisfactory because the estimation brings in the additional uncertainty of the crop coefficient as previously discussed. Wright has developed unique crop coefficients which are particularly applicable to reference ET for alfalfa computed using the Wright Penman and generally to any alfalfa-based reference ET (Wright, 1982). The development of these coefficients is different from the procedure used for the FAO coefficients, and the procedure used gives additional insight into water use on the irrigated field. A disadvantage is that the coefficients have not been developed on as wide a range of crops as the FAO coefficients. However, for those crops which are shared by both methods, the final results are very similar in magnitude (Wright, 1982).

The Wright crop coefficient method divides the actual coefficient into three parts. The basal crop coefficient is related to crop water use measured when the soil surface is dry. An additional coefficient is given to account for soil moisture conditions in the root zone. A third coefficient is related to increasing evaporation rates due to wet soil surface conditions. All of the coefficients are based on lysimeter measurements made at Kimberly, Idaho (Wright, 1982). The expression for the Wright crop coefficient is given as

$$K_c = K_{cb}(K_a) + K_s \tag{5-57}$$

where

K_{cb} = basal crop coefficient, i.e. crop coefficient related to a dry soil surface

K_a = coefficient to adjust for available soil moisture in the root zone

K_s = coefficient to adjust for evaporation due to wet soil surface conditions

The Wright crop coefficient is graphically depicted in Fig. 5-15. The figure demonstrates that with this formulation of the crop coefficient it is possible to separate out the effects of surface wetness and limited soil water in the root zone from the basal crop coefficient. This method of accounting for surface wetness and limited soil water content may be useful in other models of the soil-water-atmosphere-plant system (Cuenca, 1988). For example, an equivalent FAO coefficient (i.e., relative to the same reference ET) would tend to be higher than Wright's basal crop coefficient because the effects of surface wetness after irrigation or rainfall are incorporated into the general FAO crop coefficient.

Application of the Wright coefficients requires dividing the growing season between the dates of planting, effective full cover, and harvest. The dates and number of days used for these periods at Kimberly, Idaho are given in Table 5-15. These

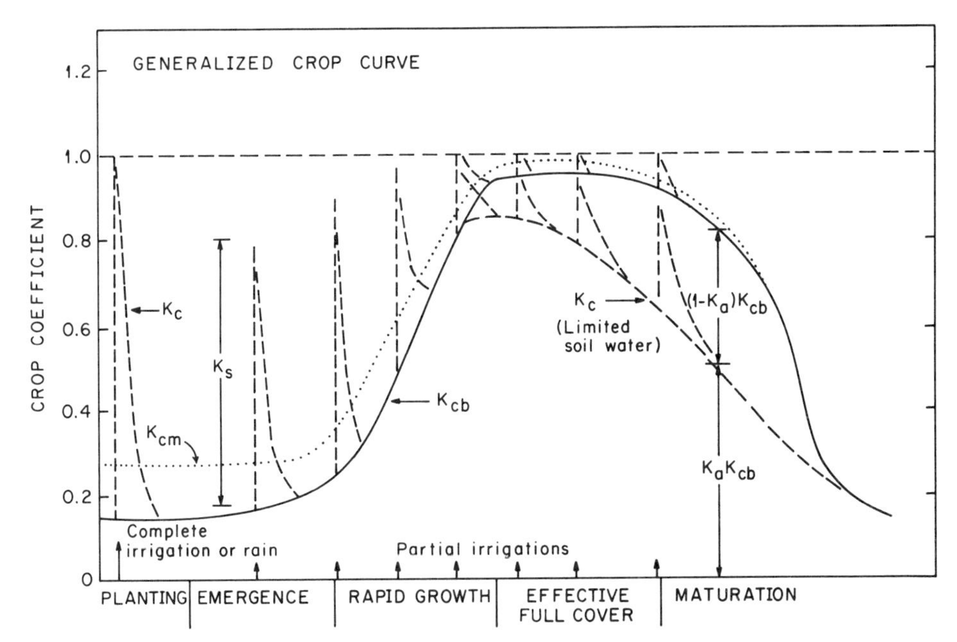

Figure 5-15 Generalized crop coefficient curves showing the effects of growth stage, wet surface soil due to rain or irrigation, and limited available soil water; K_{cb} represents the basal curve; K_c is the adjusted curve based on the surface wetness coefficient, K_s, and available water coefficient, K_a; and K_{cm} is the mean crop curve. (After Wright, 1982.)

dates are applicable to Kimberly and probably much of the intermountain western United States, but for other locations local information would be preferred. Table 5-15 indicates the limited number of crops for which the Wright coefficients have been developed.

Basal crop coefficient data are given in Table 5-16. The basal crop coefficient is given as a function of percentage of time in increments of 10 percent from planting to effective full cover and thereafter for the number of days after full cover. For alfalfa, the basal crop coefficient is given from the time of new growth until harvest as a percent to account for reduced water requirements after cutting.

The coefficient to adjust for available soil moisture in Eq. (5-57) is given as

$$K_a = \ln[(AVM/TAM) \times 100 + 1]/\ln(101.0) \tag{5-58}$$

where

AVM = available soil moisture content at time K_c is evaluated, mm

and

$$AVM = SMC - CEW \tag{5-59a}$$

TABLE 5-15 Date of various crop growth stages identifiable for crops studied at Kimberly, Idaho, 1968–1979. (After Wright, 1982.)

	Date of Occurrence (month/day)							Days	
Crop (1)	Planting (2)	Emergence (3)	Rapid growth (4)	Full Cover (5)	Heading or Bloom (6)	Ripening (7)	Harvest (8)	Planting to Full Cover (9)	Full Cover to Harvest (10)
Spring Grain*	4/01	4/15	5/10	6/10	6/10	7/20	8/10	70	61
Peas	4/05	4/25	5/10	6/05	6/15	7/05	7/25	60	50
Sugar Beets	4/15	5/10	6/01	7/10	—	—	10/15	85	95
Potatoes	4/25	5/25	6/10	7/10	7/01	9/20	10/10	75	90
Field Corn	5/05	5/25	6/10	7/15	7/30	9/10	9/20	72	67
Sweet Corn	5/05	5/25	6/10	7/15	7/20	—	8/15	72	30
Beans	5/22	6/05	6/15	7/15	7/05	8/15	8/30	55	45
Winter Wheat#	(2/15)	(3/01)	3/20	6/05	6/05	7/15	8/10	(110)	60
Alfalfa## (1s)	4/01		4/20				6/15		75
(2n)	6/15		6/25				7/31		46
(3r)	7/31		8/10				9/15		46
(4t)	9/15		10/01				10/30		46

*Spring Grain includes barley and wheat.

#Effective dates in parentheses. Crop was planted on 10/10 and emerged 10/25.

##Effective planting date for established alfalfa is date growth begins in spring or harvest of preceding crop. Final harvest is date crop becomes dormant.

TABLE 5-16 Basal crop coefficients, (K_{cb}), for dry surface soil for use with alfalfa reference for irrigated crops grown in an arid region with a temperate intermountain climate. (After Wright, 1982.)

Basal ET Crop Coefficients, K_{cb}											
	PCT, time from planting to effective cover (%)										
Crop	0	10	20	30	40	50	60	70	80	90	100
(1)	(2)	(3)	(4)	(5)	(6)	(7)	(8)	(9)	(10)	(11)	(12)
Spring Grain*	0.15	0.15	0.16	0.20	0.25	0.40	0.52	0.65	0.81	0.96	1.00
Peas	0.15	0.15	0.16	0.18	0.20	0.29	0.38	0.47	0.65	0.80	0.90
Sugar Beets	0.15	0.15	0.15	0.15	0.15	0.17	0.20	0.27	0.40	0.70	1.00
Potatoes	0.15	0.15	0.15	0.15	0.15	0.20	0.32	0.47	0.62	0.70	0.75
Corn	0.15	0.15	0.15	0.16	0.17	0.18	0.25	0.38	0.55	0.74	0.93
Beans	0.15	0.15	0.16	0.18	0.22	0.34	0.45	0.60	0.75	0.88	0.92
Winter Wheat	0.15	0.15	0.15	0.30	0.55	0.80	0.95	1.00	1.00	1.00	1.00
	DT, days after effective cover										
Crop	0	10	20	30	40	50	60	70	80	90	100
(1)	(2)	(3)	(4)	(5)	(6)	(7)	(8)	(9)	(10)	(11)	(12)
Spring Grain*	1.00	1.00	1.00	1.00	0.90	0.40	0.15	0.07	0.05	—	—
Peas	0.90	0.90	0.72	0.50	0.32	0.15	0.07	0.05	—	—	—
Sugar Beets	1.00	1.00	1.00	1.00	0.98	0.91	0.85	0.80	0.75	0.70	0.65
Potatoes	0.75	0.75	0.73	0.70	0.66	0.63	0.59	0.52	0.20	0.10	0.10
Field Corn	0.93	0.93	0.93	0.90	0.87	0.83	0.77	0.70	0.30	0.20	0.15
Sweet Corn	0.93	0.91	0.90	0.88	0.80	0.70	0.50	0.25	0.15	—	—
Beans	0.92	0.92	0.86	0.65	0.30	0.10	0.05	—	—	—	—
Winter Wheat	1.00	1.00	1.00	1.00	0.95	0.50	0.20	0.10	0.05	—	—
	Time from new growth or harvest to harvest (%)										
Crop	0	10	20	30	40	50	60	70	80	90	100
(1)	(2)	(3)	(4)	(5)	(6)	(7)	(8)	(9)	(10)	(11)	(12)
Alfalfa(1st)#	0.40	0.50	0.62	0.80	0.90	0.95	1.00	1.00	0.98	0.96	0.94
(Intermed.)	0.25	0.30	0.40	0.70	0.90	0.95	1.00	1.00	0.98	0.96	0.94
(Last)	0.25	0.30	0.40	0.50	0.55	0.50	0.40	0.35	0.30	0.27	0.25

*Spring Grain includes wheat and barley.

#1st denotes first harvest; intermediate harvests may be 1 or more depending on length of season; last harvest is when crop becomes dormant in cool weather; see text for further discussion. Cultivar used was Ranger.

where

SMC = soil moisture content in the root zone at time K_c is evaluated expressed as depth, mm

CEW = limit of crop extractable water in root zone expressed as depth, mm

TAM = total available moisture expressed as depth, mm

and

$$TAM = FC - CEW \tag{5-59b}$$

where

$$FC = \text{field capacity of root zone expressed as depth, mm}$$

The principle of the available moisture content is expressed schematically in Fig. 5-16. The magnitude of the ratio AVM/TAM goes from 1 when the soil water content is at field capacity to 0 when the soil water content is at the limit of crop extractable water.

The coefficient to adjust for surface evaporation is given as

$$K_s = (K_t - K_{cb}) \exp(-\beta t) \tag{5-60}$$

where

K_t = threshold coefficient reflecting level of crop canopy development beyond which soil evaporation effects are negligible

β = empirical coefficient representing soil characteristics and climatic conditions

t = time after wetting of soil surface, d

The K_t and β constants originally applied by Wright (1982) were developed for soil and climatic conditions in Kimberly, Idaho (Burman et al., 1982; Jensen et al., 1988). However, they continue to be widely applied due to the difficulty in independently deriving the empirical coefficients. Based on Wright's analysis, K_t is equal to 0.9 and $\exp(-\beta t)$ is equal to 0.8, 0.5, and 0.3 for t equal to 1, 2, and 3 days, respectively. For t greater than 3 days, K_s is assumed equal to zero.

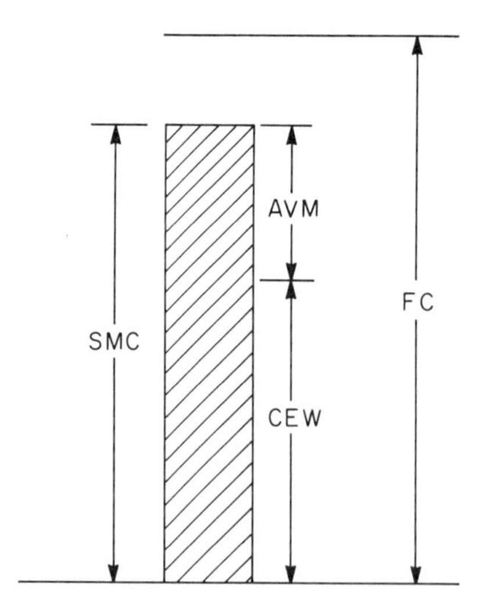

Figure 5-16 Schematic indicating representative depths of field capacity (FC), soil moisture content (SMC), limit of crop extractable water (CEW), and available moisture (AVM) in the root zone.

5.7 Irrigation Project Planning

Crop water requirement estimating methods indicated in the previous sections generally have two major applications. The first is for irrigation project planning purposes and the second is for real-time irrigation system management. This section will discuss applications of crop water requirement estimates for project planning. Irrigation system management will be presented in the next section.

Project planning can be considered on the scale of a total farm or total irrigation project. The principal feature to consider is that the crop water requirement is applied over a range of different crops and on a growing season or annual basis. Thus irrigation planning pertains to one farm with multiple crops as well as to an irrigation project made up of different farms.

Reference Evapotranspiration as a Probabilistic Variable

Long-term records of daily evapotranspiration indicate that a parameter such as reference evapotranspiration is a probabilistic variable with a normal distribution (Wright and Jensen, 1972; Nixon et al., 1972). This point is demonstrated in Fig. 5-17 which indicates the distribution of ryegrass evapotranspiration at a coastal site in California. If enough years of daily data are available, the daily evapotranspiration rate can be shown as normally distributed on any given day of the year. Therefore the monthly or seasonal rate is also normally distributed. This can be an important factor in system design.

It may not be economically feasible to construct a system with enough capacity to meet the expected crop water requirement in 10 out of 10 years. However, it may be feasible to design and construct the system which meets the requirement in eight out of 10 years. If the grower can accept the related risk—that is, that on the average

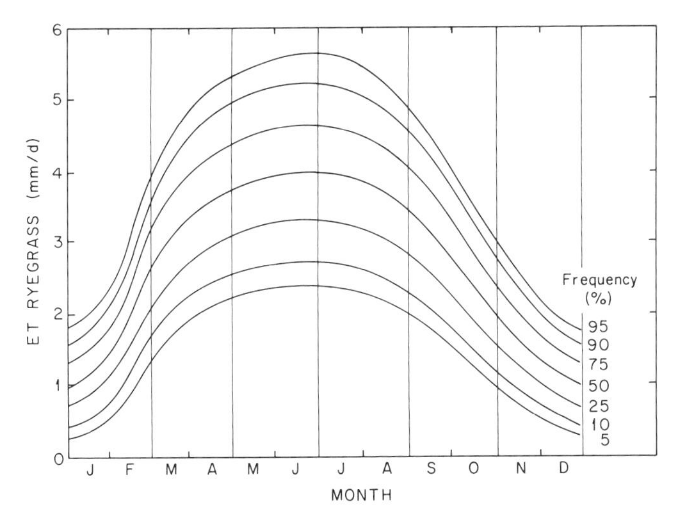

Figure 5-17 Frequency distribution of daily ET_r throughout the year measured in a coastal California valley. (Adapted from Nixon et al., 1972.)

in two out of 10 years he or she will have an inadequate water supply causing reduction in yields or possibly crop failure—then the design based on 80 percent probability of evapotranspiration may be the most cost-effective.

There is an inverse relationship between the capacity of the distribution and irrigation system and the degree of risk the grower faces. The most cost-effective system over the long term will generally be a system which requires the grower to accept some level of risk. A complete design will allow the grower to evaluate the trade-off between system cost and the level of risk. This requires the design engineer to make the necessary effort to collect sufficient data to develop the statistical distribution function for crop water requirements.

Figure 5-18 indicates another interesting feature of crop water requirements which may lead to maximizing the effectiveness of a design. The figure demonstrates that the variation of the measured evapotranspiration decreases as the sequential number of days included in the estimating period increases. In other words, at any level of probability, the evapotranspiration rates will tend to balance out to a value with less variation as more days are included in the irrigation interval. Designing for longer irrigation intervals may therefore allow the grower to meet the same level of risk with a system of smaller capacity.

The difficulty in using the probability concept in irrigation system design is the requirement for long-term data to use as input into the estimating methods. Studies have also shown that when meteorological data are applied in an estimating method, the resulting variation in estimated reference evapotranspiration over a given time is less than that of actual evapotranspiration measured using a lysimeter (Allen and Wright, 1983). Adjustment factors have been developed to modify the standard devia-

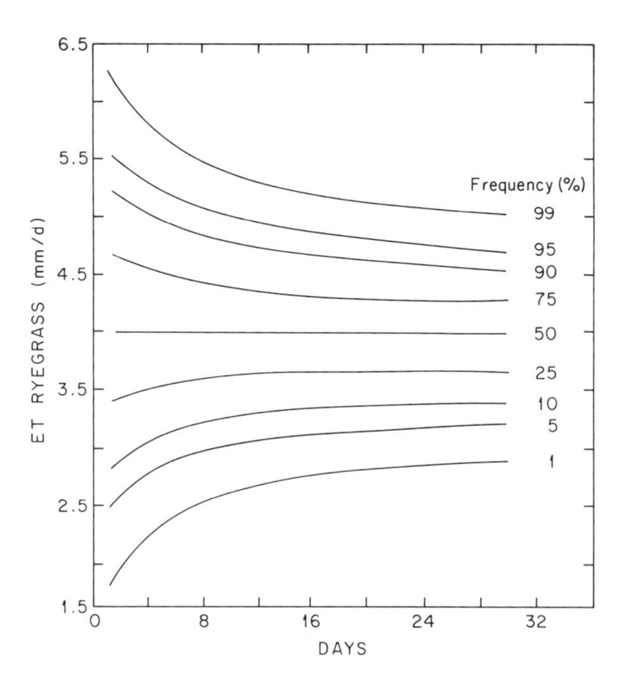

Figure 5-18 Frequency distribution of mean ET_r during peak months (June-July) over time periods of 1 to 30 days. (Adapted from Nixon et al., 1972.)

tion computed for a particular estimating method to bring it into agreement with the variation of measured evapotranspiration (Allen and Wright, 1983).

Statistical theory indicates that a minimum of approximately 30 values are needed to correctly compute the statistics of a normally distributed variable (Devore and Peck, 1986). Normal probability tables may be applied when the mean and standard deviation of the variable are computed based on this minimum number of samples. For less than 30 samples, the Student's t distribution may be applied for a variable which in reality is normally distributed. Application of these statistical procedures is indicated in other references (Devore and Peck, 1986).

Effective Precipitation

Effective precipitation is defined as that amount of precipitation which can be applied to meet the evapotranspiration requirement of a particular crop. The irrigation requirement is the difference between the crop water requirement and the effective precipitation. This is given by the equation

$$IR = CWR - P_{ef} \tag{5-61}$$

where

IR = irrigation requirement, mm

CWR = crop water requirement including inefficiencies of the application system and leaching requirements, mm

P_{ef} = effective precipitation, mm

The inefficiencies of the application system embedded in the crop water requirement of Eq. (5-61) include losses to deep percolation due to overirrigation and/or nonuniform application of water to the field, spray and drift losses in sprinkler systems, and surface runoff. Leaching requirements for salinity control were described in Chapter 4.

In reality, the amount of effective precipitation depends on the precipitation rate, soil moisture conditions which control infiltration rate, and depth of the root zone. That amount of precipitation which runs off a field or percolates past the bottom of the root zone is not considered effective in reducing the irrigation requirement. Detailed hydrologic balance models which operate on a daily basis may be useful in simulating effective precipitation. However, the detail required for this type of model is not normally applied in irrigation system design.

A simplified method which has been usefully applied in many irrigation studies is that developed by the Soil Conservation Service. It is an empirical method based on analysis of 50 years of data for soil water storage, precipitation, and evapotranspiration at 22 locations in the United States ranging from humid to arid climates. The SCS method produces a monthly estimate of effective precipitation and is not theoretically valid for shorter periods of time. This relationship is given in tabular form in Table 5-17. The values given in the body of Table 5-17 are the monthly effective precipitation if the depth of soil moisture depletion at the time of irrigation is normally equal to 75 mm. For other depths of soil moisture depletion at the time of irrigation, the values given in the body of the table are multiplied by the storage factor at the bottom of the table.

TABLE 5-17 Average monthly effective rainfall as related to average monthly evapotranspiration and mean monthly rainfall. (From Anonymous, 1970.)

P_t – mm	12.5	25	37.5	50	62.5	75	87.5	100	112.5	125	137.5	150	162.5	175	187.5	200
	Average monthly effective rainfall in mm D = 75 mm															
ET_c – mm																
25	8	16	23													
50	8	17	24	32	39	46										
75	8	18	26	34	41	48	55	62	69							
100	9	18	27	36	44	51	58	66	73	80	87	93	100			
125	9	20	29	38	46	54	62	70	77	84	92	99	106	112	119	
150	10	21	30	40	48	57	65	73	81	89	97	104	112	119	126	133
175	10	22	32	42	51	60	69	78	86	94	102	110	118	126	133	141
200	11	23	34	44	54	64	73	82	91	99	108	116	124	133	141	149
225	12	24	36	47	57	67	77	87	96	105	114	123	132	140	149	157
250	12	26	38	50	60	71	81	92	101	111	120	130	139	148	157	166

Correction factor (f) for net depths of water depletion (D) other than 75 mm

D – mm	20	25	37.5	50	60	75	100	125	160	175	200
f(D)	.73	.77	.85	.92	.96	1.00	1.03	1.04	1.05	1.07	1.14

Regression analysis has been used to force a function to fit through the data of Table 5-17. This function is given as

$$P_{ef} = f(D)\,[1.25(P_t)^{0.824} - 2.93] \times 10^{(0.000955ET_c)} \tag{5-62}$$

where

P_{ef} = effective precipitation, mm/month
$f(D)$ = function to account for depth of soil moisture depletion other than 75 mm
P_t = total precipitation, mm/month
ET_c = crop evapotranspiration, mm/month

The function f(D) is given by

$$f(D) = 0.53 + 0.0116\,D - 8.94 \times 10^{-5}(D)^2 + 2.32 \times 10^{-7}(D)^3 \tag{5-63}$$

where

D = normal depth of soil moisture depletion prior to irrigation, mm

The value of effective precipitation is limited to the lesser of P_t, ET_c, or P_{ef}, computed using Eq. (5-62).

Application of Eqs. (5-62) and (5-63) can be used to estimate that portion of fall and winter precipitation which replenishes soil moisture depletion after harvest of a crop. This fall-winter replenishment can be used to reduce the amount of preseason irrigation required at the start of a growing season to bring a soil to field capacity. If the level of fall-winter effective precipitation is significant, soil moisture content at the end of the growing season may be allowed to diminish to relatively low levels since nature can be depended upon to refill the soil-moisture reservoir before the next growing season begins.

Peak Project or Farm Irrigation Requirements

The peak irrigation requirement for a project or farm will depend on both the pattern of the reference evapotranspiration rate and the distribution of the crop coefficient curves for the crops of interest during the growing season. The irrigation system capacity must be based on the overall peak rate for the entire project or farm.

It was previously indicated that evapotranspiration is a normally distributed variable and that the most complete design would indicate required system capacity as a function of probability and associated risk. This level of design will not be possible if long-term meteorological data are not available. One is often constrained to use mean climatic data to estimate crop water requirements. Doorenbos and Pruitt (1977) give a method to adjust the maximum monthly crop water requirement computed using mean climatic data to a value of ET_c which is valid at the 75 percent probability level. This adjustment is given as a function of climate, which is indicative of degree of weather variation, and increasing depths of available soil moisture, which tend to moderate the effect of short duration, high ET_c rates.

The relationship presented by Doorenbos and Pruitt (1977) is shown in Fig. 5-19. It indicates the degree of adjustment required in the maximum monthly ET_c calculated using mean climatic data to produce a value of ET_c which is valid at the 75 per-

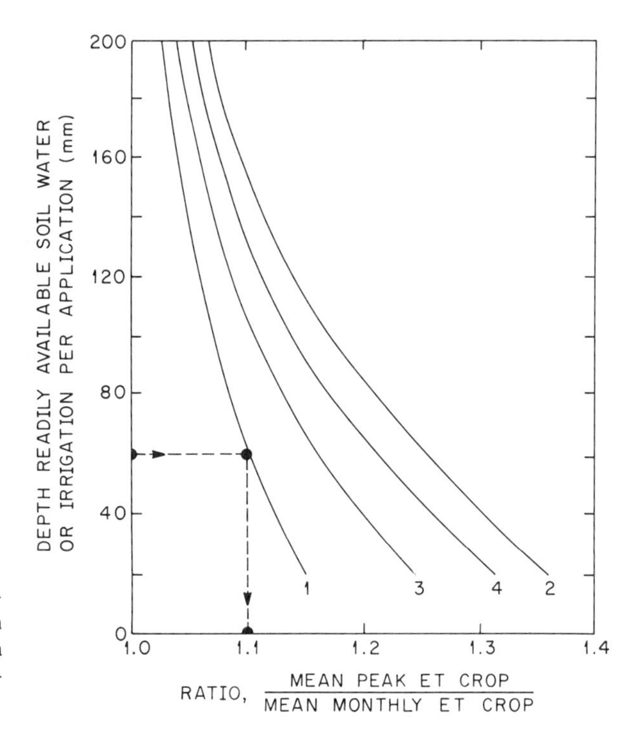

Figure 5-19 Ratio of 75 percent probability peak ET_c to ET_c for maximum month computed using mean climatic data as a function of climate. (Adapted from Doorenbos and Pruitt, 1977.)

cent probability level. This adjustment is given as a ratio of the mean peak month ET_c—that is, valid at the 75 percent probability level—to the maximum monthly ET_c computed using mean climatic data. The functions are given for four climatic conditions. Application of this figure is demonstrated in example problem 5-7.

Example Problem 5-7

Compute the adjustment to the maximum monthly ET_c computed using mean climatic data for maize in Cairo in example problem 5-6 to have a result valid at the 75 percent probability level. Assume the depth of soil moisture depletion at time of irrigation equals 60 mm.

Solution Apply the curve for an arid climate with clear weather conditions during the peak month from Fig. 5-19—that is, curve #1. Project horizontally across from a depth of 60 mm to curve #1 and down to a ratio of 1.1. The corrected value for the peak month of July is

$$ET_{c-peak-75\%} = 1.1(291 \text{ mm}) = 320 \text{ mm}$$

This value may now be applied to compute the peak project or farm irrigation requirement which is valid for 15 out of 20 years—that is, 75 percent of the time.

The overall peak irrigation requirement for a project or farm is obtained by summing the irrigation requirements for individual crops over the growing season as a function of time and choosing that time period with the maximum requirement. The length of the peak period is a function of the irrigation interval selected for operations during mid-season. The peak period irrigation requirement is calculated by

$$IR_{peak} = \sum_{d=d_s}^{d_e} \left[\sum_{c=1}^{N_c} [IR_c]_d \right] \tag{5-64}$$

where

IR_{peak} = total peak period irrigation requirement, mm
d = date
d_s = starting day of peak period
d_e = ending day of peak period
c = crop number
N_c = total number of crops
IR_c = irrigation requirement of crop c, mm

Application of Eq. (5-64) can be used to indicate the effects on the peak period irrigation requirement of varying planting dates or selecting early or late season crops.

5.8 Irrigation System Management

Application of Crop Water Use Estimates

One application of estimates of crop water use is in scheduling irrigations. This may be considered at two time scales. The first is before the season begins as part of the seasonal planning effort. Assuming the meteorological data-based methods previously demonstrated in this chapter are to be applied for crop water use estimates, the data used before the season begins would have to be historical data. The results are used to develop the general operating procedure for the season, assuming the weather conditions would be more or less normal.

The second time scale is at the real-time or operational level. In this case current meteorological data are applied in a crop water use estimating method. Automated meteorological stations which measure all parameters required for the Penman method, such as that depicted in Fig. 5-20, are often used in such applications. These stations are driven by microprocessors and can relay data to cassette tapes within the stations, over telephone lines, or using satellite communication. Data can be taken on time scales of a few minutes and results applied to daily estimates of crop water use for irrigation management. The predicted cumulative crop water use is compared with the amount of water stored in the root zone and considered to be available to the crop. When the level of estimated crop water use approaches the management allowed depletion of the soil profile, it is time to prepare for an irrigation.

The procedure to trace the pattern of crop water use versus depth of available water in the root zone is the same whether historical or current meteorological data are applied—that is, whether the results are to be applied for seasonal planning or real-time operation purposes. Three factors must be determined for the specific site-

Figure 5-20 Automated meteorological station used to collect solar radiation, windspeed, relative humidity, air and soil temperature data on a time scale of one hour or less.

soil-crop situation being analyzed. The first of these is the management allowed depletion (MAD) of soil moisture in the root zone. The selection of the MAD is based on the sensitivity of the crop to damage due to stress, the expected adequacy of seasonal water supplies, and the economic value of the crop. The economic value may be considered as the marginal value in crop production and revenue generation compared to the marginal cost of applying water to the field at different time intervals. Marginal cost and benefit concepts were discussed in the second chapter.

Different crops are susceptable to stress at different levels of soil moisture depletion. The stress to the crop is actually associated with levels of soil moisture tension, as discussed in the chapter on soil physics. But the depletion in terms of volumetric soil moisture content is a readily measurable and more convenient concept. It is clear that the volume of water applied to the field through the irrigation system goes directly to increasing the volumetric water content of the soil profile. It is generally not so clear what the quantitative impact on the soil-water tension is due to a specific volume of applied water.

Crops which are more susceptable to stress due to impacts on yield deficit or degradation of quality require more frequent irrigation than those which are more tolerant to high stress levels. In general, horticultural crops can sustain 40 percent depletion of available water and cereal crops up to 60 percent without excessive economic loss. Particular crops may have more critical levels. Potatoes, for example, suffer rapid reduction in quality at soil moisture depletion levels greater than 30 percent. Typical values for management allowed depletion are 33 percent for shallow rooted, high value crops, 50 percent for moderate value crops with medium root depths, and 67 percent for crops of relatively low value with deep roots.

The final value of acceptable MAD requires a decision based on balancing the reduction of marketable product with the increased cost of a more intensive design. As with all aspects of agricultural system design, risk is also an element. As the level of MAD is decreased, there is less likely to be serious yield reduction due to unexpected equipment failure or extremely dry meteorological conditions. However, the reduction of risk comes at the cost of a more intensive irrigation system.

Another variable which is required to schedule irrigations is the depth of the root zone, D_r. This is actually a dynamic variable since the depth of the root zone progresses during the growing season. A number of studies have shown the depth of the active root zone is a function of the cumulative reference evapotranspiration since the start of the growing season. An example of such an expression for the UC82 tomato is (Cuenca, 1978):

$$D_r = 0.59\left[\sum ET_r\right] - 121 \tag{5-65}$$

where

D_r = depth of active root zone, cm

ET_r = reference evapotranspiration for grass, mm

This function is depicted in Fig. 5-21.

For irrigation system design based on peak period water requirements, the depth of root zone required is at or close to the maximum. Table 5-18 indicates soil

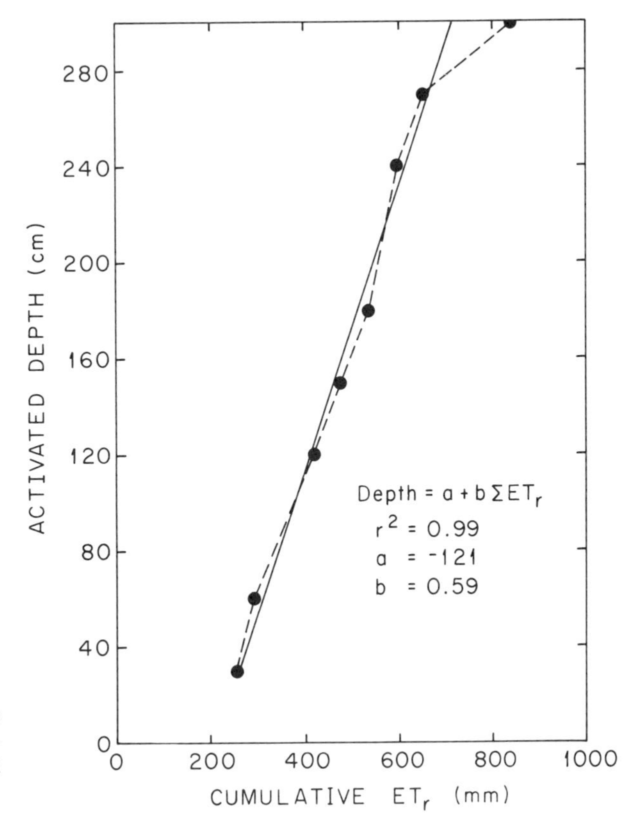

Figure 5-21 Depth of active root zone as a function of cumulative reference evapotranspiration for UC82 tomato. (Adapted from Cuenca, 1978.)

TABLE 5-18 Depths containing majority of feeder roots for various crops. (Modified from Rainbird, 1982.)

Crop	Depth, m	Crop	Depth, m
Alfalfa	1.8	Onions	0.5
Beans	0.6	Orchard	1.5
Beets	0.9	Pasture	0.5
Berries (Cane)	0.9	Grasses only	
Cabbage	0.6	Pasture	0.6
Carrots	0.6	with Clover	
Cotton	1.2	Peanuts	0.5
Cucumbers	0.6	Peas	0.8
Grain	0.8	Potatoes	0.6
Grapes	1.8	Soy Beans	0.6
Lettuce	0.3	Strawberries	0.5
Maize	0.8	Sweet Potatoes	0.9
Melons	0.9	Tobacco	0.8
Nuts	1.8	Tomatoes	0.6

depths within which are found the majority of roots for various crops. In deep, uniform soils without constrictive layers, these depths may easily be exceeded. In soils with limited depths before a constricting or rock layer is encountered, the maximum depth of the root zone will be the limit of available soil.

The final data requirement is the limit of crop extractable water, CEW. As discussed in the chapter on soil physics, the tension at the limit of CEW appears to be overestimated in greenhouse experiments compared to field trials. The soil water tension of approximately 100 m of water is used to indicate a conservative limit of CEW in this book. Representative limits of crop extractable water and total available moisture expressed as percent by volume and millimeters per meter of soil depth were indicated in Table 3-2 for various soil textures.

The product of the three variables is equal to the water which must be replenished in the root zone, or net depth of irrigation, under the given management conditions. This is expressed as

$$i_n = MAD(D_r)TAM \tag{5-66}$$

where

$$\begin{aligned} i_n &= \text{net depth of irrigation, mm} \\ MAD &= \text{management allowed depletion, fraction} \\ D_r &= \text{depth of root zone, m} \\ TAM &= \text{total available moisture, mm/m} \end{aligned}$$

Figure 5-22 indicates a plot of field capacity, management allowed depletion, limit of crop extractable water, and net depth of irrigation, all expressed as depth in mm for a given depth of root zone. Superimposed on the figure is the soil moisture depletion pattern resulting from crop water use estimates based on meteorological data averaged over five-day time intervals. The concept of irrigation management using scheduling requires that water be applied to the field when the soil moisture

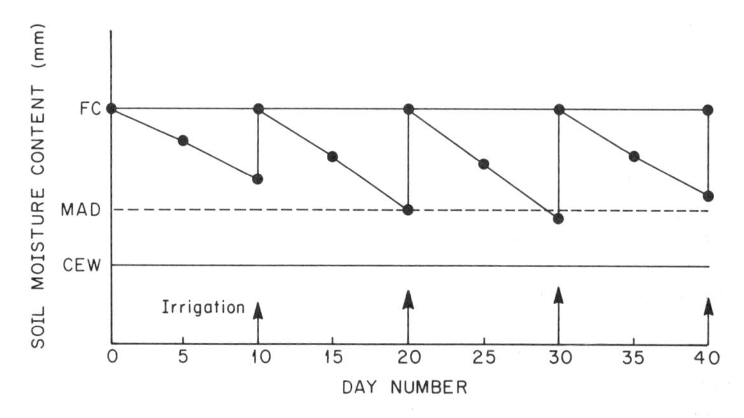

Figure 5-22 Pattern of soil moisture content with time due to crop water use and irrigation at fixed time interval to replenish root zone to field capacity. (Note: Scale of depth of irrigation is approximately one-half scale of soil moisture content.)

depletion level intersects or is close to the management allowed depletion line. Figure 5-22 indicates the results of irrigation on a fixed 10-day interval with the depth of irrigation adjusted to bring the water content up to the level of field capacity. Irrigating on a fixed time interval is often most convenient for growers. From the field capacity point, the soil moisture depletion pattern continues at the average rate for the five-day time interval based on crop water use estimates.

Application of Measured Soil Moisture Status

Numerous methods are used to measure soil moisture status in the field. Such measurements made before an irrigation can allow growers to evaluate the dryness of the soil and to evaluate whether an irrigation should be prepared for or whether soil moisture conditions are high enough that irrigation can wait a few days. Soil moisture measurements made after an irrigation can be used to evaluate if the time of irrigation was adequate and if the soil profile was wet to the expected depth in the root zone. In general, commercial irrigation scheduling services use some type of soil moisture measurement to verify the soil water status.

Detailed measurements of soil moisture content allow for verification of the soil moisture levels predicted through a scheduling scheme such as shown in Fig. 5-22. With sufficiently accurate soil moisture measurements, knowledge can be gained regarding required adjustments to procedures used to estimate crop water use rates applied in a graph such as that shown in Fig. 5-22. If the predicted soil moisture content is consistently higher or lower than the measured content, there may be a bias in the estimating method which should be adjusted.

The simplest type of soil moisture measurement tool is a shovel. Another device almost as basic is an Oakfield soil probe shown in Fig. 5-23. The Oakfield probe allows for checking the soil moisture content at depth more conveniently than a shovel. Although a shovel or samples taken with an Oakfield probe and analyzed

Figure 5-23 Soil moisture sample taken with Oakfield soil probe, bottom of photograph. Also shown are two white gypsum blocks.

by hand by squeezing in the field may not appeal to the high technologist, they can be a good measure of the adequacy of irrigation to an experienced field person. Photographic guidelines have also been developed to aid in evaluation of field samples by hand squeezing (Hurd, 1969).

Tensiometers and gypsum blocks (see Fig. 5-23) measure soil-water tension which can be related to soil moisture content through the soil-water characteristic curve. Tensiometers can measure in the moist range of soil moisture content between zero and approximately 8 m of water tension. Gypsum blocks can be selected to measure tensions up to 150 m of water. Neither device is difficult to place at the required depth of measurement in the field and they are normally installed with the aid of an Oakfield soil probe.

Neither a tensiometer nor a gypsum block can normally measure the full range of tensions possible in a field situation, but they can give an adequate measure of when moisture due to irrigation reaches certain depths in the soil profile. An automated hydraulic controller which operates on this principle is shown in Fig. 5-24. The tensiometer connected to the controller senses the soil moisture levels. Two such sensors placed at different depths can be used to automatically turn on and off water application devices. The devices are turned on when an upper-level tensiometer reaches a critical tension and are turned off when the wetting front reaches a lower-depth tensiometer. In this manner the entire root zone is adequately irrigated. Such a system is particularly applicable in trickle irrigation systems which are typically operated at high moisture content levels. The range of tensions measured by the tensiometer could be used to control irrigation in such systems.

Figure 5-24 Hydraulic controller used to turn on or off valves based on programmed level of soil moisture tension sensed by tensiometer. (Photo courtesy of Moisture Dynamics, Inc., Corvallis, Oregon.)

The difficulty with use of tension measuring devices to measure levels of water content is that the soil-water characteristic curve must be applied. This causes one additional level of uncertainty. A device which is calibrated to measure moisture content directly is the neutron probe depicted in Fig. 5-25. The neutron probe actually measures the number of hydrogen ions within a sphere of approximately 40 cm in diameter. The actual radius of measurement depends on the soil texture and soil-moisture content and is described by the following equation (Burman et al., 1982):

$$R = 15.0(100.00/\theta_v)^{1/3} \tag{5-67}$$

where

R = radius of influence, cm

θ_v = volumetric moisture content, percent

The neutron probe is typically calibrated in the United States by taking gravimetric samples and readings with the neutron probe at the same time. This is normally done at the time of installation of the aluminum access tube placed throughout the depth of the soil profile to be monitored. The neutron probe itself is slid down into the access tube and connected to the unit housing the electronics (sitting on top of the tube) by an electrical cable (see Fig. 5-25). The entire soil profile can be monitored using the neutron probe. Measurements are normally made every 10 to 30 cm of depth.

Calibration is usually done for each soil type to be monitored so the effects of organic matter can be separated from the hydrogen ions bound in the water molecule.

Figure 5-25 Neutron probe which can be calibrated to measure soil moisture content at depth. Probe housing containing microprocessor is shown sitting on aluminum access tube.

In some countries, such as France, a calibration procedure is followed which requires bombarding the soil sample with neutrons and making certain measurements of atomic absorption and desorption (Couchat, 1974). Measurements for soil moisture content at depths less than 15 cm require different calibration equations than the rest of the soil profile. This is because conditions within the sphere of influence are different near the soil surface than at depth. Near the surface, fast neutrons are able to escape to the atmosphere since the soil surface is at a distance less than the radius of influence from the neutron source.

The most convenient form of calibration equation for volumetric moisture content using the neutron probe is given as

$$\theta_v = A(NMC/STD) + B \tag{5-68}$$

where

NMC = neutron meter count

STD = standard count for the particular neutron probe

A, B = statistical calibration coefficients

Note that a calibration equation given in the preceding form must be evaluated for each change in bulk density of the soil profile (Cuenca, 1988). If the gravimetric calibration samples are carefully taken and the probe operated correctly, the calibration equation can be statistically verified to clearly describe the variation in soil moisture content (e.g., coefficient of determination equal to or greater than 90 percent).

REFERENCES

ALLEN, R. G. and C. E. BROCKWAY, "Estimating Consumptive Use on a Statewide Basis," Advances in Irrigation and Drainage, Proceedings of the American Society of Civil Engineers Irrigation and Drainage Division Specialty Conference, Jackson, Wyoming, 1983, pp. 79–89.

ALLEN, R. G., "A Penman for All Seasons," American Society of Civil Engineers, *Journal of the Irrigation and Drainage Division,* vol. 112, no. 4, 1986, pp. 348–368.

ALLEN, R. G. and W. O. PRUITT, "Rational Use of the FAO Blaney-Criddle Formula," American Society of Civil Engineers, *Journal of Irrigation and Drainage Engineering,* vol. 112, no. 2, 1986, pp. 139–156.

BLANEY, H. F. and W. D. CRIDDLE, "Determining Water Requirements in Irrigated Areas from Climatological and Irrigation Data," U.S. Dept. of Agriculture, Soil Conservation Service, Technical Report No. 96, 1950.

BURMAN, R. D., R. H. CUENCA, and A. WEISS, "Techniques for Estimating Irrigation Water Requirements," in *Advances in Irrigation,* vol. 2, ed. D. Hillel. New York: Academic Press, 1983, pp. 336–394.

COUCHAT, P., "Mesure Neutronique de l'Humidite des Sols," (In French), Doctor of Science Thesis, University of Paul Sabatier, Toulouse, France, 1974.

CUENCA, R. H., "Transferable Simulation Model for Crop Soil Water Depletion," Ph.D. Thesis, University of California, Davis, 1978.

CUENCA, R. H., J. ERPENBECK, and W. O. PRUITT, "Advances in Computation of Regional Evapotranspiration," Proceedings of Water Forum 1981, American Society of Civil Engineers Specialty Conference, San Francisco, California, vol. 1, 1981, pp. 73–80.

CUENCA, R. H. and M. T. NICHOLSON, "Application of Penman Equation Wind Function," *Journal of the Irrigation and Drainage Division,* American Society of Civil Engineers, vol. 108, no. IR 1, March, 1982, pp. 13–24.

CUENCA, R. H., "Hydrologic Balance Model Using Neutron Probe Data," American Society of Civil Engineers, *Journal of Irrigation and Drainage Engineering*. (In Press)

CUENCA, R. H. and M. E. JENSEN, "Approximating the FAO Coefficients—A Second Look," work in progress.

DEVORE, J. and R. PECK, *Statistics, The Exploration and Analysis of Data*. St. Paul, Minnesota: West Publishing Co., 1986.

DOORENBOS, J. and W. O. PRUITT, "Crop Water Requirements," Food and Agriculture Organization of the United Nations, Irrigation and Drainage Paper no. 24. Rome, Italy, 1975.

DOORENBOS, J. and W. O. PRUITT, "Crop Water Requirements," Food and Agriculture Organization of the United Nations, Irrigation and Drainage Paper no. 24, Revised, Rome, Italy, 1977.

EL KAYAL, A., "Evaluating the Design and Operation of Irrigation Canals in Egypt," unpublished M.S. Thesis, Department of Civil Engineering, Utah State University, Logan, Utah, 1983.

FERVERT, D. K., R. W. HILL, and B. C. BRAATEN, "Estimation of FAO Evapotranspiration Coefficients," *Journal of the Irrigation and Drainage Division,* American Society of Civil Engineers, vol. 109, no. 2, June, 1983, pp. 265–269.

GREBET, P., "Evapotranspiration Mesure et Calcul," Ph.D. Dissertation, l'Université Pierre et Marie Curie et l'Ecole Nationale Superieure des Mines de Paris, 1982. (In French)

HOWELL, T. A., R. L. MCCORMICK, and C. J. PHENE, "Design and Installation of Large Weighing Lysimeters," Transactions of the American Society of Agricultural Engineering, vol. 28, no. 1, 1985, pp. 106–112, 117.

HURD, C., *Sprinkler Irrigation Guidebook*. Washington, D.C.: U.S. Agency for International Development, 1969.

JENSEN, M. E., ed., *Consumptive Use of Water and Irrigation Water Requirements*. New York, NY: American Society of Civil Engineers, 1974.

JENSEN, M. E., ed., "Design and Operation of Farm Irrigation Systems," Monograph no. 3, American Society of Agricultural Engineers, St. Joseph, Michigan, 1980.

MAHRT, L. and M. EK, "The Influence of Atmospheric Stability on Potential Evaporation," *Journal of Climate and Applied Meteorology,* vol. 23, February, 1984, pp. 222–234.

MUKAMMAL, E. I., G. A. MCKAY, and V. R. TURNER, "Mechanical Balance-Electrical Readout Weighing Lysimeter," *Boundary-Layer Meteorology,* vol. 2, 1971, pp. 207–217.

MULLER, M. J., *Selected Climatic Data for a Global Set of Standard Stations for Vegetation Science*. The Hague, the Netherlands: Dr. W. Junk Publishers, 1982.

NIXON, P. R., G. P. LAWLESS, and G. V. RICHARDSON, "Coastal California evapotranspiration frequencies," Proceedings American Society of Civil Engineers, Journal of Irrigation and Drainage Division, IR 2, 1972, pp. 185–191.

PENMAN, H. L., "Natural Evaporation for Open Water, Bare Soil, and Grass," Proceedings Royal Society of London, A193, 1948, pp. 120–146.

PENMAN, H. L., "The Physical Bases of Irrigation Control," Proceedings 13th International Horticulture Congress, London, England, 1952, pp. 913–924.

PRUITT, W. O. and D. E. ANGUS, "Large Weighing Lysimeter for Measuring Evapotranspiration," Transactions of the ASAE, vol. 3, no. 2, 1960, pp. 13–15, 18.

PRUITT, W. O. and J. DOORENBOS, "Empirical Calibration, A Requisite for Evapotranspiration Formulae Based on Daily or Longer Mean Climatic Data," presented at International Round Table Conference on Evapotranspiration, International Commission on Irrigation and Drainage, Budapest, Hungary, 1977.

ROSENBERG, N. J., B. L. BLAD, and S. B. VERMA, *Microclimate, The Biological Environment*. New York: John Wiley and Sons, 1983.

RYAN, P. K. and R. H. CUENCA, "Feasibility and Significance of Revising Consumptive Use and Net Irrigation Requirements for Oregon," Project Completion Report, Department of Agricultural Engineering, Oregon State University, 1984.

SELLERS, W. D., *Physical Climatology*. Chicago, Illinois: University of Chicago Press, 1972.

U.S. Dept. of Agriculture, Soil Conservation Service, Technical Release no. 21, (Revision 2), 1970.

WRIGHT, J. L., "New Evapotranspiration Crop Coefficients," *Journal of the Irrigation and Drainage Division,* American Society of Civil Engineers, vol. 108, IR 1, June, 1982, pp. 57–74.

PROBLEMS

5-1. **(a)** Compute the saturation deficit ($e_s - e_a$) in mb for measurements made at a climatic station given the following data:

$$\text{Elevation} = 310 \text{ m}$$
$$T_{dry} = 28.5°C$$
$$T_{wet} = 22.6°C$$

(b) What is the relative humidity in percent at the time of measurement?

5-2. A class A evaporation pan is maintained in a pasture environment in which the pasture surrounds the pan for 115 m in the upwind direction. For a seven-day period, 24-hr wind run averages 235 km/d and the mean daily relative humidity is 35 percent. Total pan evaporation for the seven-day period is 54 mm. What would you estimate as the equivalent water use for grass in the same area over the same period in mm/d?

5-3. Measurements are made over a cropped field for data in the energy balance equation to be used for ET estimates using climatic data. The following values are given in equivalent mm/d:

$$\text{Incoming solar radiation} = 10.3$$
$$\text{Atmospheric long-wave radiation} = 0.7$$
$$\text{Terrestrial long-wave radiation} = 1.1$$

What is the net radiation in mm/d?

5-4. The following monthly mean climatic data is given for June for a site located at 40°N latitude:

$$T_{max} = 28°C \qquad \frac{n}{N} = 0.72$$

$$T_{min} = 15°C \qquad U_{2\ day} = 6\ m/s$$

$$RH_{min} = 43°C$$

Compute the monthly mean reference ET in mm/d using the FAO modified Blaney-Criddle.

5-5. A site instrumented with a recording climatic station collects the following data for one day:

$U_2 = 165$ km $\qquad$ $U_{day}/U_{night} = 3.0$

$RH_{max} = 55\%$ $\qquad$ $R_s = 12.4$ mm

$T_{mean} = 21.5°C$ $\qquad$ $R_n = 7.5$ mm

Mean actual vapor pressure = 12.4 mb

$$\text{Weighting function } \frac{\Delta}{\Delta + \gamma} = 0.705$$

Compute the reference ET for the day in mm using the FAO corrected Penman.

5-6. A grower wants to use estimates of reference ET to schedule irrigations on carrots on a seven-day interval in an area in which no rainfall is expected. The mean reference ET at time of planting and for 20 days after is 6 mm/d. What is the total estimated crop water use for these 20 days in mm?

5-7. A grower uses one of the FAO methods to schedule irrigation on tomatoes for the first 30 days after planting. The average computed reference ET rate during this period is 5 mm/d. The total crop ET for this period is 97.5 mm. What is the frequency of irrigation and/or rainfall in days?

5-8. An evaporation pan is used to schedule irrigations on peas during mid-season. The pan is situated in a green pasture with grass 1000 m in the upwind direction. The following climatic data are for the 10-day period during mid-season for which the depth of irrigation is to be computed:

$$RH_{max} = 80\%$$

$$RH_{min} = 35\%$$

$$\text{Wind} = 315\ km/d$$

$$E_{pan} = 6.1\ mm/d$$

What is the net depth of irrigation in mm for the 10-day period?

5-9. A class A evaporation pan is maintained at Burns, Oregon immediately surrounded by a dry fallow area with pasture at some distance upwind of the dry fallow area. Total pan evaporation for a 10-day period in August is measured at 80 mm. During the period mean relative humidity is 35 percent and mean daily wind run at 2 m height is 481 km/d. If mean reference ET for grass is estimated as 4 mm/d for the 10-day period, at what distance upwind of the pan does the pasture begin?

5-10. Small grain is to be planted in spring in a Mediterranean area. Planting date is 01 May and at this time reference ET for grass is 6 mm/d and rainfall and/or irrigation is expected every 10 days. During the majority of the growing season, minimum relative humidity is 65 percent and wind speeds at 2 m height average approximately 6 m/s. Draw a crop coefficient curve for this situation and indicate the sources of your data.

5-11. Compute the mean monthly reference ET for grass in mm/d using the FAO modified Blaney-Criddle method given the following data:

Latitude = 10°N	Month = June
T_{max} = 27°C	T_{min} = 13°
RH_{max} = 95%	RH_{min} = 60%
U_{day} = 5.4 m/s	R_s = 610 cal/cm^2

Where U_{day} refers to daytime wind speed measured at 2 m height between 0700–1900 hours and R_s is measured mean daily solar radiation.

5-12. The following data are given for a single day at a particular tropical site:

T_{mean} = 18°C	RH_{max} = 90%	RH_{min} = 65%
Δ = 1.0968 mb/°C	γ = 0.6675 mb/°C	
R_n = 6.1 mm	R_s = 9.0 mm	
U_{2m} = 256 km/d	U_{day}/U_{night} = 4.0	

Compute the FAO Modified Penman estimate of reference ET for grass in mm/d.

5-13. Class A pan evaporation measured at a site in which the pan is surrounded by grass 100 m in the daytime upwind direction is 9.2 mm/d. Reference ET for grass, correctly adjusting for pan environment using the FAO method is 6.0 mm/d. The mean relative humidity is 35 percent. Indicate the range of the daily wind run at 2 m height in km/d.

5-14. The FAO Penman method is applied to compute reference evapotranspiration for 15 July at a site with latitude 50 degrees north and elevation 630 m. The following temperature, relative humidity, and Cambell-Stokes sunshine recorder data are available:

T_{max} = 30.5°C	RH_{min} = 53 percent
T_{min} = 22.8°C	RH_{max} = 84 percent
n = 12.1 h	

Compute the net radiation in equivalent mm of evaporated water.

5-15. Reference evapotranspiration is computed for the same location given in problem 5-14 and for the same date using the Wright Penman method. Latitude is 50 degrees north, date is July 15 (not a leap year), and elevation is 630 m. The following data are given with wind measured at 3.0 m height:

T_{max} = 30.5°C	RH_{min} = 53 percent
T_{min} = 22.8°C	RH_{max} = 84 percent
n = 12.1 h	$U_{3.0}$ = 480 km/d

Compute the advection term of the Wright Penman—that is, E_a—in equivalent mm evaporation of water, when the Wright Penman is given as

$$ET_r = \frac{\Delta}{\Delta + \gamma}(R_n - G) + \frac{\gamma}{\Delta + \gamma}(E_a)$$

5-16. A sunflower crop is planted on 15 May in a very arid region of California which experiences low relative humidities and low winds. The initial period crop coefficient is 0.50. Compute the crop coefficient for 20 June.

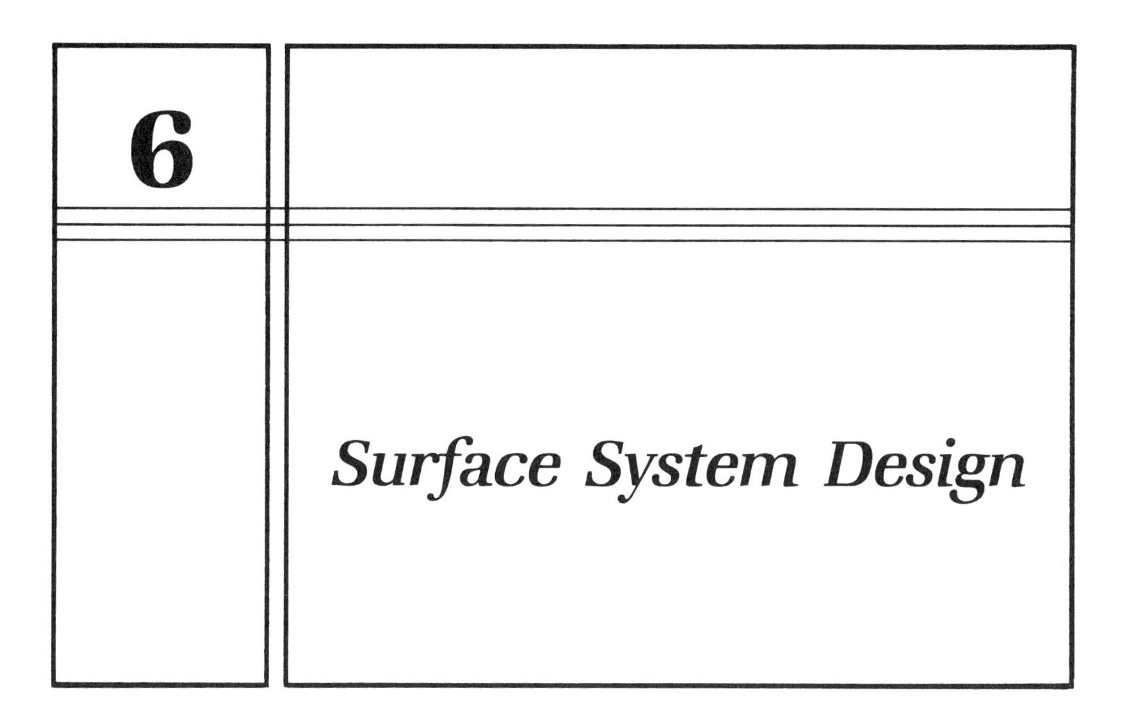

6.1 Introduction

Application

The history of irrigation commences with the application of water to the land in some type of surface scheme. Today surface irrigation takes on many different forms around the world and is the most widely practiced irrigation method. Figures 6-1 through 6-3 indicate examples of different types of furrow irrigation. Other surface irrigation methods will be demonstrated later in this chapter. From Figs. 6-1 through 6-3 we can see the wide applicability of surface systems and the field skills which are required to obtain the maximum efficiency from the available water resource.

The adaptability of different soil types and topography to methods of surface irrigation is a function of many different parameters. Table 6-1 presents information regarding applicability of different types of surface irrigation methods to various crops and topographic conditions. This table also specifies ranges of required water flows for the different methods, recommended soil characteristics, and comments regarding installation costs and labor requirements. By reviewing the type of crop to be grown, topographic and physical characteristics of the soil and available water supply, Table 6-1 can be used to determine which types of surface irrigation methods have the best potential for relatively high irrigation efficiencies with low labor and leveling requirements.

Figure 6-1 Water being carefully directed to individual furrows on an irrigation project in central Tunisia.

Figure 6-2 Graded furrows used to wet up double row tomato seed beds in the Central Valley of California.

Perspectives on Design

Contour levee, level basin, and graded border surface systems were briefly described in the first chapter, as were furrow and corrugated systems. There are other types of surface irrigation systems, but those indicated are the major types. The design

Figure 6-3 Siphon tubes being started to deliver water from elevated head ditch to furrows. (Photo courtesy of Marvin Shearer.)

of level basin, graded border, and furrow systems are discussed in this chapter in some detail.

There are two viewpoints on the design of surface irrigation systems. These may be divided into the school which prefers operation based on field tested empirical guidelines ("rules of thumb") and the school which prefers to apply sophisticated hydraulic analysis to the design of surface systems. In general, the empirical guideline group has not clearly indicated why an accurate hydraulic analysis should not be superior. Neither has the hydraulic analysis group indicated how the typical spatial variability of soil-water properties on a field scale warrants extremely precise hydraulic analysis using input which does not account for the stochastic variation of these properties.

This book will attempt to follow a middle course. The empirical guidelines will be indicated. These have proven useful under a wide variety of field conditions and at least serve as a first approximation of design feasibility which can be extremely useful in initial discussions with clients. When more sophisticated hydraulic analyses have proven to be useful in field applications, these methods will be thoroughly covered. The governing equations for the hydraulic analysis will be indicated and reviewed. It is felt that with better soil parameter determination methods and increased application of computerized analysis on a routine basis, hydraulic methods will see increased application in the future.

Irrigation Efficiency

In discussing any type of irrigation system, it is useful to have concepts of efficiency to enable comparison of different systems or different management strategies for a particular system. There have been over 20 different ways of quantifying efficiency proposed for irrigation systems. Most of these methods are useful, although some

TABLE 6.1 Surface irrigation methods and conditions of use. (After Booher, 1974.)

Irrigation Method	Suitabilities and Conditions of Use				Remarks
	Crops	Topography	Water Supply	Soils	
Small rectangular-basins	Grain, field crops, orchards, rice	Relatively flat land; area within each basin should be leveled	Can be adapted to streams of various size	Suitable for soils of high or low intake rates; should not be used on soils that tend to puddle	High installation costs. Considerable labor required for irrigating. When used for close-spaced crops, a high percentage of land is used for levees and distribution ditches. High efficiencies of water use possible.
Large rectangular basins	Grain, field crops, rice	Flat land; must be graded to uniform plane	Large flows of water	Soils of fine texture with low intake rates	Lower installation costs and less labor required for irrigation than with small basins. Substantial levees needed.
Contour checks	Orchards, grain, rice, forage crops	Irregular land; slopes less than 2 percent	Flows greater than 30 liters (1 cubic foot) per second	Soils of medium to heavy texture which do not crack on drying	Little land grading required. Checks can be continuously flooded as for rice, water ponded as for orchards, or intermittently flooded as for pastures.
Narrow borders up to 5 meters (16 feet) wide	Pasture, grain, lucerne, vineyards, orchards	Uniform slopes less than 7 percent	Moderately large flows	Soils of medium to heavy texture	Borders should be in direction of maximum slope. Accurate cross leveling required between guide levees.
Wide borders up to 30 meters (100 feet) wide	Grain, lucerne, orchards	Land graded to uniform plane with maximum slope less than 0.5 percent	Large flows, up to 600 liters (20 cubic feet) per second	Deep soils of medium to fine texture	Very careful land grading necessary. Minimum of labor required for irrigation. Little interference with use of farm machinery.
Wild flooding	Pasture, grain	Irregular surfaces with slopes up to 20 percent	Can use small continuous flows on steeper land or large flows on flatter land	Soils of medium to fine texture with stable aggregate which do not crack on drying	Little land grading required. Low initial cost for system. Best adapted to shallow soils since percolation losses may be high on deep permeable soils.
Benched terraces	Grain, field crops, forage crops, orchards, vineyards	Slopes up to 20 percent	Streams of small to medium size	Soils must be sufficiently deep that grading operations will not impair crop growth	Care must be taken in constructing benches and providing adequate drainage channels for excess water. Irrigation water must be properly managed. Misuse of water can result in serious soil erosion.

TABLE 6.1 Surface irrigation methods and conditions of use. (After Booher, 1974.) (continued)

Irrigation Method	Suitabilities and Conditions of Use: Crops	Topography	Water Supply	Soils	Remarks
Straight furrows	Vegetables, row crops, orchards, vineyards	Uniform slopes not exceeding 2 percent for cultivated crops	Flows up to 350 liters (12 cubic feet) per second	Can be used on all soils if length of furrows is adjusted to type of soil	Best suited for crops which cannot be flooded. High irrigation efficiency possible. Well adapted to mechanized farming.
Graded contour furrows	Vegetables, field crops, orchards, vineyards	Undulating land with slopes up to 8 percent	Flows up to 100 liters (3 cubic feet) per second	Soils of medium to fine texture which do not crack on drying	Rodent control is essential. Erosion hazard from heavy rains or water breaking out of furrows. High labor requirement for irrigation.
Corrugations	Close-spaced crops such as grain, pasture, lucerne	Uniform slopes of up to 10 percent	Flows up to 30 liters (1 cubic foot) per second	Best on soils of medium to fine texture	High water losses possible from deep percolation or surface runoff. Care must be used in limiting size of flow in corrugations to reduce soil erosion. Little land grading required.
Basin furrows	Vegetables, cotton, maize and other row crops	Relatively flat land	Flows up to 150 liters (5 cubic feet) per second	Can be used with most soil types	Similar to small rectangular basins, except crops are planted on ridges.
Zigzag furrows	Vineyards, bush berries, orchards	Land graded to uniform slopes of less than 1 percent	Flows required are usually less than for straight furrows	Used on soils with low intake rates	This method is used to slow the flow of water in furrows to increase water penetration into soil.

appear more cumbersome or abstract than others. Basically the different systems proposed merely represent the various authors' preferences based on their own experiences. The definitions for efficiency which follow are no exception. The important point is that if the efficiencies of different systems are being compared, it is imperative that it is clearly stated how the efficiencies were derived to see that they actually measure the same value of efficiency. The efficiencies which follow have been chosen because they are straightforward and result in a consistent and related set of efficiencies which can be applied to various types of irrigation systems.

General efficiency

In general, efficiency is defined as the ratio of output to input. For irrigation systems, this general efficiency has been defined by the following:

$$\text{Efficiency} = \frac{\text{Output}}{\text{Input}} = \frac{\text{Quantity Beneficially Used}}{\text{Total Quantity from Supply}} \tag{6-1}$$

The quantity in the case of irrigation systems refers to volumes of water. Water applied for a wide variety of reasons may be considered beneficially used in irrigation. The normal beneficial uses considered include meeting plant transpiration demands, evaporation from the plant and soil, quantities used for leaching dissolved salts out of the root zone, frost protection, and crop cooling.

Specific efficiency

In this book, a system of efficiencies will be applied which result in a consistent set of values which are all related (Painter and Carran, 1978). These efficiencies given in percent are defined as follows:

Extraction Efficiency

$$e_x = \frac{\text{volume delivered to distribution system}}{\text{volume extracted from supply}} \times 100 \tag{6-2}$$

Conveyance Efficiency

$$e_c = \frac{\text{volume delivered to application devices}}{\text{volume delivered to distribution system}} \times 100 \tag{6-3}$$

Application Efficiency

$$e_a = \frac{\text{volume delivered to application surface}}{\text{volume delivered to application devices}} \times 100 \tag{6-4}$$

Distribution Pattern Efficiency

$$e_d = \frac{\text{volume stored in crop root zone}}{\text{volume delivered to application surface}} \times 100 \tag{6-5}$$

Irrigation System Efficiency

$$e_i = \frac{\text{volume stored in crop root zone}}{\text{volume extracted from supply}} \times 100 \qquad (6\text{-}6)$$

$$e_i = \frac{e_x}{100}\frac{(e_c)}{100}\frac{(e_a)}{100}\frac{(e_d)}{100} \times 100 \qquad (6\text{-}7)$$

It should be noted that efficiencies of various parts of a water resources system are described by the foregoing set of relationships. The system of efficiencies proceeds from the water source, through the distribution system, down to the farm level. The irrigation system efficiency given by Eqs. (6-6) and (6-7), sometimes termed the overall efficiency, indicates the efficiency for the complete system from water supply to field.

Example Problem 6-1

Farmer Harris is a member of a large irrigation district which has as its water source a storage reservoir 60 km upstream from the entrance to the district distribution system. He wants to irrigate a 65 ha parcel of maize by center pivot. The crop water requirement since the last irrigation is estimated as 50 mm. Due to the expected uniformity of application of the center pivot system, an areal average of 55 mm of water will have to be applied to the field to insure that all parts of the field receive a minimum of 50 mm. Estimated spray and windrift losses are 8 percent of the water discharge through the sprinkler nozzles. Distribution system losses from where the water enters the irrigation district in a canal to the head of Farmer Harris's field is estimated at 15 percent. Seepage and evaporation losses in the unlined canal between the storage reservoir and the entrance to the district distribution system are 45 percent.

Compute the applicable irrigation efficiencies and volume of water (m^3) which must be released from the storage reservoir to meet Farmer Harris's crop water requirement.

Solution Compute required volumes:

(a) Root zone:

$$V_{\text{root zone}} = 65\text{ ha}\left(10{,}000\frac{\text{m}^2}{\text{ha}}\right)50\text{ mm}\left(\frac{1\text{ m}}{1000\text{ mm}}\right)$$

$$V_{\text{root zone}} = 32{,}500\text{ m}^3$$

(b) Application surface:

$$V_{\text{app sfc}} = 32{,}500\text{ m}^3\left(\frac{55\text{ mm}}{50\text{ mm}}\right)$$

$$V_{\text{app sfc}} = 35{,}750\text{ m}^3$$

(c) Application device considering 8 percent spray and drift losses:

$$V_{\text{app dev}} = 35{,}750\text{ m}^3/(1 - 0.08)$$

$$V_{\text{app dev}} = 38{,}859\text{ m}^3$$

(d) Distribution system considering 15 percent distribution losses:

$$V_{\text{dist sys}} = 38{,}859\text{ m}^3/(1 - 0.15)$$

$$V_{\text{dist sys}} = 45{,}716\text{ m}^3$$

(e) Extraction considering 45 percent seepage and evaporation:

$$V_{extract} = 45{,}716 \text{ m}^3/(1 - 0.45)$$
$$V_{extract} = 83{,}120 \text{ m}^3$$

Compute efficiencies:

(a) Extraction:

$$e_x = \frac{V_{dist\ sys}}{V_{extract}} \times 100 = \frac{45{,}716 \text{ m}^3}{83{,}120 \text{ m}^3} \times 100$$
$$e_x = 55\%$$

(b) Conveyance:

$$e_c = \frac{V_{app\ dev}}{V_{dist\ sys}} \times 100 = \frac{38{,}859 \text{ m}^3}{45{,}716 \text{ m}^3} \times 100$$
$$e_c = 85\%$$

(c) Application:

$$e_a = \frac{V_{app\ sfc}}{V_{app\ dev}} \times 100 = \frac{35{,}750 \text{ m}^3}{38{,}859 \text{ m}^3} \times 100$$
$$e_a = 92\%$$

(d) Distribution pattern:

$$e_d = \frac{V_{root\ zone}}{V_{app\ sfc}} \times 100 = \frac{32{,}500 \text{ m}^3}{35{,}750 \text{ m}^3} \times 100$$
$$e_d = 91\%$$

(e) Irrigation system:

$$e_i = \frac{V_{root\ zone}}{V_{extract}} \times 100 = \frac{32{,}500 \text{ m}^3}{83{,}120 \text{ m}^3} \times 100$$
$$e_i = 39\%$$

Volume which must be released from the district storage reservoir is 83,120 m^3.

6.2 Definitions of Surface System Terms

Physical Description of Distribution System

The components of a typical surface irrigation system are indicated in Fig. 6-4. We begin with the delivery part which is from the source of water to the location of the field. The delivery part may include flow measuring devices, such as the Parshall flume shown, and perhaps drop structures to control dissipation of energy in controlled locations rather than in more easily erodible channel sections.

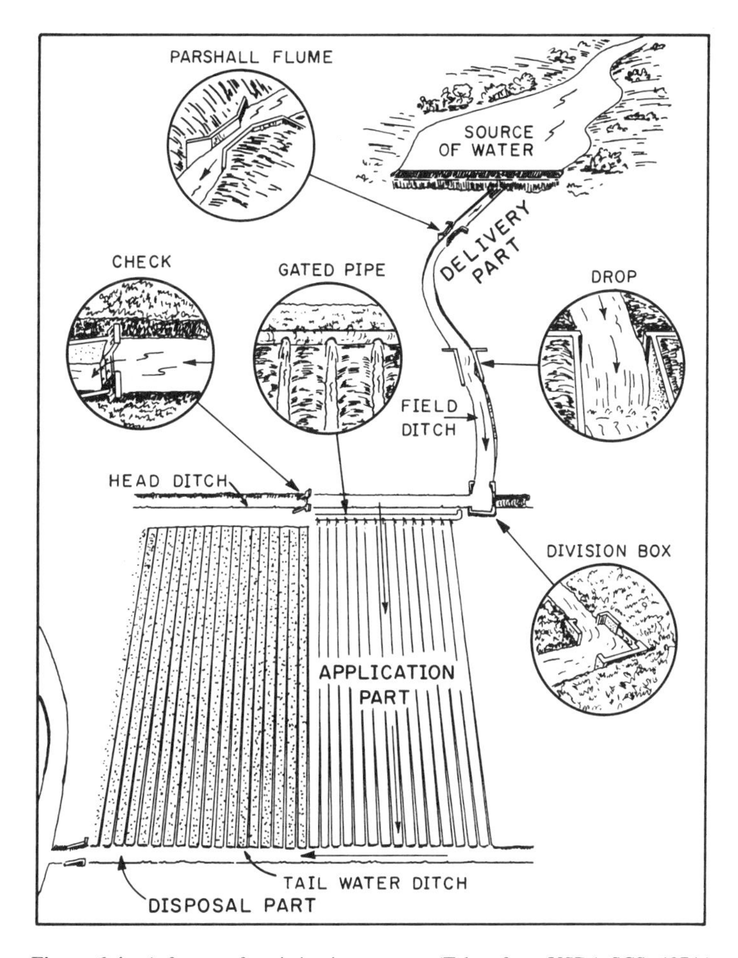

Figure 6-4 A farm surface irrigation system. (Taken from USDA-SCS, 1974.)

The last section of the delivery part in the example shown is the field ditch which supplies water to the head ditch. The head ditch is the means by which water is actually applied to the field to be irrigated. In the case shown, water is turned through a division box which could be used to divide different amounts of the flow stream for distribution to different parts of the field. The head ditch is the first component of the application part which is also made up of the field to be irrigated.

The final component of the system is the disposal part which includes a tail water ditch to deliver excess flow to an adjacent water course in the case illustrated. Although it is common, not all surface systems require a disposal part. Some systems, such as level borders, are closed and do not require a disposal section. Others, such as graded furrows, normally require a disposal part for efficient operation and to insure water distribution to all parts of the field.

Phases of Water Distribution over Soil

Figure 6-5 depicts the phases of water distribution on soils which are applicable to all types of surface systems. In this figure, time increases along the vertical axis and distance down the field increases along the horizontal axis. Water is assumed to be applied to the head of the field at time t = 0. The advance phase occurs as water is moving down the field and data for the advance curve is obtained by noting the time required for the stream of water to reach various distances down the field.

During the storage phase, water continues to be applied to the field even though the advance curve has reached the end of the field. The storage phase is the time between when the water reaches the end of the field and when water is cut off at the head of the field. The time of cut-off is denoted by T_{co} on the time axis. The time of cut-off is the beginning of the depletion phase. The depletion phase continues until the soil surface at the head of the field is again visible—that is, is no longer submerged. This time marks the beginning of the recession phase which continues until water has drained completely off the surface of the field. The plot of time versus the distance down the field of the exposed soil surface describes the recession curve. The difference in time between the recession and advance curves at any distance down the field is termed the infiltration opportunity time. This is the total time available for infiltration at that point in the field.

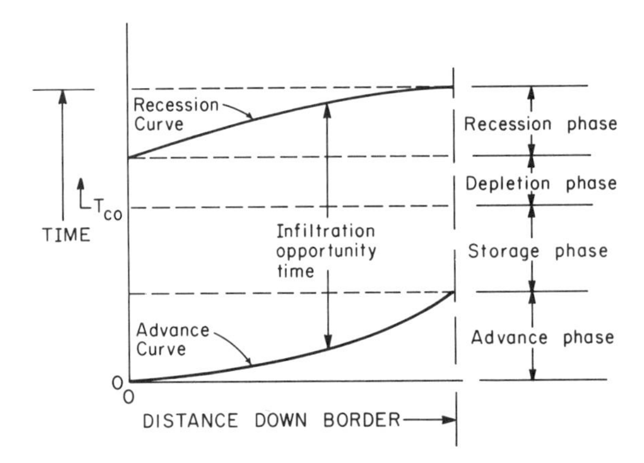

Figure 6-5 Definition sketch showing surface irrigation terms. (Taken from Jensen, 1980.)

6.3 Furrow System Design

Physical Relationship between Advance Time, Infiltration Rate, and Deep Percolation

The Kostiakov equation for infiltration was previously given as

$$i = c(t)^{\alpha} \tag{6-8}$$

where

$$i = \text{depth of infiltration}$$
$$t = \text{time of infiltration}$$
$$c \text{ and } \alpha = \text{empirical constants}$$

Equation (6-8) may be rearranged as

$$t = \left(\frac{i}{c}\right)^{1/\alpha} \tag{6-9}$$

The t in Eq. (6-9) may be thought of as the required time of application for depth of irrigation i.

If we assume a homogeneous soil with uniform initial water content, the distance z from the soil surface to the depth of the wetting front below the surface is given as:

$$z = k(i) \tag{6-10}$$

where

$$k = \text{proportionality constant}$$

Substituting for i in Eq. (6-8),

$$z = kc(t)^{\alpha} = K'(t)^{\alpha} \tag{6-11}$$

Consider Fig. 6-6 in which the advance phase is assumed to equal one-fourth the total time required for irrigation:

$$T_t = \frac{1}{4}T_n \tag{6-12}$$

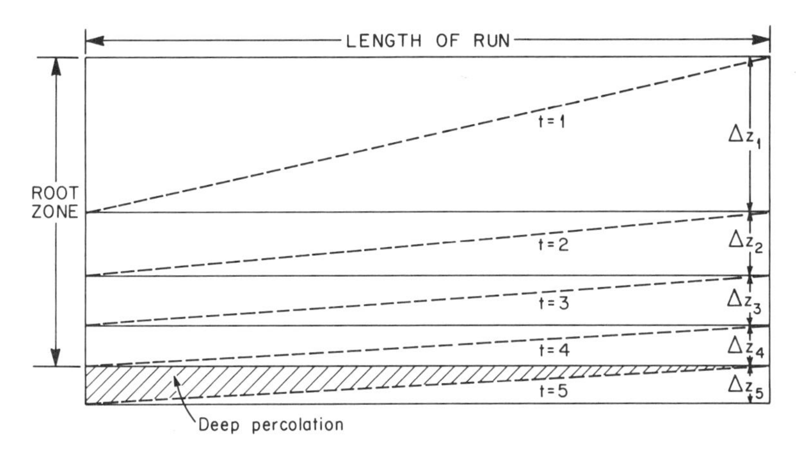

Figure 6-6 Depth profile for a field with water running from left to right. Dashed lines represent wetting fronts at the times indicated. The Δ z's represent the depth of wetting front advance during each equal time interval.

where

T_t = time required for advance curve to reach end of field

T_n = time required to infiltrate net irrigation requirement, i_n

Dividing the total infiltration time into four equal intervals,

$$\Delta t_1 = \Delta t_2 = \Delta t_3 = \Delta t_4 = \frac{T_n}{4} \tag{6-13}$$

The depth of water infiltrated in interval Δt_1 is given by

$$\Delta i_1 = i_1 - i_o \tag{6-14}$$

The depth of soil wet during interval Δt_1 is

$$\Delta z_1 = z_1 - z_o \tag{6-15}$$

As an example, let $\alpha = 0.5$ and $z_o = 0$. Then,

$$\Delta z_1 = K'(t_1)^{0.5} - 0 = K'(t_1)^{0.5} \tag{6-16}$$

or

$$\Delta z_1 = z_1 \tag{6-17}$$

For time interval, Δt_2,

$$\Delta z_2 = z_2 - z_1 = K'(t_2)^{0.5} - \Delta z_1 \tag{6-18}$$

Since $t_2 = 2t_1$,

$$\Delta z_2 = K'(2t_1)^{0.5} - \Delta z_1 \tag{6-19}$$

but

$$K'(2t_1)^{0.5} = \Delta z_1(2)^{0.5} \tag{6-20}$$

so

$$\Delta z_2 = \Delta z_1[(2)^{0.5} - 1] \tag{6-21}$$

Similarly,

$$\Delta z_3 = \Delta z_1[(3)^{0.5} - (2)^{0.5}] \tag{6-22}$$

$$\vdots \qquad \vdots$$

$$\Delta z_5 = \Delta z_1[(5)^{0.5} - (4)^{0.5}] \tag{6-23}$$

Referring to Fig. 6-6, to infiltrate depth i_n into the lower end of the field requires an infiltration opportunity time of T_n at the lower end. Since

$$t_4 = 4t_1 \tag{6-24}$$

$$z_4 = K'(4t_1)^{0.5} \tag{6-25}$$

or

$$z_4 = \Delta z_1(4)^{0.5} \tag{6-26}$$

At opportunity time T_n at the lower end, the infiltration opportunity at the head of the field is

$$t_5 = T_n + \frac{T_n}{4} = 5t_1 \tag{6-27}$$

and

$$z_5 = \Delta z_1(5)^{0.5} \tag{6-28}$$

Assuming a linear relationship between the depth of wetting front and distance down field, the average depth of soil wet by deep percolation is

$$\frac{\Delta z_5}{2} = \frac{z_5 - z_4}{2} = \frac{[(5)^{0.5} - (4)^{0.5}]\Delta z_1}{2} \tag{6-29}$$

The average depth of wet soil is given by

$$\frac{z_4 + z_5}{2} = \frac{[(4)^{0.5} + (5)^{0.5}]\Delta z_1}{2} \tag{6-30}$$

Now we are able to calculate the ratio of the average depth of deep percolation to average depth of wet soil:

$$\frac{\Delta z_5/2}{(z_4 + z_5)/2} = \frac{(5)^{0.5} - (4)^{0.5}}{(4)^{0.5} + (5)^{0.5}} = 0.056 \tag{6-31}$$

The result of Eq. (6-31) indicates that if the time of advance, T_t, is one-fourth the net infiltration time, T_n, and $\alpha = 0.5$, then the average depth of deep percolation is equal to 5.6 percent.

Table 6-2 gives expected percentage of deep percolation for various fractional advance ratios and values of α subject to the assumptions made in the previous anal-

TABLE 6-2 The expected percentage of percolation losses for different fractional advance ratios and several values of α.

Fractional Advance Ratios	α								
	0.1	0.2	0.3	0.4	0.5	0.6	0.7	0.8	0.9
t	3.5	6.9	10.4	13.8	17.2	20.5	23.8	27.0	30.2
t/2	2.0	4.1	6.1	8.1	10.1	12.1	14.1	16.1	18.0
t/3	1.4	2.9	4.3	5.7	7.2	8.6	10.1	11.5	12.9
t/4	1.1	2.2	3.3	4.5	5.6	6.7	7.8	8.9	10.0
t/5	0.9	1.8	2.7	3.6	4.6	5.5	6.4	7.3	8.2
t/10	0.5	1.0	1.4	1.9	2.4	2.9	3.3	3.8	4.3

ysis. The fractional advance ratio, FAR, is given by the ratio of advance time to net time of irrigation at the end of the field:

$$FAR = \frac{T_t}{T_n} \tag{6-32}$$

Table 6-2, subject to the assumptions applied, can be used to solve for the required fractional advance ratio by specifying tolerable deep percolation for a soil of known α.

Example Problem 6-2

Assume that for a given soil the empirical constants for the Kostiakov equation with depth in cm and time in minutes are $\alpha = 0.7$ and $c = 0.21$ and that 10 percent deep percolation is acceptable. If the net irrigation requirement, i_n, is 8.0 cm, determine the net time of irrigation, T_n, and the advance time required for the water to reach the end of the field, T_t.

Solution From Table 6-2,

$$FAR = \frac{T_t}{T_n} = \frac{1}{3}$$

or

$$T_t = \frac{T_n}{3}$$

Substituting the required variables into Eq. (6-9),

$$T_n = \left[\frac{i_n}{c}\right]^{1/\alpha} = \left[\frac{8.0}{0.21}\right]^{1/0.7}$$

$$T_n = 181 \text{ min}$$

$$T_t = \frac{T_n}{3} = \frac{181 \text{ min}}{3} = 60 \text{ min}$$

Note that the result has been rounded off to the nearest minute to represent a reasonable answer. There is no point in listing an answer which is more precise than a soil is uniform. In fact a result to the nearest 10 minutes is reasonable for this problem.

The results of this section indicate that there are quantifiable relationships between soil intake rates, time of advance, and deep percolation. The advance time is a variable affected by both system design and management, and the amount of deep percolation is directly related to overall system efficiency. The design engineer needs to be aware of these relationships in choosing a design configuration to maximize efficiency on a given soil type. Although these relationships have been described relative to furrow systems, the same general approach is applicable to all types of surface irrigation systems.

Experience Relationships

The relationships described in this section have been derived from numerous field trials on furrow systems. In general, the relationships indicated do not require extensive data compared to more sophisticated methods based on hydraulic analysis.

Because of the general and limited amount of input data required, these relationships must be considered as giving only approximate solutions to design problems. Nevertheless, these relationships are based on vast field experience and give estimates which are useful in the case of a limited database or as a first estimate. When more detailed databases become available or when more refined analysis is required to serve as the basis of extensive economic decisions, it is necessary to proceed to the more detailed hydraulic analysis indicated later in this chapter.

Nonerosive stream size

To maintain proper furrow shape and reduce sediment loss from the head of the field and deposition at the tail of the field or in an adjacent waterway, it is desirable to operate the furrow at a velocity that is nonerosive. Since the flow velocity is related to the flow rate of water into the furrow, an empirical relationship has been developed by the USDA-SCS for the maximum nonerosive stream size. This relationship is given by

$$Q_{max} = \frac{C}{S} \qquad (6\text{-}33)$$

where

S = ground slope down the furrow, percent

C = empirical constant given in Table 6-3

The relationship given in Eq. (6-33) does not account for soil type and therefore is limited in accuracy. A complete design for a furrow system would indicate a field test to allow for evaluation of the Q_{max} parameter. The field test procedure is explained later in this chapter. Table 6-4 indicates the solution of Eq. (6-33) for some important field slopes and presents some cautionary comments regarding these slopes. Note that this table taken from Booher (1974) indicates that a 2.0 percent slope is considered maximum for cultivated furrows.

Suggested field grade

To a certain degree, field slopes will be controlled by the natural grade of the land to be irrigated. Grades which are extremely different from the natural grade will require excessive time and costs for earth movement. In most locations, only a limited amount of material may be removed before the most productive portion of the topsoil has been taken away. The soil profile must therefore be investigated to assure that the depth of material to be removed will not severely limit a productive root zone.

TABLE 6-3 Units and value for C to compute maximum nonerosive stream size.

System	Units for Q_{max}	C
International	L/s	0.6
English	gal/min	10.0

TABLE 6-4 Relation of maximum nonerosive flow rates to critical slopes in furrows. (Taken from Booher, 1974.)

Furrow Slope, S	Maximum Flow Rate, Q_{max}		Comments
Percent	Liters per second	U.S. gallons per minute	
0.1	6.0	100	The flow rate indicated is about double the carrying capacity of most furrows in normal use on a 0.1 percent slope. Erosion is negligible with furrows flowing to capacity on this slope.
0.3	2.0	33	A slope of 0.3 percent is near the upper limit where furrows flowing at capacity will not cause serious erosion.
0.5	1.2	20	Cultivated furrows with 0.5 percent slope will erode unless the flow rate is considerably less than furrow capacity.
2.0	0.3	5	This indicates the reduction in flow rate needed to prevent serious erosion on a 2 percent slope. This is considered the maximum slope allowable for cultivated furrows.

Some general guidelines for grade have been developed by the United States Soil Conservation Service (1983) based on field experience. These suggested grades vary with climatic conditions and have been developed to a certain degree based on minimizing soil erosion. In general, it is recommended that furrow grades be 1.0 percent or less. A minimum grade of 0.1 percent has been found effective in moving water through a furrow. In arid areas where erosion from rainfall is not expected to be a problem, the grade can be as much as 3.0 percent. In humid areas which experience high rates of rainfall, the furrow grade should usually not exceed 0.3 percent. Grades up to 0.5 percent may be acceptable in such areas if the lengths of runs are kept short enough that water does not accumulate to the point that it causes erosion. In humid and subhumid areas, a minimum grade of 0.05 percent is necessary to provide for surface drainage. Maximum recommended grades in subhumid and semiarid areas are 1.0 and 2.0 percent, respectively. The cross slope perpendicular to the direction of flow for furrows with grades of 0.5 percent or more should be limited to 1.0 percent or to the furrow grade, whichever is the lesser.

Suggested field size

As can be noted from inspection of Table 6-2, increasing field length, thereby increasing FAR, increases the amount of deep percolation which subsequently decreases the irrigation efficiency. General relationships have been developed to indicate the maximum recommended length of furrows to maintain an adequate efficiency and therefore a reasonably economical irrigation operation. These lengths are a function of field slope down the furrow, average depth of water to be applied, and soil type. Tabulated values for maximum recommended length of run are indicated in Table 6-5. The values given in this table are for the broad soil categories of clays, loams, and sands. Intermediate values would be applicable for a mixture of these main soil types. The values given in Table 6-5 are only a first approximation but are nevertheless useful for preliminary estimates of field layout.

Hydraulic Relationships

This section indicates the application of principles of fluid dynamics to the design of furrow irrigation systems. The relationships indicated must be applied with a balanced approach to system design. Engineers tend to prefer to be able to arrive at exact solutions to engineering problems. It must be kept in mind that irrigation systems are made up partly of mechanical components such as pumps, pipelines, and valves. The remainder of the system is made up of natural components such as soil, plants, and climate. Natural systems do not lend themselves to exact solutions, and a system designed at the drawing table should not be expected to perform exactly as predicted in the field.

TABLE 6-5 Suggested maximum lengths of cultivated furrows for different soils, slopes, and depths of water to be applied. (Taken from Booher, 1974.)

A. Lengths in meters; depths in centimeters

Furrow slope	Average depth of water applied (centimeters)											
	7.5	15	22.5	30	5	10	15	20	5	7.5	10	12.5
	Clays				Loams				Sands			
Percent	Meters											
0.05	300	400	400	400	120	270	400	400	60	90	150	190
0.1	340	440	470	500	180	340	440	470	90	120	190	220
0.2	370	470	530	620	220	370	470	530	120	190	250	300
0.3	400	500	620	800	280	400	500	600	150	220	280	400
0.5	400	500	560	750	280	370	470	530	120	190	250	300
1.0	280	400	500	600	250	300	370	470	90	150	220	250
1.5	250	340	430	500	220	280	340	400	80	120	190	220
2.0	220	270	340	400	180	250	300	340	60	90	150	190

B. Lengths in feet; depths in inches

Furrow slope	Average depth of water applied (inches)											
	3	6	9	12	2	4	6	8	2	3	4	5
	Clays				Loams				Sands			
Percent	Feet											
0.05	1000	1300	1300	1300	400	900	1300	1300	200	300	500	600
0.1	1100	1400	1500	1600	600	1100	1400	1500	300	400	600	700
0.2	1200	1500	1700	2000	700	1200	1500	1700	400	600	800	1000
0.3	1300	1600	2000	2600	900	1300	1600	1900	500	700	900	1300
0.5	1300	1600	1800	2400	900	1200	1500	1700	400	600	800	1000
1.0	900	1300	1600	1900	800	1000	1200	1500	300	500	700	800
1.5	800	1100	1400	1600	700	900	1100	1300	250	400	600	700
2.0	700	900	1100	1300	600	800	1000	1100	200	300	500	600

No field will have all the parameters required for a hydraulic solution defined to exactly the same degree of accuracy. In fact, values of required variables will stochastically vary from place to place in the field (Warrick and Nielsen, 1980). The results of hydraulic solutions to different design options should therefore not be considered exact, but rather as indicating the relative magnitude or direction of the solution. Hydraulic solutions require more field data than application of empirical relationships. The design engineer should therefore have a greater degree of confidence in hydraulic solutions than results using simplified empirical relationships provided field data are accurately collected and represent average field conditions.

The hydraulic relationships used throughout this chapter are basically those developed by the Soil Conservation Service (USDA, 1983, 1974). These rely on the infiltration concepts developed in the Soil Physics chapter. Infiltration constants in addition to those listed in the Soil Physics chapter are required for design of surface irrigation systems. These constants are listed for each intake family in Table 6-6 and will be applied in equations later in this chapter. Note that the intake constants listed in Table 6-6 are for the depth of infiltration in mm, whereas the corresponding constants given in the Soil Physics chapter were for the depth in cm.

It should be noted that the hydraulic relationships described in this chapter, derived by the Soil Conservation Service (SCS), contain certain assumptions which make them easier to solve than other hydraulic models. Other models have been developed, oftentimes employing parts of the SCS methods, which are based on more general flow equations. Such equations are not subject to the same simplifying assumptions as the SCS method, but at the same time normally require more complex

TABLE 6-6 Intake family and advance coefficients for depth of infiltration in mm, time in minutes, and length in meters.

Intake Family	a	b	c	f	g
0.05	0.5334	0.618	7.0	7.16	1.088×10^{-4}
0.10	0.6198	0.661	7.0	7.25	1.251×10^{-4}
0.15	0.7110	0.683	7.0	7.34	1.414×10^{-4}
0.20	0.7772	0.699	7.0	7.43	1.578×10^{-4}
0.25	0.8534	0.711	7.0	7.52	1.741×10^{-4}
0.30	0.9246	0.720	7.0	7.61	1.904×10^{-4}
0.35	0.9957	0.729	7.0	7.70	2.067×10^{-4}
0.40	1.064	0.736	7.0	7.79	2.230×10^{-4}
0.45	1.130	0.742	7.0	7.88	2.393×10^{-4}
0.50	1.196	0.748	7.0	7.97	2.556×10^{-4}
0.60	1.321	0.757	7.0	8.15	2.883×10^{-4}
0.70	1.443	0.766	7.0	8.33	3.209×10^{-4}
0.80	1.560	0.773	7.0	8.50	3.535×10^{-4}
0.90	1.674	0.779	7.0	8.68	3.862×10^{-4}
1.00	1.786	0.785	7.0	8.86	4.188×10^{-4}
1.50	2.284	0.799	7.0	9.76	5.819×10^{-4}
2.00	2.753	0.808	7.0	10.65	7.451×10^{-4}

numerical methods for solution. The more general flow equations are usually combined with hydraulic theories such as zero-inertia or kinematic-wave for solution. Some of the original modeling work, more complex than the SCS methods, was done by Katopodes and Strelkoff (1977). Strelkoff and Clemmens (1984) have written an overview of development of the more sophisticated hydraulic methods. Allen et al. (1978) indicate application of some of the complex hydraulic models and show sample program listings. Although the more complex hydraulic solutions do not normally take into account the stochastic variation of soil-water properties, Walker (1984) has indicated good agreement of various complex hydraulic models with field measurements of water movement.

This text will emphasize the SCS method of solution for surface water hydraulics since the hydraulic and numerical methods' background required is more in line with the expected level of training for those using this textbook. However, as the capacity, speed, and affordability of computing systems improves, it can be expected that more engineering design firms will make use of the more general and therefore more rigorous hydraulic models. Some of these models are described in the surface irrigation text by Walker and Skogerboe (1987).

Infiltration in furrow systems and calculations of required infiltration time must be handled differently than for other types of surface systems. This is because infiltration only takes place on the wetted perimeter of the furrow yet it must be expressed as an equivalent depth of infiltration over the surface area of the field. Furrow intake rates are measured by instrumenting a furrow for inflow and outflow over a minimum distance of 60 m to 90 m for high intake rate soils and 150 m to 180 m for low intake rate soils (Jensen, 1980). It is desirable to run the test for the expected time of irrigation, but multiples of one-half or one-quarter time are acceptable. It is necessary to compute the volume of inflow and outflow at a minimum of three intermediate times. The average intake over the length of the furrow is given by

$$i = \frac{1}{LP}(V_{in} - V_{out} - V_s) \tag{6-34}$$

where

i = equivalent depth infiltrated over wetted surface area of field, mm
L = distance between inflow and outflow measurements, m
P = adjusted wetted perimeter, m
V_{in} = inflow volume, L
V_{out} = outflow volume, L
V_s = volume of water in storage, L

The adjusted wetted perimeter is given by the following equation:

$$P = 0.265\left[\frac{Qn}{S^{0.5}}\right]^{0.425} + 0.227 \tag{6-35}$$

where

$$P = \text{adjusted wetted perimeter, m}$$
$$Q = \text{volumetric inflow rate, L/s}$$
$$n = \text{Manning's roughness coefficient}$$
$$S = \text{furrow slope or hydraulic gradient, m/m}$$

Note that in most cases of infiltration tests in furrows—that is, after the flow has stabilized and is uniform, the hydraulic gradient is equal to the furrow slope. A roughness coefficient of 0.04 is normally used for design of furrow irrigation systems. Other values of the roughness coefficient for various cropping conditions will be given in later sections of this chapter. The volume of channel storage, V_s, is given by

$$V_s = \frac{L}{0.305}\left[2.947\left[\frac{Qn}{S^{0.5}}\right]^{0.735} - 0.0217\right] \tag{6-36}$$

where the variables and units are as previously defined. Note that the Q in Eq. (6-36) refers to the volumetric inflow rate.

The required depth of infiltration for a furrow system must be expressed as an equivalent depth over the total field area. For this reason, the infiltration depth given by Eq. (3-28) must be modified by the ratio of adjusted wetted perimeter to furrow spacing using the following:

$$i = [a(t)^b + c]\frac{P}{W} \tag{6-37}$$

where

$$W = \text{furrow spacing, m}$$

The advance time for a stream of water moving down the furrow is given by

$$T_t = \frac{x}{f}\exp\left[\frac{gx}{Q(S)^{0.5}}\right] \tag{6-38}$$

where

$$T_t = \text{advance time, min}$$
$$x = \text{distance down the furrow, m}$$
$$f = \text{advance coefficient, Table 6-6}$$
$$g = \text{advance coefficient, Table 6-6}$$
$$Q = \text{volumetric inflow rate, L/s}$$
$$S = \text{furrow slope or hydraulic gradient, m/m}$$

As previously described, the infiltration opportunity time is equal to the time of water application minus the advance time plus the recession time. This is described by

$$T_o = T_{co} - T_t + T_r \tag{6-39}$$

where

$$T_o = \text{infiltration opportunity time, min}$$
$$T_{co} = \text{cut-off time, min}$$
$$T_t = \text{advance time, min}$$
$$T_r = \text{recession time, min}$$

The cut-off time, T_{co}, reflects an irrigation management decision made by the farmer and designer. It should be an adequate length of time to infiltrate a satisfactory depth of water over the length of the furrow without causing excessive deep percolation. T_{co} is normally set equal to the time to advance to the end of the furrow plus the required net infiltration time less recession time. Letting i_n equal the desired net depth of infiltration, the net infiltration time is determined by rearranging Eq. (6-37):

$$T_n = \left[\frac{i_n\left(\frac{W}{P}\right) - c}{a}\right]^{1/b} \tag{6-40}$$

The recession time is assumed zero for open-ended gradient furrows (i.e., furrows for which S is not equal to zero) without loss of accuracy. For gradient furrows, Eq. (6-39) can therefore be revised to

$$T_o = T_{co} - T_t \tag{6-41}$$

Using the preceding assumptions, T_{co} would be equal to

$$T_{co} = T_t + T_n \tag{6-42}$$

where

T_t = advance time required to reach end of the field at distance L, min

For simplification, let us use the symbol β for the term in brackets in Eq. (6-38). Therefore,

$$\beta = \frac{gx}{Q(S)^{0.5}} \tag{6-43}$$

The average infiltration opportunity time over distance x down the furrow is given by

$$T_{O-x} = T_{co} - \frac{0.0929}{f(x)\left[\frac{0.305(\beta)}{x}\right]^2}[(\beta - 1)\exp(\beta) + 1] \tag{6-44}$$

The average infiltration time for the full furrow length, T_{O-L}, is obtained by substituting L into Eq. (6-44) for x. The average depth of infiltration for the entire furrow length, i_{avg}, is therefore determined by substituting T_{O-L} into Eq. (6-37) for t.

The gross depth of water application, i_g, is defined as the required net depth of irrigation, i_n, divided by the product of the application and distribution pattern efficiencies:

$$i_g = \frac{i_n}{\frac{e_a}{100}\left(\frac{e_d}{100}\right)} \tag{6-45}$$

In surface systems, the application efficiency, e_a, is normally assumed equal to 100 percent—that is, evaporation losses are negligible. Equation (6-45) therefore reduces to

$$i_g = \frac{i_n}{e_d/100} \tag{6-46}$$

The equivalent gross depth of application as a function of inflow rate and field geometry is

$$i_g = \frac{60(Q)(T_{co})}{WL} \tag{6-47}$$

where

i_g = gross depth of application, mm
Q = inflow rate, L/s
W = furrow spacing, m

The equivalent depth of surface runoff from the furrow, d_{ro}, can be estimated as the difference between the gross and average depths of application:

$$d_{ro} = i_g - i_{avg} \tag{6-48}$$

The equivalent depth of deep percolation, d_{dp}, is the average minus the net depth of application:

$$d_{dp} = i_{avg} - i_n \tag{6-49}$$

Example Problem 6-3

Given the following information,

Intake family	$I_f = 0.3$
Furrow length	$L = 275$ m
Furrow slope	$S = 0.004$ m/m
Furrow spacing	$W = 0.75$ m
Roughness coefficient	$n = 0.04$
Net irrigation depth	$i_n = 75$ mm
Inflow rate	$Q = 0.6$ L/s

Compute the desired time to cut-off, T_{co}, the equivalent depths of surface runoff and deep percolation, d_{ro} and d_{dp}, and the distribution pattern efficiency.

Solution Advance time, T_t:

$$\beta = \frac{(1.904 \times 10^{-4})275}{0.6(0.004)^{0.5}} = 1.38$$

$$T_t = \frac{275}{7.61}\exp(1.38) = 144 \text{ min}$$

Adjusted wetted perimeter:

$$P = 0.265\left[\frac{(0.6)(0.04)}{(0.004)^{0.5}}\right]^{0.425} + 0.227 = 0.40 \text{ m}$$

Net infiltration time:

$$T_n = \left[\frac{75\left(\frac{0.75}{0.40}\right) - 7.0}{0.925}\right]^{1/0.720} = 999 \text{ min}$$

Design cut-off time:

$$T_{co} = 144 + 999 = 1143 \text{ min}$$

Gross application depth:

$$i_g = \frac{60(0.6)(1143)}{(0.75)(275)} = 200 \text{ mm}$$

Average infiltration time:

$$T_{O-L} = 1143 - \frac{0.0929}{7.61(275)\left[\frac{(0.305)(1.38)}{275}\right]^2}[(1.38 - 1)\exp(1.38) + 1]$$

$$= 1143 - 47.6 = 1095 \text{ min}$$

Average infiltration:

$$i_{avg} = [0.925(1095)^{0.720} + 7.0]\frac{0.40}{0.75} = 80 \text{ mm}$$

Surface runoff depth:

$$d_{ro} = 200 - 80 = 120 \text{ mm}$$

Deep percolation depth:

$$d_{dp} = 80 - 75 = 5 \text{ mm}$$

Distribution pattern efficiency:

$$e_d = \frac{75 \times 100}{200} = 37.5 \text{ percent}$$

Graphical procedures

Design charts have been developed for graphical solution of the foregoing series of equations. These charts are specified for conditions of intake family, net depth of irrigation, field slope, roughness coefficient and furrow spacing. Figure 6-7

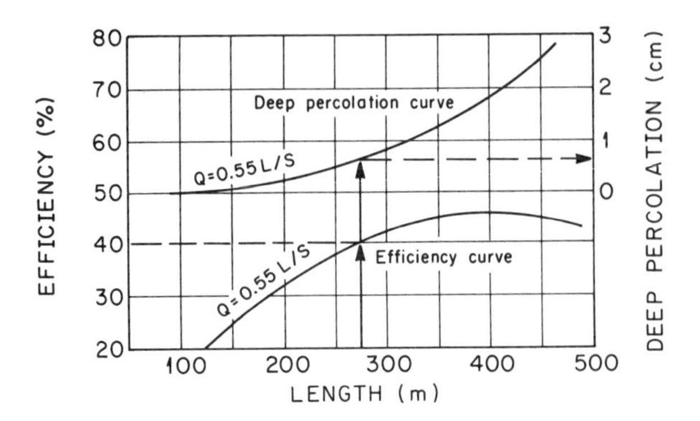

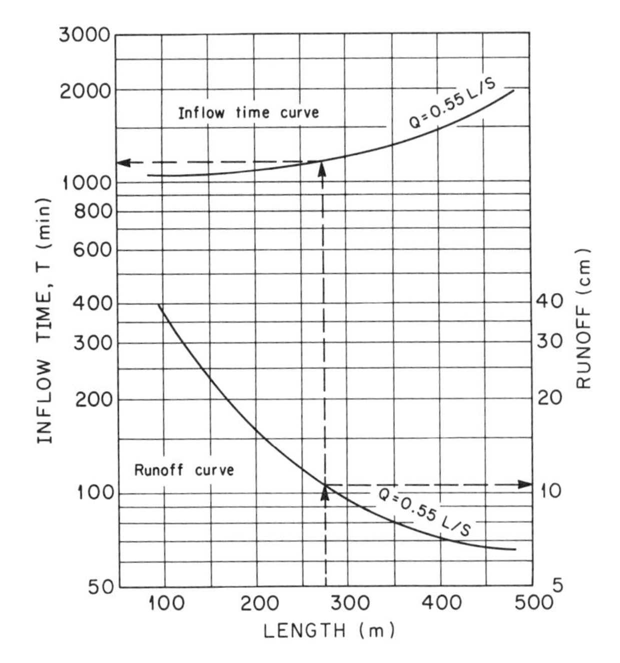

Figure 6-7 Sample furrow irrigation design chart. Intake family = 0.3, i_n = 7.5 cm, S = 0.004 m/m, n = 0.04, and furrow spacing = 75 cm. (Taken from Jensen, 1980.)

indicates a solution to a problem similar to the preceding example. Figure 6-8 indicates a sample design chart for the same conditions but covering a wide range of inflow rates. Such design charts are simply graphical solutions of the equations in this section. They may be rapidly applied in design work, but a wide variety of charts is required to cover the range of possible field conditions. With the increasing capacity of desk-top computers and hand-held calculators, the equations in this section are easily programmed. The design charts do give an engineer a physical feel for the expected response in changing different system parameters. Both methods of solution therefore have advantages and disadvantages.

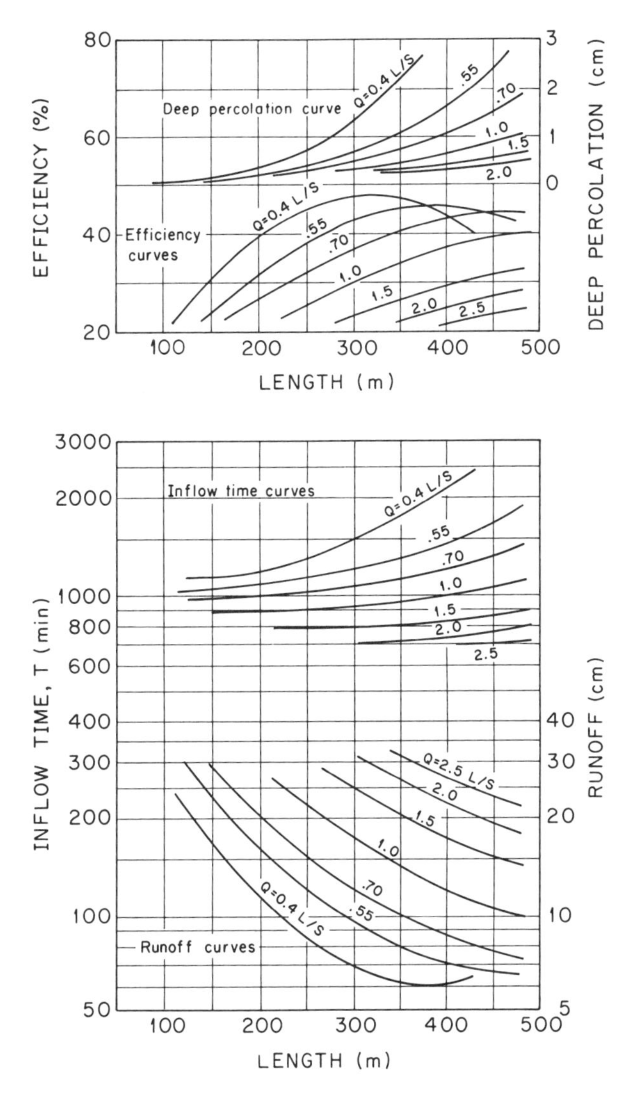

Figure 6-8 Furrow irrigation design chart. Intake family = 0.3, i_n = 7.5 cm, S = 0.004 m/m, n = 0.04, and furrow spacing = 75 cm. (Taken from Jensen, 1980.)

Modifications for Cut-Back Systems

It can be recognized from the infiltration curves in the Soil Physics chapter that infiltration rate decreases with time. It can also be recognized from Table 6-2 that the smaller the fractional advance ratio—that is, the larger the inflow rate—the less will be the amount of deep percolation. This combination has led to the principle of the cut-back system in surface irrigation. Using this method, once the water reaches the end of the furrow, the stream size at the head of the furrow is reduced. The actual time and degree of cut-back is the option of the designer. The following equations are given for conditions in which the time of cut-back, T_{cb}, is equal to the advance time, T_t, and the final flow rate is equal to one-half the initial flow rate. In the following equations, the subscript 1 will be used to indicate initial conditions, and subscript 2 to indicate cut-back conditions.

The following modifications are necessary to solve the hydraulic equations for the cut-back conditions. The adjusted wetted perimeter under cut-back conditions is computed by substituting Q_2 into Eq. (6-35). The required net infiltration time at length L is solved for by substituting P_2 into Eq. (6-40). The average opportunity time for infiltration during the advance period is given by the absolute value of the second term on the right-hand side of Eq. (6-44) with x set equal to L:

$$T_{avg} = \frac{0.0929}{f(L)\left[\frac{0.305(\beta)}{L}\right]^2}[(\beta - 1)\exp(\beta) + 1] \tag{6-50}$$

The average infiltration under cut-back conditions is given by

$$i_{avg} = [a(T_{co} - T_{avg})^b + c]\frac{P_2}{W} + [(a(T_{avg})^b + c)]\frac{P_1 - P_2}{W} \tag{6-51}$$

The gross depth of application is given by

$$i_g = \frac{60}{WL}[Q_1(T_t) + Q_2(T_n)] \tag{6-52}$$

Example Problem 6-4

Given the same conditions as Example Problem 6-3, compute the same information required for that problem if a cut-back system is used and Q is reduced by one-half.

Solution Time of cut-back, T_{cb}, is the time of advance at full flow, T_t, and is equal to that calculated in the previous example.

$$T_{cb} = T_t = 144 \text{ min}$$

Adjusted wetted perimeter during advance is P as calculated in the previous example.

$$P_1 = 0.40 \text{ m}$$

Adjusted wetted perimeter during reduced flow is calculated with flow Q_2.

$$P_2 = 0.265\left[\frac{(0.3)(0.04)}{(0.004)^{0.5}}\right]^{0.425} + 0.227 = 0.36 \text{ m}$$

Net application time is the time water must remain on the surface at the end of the field and is equal to T_n under reduced flow conditions.

$$T_n = \left[\frac{\frac{(75)(0.75)}{0.36} - 7.0}{0.925}\right]^{1/0.720} = 1165 \text{ min}$$

Time of cut-off is the sum of T_t and T_n

$$T_{co} = 144 + 1165 = 1309 \text{ min}$$

Average infiltration time during the advance period is the absolute value of the second term of Eq. (6-44) and was calculated in the previous example as part of T_{O-L}.

$$T_{avg} = 47.6 \text{ min}$$

Average infiltration calculated from Eq. (6-51)

$$i_{avg} = [0.925(1309 - 47.6)^{0.720} + 7.0]\frac{0.36}{0.75} + [0.925(47.6)^{0.720} + 7.0]\frac{(0.40 - 0.36)}{0.75}$$

$$= 79 + 1.2 = 80 \text{ mm}$$

Gross application depth:

$$i_g = \frac{60}{(0.75)(275)}[(0.6)(144) + (0.6/2)(1165)]$$

$$= 127 \text{ mm}$$

Surface runoff depth:

$$d_{ro} = (127 - 80) = 47 \text{ mm}$$

Deep percolation depth:

$$d_{dp} = (80 - 75) = 5 \text{ mm}$$

Distribution pattern efficiency:

$$e_d = 100(75/127) = 59 \text{ percent}$$

Table 6-7 illustrates a comparison of the major parameters of example problems 6-3 and 6-4 with the standard and cut-back designs. It can be observed that there is a significant improvement in distribution pattern efficiency due basically to reduction in runoff with the cut-back design.

Graphical procedures

Graphical solutions can also be developed for cut-back systems. An example of a design chart for a given intake family, net depth of irrigation, slope, roughness coefficient, and furrow spacing is indicated in Fig. 6-9.

TABLE 6-7 Comparison of standard and cut-back solutions from example problems 6-3 and 6-4.

Variable	Standard Solution	Cut-Back Solution
T_t (min)	144	144
T_n (min)	999	1165
T_{co} (min)	1143	1309
i_{avg} (mm)	80	80
i_g (mm)	200	127
d_{ro} (mm)	120	47
d_{dp} (mm)	5	5
e_d (%)	37.5	59

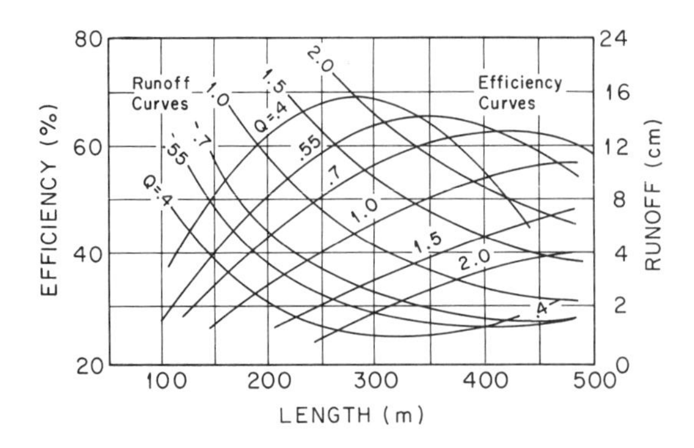

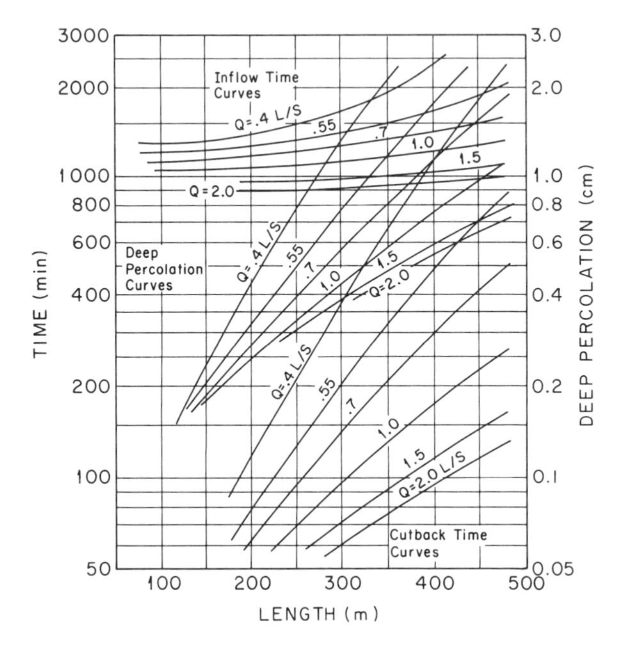

Figure 6-9 Cutback furrow irrigation design chart. Intake family = 0.3, i_n = 7.5 cm, S = 0.004 m/m, n = 0.04, and furrow spacing = 75 cm. (Taken from Jensen, 1980.)

Field Trial Verification of Design Parameters

Because of the previously mentioned stochastic variation of soil-water properties on a field scale as well as variation of hydraulic parameters such as the roughness coefficient, it has been found desirable to conduct field trials on three or four typical furrows with measured slope to verify design variables. This procedure is initiated by estimating a Q_{max} using Eq. (6-33). The four trial furrows are then operated simultaneously with some inflow rates less than and some greater than Q_{max}. The low flow rate should be of such magnitude that water never reaches the end of the furrow and the trial high flow rate should be clearly erosive. Advance curves are plotted for each trial furrow as shown in Fig. 6-10. The highest allowable nonerosive application rate is then chosen for operations so as to increase uniformity of ap-

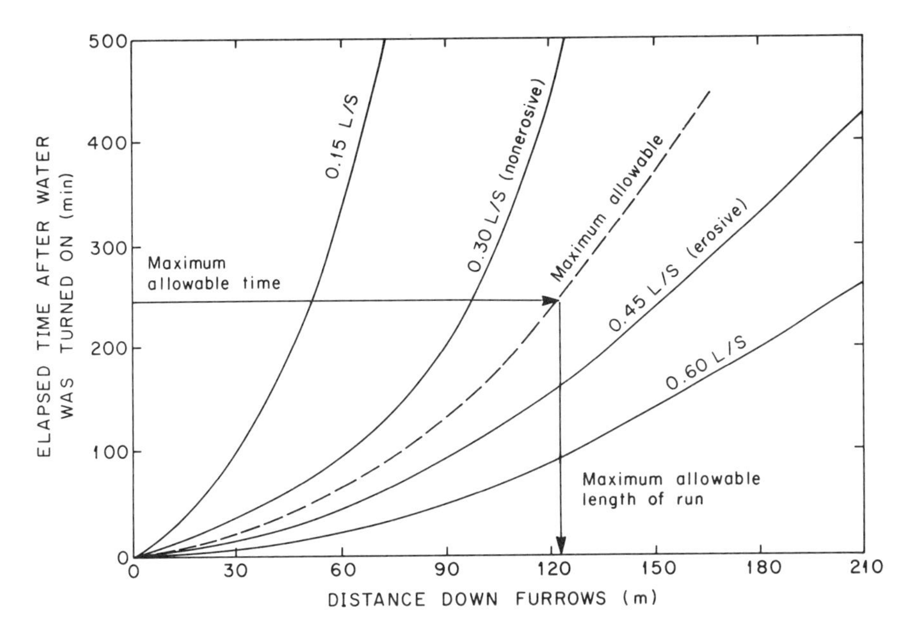

Figure 6-10 Rate of advance curves for determining maximum length of furrows. (Adapted from Criddle et al., 1956.)

plication and decrease deep percolation. With the field verified value of Q_{max} and measured slope S, the length of run and advance time are solved for by a trial and error solution of Eq. (6-38).

Example Problem 6-5

Refer to Fig. 6-10 for field trial results giving a Q_{max} of 0.4 L/s on a slope of 0.0005 m/m on a soil with an intake family of 0.30.

As an initial trial, assume furrow length equal to 100 m.

$$\beta = \frac{gx}{Q(S)^{0.5}} = \frac{1.904 \times 10^{-4}(100)}{0.4(0.0005)^{0.5}} = 213$$

$$T_{t1} = \frac{x}{f}\exp(\beta) = \frac{100}{7.61}\exp(2.13)$$

$$T_{t1} = 110 \text{ min}$$

Projecting across from T_{t1} = 110 min on Fig. 6-10 to the intersection with L = 100 m this point falls below the Q_{max} curve.

Assume a new value of L equal to 140 m to try to improve the solution. Substituting this new L into the preceding set of equations,

$$\beta = 2.98$$

$$T_{t2} = 362 \text{ min}$$

Projecting this result on Fig. 6-10 to the intersection with L = 140 m gives a point which falls above the Q_{max} curve. At this point a satisfactory solution has been bracketed. For the third trial assume L equals 125 m.

The analysis now yields

$$\beta = 2.66$$

$$T_{t3} = 235 \text{ min}$$

Projecting this result on Fig. 6-10 indicates a solution which falls reasonably close to the field trial Q_{max} curve. For design purposes, a system with a maximum nonerosive flow rate of 0.4 L/s will advance 125 m in 235 min. Other system parameters may require recalculation if the projected solution was too far from that verified by field trials.

6.4 Level Basin System Design

Description

Fields to be irrigated by a level basin system are divided into level rectangles of limited extent by ridges of adequate height to retain the depth of flow. The entire field is flooded and the water is allowed to infiltrate into the root zone after ponding on the soil surface. The greater the uniformity of application and infiltration over the field, the greater will be the distribution pattern and overall irrigation efficiency. Level basin systems are designed on the basis of water application rate, soil intake family, and field dimensions.

Level basin systems are applicable to a wide variety of crops. It is possible to grow crops ranging from orchards to small grains using such systems. Level basin systems are especially applicable to soils with low to moderate intake rates, as reasonable efficiencies are attainable on such soils. Sometimes furrows are used in conjunction with level basins by installing a set of furrows within the boundary formed by the ridge lines. Level basin systems are found around the world ranging in areas of from a few square meters to tens of hectares depending on the available stream size and soil intake rate. Growers in many countries have found level basins an efficient way to irrigate basins of small size when limited flow rates are available.

Figures 6-11 and 6-12 indicate applications of level basin systems to fields of various sizes. In the first photograph, a limited water source is carefully applied to a very small parcel. Figure 6-12 indicates a more extensive level basin system in the Nile Delta.

Experience Relationships

As with furrow systems, empirical relationships have been developed for the design of level basin systems based on reasonably successful designs in field situations. These relationships are a compromise between available stream size, soil intake family, basin size, and irrigation efficiency. Table 6-8 indicates suggested basin areas as a function of soil and water supply parameters necessary to achieve a reasonable irrigation efficiency. It should be repeated that the information available in a table of this type gives a general idea of design criteria. Evaluation of important economic alternatives and final system design should be based on the more detailed analysis described in the following section.

Figure 6-11 Water being spread over level basins of very limited size in central Tunisia.

Figure 6-12 Head ditch in foreground serving water to supply ditches used in relatively large level basin system for cereal production in Nile River Valley, Egypt.

TABLE 6-8 Suggested basin areas for different soil types and rates of water flow. (Taken from Booher, 1974.)

A. Area in hectares

Flow rate		Soil type			
		Sand	Sandy loam	Clay loam	Clay
Liters per second	Cubic meters per hour	Hectares			
30	108	0.02	0.06	0.12	0.2
60	216	0.04	0.12	0.24	0.4
90	324	0.06	0.18	0.36	0.6
120	432	0.08	0.24	0.48	0.8
150	540	0.10	0.30	0.60	1.0
180	648	0.12	0.36	0.72	1.2
210	756	0.14	0.42	0.84	1.4
240	864	0.16	0.48	0.96	1.6
270	972	0.18	0.54	1.08	1.8
300	1080	0.20	0.60	1.20	2.0

B. Area in acres

Flow rate		Soil type			
		Sand	Sandy loam	Clay loam	Clay
Cubic feet per second	U.S. gallons per minute	Acres			
1	450	0.05	0.15	0.3	0.5
2	900	0.10	0.30	0.6	1.0
3	1350	0.15	0.45	0.9	1.5
4	1800	0.20	0.60	1.2	2.0
5	2250	0.25	0.75	1.5	2.5
6	2700	0.30	0.90	1.8	3.0
7	3150	0.35	1.05	2.1	3.5
8	3600	0.40	1.20	2.4	4.0
9	4050	0.45	1.35	2.7	4.5
10	4500	0.50	1.50	3.0	5.0

Hydraulic Relationships

The hydraulic relationships described in this section are based on design procedures developed by the Soil Conservation Service and will use the intake family concept previously discussed. Infiltration tests for level basin design are made using concentric ring infiltrometers or by flooding small representative basins if possible. The equations in this section can be derived by application of the continuity equation, infiltration equation, and Manning's equation for channel flow applied to relatively wide basins with limited depth of flow.

The net time of infiltration, T_n, in a level basin system is computed using an equation similar to (6-40) without the adjustment for wetted perimeter and furrow spacing required for furrow design:

$$T_n = \left[\frac{i_n - c}{a}\right]^{1/b} \tag{6-53}$$

The required advance time, T_t, is determined by multiplying the net infiltration time by the fractional advance ratio, T_t/T_n, which is a function of distribution pattern efficiency. Tabulated values for the fractional advance ratio for different efficiencies are indicated in Table 6-9 and a graphical solution is presented in Fig. 6-13.

TABLE 6-9 Ratio of T_t to T_n for various distribution efficiency values.

Distribution Pattern Efficiency	Ratio T_t to T_n
95	0.16
90	.28
85	.40
80	.58
75	.80
70	1.08
65	1.45
60	1.90
55	2.45
50	3.20

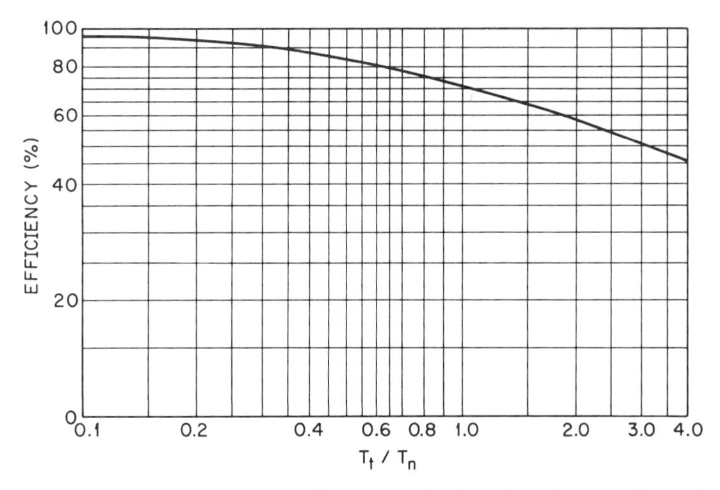

Figure 6-13 Chart for estimating efficiency of level border irrigation. (Taken from USDA-SCS, 1974.)

TABLE 6-10 Common roughness coefficients used in border design.

Smooth, bare soil surfaces noncultivated, oil-mulch-treated citrus	0.04
Small grain, drill rows parallel to border strip	0.10
Alfalfa, mint, broadcast small grain, and similar crops	0.15
Dense sod crops, small grain with drill rows across the border strip	0.25

The data given in Table 6-9 may also be represented with reasonable accuracy by the following regression equation:

$$e_d = 105.81 - 32.676(T_t/T_n)^{0.5} \tag{6-54}$$

where

e_d = distribution pattern efficiency, percent

The relationship between the advance time, basin length, L, and inflow rate will apply the unit inflow rate concept. The unit inflow rate, Q_u, is computed by dividing the level basin inflow by the basin width and has units of cubic meters per second per meter width, equivalent to square meters per second. The relationship between basin length and the other parameters is given by

$$L = \frac{6 \times 10^4(Q_u)(T_t)}{\dfrac{a(T_t)^b}{1 + b} + c + 1798(n)^{3/8}(Q_u)^{9/16}(T_t)^{3/16}} \tag{6-55}$$

Values of the roughness coefficient applied in design of both level basin and graded border systems are indicated in Table 6-10.

Basin length may be solved for directly using Eq. (6-55) for a given unit inflow rate and advance time associated with a given irrigation efficiency. To solve for the required unit inflow rate for a given basin length and advance time requires a trial and error or numerical method solution applied to the highly nonlinear Eq. (6-55). Application of numerical techniques, such as Newton's method, indicates an efficient convergence to the required unit inflow rate and can be conveniently programmed on calculators of adequate capacity.

The time to cut-off, T_{co}, is the time required to put the gross depth of irrigation, i_g, onto the basin. This time can be computed from the following:

$$T_{co} = \frac{i_n L}{600 Q_u e_d} \tag{6-56}$$

where

T_{co} = time to cut-off, min

i_n = net depth of irrigation, mm

e_d = distribution pattern efficiency, percent

Note that Eq. (6-56) assumes that the application efficiency, e_a, is equal to 100 percent—that is, that all the water turned into the basin is infiltrated into the soil. If for some reason e_a is not equal to 100 percent, Eq. (6-56) must be revised as follows:

$$T_{co} = \frac{i_n L}{6(Q_u)(e_a)(e_d)} \quad (6\text{-}57)$$

where e_a and e_d are both given in percent.

The maximum depth of flow in the basin, d_{max}, is an important parameter in basin design in that it governs the minimum ridge height. The ridge height should be equal to 1.25 times the maximum depth of flow and the ridge should have a maximum side slope ratio of 2.5 to 1. The maximum flow depth is derived from application of Manning's equation with Eq. (6-56) and the empirical assumption that the average flow depth is 80 percent of the maximum flow depth. The maximum flow depth is given by

$$d_{max} = 2250(n)^{3/8}(Q_u)^{9/16}(T_{co})^{3/16} \quad (6\text{-}58)$$

where

d_{max} = maximum flow depth, mm

The advance time must be used in place of the time to cut-off in Eq. (6-58) if T_t is greater than T_{co}.

Graphical solutions for the hydraulic equations presented in this section have been developed. These solutions are given for various basin lengths and distribution pattern efficiencies as a function of intake family, roughness coefficient, and net depth of irrigation. They have similar advantages and disadvantages as previously discussed under furrow design charts. An example design chart for level basins is indicated in Fig. 6-14.

Example Problem 6-6

Given the following information,

Intake family	$I_f = 0.5$
Target distribution pattern efficiency	e_d = 80 percent
Unit flow rate	$Q_u = 0.005$ m²/s
Net irrigation depth	i_n = 100 mm
Roughness coefficient	$n = 0.15$

Assuming 100 percent application efficiency, compute the net infiltration time, basin length, time to cut-off, and maximum depth of flow.

Solution Net infiltration time:

$$T_n = \left[\frac{100 - 7.0}{1.196}\right]^{1/0.748} = 337 \text{ min}$$

Advance time:

Ratio T_t/T_n at 80 percent efficiency = 0.58

$$T_t = (0.58)(337) = 195 \text{ min}$$

Basin length:

$$L = \frac{(6 \times 10^4)(0.005)(195)}{\frac{(1.196)(195)^{0.748}}{1 + 0.748} + 7.0 + (1798)(0.15)^{3/8}(0.005)^{9/16}(195)^{3/16}}$$
$$= 359 \text{ m}$$

Time to cut-off:

$$T_{co} = \frac{(100)(359)}{(600)(0.005)(80)} = 150 \text{ min}$$

Maximum depth of flow (note $T_t > T_{co}$):

$$d_{max} = 2250(0.15)^{3/8}(0.005)^{9/16}(195)^{3/16}$$
$$= 151 \text{ mm}$$

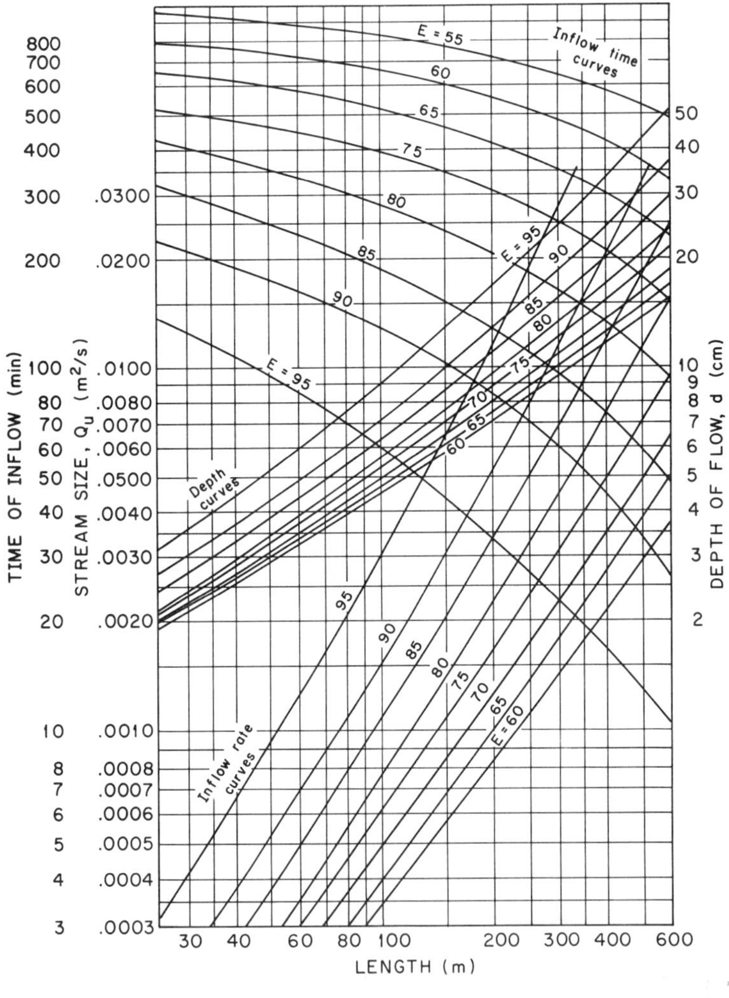

Figure 6-14 Level basin design chart for intake family = 0.5, n = 0.15, and i_n = 100 mm.

6.5 Graded Border System Design

Description

Graded border systems are similar in concept to level basin systems except that there is a slope down the border and there may be limited cross slope. Graded border systems do not have to be, and oftentimes are not, blocked on the end. Water is therefore allowed to pass through the border in a similar manner as in operation of a gradient furrow system. Graded border systems may be more conveniently applied to soils of limited depth than level basin systems because of reduced leveling requirements.

Graded border systems are most applicable to soils with moderately low to moderately high intake rates. Such systems on soils with high intake rates are not feasible because borders too limited in length are required to have a reasonable irrigation efficiency. Soils with low intake rates tend to need excessive time to infiltrate the required depth of irrigation into the soil, thereby causing excessive runoff. This method is best suited to lands with slopes less than 0.5 percent. It can be used on lands of slopes up to 2 percent for non-sod crops and up to approximately 4 percent for sod crops.

Figures 6-15 and 6-16 show different size graded border systems. The field sizes indicated require a substantial water supply to produce reasonable application uniformities. Skill is required to bring this size field to proper grade for efficient distribution of the water commonly applied at one point on the field border.

Graded border systems have similar advance and recession characteristics as those indicated by Fig. 6-5. Highest efficiency is obtained if it is possible to balance

Figure 6-15 Land brought to grade for moderate size graded border systems in Aragon Province, Spain. Gate from supply ditch is visible in foreground.

Figure 6-16 Initial stages of graded border irrigation in Aragon Province, Spain. Inlet from water supply is at corner of rectangular field.

the advance and recession curves so each point along the border has equal opportunity for infiltration. This condition can be met if the depth of water applied is equal to the gross depth of irrigation required and if the infiltration opportunity time at the end of the border is equal to the net time of irrigation.

Experience Relationships

As with other types of surface irrigation systems, a number of guidelines have been developed for the design of graded border systems based on accumulated field experience. These relationships allow for a first approximation of system layout. Tables 6-11 and 6-12 indicate relationships developed for design of graded border systems for shallow and deep rooted crops, respectively. These tables indicate recommended borderstrip widths and lengths as a function of soil type and percent slope. The soil types covered for shallow rooted crops are clay loam, clay, and loam. For deep rooted crops, sand, loamy sand, sandy loam, clay loam, and clay are covered. These soils are representative of soils for which graded border systems may be effectively used. The tables also indicate recommended available flow rates and gross depth of water to be applied with each irrigation. These guidelines can aid the grower and designer in determining system feasibility. Final decisions and examination of economic alternatives should be based on hydraulic analysis.

Hydraulic Relationships

As with level basin systems, design procedures in this section are based on those developed by the Soil Conservation Service and apply the intake family concept. Graded border system design also applies the concept of discharge per unit width.

TABLE 6-11 Suggested standards for the design of border-strips for shallow-rooted crops. (Taken from Booher, 1974.)

A. Metric units

Soil profile	Percent of slope	Unit flow per meter of strip width	Average depth of water applied	Border-strip Width	Border-strip Length
	Meters per 100 meters	Liters per second	Milli-meters	Meters	
CLAY LOAM	0.15–0.6	6–8	50–100	5–18	90–180
0.6 meter deep over permeable	0.6 –1.5	4–6	50–100	5–6	99–180
subsoil	1.5 –4.0	2–4	50–100	5–6	90
CLAY	0.15–0.6	3–4	100–150	5–18	180–300
0.6 meter deep over permeable	0.6 –1.5	2–3	100–150	5–6	180–300
subsoil	1.5 –4.0	1–2	100–150	5–6	180
LOAM	1.0 –4.0	1–4	25–75	5–6	90–300
0.15 to 0.45 meter deep over hardpan					

B. British units

Soil profile	Percent of slope	Unit flow per foot of strip width	Average depth of water applied	Border-strip Width	Border-strip Length
	Feet per 100 feet	Cubic feet per second	Inches	Feet	
CLAY LOAM	0.15–0.6	0.06–0.08	2–4	15–60	300–600
24 inches deep over permeable	0.6 –1.5	0.04–0.07	2–4	15–20	300–600
subsoil	1.5 –4.0	0.02–0.04	2–4	15–20	300
CLAY	0.15–0.6	0.03–0.04	4–6	15–60	600–1000
24 inches deep over permeable	0.6 –1.5	0.02–0.03	4–6	15–20	600–1000
subsoil	1.5 –4.0	0.01–0.02	4–6	15–20	600
LOAM	1.0 –4.0	0.01–4.0	1–3	15–20	300–1000
6 to 18 inches deep over hardpan					

The hydraulic relationships applied are complicated relative to level basins in that the water applied is continuously moving downslope. The hydraulic relationships are derived by consideration of the continuity relationship, the Manning equation, and the assumption that the amount of water infiltrated into the soil may be approximated by the volume of a section with a triangular cross-sectional shape as the recession curve moves down the field.

Graded border systems are designed on the principle that any point in the field should have water applied to it for a time equal to that required to infiltrate the net depth of irrigation. In a graded border system, the time between cut-off of water at the head of the field and the disappearance of water at the head of the field is termed the recession lag time, T_{rl}. The recession lag time is equivalent to the depletion

TABLE 6-12 Suggested standards for the design of border-strips for deep-rooted crops. (Taken from Booher, 1974.)

A. Metric Units

Soil type	Percent of slope	Unit flow per meter of strip width	Average depth of water applied	Border-strip Width	Border-strip Length
	Meters per 100 meters	Liters per second	Milli-meters	Meters	
SAND	0.2–0.4	10–15	100	12–30	60–90
Infiltration rate of 2.5 + cm per	0.4–0.6	8–10	100	9–12	60–90
hour	0.6–1.0	5–8	100	6–9	75
LOAMY SAND	0.2–0.4	7–10	125	12–30	75–150
Infiltration rate of 1.8 to 2.5 cm	0.4–0.6	5–8	125	9–12	75–150
per hour	0.6–1.0	3–6	125	6–9	75
SANDY LOAM	0.2–0.4	5–7	150	12–30	90–250
Infiltration rate of 1.2 to 1.8 cm	0.4–0.6	4–6	150	6–12	90–180
per hour	0.6–1.0	2–4	150	6	90
CLAY LOAM	0.2–0.4	3–4	175	12–30	180–300
Infiltration rate of 0.6 to 0.8 cm	0.4–0.6	2–3	175	6–12	90–180
per hour	0.6–1.0	1–2	175	6	90
CLAY	0.2–0.3	2–4	200	12–30	350
Infiltration rate of 0.25 to 0.6 cm per hour					

B. British units

Soil profile	Percent of slope	Unit flow per foot of strip width	Average depth of water applied	Border-strip Width	Border-strip Length
	Feet per 100 feet	Cubic feet per second	Inches	Feet	
SANDY	0.2–0.4	0.11–0.16	4	40–100	200–300
Infiltration rate of 1 + inch per	0.4–0.6	0.09–0.11	4	30–40	200–300
hour	0.6–1.0	0.06–0.09	4	20–30	250
LOAMY SAND	0.2–0.4	0.07–0.11	5	40–100	250–500
Infiltration rate of 0.75 to 1 inch	0.4–0.6	0.06–0.09	5	25–40	250–500
per hour	0.6–1.0	0.03–0.06	5	25	250
SANDY LOAM	0.2–0.4	0.06–0.08	6	40–100	300–800
Infiltration rate of 0.5 to 0.75	0.4–0.6	0.04–0.07	6	20–40	300–600
inch per hour	0.6–1.0	0.02–0.04	6	20	300
CLAY LOAM	0.2–0.4	0.03–0.04	7	40–100	600–1000
Infiltration rate of 0.25 to 0.5	0.4–0.6	0.02–0.03	7	20–40	300–600
inch per hour	0.6–1.0	0.01–0.02	7	20	300
CLAY	0.2–0.3	0.02–0.04	8	40–100	1200
Infiltration rate of 0.10 to 0.25 inch per hour					

phase shown in Fig. 6-5. If the infiltration opportunity time is to be equal to the required net infiltration time, it can be shown that the time to cut-off is equal to the net infiltration time minus the recession lag time:

$$T_{co} = T_n - T_{rl} \tag{6-59}$$

The term *high gradient borders* is used to denote borders with a surface slope greater than approximately 0.004 m/m. In such borders, the water surface slope is assumed equal to the field slope and the normal flow depth—that is, the depth of flow under conditions of uniform flow—is assumed equal to the depth of flow at the head of the border. Under such conditions, the recession lag time is given by:

$$T_{rl} = \frac{(Q_u)^{0.2}(n)^{1.2}}{120(S)^{1.6}} \tag{6-60}$$

where

T_{rl} = recession lag time, min
Q_u = unit inflow rate, m^2/s
n = Manning's roughness coefficient
S = surface slope, m/m

Manning's roughness coefficients used in design of graded border systems are the same as those given in Table 6-10.

For low gradient borders with surface slopes of less than 0.004 m/m, the depth of flow at the head of the border is less than the normal depth, and the hydraulic analysis is more complicated. The hydraulic slope required in Manning's equation is no longer equal to the surface slope as in the case of uniform flow at normal depth. The hydraulic slope is approximated by the surface slope plus the depth of flow at the head of the field divided by the length of advance. In such cases, the recession lag time is given by

$$T_{rl} = \frac{(Q_u)^{0.2}(n)^{1.2}}{120\left[S + \dfrac{0.0094n(Q_u)^{0.175}}{(T_n)^{0.88}(S)^{0.5}}\right]^{1.6}} \tag{6-61}$$

where

T_n = net infiltration time, min

Tabulated results for the recession lag time as a function of net infiltration time, surface slope, and unit inflow rate are indicated in Table 6-13. This table is a necessary guide for solution of design parameters for a low gradient border by trial and error methods as will be demonstrated later in this chapter.

The unit inflow rate is derived from application of a balance between the volume of water applied to the border and the required net depth of irrigation divided by irrigation efficiency. The inflow rate per unit width of border strip in square meters per second is given by

$$Q_u = \frac{0.00167 i_n L}{(T_n - T_{rl})e_d} \tag{6-62}$$

TABLE 6-13 Recessional-lag times T_{rl} (min) in low gradient borders. (Taken from USDA-SCS, 1974.)

Oppor. time, T_n (min)	Border slope, S_o (m/m)															
	0.0005				0.001				0.002				0.004			
	Inflow rate, Q_u (m^2/s)				Inflow rate, Q_u (m^2/s)				Inflow rate, Q_u (m^2/s)				Inflow rate, Q_u (m^2/s)			
	0.0001	0.001	0.01	0.02	0.0001	0.001	0.01	0.02	0.0001	0.001	0.01	0.02	0.0001	0.001	0.01	0.02
	Manning's n = 0.04															
10	1.9	2.2	2.3	2.3	1.1	1.5	1.9	2.0			1.1	1.1				
25	3.1	4.0	4.8	5.1	1.4	2.0	2.8	3.1			1.2	1.4				
50	3.9	5.4	7.1	7.7	1.6	2.3	3.4	3.8			1.3	1.5				
100	4.4	6.5	9.2	10.1	1.6	2.5	3.8	4.3			1.4	1.6				
200	4.8	7.3	10.8	12.1	1.7	2.6	4.1	4.6			1.4	1.6				
500	5.1	7.9	12.1	13.7	1.7	2.7	4.2	4.9			1.4	1.6				
1000	5.2	8.1	12.6	14.4	1.7	2.7	4.3	4.9			1.4	1.7				
2000	5.2	8.2	12.9	14.8	1.7	2.8	4.4	5.0			1.4	1.7				
	Manning's n = 0.15															
10	2.5	2.4	2.2	2.1	2.5	2.7	2.7	2.7	1.6	2.1	2.5	2.6		1.1	1.5	1.6
25	6.1	6.3	6.3	6.2	4.4	5.4	6.2	6.4	2.2	3.0	4.1	4.4		1.3	1.9	2.1
50	10.1	11.6	12.5	12.7	5.7	7.7	9.8	10.4	2.4	3.6	5.1	5.7		1.4	2.1	2.3
100	14.5	18.4	21.9	22.7	6.8	9.7	13.4	14.6	2.6	3.9	5.9	6.6		1.4	2.2	2.5
200	18.4	25.3	32.9	35.2	7.5	11.2	16.3	18.1	2.7	4.2	6.4	7.3		1.4	2.3	2.6
500	22.1	32.5	46.3	51.2	8.1	12.4	18.9	21.4	2.8	4.3	6.8	7.7		1.5	2.3	2.6
1000	23.7	36.0	53.6	60.2	8.3	12.9	20.0	22.8	2.8	4.4	6.9	7.9		1.5	2.3	2.7
2000	24.7	38.2	58.4	66.2	8.4	13.2	20.7	23.6	2.8	4.4	7.0	8.0		1.5	2.3	2.7
	Manning's n = 0.25															
10	2.4	2.2	1.9	1.8	2.8	2.8	2.7	2.6	2.2	2.7	2.9	3.0	1.2	1.7	2.1	2.3
25	6.5	6.4	6.0	5.8	5.8	6.6	7.0	7.1	3.4	4.4	5.6	6.0	1.5	2.1	3.0	3.4
50	12.3	13.0	13.1	12.9	8.5	10.6	12.5	12.9	4.1	5.8	7.9	8.5	1.6	2.4	3.5	4.0
100	19.9	23.3	25.6	26.0	10.9	14.9	19.1	20.4	4.5	6.7	9.7	10.8	1.6	2.5	3.9	4.4
200	28.1	36.0	43.5	45.5	12.8	18.5	25.6	28.1	4.8	7.4	11.1	12.4	1.7	2.6	4.1	4.6
500	36.9	52.1	70.5	76.5	14.3	21.7	32.2	36.0	5.0	7.8	12.1	13.8	1.7	2.7	4.2	4.8
1000	41.3	61.2	88.0	97.5	15.0	23.1	35.3	39.9	5.1	8.0	12.5	14.3	1.7	2.7	4.3	4.9
2000	44.1	67.3	100.7	113.3	15.3	24.0	37.2	42.4	5.2	8.1	12.8	14.6	1.7	2.7	4.3	4.9

Note:
Recession-lag times of less than one minute are omitted

where

i_n = net depth of irrigation, mm

L = border length, m

e_d = distribution pattern efficiency, percent

As in Eq. (6-56) for level basin systems, Eq. (6-62) assumes that the application efficiency, e_a, is 100 percent. If not, Eq. (6-62) must be revised by insertion of e_a as a product term in the denominator as shown in Eq. (6-57). Since Q_u appears on the right-hand side of Eqs. (6-60) and (6-61) for the recession lag time, solution of either the recession lag time or unit inflow rate is seen to require a trial and error or numerical methods solution.

The distribution pattern efficiency in a graded border system accounts for water loss due to both deep percolation and runoff. It is a function of intake family, surface slope, net irrigation depth, border length, and unit inflow rate. Higher efficiencies under common operating conditions are normally obtained on soils with moderate to high intake rates—that is, intake families 1.5 to 3.0. For intake families below and above this range, irrigation efficiency is reduced. For gently sloping, well-leveled fields with adequate water control, distribution pattern efficiencies in the range of 60 to 75 percent are feasible. It is generally not economically justified to design graded border systems for efficiencies of less than 50 percent.

The maximum depth of flow in the border strip is determined by the border ridge height. The border ridge height is normally established at 1.25 times the maximum flow depth. Maximum flow depths of less than 150 mm are generally acceptable. In certain soils resistant to erosion, flow depths in the range of 200 mm may be acceptable.

The depth of flow at the head of the border in high gradient borders is the normal depth for uniform flow given by

$$d_h = \frac{1000(Q_u)^{0.6}(n)^{0.6}}{S^{0.3}} \qquad (6\text{-}63)$$

where

d_h = depth of flow at head of border, mm

For low gradient borders, flow is at a depth less than the normal depth which occurs under uniform flow. Under such conditions, d_h may be calculated by:

$$d_h = 2454(T_{rl})^{0.1875}(Q_u)^{0.5625}(n)^{0.1875} \qquad (6\text{-}64)$$

Design Limitations

Certain design limitations have been developed for graded border systems to insure that the design is reasonably efficient and erosion and hydraulic constraints are adhered to. These limitations are aimed at achieving reasonable efficiencies without excessive erosion based on the hydraulic considerations presented in this section. Some of the limitations are empirical in nature while others are based on the hydraulics of flow. The final design should be checked to see that it falls within the constraints described in this section.

A maximum flow rate criterion has been established to aid in producing a design with a nonerosive stream size. This criterion is empirical and needs to be verified by field checks. For non-sodforming crops such as alfalfa and small grains, the stream size per unit width in square meters per second is given by

$$Q_{u_{max}} = \frac{1.765 \times 10^{-4}}{S^{0.75}} \tag{6-65}$$

For well-established, dense sod crops such as pasture and other grasses, the relationship is

$$Q_{u_{max}} = \frac{3.53 \times 10^{-4}}{S^{0.75}} \tag{6-66}$$

Notice that the relationships in Eqs. (6-65) and (6-66) do not account for soil type and therefore must be applied as an initial approximation of nonerosive stream size.

A minimum depth of flow criterion is required to insure that the water stream is large enough to spread over the entire border. This minimum depth of flow is maintained by specifying a minimum unit inflow rate given by the following:

$$Q_{u_{min}} = \frac{5.95 \times 10^{-6} L(S)^{0.5}}{n} \tag{6-67}$$

Note that a lower flow rate is required on land with a rougher surface which would have a larger value of n compared to a smooth surface.

The maximum field slope is given as a function of surface roughness, net depth of irrigation, intake family, and desired irrigation efficiency. It is based on the principle of the minimum depth of flow. Normally, surface slopes of greater than 0.04 m/m are not practical due to erosion hazards. The theoretical relationship for maximum slope is given by:

$$S_{max} = \left[\frac{n}{0.0117 e_d} \frac{i_n}{T_n - T_{rl}}\right]^2 \tag{6-68}$$

The maximum border length is limited by the maximum unit inflow rate. This rate is in turn limited by maximum nonerosive stream size on steeper borders and by maximum depth of flow on flatter borders. On borders of low field slope made up of soils with low intake rates, the theoretical maximum may be too long for practical irrigation operations. Border lengths in excess of 400 m are generally considered excessive. The theoretical maximum length is given by

$$L_{max} = \frac{Q_{u_{max}} e_d (T_n - T_{rl})}{0.00167 i_n} \tag{6-69}$$

Note that in both Eqs. (6-68) and (6-69) the application efficiency is assumed equal to 100 percent. If not, it must be inserted as a product term times e_d in both equations.

Example Problem 6-7

Given the following information,

Intake family	$I_f = 0.5$
Net irrigation depth	$i_n = 100$ mm
Surface slope	$S = 0.001$ m/m
Roughness coefficient	$n = 0.15$
Desired distribution pattern efficiency	$e_d = 70$ percent
Allowable flow depth	$d_{max} = 150$ mm
Border length	$L = 250$ m
Crop	Alfalfa

Assuming 100 percent application efficiency, compute the net infiltration time, recession lag time, unit inflow rate, time to cut-off, and depth of flow.

Solution Net infiltration time:

$$T_n = \left[\frac{100 - 7.0}{1.196}\right]^{1/0.748} = 337 \text{ min}$$

Solve for Q_u using trial and error solution of Eqs. (6-62) and (6-61). From Table 6-13 the recession lag time for the low gradient border in this example must be within the following limits:

$$7.5 \text{ min} < T_{rl} < 21.4 \text{ min}$$

Initial trial: $T_{rl} = 17$ min.

$$Q_u = \frac{0.00167(100)(250)}{(337 - 17)(70)} = 0.00186 \text{ m}^2\text{/s}$$

Referring to Table 6-13, revise T_{rl} down to 13 min for second trial.

$$Q_u = \frac{0.00167(100)(250)}{(337 - 13)(70)} = 0.00184 \text{ m}^2\text{/s}$$

Check T_{rl} using Eq. (6-61).

$$T_{rl} = \frac{(0.00184)^{0.2}(0.15)^{1.2}}{120\left[0.001 + \dfrac{0.0094(0.15)(0.00184)^{0.175}}{(337)^{0.88}(0.001)^{0.5}}\right]^{1.6}}$$

$$T_{rl} = 13.4 \text{ min} \cong 13 \text{ min}$$

The foregoing solution is adequate considering natural variation in field conditions.
Time to cut-off:

$$T_{co} = T_n - T_{rl} = 337 - 13 = 324 \text{ min}$$

Depth of flow:

$$d_h = 2454(13)^{0.1875}(0.00184)^{0.5625}(0.15)^{0.1875}$$

$$d_h = 80 \text{ mm}$$

Check design limitations:

Maximum inflow rate:

$$Q_{u_{max}} = \frac{1.765 \times 10^{-4}}{(0.001)^{0.75}} = 0.0314 \text{ m}^2/\text{s} > 0.00184 \text{ m}^2/\text{s}$$

Minimum inflow rate:

$$Q_{u_{min}} = \frac{5.95 \times 10^{-6}(250)(0.001)^{0.5}}{0.15}$$

$$Q_{u_{min}} = 0.00031 \text{ m}^2/\text{s} < 0.00184 \text{ m}^2/\text{s}$$

Maximum slope:

$$S_{max} = \left[\frac{0.15}{0.0117(70)} \frac{100}{337 - 13}\right]^2$$

$$S_{max} = 0.0032 \text{ m/m} > 0.001 \text{ m/m}$$

Maximum length:

$$L_{max} = \frac{(0.0314)(70)(337 - 21.7)}{0.00167(100)}$$

The value of 21.7 min in the preceding equation is T_{r1} computed using Eq. (6-61) with $Q_{u_{max}}$ equal to 0.0314 m²/s.

$$L_{max} = 4150 \text{ m} > 250 \text{ m}$$

Maximum flow depth:

$$d_h = 80 \text{ mm} < 150 \text{ mm}$$

The design values for T_n, T_{r1}, Q_u, T_{co}, and d_h as specified in the preceding therefore fall within all the design limitations.

Application of End Blocks

Greater irrigation efficiencies can be realized in graded border systems if runoff from the end of the border is eliminated. This can be achieved using two methods which require blocking the end of the border. The first method requires extending the border length and impounding the runoff on the length extension. The second method entails reducing the inflow rate on a blocked border of original length. Both of these methods are discussed in this section.

Border extensions

The length of a border extension using end blocks is limited by the lesser of the following two conditions.

1. Length covered by an impoundment whose maximum depth equals the net infiltration depth.

$$L_e = \frac{I_n}{1000(S)} \tag{6-70}$$

where

$$L_e = \text{length extension, m}$$

2. Length that can be adequately irrigated by the volume of run-off from an open-ended border. This length is determined by the following empirical relationship,

$$L_e = \left(1 - \frac{e_d}{100}\right) r_i r_n (L) \tag{6-71}$$

where

$$r_i = \text{factor for effect of intake family on runoff}$$
$$r_n = \text{factor for effect of roughness on runoff}$$
$$L = \text{design border length, m}$$

In Eqs. (6-70) and (6-71), L_e is the allowable length extension using end blocks to impound the runoff. Empirical values for r_i and r_n to apply in Eq. (6-71) are given in Table 6-14.

Inflow reduction

On fields which are limited in geometric shape such that length extension is not possible, runoff may be eliminated by use of end blocks and reduction of inflow stream size. The reduced inflow rate with end blocks can be computed from

$$Q_{ue} = \frac{Q_u}{1 + r_i r_n \left[1 - \dfrac{e_d}{100}\right]} \tag{6-72}$$

where

$$Q_{ue} = \text{inflow rate per unit width with end blocks, m}^2/\text{s}$$
$$Q_u = \text{design inflow rate, m}^2/\text{s}$$

As previously indicated for other cases, Eqs. (6-71) and (6-72) assume application efficiency equals 100 percent. The distribution pattern efficiency therefore accounts for both deep percolation and runoff losses. If e_a is not 100 percent, it must be ap-

TABLE 6-14 Intake and roughness factors for estimating potential runoff.

Intake family	Intake factor, r_i (dimensionless)	Manning coefficient	Roughness factor, r_n (dimensionless)
0.3	0.90	0.10	0.80
0.5	0.80	0.15	0.75
1.0	0.70	0.20	0.70
1.5	0.65	0.25	0.65
2.0	0.60		
3.0	0.50		
4.0	0.40		

plied as a multiplier of e_d in both equations. Application of Eq. (6-72) assumes that reduction of inflow stream size to Q_{ue} will not result in a significant change in recession lag time.

Example Problem 6-8

Rework Example Problem 6-7 with the addition of end blocks. Compute the reduced flow rate with end blocks and no length extension and, alternatively, the length extension with end blocks.

Solution Reduced flow rate:
From Table 6-14,

$$r_i = 0.80$$

$$r_n = 0.75$$

$$Q_{ue} = \frac{0.00184}{1 + 0.80(0.75)\left[1 - \frac{70}{100}\right]} = 0.00156 \text{ m}^2/\text{s}$$

Length extension:
The allowable length extension is the lesser of the results from applying Eqs. (6-70) and (6-71). Using Eq. (6-70),

$$L_e = \frac{100}{1000(0.001)} = 100 \text{ m}$$

Applying Eq. (6-71),

$$L_e = \left[1 - \frac{70}{100}\right](0.80)(0.75)(250)$$

$$L_e = 45 \text{ m}$$

The extended length is thus equal to

$$L = L + L_e = 250 + 45 = 295 \text{ m}$$

6.6 Computer Program

The sample computer program which follows can be applied in computer aided design (CAD) of furrow irrigation systems. The program solves the equations indicated in this chapter for furrow system design and is written in BASIC. The following input data are required to execute the program

(a) Intake family constants a, b, c, f, and g
(b) Furrow length, m
(c) Furrow slope, m/m
(d) Furrow spacing, m
(e) Manning's roughness coefficient
(f) Net irrigation depth, mm
(g) Inflow rate, L/s

The program allows for iteration in length, slope, net depth of irrigation, or inflow rate and produces output for each iteration. The program yields the following output

(a) Advance time, min
(b) Adjusted wetted perimeter, m
(c) Net infiltration time, min
(d) Design cut-off time, min
(e) Gross application depth, mm
(f) Average infiltration time, min
(g) Average infiltration depth, mm
(h) Surface runoff depth, mm
(i) Deep percolation depth, mm
(j) Distribution pattern efficiency, percent

There are a number of noteworthy features demonstrated in the program. They will be described in their order of appearance in the program.

(a) Two default data sets are built into the program, one for the intake family constants and the other for the problem physical description. There are two major advantages to default data sets. First, in program development or modification, it is not necessary to re-enter input data for each program run. Second, the user can run the program with the default data set to rapidly become more familiar with the program and to check that the program is running as expected.
(b) The units associated with each required input variable are indicated when input is requested.
(c) REM (remark) statements are used throughout the program to indicate the action taking place in various program segments.
(d) Iterations for length, slope, net depth, and flow rate are handled in independent subroutines.
(e) Program output is first directed to the screen. The user then has the option of directing the output to a printer if it is satisfactory. This greatly facilitates running through a number of different "what if" trails and saving only those which are feasible or noteworthy.
(f) The input data are repeated in the output listing. This allows for convenient verification of the physical setup for which the output is given. Examples of the repeat of input data are shown in the output listing which follows the program listing. The output indicated is for the default data set.

The program indicated does not have physical constraints; that is, the program can produce output which is not physically possible. For example, a negative value for the depth of runoff is not physically possible, but instead indicates that the length of the furrow was too long or inflow rate too small for the water to reach the end of the furrow given the intake characteristics of the soil. Invalid results such as this are obvious upon inspection and support the concept of outputting results to the screen for rapid inspection before producing a permanent copy.

The form and understandability of the output are important both to the design engineer and to the client who is paying for the designer's time. It is therefore worth the effort to carefully organize the output.

```
10 REM ***********************************************************************
20 REM
30 REM      *****  INPUT INTAKE FAMILY CONSTANTS   *****
40 REM
50 CLS:PRINT:PRINT:PRINT
60 INPUT "DO YOU WISH TO LOAD INTAKE FAMILY DEFAULT DATA FILE (Y/N)"; FILE1$
70 IF FILE1$ = "Y" THEN GOSUB 1170
80 IF FILE1$ = "Y" THEN 180
90 REM INPUT INTAKE FAMILY CONSTANTS
100 INPUT "a = "; A
110 INPUT "b = "; B
120 INPUT "c = "; C
130 INPUT "f = "; F
140 INPUT "g = "; G
150 REM
160 REM **********************************************************************
170 REM
180 REM      *****  INPUT PHYSICAL DESCRIPTION   *****
190 REM
200 PRINT:PRINT
210 INPUT "DO YOU WISH TO LOAD PHYSICAL SYSTEM DEFAULT DATA FILE (Y/N)"; FILE2$
220 IF FILE2$ = "Y" THEN GOSUB 1280
230 IF FILE2$ = "Y" THEN 310
240 PRINT: INPUT "FURROW LENGTH (m) = "; L
250 PRINT: INPUT "FURROW SLOPE (m/m) = "; S
260 PRINT: INPUT "FURROW SPACING (m) = "; W
270 PRINT: INPUT "ROUGHNESS COEFFICIENT = "; N
280 PRINT: INPUT "NET IRRIGATION DEPTH (mm) = "; IN
290 PRINT: INPUT "INFLOW RATE (l/s) = "; Q
300 CLS:PRINT:PRINT
310 PRINT: PRINT: INPUT "DO YOU WISH TO MAKE ITERATIVE SOLUTION FOR CHANGES IN L
ENGTH, SLOPE, NET        IRRIGATON DEPTH, OR INFLOW RATE VARIABLES (Y/N)"; CHK1$

320 IF CHK1$ = "N" THEN 430
330 PRINT: PRINT "INPUT VARIABLE CODE: ": PRINT
340 PRINT "     a) L = LENGTH (m) "
350 PRINT "     b) S = SLOPE (m/m) "
360 PRINT "     c) IN = NET IRRIGATION DEPTH (mm) "
370 PRINT "     d) Q = INFLOW RATE (l/s) "
380 PRINT: INPUT "ENTER VARIABLE CODE"; CODE$
390 IF CODE$ = "L" THEN GOSUB 1360
400 IF CODE$ = "S" THEN GOSUB 1530
410 IF CODE$ = "IN" THEN GOSUB 1700
420 IF CODE$ = "Q" THEN GOSUB 1870
430 CLS:PRINT:PRINT
440 PRINT "FURROW LENGTH (m) = " L
450 PRINT "FURROW SLOPE (m/m) = " S
460 PRINT "FURROW SPACING (m) = " W
470 PRINT "ROUGHNESS COEFFICIENT = " N
480 PRINT "NET IRRIGATION DEPTH (mm) = " IN
490 PRINT "INFLOW RATE (l/s) = " Q
500 PRINT
510 REM
520 REM **********************************************************************
530 REM
540 REM      *****  CALCULATE SYSTEM VARIABLES   *****
550 REM
560 REM      *****  CALCULATE ADVANCE TIME   *****
570 REM
580 BETA = (G * L)/(Q * S^.5)
```

```
590 PRINT "BETA = " BETA
600 TT = (L/F) * EXP(BETA)
610 PRINT "ADVANCE TIME (min) = " TT
620 REM
630 REM      *****  CALCULATE ADJUSTED WETTED PERIMETER  *****
640 REM
650 P = .265 * ((Q * N)/S^.5)^.425 + .227
660 PRINT "ADJUSTED WETTED PERIMITER (m) = " P
670 REM
680 REM      *****  CALCULATE NET INFILTRATION TIME  *****
690 REM
700 TN = ((IN * (W/P) - C)/A)^(1/B)
710 PRINT "NET INFILTRATION TIME (min) = " TN
720 REM
730 REM      *****  CALCULATE DESIGN CUT-OFF TIME  *****
740 REM
750 TCO = TT + TN
760 PRINT "DESIGN CUT-OFF TIME (min) = " TCO
770 REM
780 REM      *****  CALCULATE GROSS APPLICATION DEPTH  *****
790 REM
800 IG = (60 * Q * TCO)/(W * L)
810 PRINT "GROSS APPLICATION DEPTH (mm) = " IG
820 REM
830 REM      *****  CALCULATE AVERAGE INFILTRATION TIME  *****
840 REM
850 TAVG = TCO - (.0929/(F * L * (.305 * BETA/L)^2!)) * ((BETA - 1) * EXP(BETA)
+ 1)
860 PRINT "AVERAGE INFILTRATION TIME (min) = " TAVG
870 REM
880 REM      *****  CALCULATE AVERAGE INFILTRATION DEPTH  *****
890 REM
900 IAVG = (A * TAVG^B + C) * (P/W)
910 PRINT "AVERAGE INFILTRATION DEPTH (mm) = " IAVG
920 REM
930 REM      *****  CALCULATE SURFACE RUNOFF DEPTH  *****
940 REM
950 DRO = IG - IAVG
960 PRINT "SURFACE RUNOFF DEPTH (mm) = " DRO
970 REM
980 REM      *****  CALCULATE DEEP PERCOLATION DEPTH  *****
990 REM
1000 DDP = IAVG - IN
1010 PRINT "DEEP PERCOLATION DEPTH (mm) = " DDP
1020 REM
1030 REM      *****  CALCULATE DISTRIBUTION PATTERN EFFICIENCY  *****
1040 REM
1050 ED = (IN/IG) * 100
1060 PRINT "DISTRIBUTION PATTERN EFFICIENCY (percent) = " ED
1070 PRINT: PRINT: INPUT "DO YOU WISH TO LIST RESULTS ON PRINTER (Y/N)"; PRNTCHK
$
1080 IF PRNTCHK$ = "Y" THEN GOSUB 2040
1090 IF ITCHK = 1 THEN 1110
1100 END
1110 GOTO 390
1120 REM
1130 REM ******************************************************************
1140 REM
1150 REM     *****  DEFAULT TABLE FOR INTAKE FAMILY CONSTANTS  *****
1160 REM
1170 A = 1.56
1180 B = .773
1190 C = 7!
1200 F = 8.5
1210 G = .0003535
1220 RETURN
1230 REM
```

```
1240 REM *******************************************************************
1250 REM
1260 REM      *****  DEFAULT TABLE FOR PHYSICAL DESCRIPTION  *****
1270 REM
1280 L = 100
1290 S = .004
1300 W = .65
1310 N = .04
1320 IN = 35
1330 Q = .4
1340 RETURN
1350 REM
1360 REM *******************************************************************
1370 REM       *****  INTERATIONS FOR LENGTH  *****
1380 REM *******************************************************************
1390 REM
1400 IF LCHK = 1 THEN 1490
1410 CLS: PRINT: PRINT
1420 INPUT "ENTER STARTING LENGTH (m)"; L1: PRINT
1430 INPUT "ENTER ENDING LENGTH (m)"; L2: PRINT
1440 INPUT "ENTER NUMBER OF INCREMENTS REQUIRED"; L3
1450 INCL = (L2 - L1)/L3
1460 ITCHK = 1
1470 LCHK = 1
1480 L = L1 - INCL
1490 L = L + INCL
1500 IF L => L2 THEN ITCHK = 0
1510 RETURN
1520 REM
1530 REM *******************************************************************
1540 REM       *****  ITERATIONS FOR SLOPE  *****
1550 REM *******************************************************************
1560 REM
1570 IF SCHK = 1 THEN 1660
1580 CLS: PRINT: PRINT
1590 INPUT "ENTER STARTING SLOPE (m/m) "; S1: PRINT
1600 INPUT "ENTER ENDING SLOPE (m/m) "; S2: PRINT
1610 INPUT "ENTER NUMBER OF INCREMENTS REQUIRED"; S3
1620 INCS = (S2 - S1)/S3
1630 ITCHK = 1
1640 SCHK = 1
1650 S = S1 - INCS
1660 S = S + INCS
1670 IF S => S2 THEN ITCHK = 0
1680 RETURN
1690 REM
1700 REM *******************************************************************
1710 REM       *****  ITERATIONS FOR NET INFILTRATION DEPTH  *****
1720 REM *******************************************************************
1730 REM
1740 IF INCHK = 1 THEN 1830
1750 CLS: PRINT: PRINT
1760 INPUT "ENTER STARTING NET IRRIGATION DEPTH (mm) "; IN1: PRINT
1770 INPUT "ENTER ENDING NET IRRIGATION DEPTH (mm) "; IN2: PRINT
1780 INPUT "ENTER NUMBER OF INCREMENTS REQUIRED"; IN3
1790 INCIN = (IN2 - IN1)/IN3
1800 ITCHK = 1
1810 INCHK = 1
1820 IN = IN1 - INCIN
1830 IN = IN + INCIN
1840 IF IN => IN2 THEN ITCHK = 0
1850 RETURN
1860 REM
1870 REM *******************************************************************
1880 REM       *****  ITERATIONS FOR FLOW RATE  *****
1890 REM *******************************************************************
1900 REM
```

```
1910 IF QCHK = 1 THEN 2000
1920 CLS: PRINT: PRINT
1930 INPUT "ENTER STARTING INFLOW RATE (l/s) "; Q1: PRINT
1940 INPUT "ENTER ENDING INFLOW RATE (l/s) "; Q2: PRINT
1950 INPUT "ENTER NUMBER OF INCREMENTS REQUIRED"; Q3
1960 INCQ = (Q2 - Q1 )/Q3
1970 ITCHK = 1
1980 QCHK = 1
1990 Q = Q1 - INCQ
2000 Q = Q + INCQ
2010 IF Q => Q2 THEN ITCHK = 0
2020 RETURN
2030 REM
2040 REM ******************************************************************
2050 REM
2060 REM      *****  LIST RESULTS ON PRINTER  *****
2070 REM
2075 LPRINT: LPRINT: LPRINT: LPRINT: LPRINT: LPRINT: LPRINT: LPRINT: LPRINT
2078 LPRINT CHR$(27)"E"
2080 LPRINT "*****************************************************************"
2090 LPRINT "                INTAKE FAMILY CONSTANTS": LPRINT
2100 LPRINT "     a = "; A
2110 LPRINT "     b = "; B
2120 LPRINT "     c = "; C
2130 LPRINT "     f = "; F
2140 LPRINT "     g = "; G
2150 LPRINT
2160 LPRINT "*****************************************************************"
2170 LPRINT "                  PHYSICAL DESCRIPTION": LPRINT
2180 LPRINT "     FURROW LENGTH (m) = " L
2190 LPRINT "     FURROW SLOPE (m/m) = " S
2200 LPRINT "     FURROW SPACING (m) = " W
2210 LPRINT "     ROUGHNESS COEFFICIENT = " N
2220 LPRINT "     NET IRRIGATION DEPTH (mm) = " IN
2230 LPRINT "     INFLOW RATE (l/s) = " Q
2240 LPRINT
2250 LPRINT "*****************************************************************"
2260 LPRINT "                     RESULTS": LPRINT
2270 LPRINT "     ADVANCE TIME (min) = " TT
2280 LPRINT "     ADJUSTED WETTED PERIMITER (m) = " P
2290 LPRINT "     NET INFILTRATION TIME (min) = " TN
2300 LPRINT "     DESIGN CUT-OFF TIME (min) = " TCO
2310 LPRINT "     GROSS APPLICATION DEPTH (mm) = " IG
2320 LPRINT "     AVERAGE INFILTRATION TIME (min) = " TAVG
2330 LPRINT "     AVERAGE INFILTRATION DEPTH (mm) = " IAVG
2340 LPRINT "     SURFACE RUNOFF DEPTH (mm) = " DRO
2350 LPRINT "     DEEP PERCOLATION DEPTH (mm) = " DDP
2360 LPRINT "     DISTRIBUTION PATTERN EFFICIENCY (percent) = " ED
2370 LPRINT
2380 LPRINT "*****************************************************************"
2385 LPRINT CHR$(12)
2387 LPRINT CHR$(27)"N"CHR$(8)
2390 RETURN
```

Output

```
*****************************************************************
                INTAKE FAMILY CONSTANTS

     a =  1.56
     b =  .773
     c =  7
     f =  8.5
     g =  .0003535
```

```
*******************************************************************
                    PHYSICAL DESCRIPTION

     FURROW LENGTH (m) =  100
     FURROW SLOPE (m/m) =  .004
     FURROW SPACING (m) =  .65
     ROUGHNESS COEFFICIENT =  .04
     NET IRRIGATION DEPTH (mm) =  35
     INFLOW RATE (l/s) =  .4

*******************************************************************
                         RESULTS

     ADVANCE TIME (min) =  47.58109
     ADJUSTED WETTED PERIMITER (m) =  .3747608
     NET INFILTRATION TIME (min) =  97.32379
     DESIGN CUT-OFF TIME (min) =  144.9049
     GROSS APPLICATION DEPTH (mm) =  53.50334
     AVERAGE INFILTRATION TIME (min) =  129.2181
     AVERAGE INFILTRATION DEPTH (mm) =  42.58534
     SURFACE RUNOFF DEPTH (mm) =  10.918
     DEEP PERCOLATION DEPTH (mm) =  7.585335
     DISTRIBUTION PATTERN EFFICIENCY (percent) =  65.41648

*******************************************************************
```

6.7 Computer Problems

1. Execute the furrow design program for the default data set and check that the results agree. Run the program for the following iterations of length holding all other factors constant and observe the variation in results: L = 75 m, 100 m, 125 m, 150 m. Run the program for other input data configurations and print out those results which are the most promising in terms of efficient design.
2. Add a subroutine to the furrow design program to write the output to a disk file for storage and later retrieval. Such output could be used in conjunction with a spreadsheet program to graph the results of the design program trials.
3. Modify the furrow design program to handle a furrow system with cutback.

REFERENCES

ALLEN, R. G., C. E. BROCKWAY, and J. R. BUSCH, "Planning Optimal Irrigation Distribution and Application Systems: Teton Flood Damaged Lands," Research Technical Completion Report, Idaho Water Resources Research Institute, University of Idaho, Moscow, Idaho, 1978.

BOOHER, L. J., *Surface Irrigation*. Rome: Food and Agriculture Organization Agricultural Development Paper No. 95, 1974.

CRIDDLE, W. D., S. DAVIS, C. H. PAIR, and D. C. SHOCKLEY, "Methods for Evaluating Irrigation Systems," U.S. Department of Agriculture, Soil Conservation Service, Agricultural Handbook no. 82, 1956.

JENSEN, M. E., ed. *Design and Operation of Farm Irrigation Systems*. St. Joseph, Michigan: American Society of Agricultural Engineers, Monograph Number 3, 1980.

KATOPODES, N. D. and T. STRELKOFF, "Dimensionless Solutions of Border-Irrigation Advance," *Journal of the Irrigation and Drainage Division,* American Society of Civil Engineers. vol. 103, 1977, pp. 401–417.

PAINTER and CARRAN, "What is Irrigation Efficiency?" *Soil and Water (New Zealand),* vol. 14, no. 5, 1978, pp. 15–17, 22.

STRELKOFF, T. and A. J. CLEMMENS, "Current Status of Irrigation Modelling," Water Today and Tomorrow, Proceedings of the American Society of Civil Engineers Irrigation and Drainage Division Specialty Conference, 1984, pp. 93–103.

United States Department of Agriculture, Soil Conservation Service, "Furrow Irrigation," section 15, chapter 5, National Engineering Handbook, 1983.

United States Department of Agriculture, Soil Conservation Service, "Border Irrigation," section 15, chapter 4, National Engineering Handbook, 1974.

WALKER, W. R., "Optimizing Surface Irrigation Performance," Water Today and Tomorrow, Proceedings of the American Society of Civil Engineers Irrigation and Drainage Division Specialty Conference, 1984, pp. 104–111.

WALKER, W. R. and G. V. SKOGERBOE, *Surface Irrigation*. Englewood Cliffs, NJ: Prentice-Hall, 1987.

WARRICK, A. W. and D. R. NIELSEN, "Spatial Variability of Soil Physical Properties in the Field," in *Application of Soil Physics,* D. Hillel, ed. New York: Academic Press, 1980, pp. 300–344.

PROBLEMS

6-1. Ring infiltrometer tests on a soil yield the following results:

$$t = 100 \text{ min} \qquad i = 33 \text{ mm}$$

$$t = 700 \text{ min} \qquad i = 107 \text{ mm}$$

What SCS intake family does this soil correspond to?

6-2. Water is ponded in a level basin for a total of 5 hours. The soil corresponds to a 0.20 intake family using the SCS method. Compute the depth of infiltration in mm.

6-3. Compute the time required in minutes to infiltrate 70 mm depth of water for a soil corresponding to the SCS 0.45 intake family.

6-4. Results of infiltration tests on a soil indicate it has an α factor of 0.85 in the Kostiakov equation and a c factor of 1.5 for depth in mm and time in minutes. The field is to be furrow irrigated. Net time of irrigation is 480 minutes and time for water to advance to the end of the furrow is 96 minutes. Compute depth of water lost as deep percolation in mm.

6-5. An irrigation system is to be supplied by a groundwater well. The well discharges into a field ditch with a Parshall flume at the end of it. After the flume, the water enters the head ditch from which it is applied to the field using siphon tubes. Over a period of

6 hours, the following average measurements are observed for well discharge, Parshall flume, and total flow from siphon tubes:

$$\text{Well discharge} = 8.5 \text{ L/s}$$

$$\text{Parshall flume} = 6.9 \text{ L/s}$$

$$\sum \text{Siphon tubes} = 6.4 \text{ L/s}$$

State which efficiencies are applicable and compute their values.

6-6. Water is applied through a cut-throat flume to a level basin system which is 0.3 ha in area. The water is applied for 120 minutes with flow measured through the flume at 11.8 L/s. The root zone has a capacity to store an equivalent depth of 26 mm of water. State which efficiencies are applicable and compute their values.

6-7. You are required to estimate the maximum furrow length in an arid area for a loam soil for a system which has a gross irrigation requirement of 15 cm per irrigation. The flow rate available per furrow is 2.2 L/s and this rate is not erosive. The fields are to be laid out on gently sloping ground ($S = 0.1$ percent) and the slope may be changed at negligible cost. What is your recommendation for length in m, field slope in percent, and furrow inflow rate in L/s?

6-8. A furrow system is to be designed for a field length of 125 m and furrow spacing of 0.60 m. The net time of irrigation is 391 min and the time for water to advance to the end of the furrow is 115 min. The soil corresponds to an SCS family of 0.6 and the adjusted wetted perimeter equals 0.398 m. If the furrow inflow rate is 0.4 L/s, compute the time to cut-off, distribution pattern efficiency, and field slope.

6-9. The following information is given for a gradient furrow system:

Furrow intake family = 0.45	Furrow length = 320 m
Slope = 0.005 m/m	Furrow spacing = 1.1 m
Roughness coefficient = 0.04	Net water requirement = 85 mm
Inflow rate to furrow = 0.65 L/s	

Compute the gross depth of application for this system in mm to the nearest integer.

6-10. A field made up of a soil of the SCS 0.30 intake family is laid out for furrow irrigation with 0.40 m between rows. The field slope down the furrow is 0.008 m/m, the net irrigation requirement is 75 mm of water, and the net time of irrigation is 336 min. Compute the volumetric inflow rate for each furrow in L/s.

6-11. A furrow system is to be laid out with a furrow slope of 0.005 m/m, flow rate into an individual furrow of 0.4 L/s, and length of the furrow equal to 210 m. The integrated average amount of runoff over 3 hours at the end of a furrow is measured with a Parshall flume at 0.06 L/s. Compute the equivalent depth of water infiltrated in mm over the surface area of the field at the end of a 3-hour irrigation. (Conversion factor: $1.0 \text{ L} = 0.001 \text{ m}^3$).

6-12. A furrow system is to be set up on a soil of the 0.25 Soil Conservation Service intake family. The design net irrigation requirement for the crop is 80 mm and the target distribution pattern efficiency is 45 percent. Furrow spacing is 0.80 m, furrow slope is 0.005 m/m, and the unit inflow rate is 0.8 L/s. Compute the roughness coefficient if the furrow length is 325 m and the time to cut-off is 963 min.

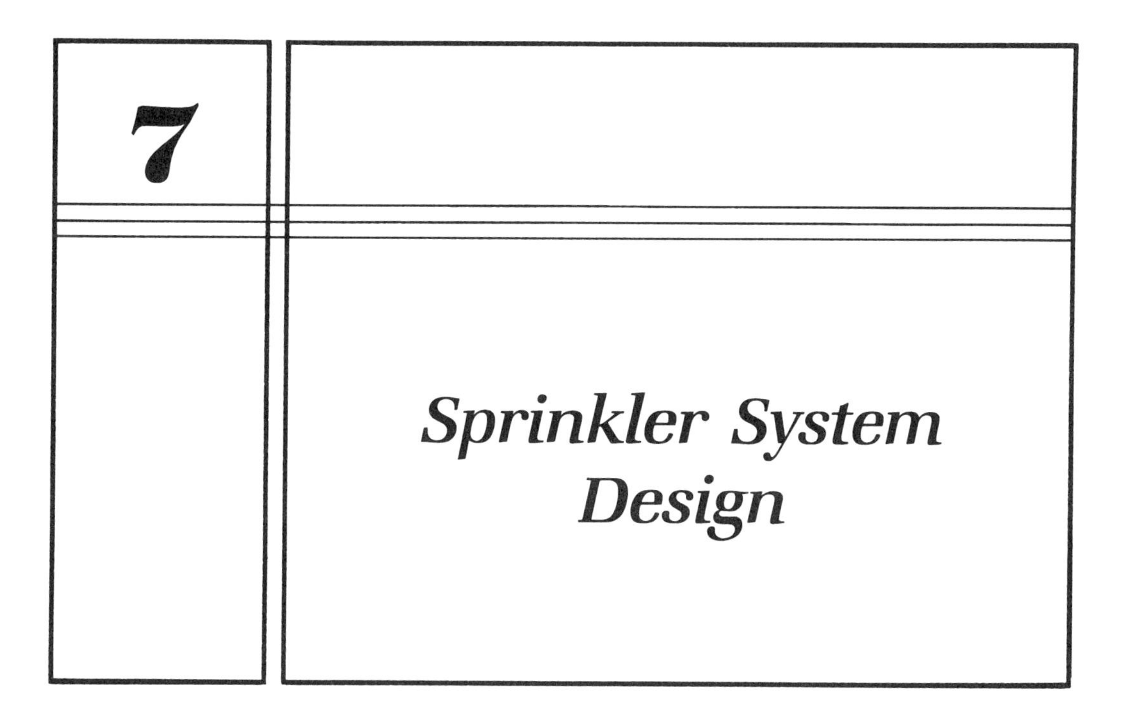

7 Sprinkler System Design

7.1 Introduction

Applications

Sprinkler systems are used to irrigate crops to increase crop production as are all irrigation systems. However, sprinkler systems also have other uses, some of them unique to this type of water application device. Sprinkler systems may be used for disposal of municipal, industrial, and agricultural wastes (Pettygrove and Asano, 1986). Sprinkler systems can also be used to lightly wet the soil surface after seeding to improve germination. Fertilizers may be applied to a crop through a sprinkler system if the uniformity of the application system is sufficient. Sprinkler systems may also be used for frost protection under adverse weather conditions in which bud freezing may bring serious economic damage to a crop. When such systems are used for frost protection, the bud is covered with a thin layer of frozen water. The energy released when the ice is formed protects the plant from air temperatures below freezing. The design of the sprinkler system must be varied to account for these special applications.

Types of Systems

Various types of sprinkler systems have been developed in response to economic and labor conditions, topographic conditions, special water application needs, and the availability of water and land resources. Sustained development of sprinkler systems began in the Pacific Northwest region of the United States where land and water resources were available for irrigation along with low cost hydroelectric power. The hydroelectric power rates brought aluminum manufacturers to the region due to the high energy requirements of aluminum fabrication (Cuenca et al., 1981). One convenient use of lightweight aluminum tubing was in irrigation sprinkler lines. Sprinkler irrigation systems therefore evolved out of the need to develop a market for fabricated aluminum while at the same time offering the possibility to improve irrigation management in the arid, high elevation intermountain plateau of the Pacific Northwest.

Several types of sprinkler irrigation systems were introduced in the first chapter. These included hand-move, side-roll, center pivot, and big gun. The original sprinkler systems were hand-move types such as that depicted in Fig. 7-1. These systems are placed in the field with the pipelines which deliver water to the sprinkler nozzles running parallel to the cropped rows. The sprinkler nozzles rotate such that each nozzle wets in a circular pattern. The uniformity of application comes from the overlapping of these circular patterns both along the sprinkler line and between successive positions of the lines. The topic of uniformity is discussed in Section 7.2.

A solid-set system is similar in principal to the type of system depicted in Fig. 7-1 except that the pipelines are left in a fixed position for the entire growing season. Because there are no lines to be moved, the entire field must be covered by application patterns of the overlapping sprinkler nozzles.

A side-roll system is shown in Fig. 7-2. It is made up of a series of aluminum wheels which serve to elevate the sprinkler line. This type of system essentially

Figure 7-1 Hand-moved sprinkler system used for vegetable crop production in the Willamette Valley, Oregon.

Figure 7-2 Side-roll sprinkler system used for mint production in the Willamette Valley, Oregon.

evolved out of a labor shortage to move the hand-move lines shown in Fig. 7-1. An entire length of side-roll line is moved by a small drive motor installed in the center of the line as shown in Fig. 7-3. After the line has finished an irrigation in one position in the field, the motor is turned on to move it into position for the next irrigation. The elevated distribution line serves as an axle about which the wheels rotate. A self-leveling nozzle, shown in Fig. 7-4, is used to keep the sprinkler in position for proper operation regardless of the number of rotations of the drive wheels.

A further development in labor-saving sprinkler systems came with the center pivot system of the type shown in Fig. 7-5. The water source for this system, whether a well or buried pipeline, is located at the center of the field and delivers water to the pivot arm. The pivot arm is rotated by hydraulic or electric drive motors

Figure 7-3 Drive motor apparatus for side-roll sprinkler line.

Figure 7-4 Self-leveling sprinkler nozzle for side-roll system. Weighted block at bottom of sprinkler maintains the nozzle in an upright position.

Figure 7-5 Center pivot system with low pressure nozzles on drop tubes used to irrigate cereals in eastern Oregon. Wheels used to change position of pivot arm are shown at base of tower.

connected to the wheels at the intermittant towers. The rotation of the pivot arm results in the pattern of circular irrigated areas such as that shown in Fig. 7-6. Modifications to the arid environment produced by center pivot systems are so predominant that their effects can be easily seen from high altitude aircraft and even in satellite photos.

Figure 7-6 Aerial view of circular patterns irrigated by center pivot systems. Two-lane road shown in right foreground of photo. One center pivot system typically irrigates from 55 to 65 ha in the western United States.

Specially designed systems are available to irrigate the corners of a center pivot system so this land is not lost from the irrigated acreage. However, the situation depicted in Fig. 7-6 in which the corners are not irrigated is typical. To avoid this loss of irrigated acreage, the linear move system was developed. One such system is shown in Fig. 7-7. This system is similar in design to the center pivot system, but moves linearly down the field thereby irrigating all available area in a square or rectangular field. A linear move system requires that the source of water be available all along one edge of the field. This is accomplished either by an open canal, such as shown in Fig. 7-7, or by using a buried mainline which is coupled to the linear move

Figure 7-7 Linear move system with low pressure nozzles on drop tubes. Water source is open channel in foreground. Water is pressurized by pump within triangular frame at end of system.

at different locations down the field by a flexible hose. The water supply system is more complex for a linear move system than a center pivot because the distribution system delivers water along the entire length of one side of the field instead of only at the center.

Another type of system which has been used in various countries is the big gun system shown in Fig. 7-8. This system has a single large diameter nozzle which sprays large volumes of water in a circular pattern. The nozzle is connected to a flexible hose which is pulled along a straight line so a strip of field is irrigated on one pass. The big gun is then moved to the next strip and the operation repeated until the entire field has been irrigated.

System Components

In spite of the diversity of sprinkler system designs, certain terminology and components are common to all systems. Water for all sprinkler systems must be pressurized, whether at high pressure [830 to 1035 kPa (120 to 150 psi)] for the big gun system in Fig. 7-8, moderate pressure [275 to 485 kPa (40 to 70 psi)] for the impact sprinklers in Figs. 7-1 and 7-4, or relatively low pressure [105 to 210 kPa (15 to 30 psi)] for rotating nozzles such as those indicated in Figs. 7-5 and 7-7. Therefore a pressurized water source, whether developed by the gravitational head available from an elevated water source or the output pressure of a pump, is an integral part of any sprinkler system.

A mainline between the source of pressurized water and the point at which water is delivered to the field is the next component. Generally the mainline is a buried or above-ground pipeline. Of the sprinkler systems depicted in this chapter, only the linear move in which the water may be supplied in an open channel does not require a typical mainline under pressure. For the linear move in Fig. 7-7, the pressurizing

Figure 7-8 Big gun system used to irrigate maize in southwest France. Rotating drum used to store hose is shown next to building.

device is a pump which is shown within the triangular framework at the end of the lineal arm.

In the most common systems such as hand-move, solid-set, side-roll, or big gun, what is termed a lateral line comes off of the mainline to deliver water to the sprinkler nozzles. The position of the lateral may be permanent, as in a solid set, or movable as in the hand-move and side-roll systems. The spacing between the successive positions of the lateral along the mainline is called the mainline spacing and is designated as s_m.

The distance between sprinkler nozzles along a lateral, including on center pivot and linear move lines, is termed the lateral spacing and designated as s_l. On all but center pivot systems, the lateral spacing is constant. On center pivot systems, the lateral spacing is often varied because the outer part of the pivot arm rotates at a much faster velocity than the inner part. To get a uniform rate of water application, the nozzle size, nozzle spacing, or both, are varied along the length of the pivot arm.

The spray area which is wet by each sprinkler nozzle at a particular operating pressure is designated as the wetted diameter, D_w. The wetted diameters are overlapped along the lateral to promote a more uniform distribution of water application. Figure 7-9 indicates the major components and typical layout of typical sprinkler systems other than center pivot and linear move.

Design Objectives

Concentrating on the use of sprinkler systems to increase crop production, the objective of a proper design is to replace the water used by the plant during the peak period of the growing season to avoid crop stress. The required depth of application is dependent upon the peak period evapotranspiration rate, the water holding capacity of the root system, and the management allowed depletion. These concepts were introduced in the chapter on Crop Water Requirements.

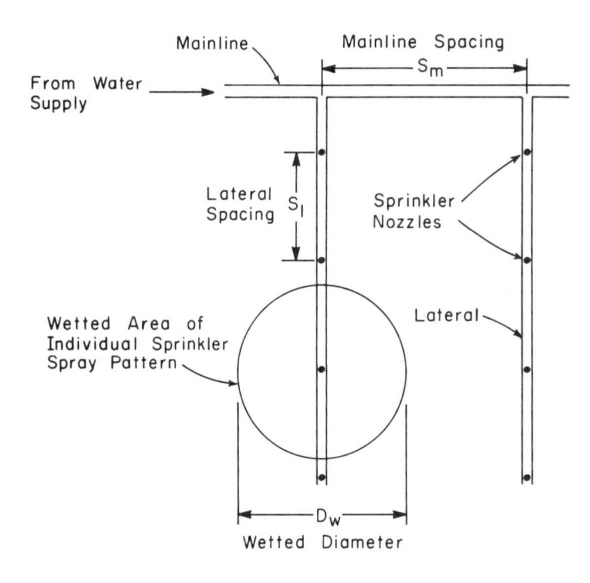

Figure 7-9 Definition sketch for components of sprinkler systems.

In addition to meeting the crop water requirement, the sprinkler system must be designed in balance with the intake rate of the soil. Sprinkler systems are normally designed to avoid all runoff from the irrigated field. The intake rate of the soil relative to sprinkler application must therefore be measured and the sprinkler application rate chosen accordingly. The sprinkler application rate will depend on the nozzle size, operating pressure, and spacing of sprinklers.

The maximum reasonable uniformity of application is required to minimize losses due to deep percolation. Deep percolation losses are accentuated by increasingly nonuniform applications of water over the field. The uniformity of application will partially depend on the velocity of the prevailing wind during sprinkler operation and sprinkler spacing. It will also depend on pressure variations along the sprinkler line. Pressure variations will be influenced by the system design criteria, topography, and pump selection.

The final design must balance the physical and biological requirements of the system with a reasonable economic cost and convenience for the grower. The required labor and availability of labor must be considered. The initial cost of equipment and installation must also be considered along with the operating cost to pressurize the system and annual maintenance costs. The experience of the grower in managing and maintaining a system which must operate within limited pressure ranges to be effective must also be accounted for.

7.2 Uniformity of Application

Pressure Effects

Uniformity of application depends on matching operating pressure with the selected sprinkler nozzle diameter, wind effects, and sprinkler spacing. The effects of operating pressure are indicated in Fig. 7-10 which gives sectional views of depth of applied water with increasing distance from the sprinkler nozzle. If the pressure is too low, the water stream is not adequately broken up and a donut-shaped application pattern results. If the pressure at the nozzle is too high, the stream is broken up into excessively small droplets and the water does not carry to the extent of the design wetted diameter. Excessive amounts of water are instead deposited in the vicinity of the nozzle.

The result of too high or too low pressure conditions, as seen in Fig. 7-10, is an inadequate depth of water being deposited in certain parts of the field. The effects of the limited amount of applied water can be visually observed in the crop growth pattern. The same conditions tend to prevail along the full length of the sprinkler lateral. Lines of reduced crop growth running parallel to the sprinkler line are therefore a possible sign of incorrect operating pressure. If the operating pressure is too high, reduced growth is observed midway between laterals because the design wetted diameter is not being attained. (This condition can also be caused by excessive spacing between adjacent positions of the lateral. Correct design criteria for spacing between laterals is discussed later in this chapter.) Reduced growth strips appear both adjacent to and between laterals if the pressure is too low.

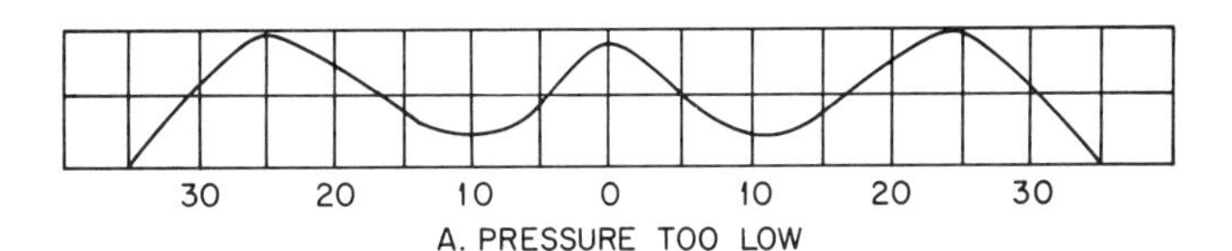

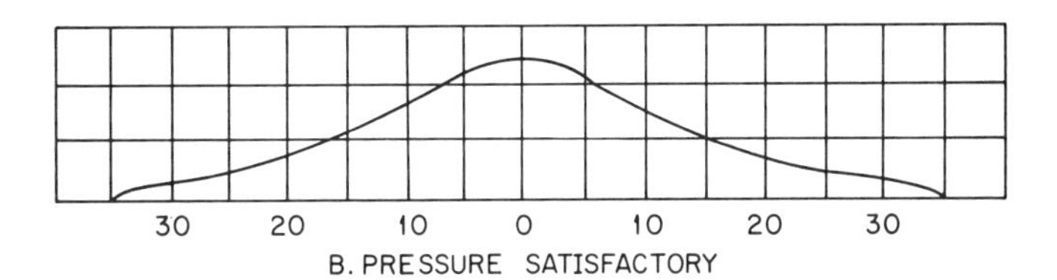

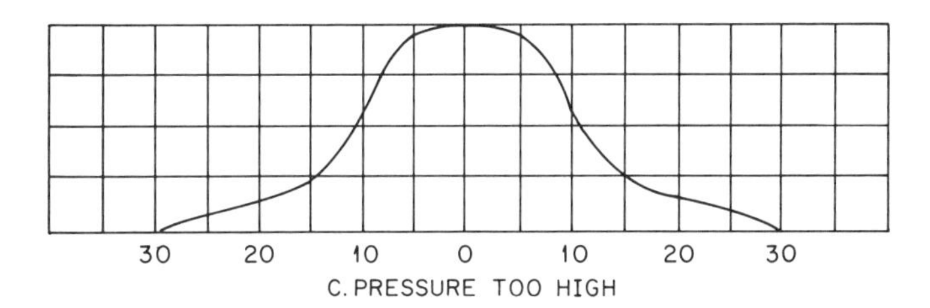

Figure 7-10 Effects of system operating pressure on sprinkler distribution pattern. Cross-sectional views of depth of applied water as function of horizontal distance from sprinkler nozzle at zero point. (Adapted from SCS, 1983.)

With proper operating pressure, a sectional view of depth of applied water indicates that an application pattern close to triangular is produced by most sprinkler nozzles. When this pattern is overlapped with similar triangular patterns from sprinklers on adjacent laterals, the fairly uniform application pattern depicted in Fig. 7-11 results. The zero position on each side of the figure indicates the position of adjacent laterals spaced at distance s_m along the mainline. It should be recognized that it is not possible to have a perfectly uniform application pattern. The design engineer has to balance the degree of uniformity with the capital cost of the system. Figure 7-12 indicates the effects of the overlap pattern of adjacent sprinklers along the lateral. In this case the zeros along the abscissa indicate the position of sprinkler nozzles with spacing s_l between nozzles along the lateral.

Wind Effects

Prevailing wind conditions can have a tremendous effect on the application pattern of a sprinkler system. Consistent high velocity winds can in fact rule out the effective use of sprinkler irrigation or limit operations to times of relatively low winds

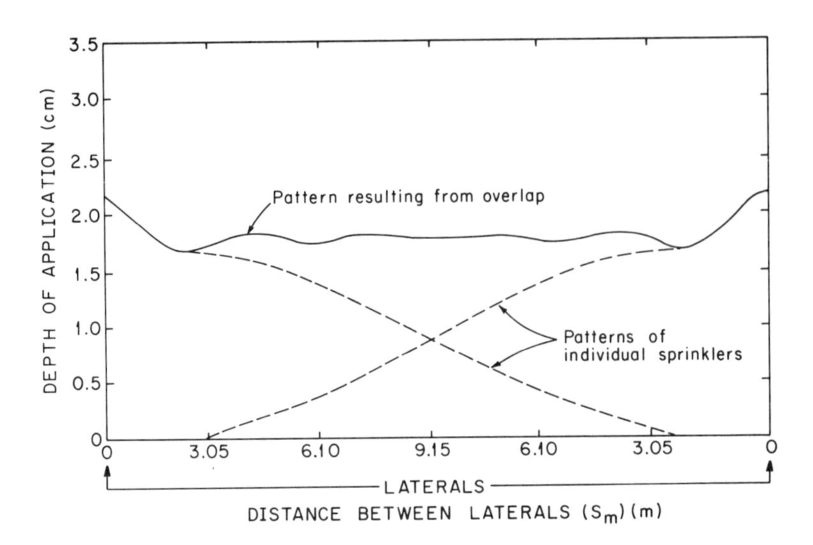

Figure 7-11 Distribution pattern of applied water resulting from overlap of sprinklers on adjacent laterals. Individual laterals at zero positions in sketch. (Adapted from SCS, 1983.)

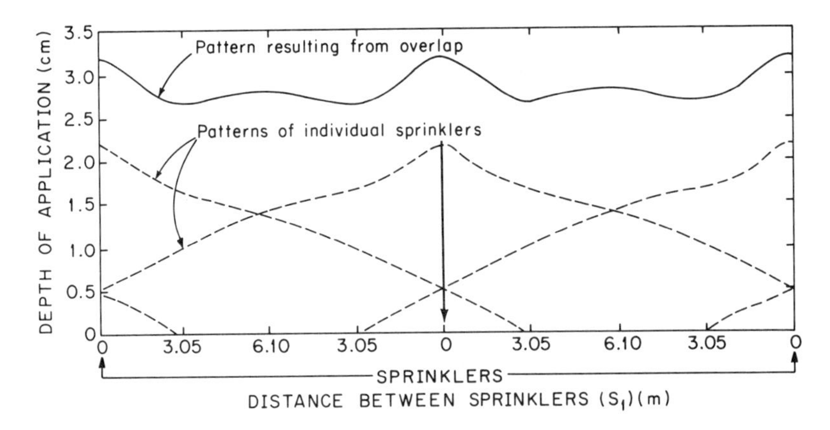

Figure 7-12 Distribution pattern of applied water resulting from overlap of sprinklers along a lateral. Individual sprinkler nozzles at zero positions in sketch. (Adapted from SCS, 1983.)

such as at night. The effects of wind on the distribution pattern of a single nozzle are indicated in Figs. 7-13 and 7-14. In these figures, the sprinkler nozzle is at coordinate position (0, 0) indicated on the sectional views of applied water. Under no or low wind conditions, the concentric pattern of depths of applied water and the triangular sectional pattern indicating proper operating pressure are evident. Under the high wind conditions shown in Fig. 7-14, the application pattern is tremendously skewed. This results in the highly nonuniform application pattern shown in the sectional views. Anticipated wind conditions are accounted for in the system design by

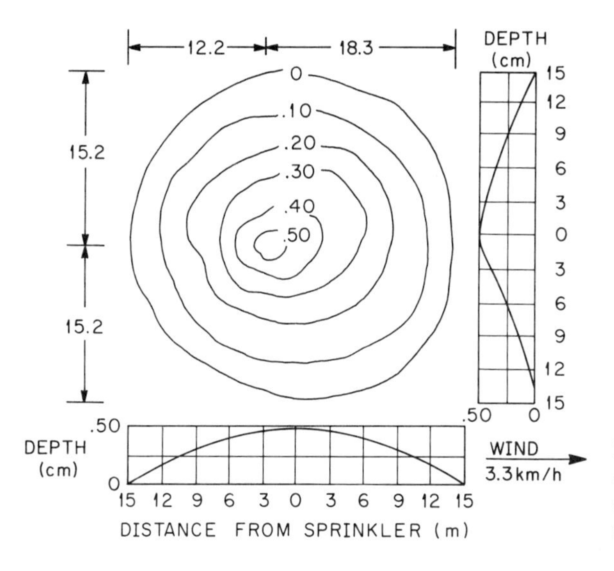

Figure 7-13 Effect of low wind conditions on sprinkler distribution pattern. Sprinkler nozzle at zero position in sectional views. (Adapted from SCS, 1983.)

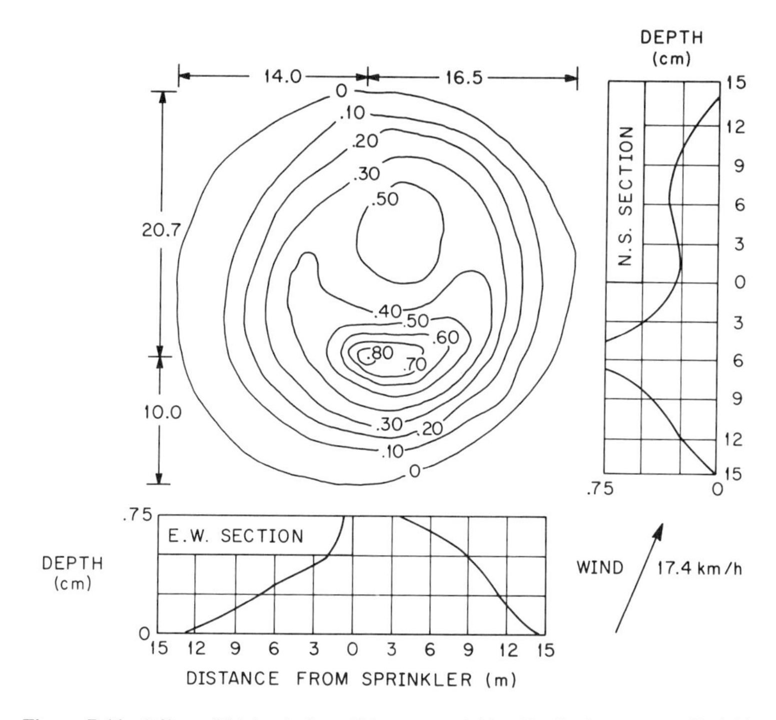

Figure 7-14 Effect of high wind conditions on sprinkler distribution pattern. Sprinkler nozzle at zero position in sectional views. (Adapted from SCS, 1983.)

decreasing the spacing between nozzles both along the lateral and between laterals with increasing wind velocities. Specific design criteria are given later in this chapter.

Uniformity Coefficients

Methods have been developed to quantify the uniformity of sprinkler application systems. These methods involve placement of catch-cans in the field. A sample of catch-can placement to evaluate uniformity of a sprinkler system with a rectangular spacing (i.e., s_m and s_l describe a rectangle) is indicated in Fig. 7-15. The depth of water caught in each can is measured at the end of the irrigation trial.

Two formulas are most commonly used to calculate the uniformity coefficient. The first is called Christiansen's uniformity coefficient and is given as

$$UC_C = 1 - \sum_{i=1}^{n} [\mathrm{abs}(x_i - \overline{x})]/(n\overline{x}) \tag{7-1}$$

where

UC_C = Christiansen's uniformity coefficient, fraction

x_i = depth caught in can i, mm

$\overline{x}$ = mean depth caught, mm

n = number of cans

A second method to quantify the uniformity is called the Hawaiian Sugar Planters Association coefficient which is given by

$$UC_H = 1 - [2/\pi]^{0.5}(s/\overline{x}) \tag{7-2}$$

where

UC_H = Hawaiian Sugar Planters Association coefficient, fraction

s = standard deviation of depth caught, mm

The standard procedure for making a uniformity test for a single sprinkler involves setting up a square grid with a minimum of 80 collectors within the wetted

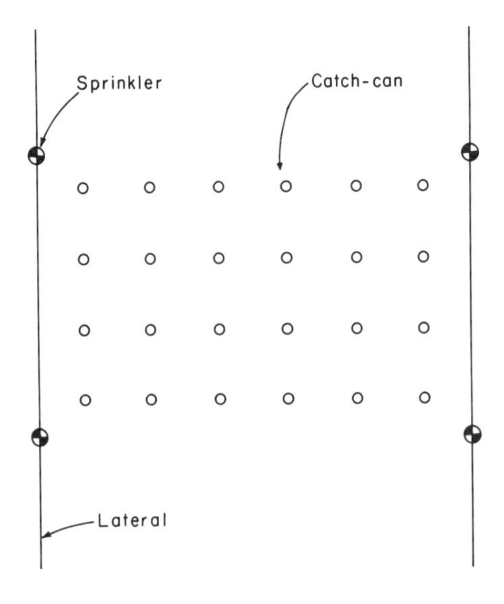

Figure 7-15 Arrangement of collectors for determination of uniformity coefficient in sprinkler system with rectangular spacing when overlapping laterals operate simultaneously.

diameter. The minimum collector diameter is 80 mm. The sprinkler to be tested is placed at the center of the grid—that is, midway between four adjacent collectors, at a height of 0.6 m above the average elevation of the tops of the four nearest collectors. The nozzle discharge pressure, pressure in the sprinkler riser, sprinkler flow rate, and speed of rotation must be accurately measured. Required meteorological data are wind speed and direction and wet and dry bulb temperatures. Recommended time of the test is one hour. Additional test details are indicated in Standard ASAE S330.1 of the American Society of Agricultural Engineers (ASAE, 1985).

The catch-can pattern shown in Fig. 7-15 should be applied for uniformity tests involving sprinkler systems at typical mainline and lateral spacing when adjacent laterals are operated simultaneously. A minimum of 24 collectors with maximum spacing between collectors of 3 m is recommended for such tests. The catch-can pattern shown in Fig. 7-16 is required to evaluate uniformity on systems in which a single lateral, with no overlapping laterals, operates at a given time. The minimum number of collectors and maximum spacing between collectors is the same as for the rectangular system. Discharge from a minimum of three sprinklers along the lateral must

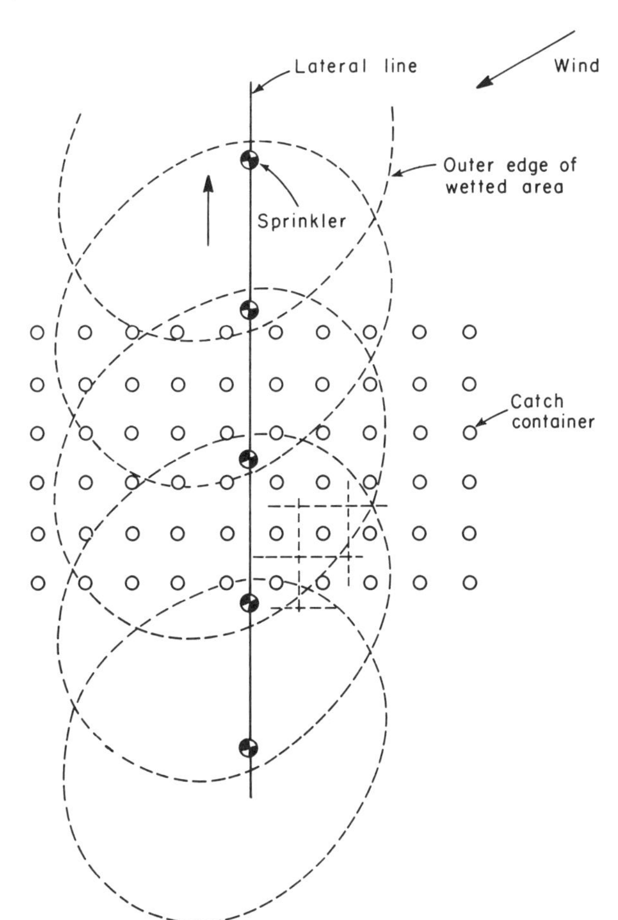

Figure 7-16 Arrangement of collectors for determination of uniformity coefficient in sprinkler system without simultaneously operating overlapping laterals. (Adapted from SCS, 1983).

be measured by the collector grid. Different test criteria, discussed later, are required for evaluation of uniformity on center pivot and linear move systems.

The water application pattern is usually assumed to have a normal statistical distribution in applications of the uniformity coefficient. The two uniformity coefficients given in Eqs. (7-1) and (7-2) are equal if the application pattern has a normal distribution. This is in fact one means of checking if the field catch-can data are normally distributed. This book will apply Christiansen's uniformity coefficient unless otherwise indicated.

7.3 Adequacy of Application

Concept of Adequacy

The net irrigation requirement at a given time is equal to the crop water requirement since the last irrigation minus the amount of precipitation which infiltrated into the soil profile—that is, effective precipitation. Figures 7-11 and 7-12 demonstrated that no sprinkler application system is perfectly uniform. The superposition of application patterns of individual sprinkler nozzles causes some parts of the field to receive greater depths of applied water than others.

The concept of adequacy addresses the question of what portion of a field received at least the net irrigation requirement. Those sections of the field which received a depth of applied water equal to or greater than the net irrigation requirement are said to be adequately irrigated. The remaining sections of the field are under-irrigated. Figures 7-11 and 7-12 indicate that some point in the field will receive the maximum amount of applied water. Any depth of applied water equal to more than the net irrigation requirement goes to deep percolation. If 25 percent of the field receives at least the net depth of irrigation, then the field is said to have an adequacy level of 25 percent. The relationship between adequacy and uniformity of application determines the amount of deep percolation for a sprinkler system.

Interrelationship between Adequacy, Uniformity, and Deep Percolation

The interrelationship between adequacy, uniformity, and deep percolation can be visualized with reference to Fig. 7-17. This figure demonstrates the portion of applied water which goes to deep percolation compared to the portion which is under-irrigated if the adequacy level is 50 percent and the uniformity coefficient is equal to 80 percent. If the net irrigation requirement is 1 cm, 91 percent of the applied water will be stored in the root zone and 9 percent will go to deep percolation for the conditions given. By definition, 50 percent of the field will be under-irrigated.

Figure 7-18 demonstrates the expected changes in the amount of deep percolation and crop stress as a function of uniformity of application for a constant adequacy level of 50 percent. For the 70 percent uniformity coefficient, the amount of deep percolation is the greatest for the three cases shown and the degree of crop stress due to under-irrigation is also maximum. At the other extreme, for a uniformity coefficient of 90 percent, the deep percolation is minimized. The degree of

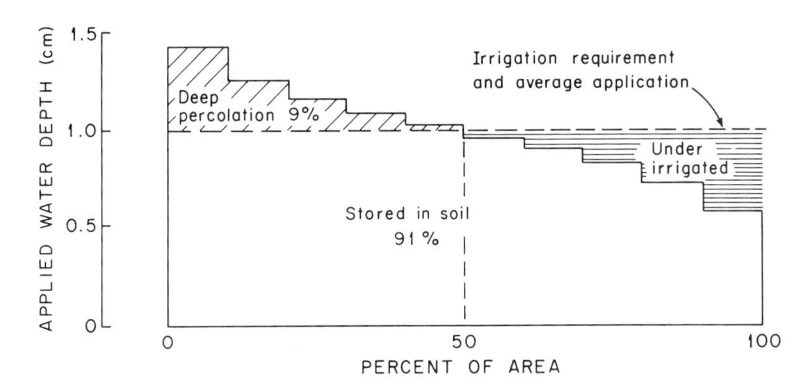

Figure 7-17 Water distribution for 80 percent uniformity and 50 percent adequacy with a mean depth of application equal to 1.0 cm. (Adapted from Shearer, 1978.)

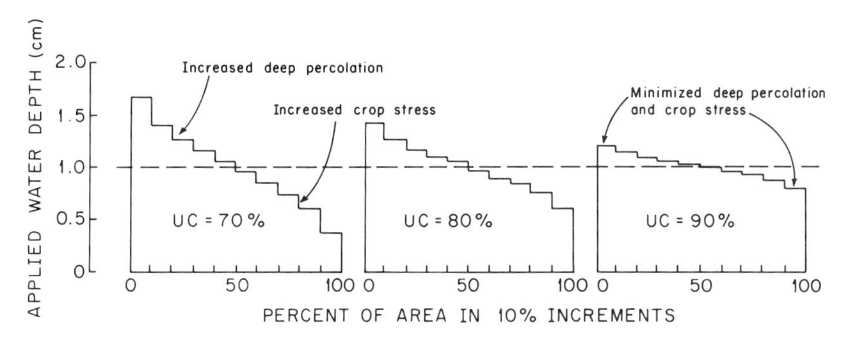

Figure 7-18 Water distribution for 50 percent adequacy and three levels of uniformity with mean depth of application equal to 1.0 cm. (Adapted from Shearer, 1978.)

crop stress is also minimized compared to the other treatments, even though the level of adequacy is equal for all treatments. At the higher uniformity levels, 50 percent of the field still remains under-irrigated, but not seriously under-irrigated.

Figure 7-19 graphically demonstrates the distribution pattern of applied water in the root zone for an adequacy level of 95 percent and a uniformity coefficient of 84 percent. For this figure, the net depth of irrigation is equivalent to 2.0 cm indicated by distance G. The amount of water applied in cross-sectional area A above the dashed line is stored in the root zone. The amount of water indicated in cross-sectional area B below the dashed line is deep percolation. Area C represents the amount of under-irrigation for the field. These areas were determined assuming the applied water pattern was normally distributed.

Table 7-1 indicates the parameters pictured graphically in Fig. 7-19. The tabulated values of the parameters were developed based on the Hawaiian Sugar Planters Association value of the uniformity coefficient but are equally valid for application with Christiansen's uniformity coefficient if the depth of applied water is normally distributed. The table indicates the relationship between the uniformity coefficient, level of adequacy, water storage coefficient, E, and distribution coefficient, H. The

TABLE 7-1 Sprinkler performance based on a normal distribution.

Unif. Coef. percent	Std. Dev. ÷ Mean	Parameter	Fraction of Area Adequately Irrigated, Percent										
			99.9	95	90	85	80	75	70	65	60	55	50
99.9	.00125	H	.996	.998	.998	.999	.999	1.000	1.000	1.000	1.000	1.000	1.000
		E	.996	.998	.998	.999	.999	.999	.999	.999	.999	.999	.999
98	.0251	H	.923	.959	.968	.974	.979	.983	.987	.990	.994	.997	1.000
		E	.923	.958	.967	.972	.976	.979	.982	.984	.986	.988	.990
96	.0501	H	.845	.917	.936	.948	.958	.966	.974	.981	.987	.994	1.000
		E	.845	.916	.933	.944	.952	.959	.964	.969	.973	.977	.980
94	.0752	H	.768	.876	.903	.922	.937	.949	.961	.971	.981	.991	1.000
		E	.768	.875	.900	.916	.928	.938	.946	.953	.959	.965	.970
92	.1003	H	.690	.835	.871	.896	.915	.932	.947	.961	.975	.987	1.000
		E	.690	.833	.867	.888	.904	.917	.928	.938	.946	.953	.960
90	.1253	H	.613	.794	.839	.870	.894	.915	.934	.952	.968	.984	1.000
		E	.613	.791	.833	.860	.880	.897	.910	.922	.932	.941	.950
88	.1504	H	.535	.753	.807	.844	.873	.899	.921	.942	.962	.981	1.000
		E	.535	.749	.800	.832	.856	.876	.892	.906	.919	.930	.940
86	.1755	H	.458	.711	.775	.818	.852	.882	.908	.932	.956	.978	1.000
		E	.458	.707	.767	.804	.832	.855	.874	.891	.905	.918	.930
84	.2005	H	.380	.670	.743	.792	.831	.865	.895	.923	.949	.975	1.000
		E	.380	.666	.733	.776	.809	.835	.856	.875	.892	.906	.920
82	.2256	H	.303	.629	.711	.766	.810	.848	.882	.913	.943	.971	1.000
		E	.303	.624	.700	.749	.785	.814	.839	.860	.878	.895	.910

80	.2507	H	.225	.588	.679	.740	.789	.831	.869	.903	.937	.968	1.000
		E	.225	.582	.667	.721	.761	.793	.821	.844	.865	.883	.900
78	.2757	H	.148	.546	.647	.714	.768	.814	.855	.894	.930	.965	1.000
		E	.148	.541	.633	.693	.737	.773	.803	.829	.851	.871	.890
76	.3008	H	.071	.505	.614	.688	.747	.797	.842	.884	.924	.969	1.000
		E	.070	.499	.600	.665	.713	.752	.785	.813	.838	.860	.880
74	.3258	H		.464	.582	.662	.726	.780	.829	.875	.917	.959	1.000
		E		.457	.567	.637	.689	.731	.767	.797	.824	.848	.869
72	.3509	H		.423	.550	.636	.704	.763	.816	.865	.911	.956	1.000
		E		.415	.533	.609	.665	.711	.749	.782	.811	.836	.859
70	.3760	H		.381	.518	.610	.683	.747	.802	.855	.905	.953	1.000
		E		.373	.500	.581	.641	.690	.731	.766	.797	.825	.849
68	.4011	H		.340	.486	.585	.662	.730	.790	.845	.899	.949	1.000
		E		.332	.467	.553	.617	.669	.713	.751	.784	.813	.839
66	.4261	H		.299	.454	.559	.641	.713	.777	.836	.892	.946	1.000
		E		.290	.433	.525	.593	.649	.695	.735	.770	.801	.829
64	.4512	H		.258	.421	.533	.620	.696	.763	.826	.886	.943	1.000
		E		.248	.400	.497	.569	.628	.677	.719	.757	.789	.819
62	.4763	H		.217	.389	.507	.599	.679	.750	.817	.879	.940	1.000
		E		.206	.367	.469	.545	.607	.659	.704	.743	.778	.809
60	.5013	H			.357	.401	.578	.662	.737	.807	.873	.937	1.000
		E			.333	.441	.521	.587	.641	.688	.730	.766	.799

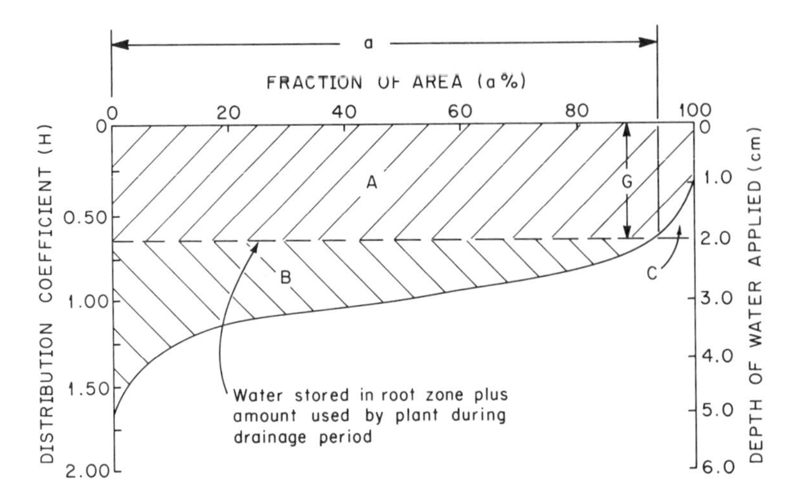

Figure 7-19 Relationship between fraction of area adequately irrigated and water stored in the root zone, A; deep percolation, B; and amount of deficit, C. Relationship shown for 84 percent uniformity, 95 percent adequacy, and net irrigation requirement of 2.0 cm. (Adapted from Hart and Reynolds, 1965.)

water storage coefficient, a fraction, is equal to one minus the deep percolation expressed as a fraction. The distribution coefficient is equal to the fraction of the mean depth of application which is equaled or exceeded over the area adequately irrigated. Table 7-1 has been developed relative to a unit depth of applied water. Application of Table 7-1 is indicated in the following example problem.

Example Problem 7-1

A sprinkler system is to be designed based on the criteria of a uniformity coefficient of 80 percent and a 75 percent adequacy level. Measurement of depth of applied water with a series of catch-cans indicates a mean depth of 2.0 cm. Compute the net depth of irrigation received over the 75 percent of the field adequately irrigated and the areal average depth of deep percolation for the field.

Solution Referring to Table 7-1 for a uniformity coefficient of 80 percent and an adequacy level of 75 percent,

$$H = 0.831 \qquad E = 0.793$$

H is equal to the fraction of the mean application depth received over the area adequately irrigated, a. Representing the depth equaled or exceeded over area a as i_{ad},

$$i_{ad} = H(\bar{x}) = 0.831(2.0 \text{ cm}) \tag{7-3}$$

$$i_{ad} = 1.66 \text{ cm}$$

The water storage coefficient is equal to 1 minus the deep percolation. Designating the fraction of deep percolation as L_d,

$$E = 1 - L_d \tag{7-4}$$

$$L_d = 1 - 0.793 = 0.207$$

Compute the depth of deep percolation, i_{dp}, for a mean depth of applied water of 2.0 cm,

$$i_{dp} = 0.207(\overline{x}) = 0.207(2.0 \text{ cm})$$
$$i_{dp} = 0.41 \text{ cm}$$

Therefore the average depth of deep percolation over the field is equivalent to 0.41 cm.

Reasonable Combinations of Uniformity and Adequacy

The design uniformity and adequacy criteria to apply for a particular sprinkler system depend on the cost of sprinkler equipment, value of the crop, and total cost of applying water to the field. This includes the cost of the volume of water applied and operation costs. The uniformity coefficient is generally increased by decreasing the mainline and lateral spacing, but this comes at the increased capital costs for more lateral lines, valves, and sprinkler nozzles. Certain crops are less susceptible to stress than others. These crops can be irrigated at a lower adequacy level without a serious economic impact to the grower. Lower value crops can also experience lower levels of adequacy without serious economic effects.

Reasonable combinations have been developed based on the interrelationships of uniformity, adequacy, the economic value of crops, and system cost. These criteria evolved from a tremendous amount of field experience in design and construction of economically successful sprinkler systems (Marvin Shearer, personal communication). The uniformity levels given are attainable and reasonable for systems designed applying the procedures given in this chapter. The criteria are given for three crop categories as follows:

Field crops:	UC_C = 80 percent	a = 75 percent
Orchards:	UC_C = 70 percent	a = 50 percent
Specialty Crops:	UC_C = 85 percent	a = 90 percent

These criteria reflect the value of the crop contrasted with the capital cost of sprinkler equipment. They also reflect the ability of the soil water profile to redistribute applied water along the gradient of total potential as discussed in the Soil Physics chapter. This is the reason the criteria for orchards, which have greater spacing between plants but plants with active and permanent root zones at depth, are lower in both categories.

7.4 Evaporation and Wind Drift

Review of Concepts

We have seen in Figs. 7-13 and 7-14 how wind velocities during irrigation can seriously affect sprinkler application patterns. If wind velocities are high, droplets from sprinkler nozzles can be blown completely off the field being irrigated before they reach the ground. In this case the water is said to be lost due to wind drift. Such losses can be fairly significant if the sprinkler nozzles are located high above the crop canopy where they are subjected to higher wind velocities than those that occur close to the canopy. The wind velocity increases logarithmically with height above the canopy as described by the log-wind law equation given in the Crop Water Requirements chapter.

It should be recognized that droplets affected by wind drift are actually only lost to the farm if they are blown completely off of the area to be irrigated. This may be an important consideration in systems such as the center pivot which typically irrigates very large areas. If the wind predominantly blows from the south on a particular farm with center pivots, the wind will probably blow the water to the northern part of the circle but not off the irrigated field when the pivot arm is in the southern part of the circle. However, when the pivot arm is in the northern portion of the circle, at least some of the droplets may be blown completely off the irrigated circle. If the grower has additional irrigated area to the north of the circle in question, this water may not be completely lost to the farm. However, it will create a very nonuniform application pattern and complicate the analysis of the correct amount of irrigation water to apply.

The question of evaporation losses is even more complex and often misinterpreted. We recognize from the Crop Water Requirement chapter that the movement of water off a cropped surface is a complex process involving the soil-plant-atmosphere continuum. Moisture moves off the leaf surface in response to the gradient of water vapor in the atmosphere. The amount of moisture which moves through the plant is in turn controlled by the total potential in the soil and the crop's perception of stress.

There is no doubt that evaporation takes place as small droplets from a sprinkler nozzle pass through the air towards the ground. Figure 7-10 demonstrated that smaller droplets which produce a greater total surface area over which evaporation can take place will be formed at higher operating pressures. The gradient to drive water vapor off the droplets increases as the relative humidity of the atmosphere decreases. Therefore evaporation will be a function of nozzle diameter, operating pressure, and relative humidity within the wetted diameter.

What is difficult to quantify is the effect of the evaporating water on moisture movement away from the crop canopy. The gradient for moisture movement off the crop is decreased during irrigation and the plant transpiration rate will decrease to low and even negligible levels. However, from the perspective of the air mass in the boundary layer above the crop, the gradient for water vapor movement is being supported by the evaporation off the sprinkler droplets. The plant itself is not feeling stress because it is within an envelope of high relative humidity just as would be the case if it were freely transpiring.

The difficulty in assessing evaporation effects is that field tests have been made using catch-cans under varying conditions of operating pressure, nozzle diameter, wind speed, air temperature, and relative humidity. Water which was gauged as going through the sprinkler system but which was never recorded in the catch-cans was assumed lost due to wind drift and evaporation. As explained earlier, the water moved by the wind may be considered lost or not depending on where it is finally deposited. The amount of water evaporated from the droplets may be considered lost or not depending on whether or not it served to meet the atmospheric demand for water vapor over the crop canopy.

These fine points are normally not taken into account in evaluating wind drift and evaporation "losses." What is taken into account is whether or not the water made it into the catch-cans. For that reason, the usual methods of evaluating evaporation and wind drift losses probably overestimate the actual loss in terms of the crop canopy. Some amount of the irrigation requirement is probably met by the calculated "losses" because they occur in the vicinity immediately above the crop canopy. This is especially true in the case of evaporation. What portion of the computed loss actually serves to reduce the water demand of the crop is not clear and is probably not easily quantifiable.

Due to the practical considerations required for design, this book will present and apply traditionally used estimates of evaporation and wind drift. These methods probably result in overestimation of losses and therefore lead to a somewhat conservative design. It is not clear to what degree the resulting design will be conservative. As indicated throughout this book, if an error must be made it is prudent to develop a conservative design which will not contribute additional risk to the already uncertain business of irrigated agriculture. The irrigation system operating conditions resulting from the assumptions made may be modified to increase overall irrigation efficiency when improved information is available to more precisely quantify evaporation and wind drift.

Frost and Schwalen Nomograph

A common method of computing evaporation and wind drift is using the nomograph developed by Frost and Schwalen (1955). This nomograph was based on over 700 sprinkler nozzle tests. Test variables included nozzle diameter, operating pressure, wind speed, relative humidity, and air temperature. The tests measured the flow rate into the nozzle and the amount of water collected in catch-cans as previously discussed. Although Frost and Schwalen did not have access to computerized data analysis systems which are common today, their series of tests remain probably the most extensive ever undertaken and therefore warrant serious attention.

The Frost and Schwalen nomograph is depicted in Fig. 7-20. Solution for the evaporation and wind drift losses in line number 6 is achieved by establishing two pivot points on lines number 4 and 8. The pivot point on line 4 is determined by first drawing a straight line between the percent relative humidity and air temperature to intersect the vapor pressure deficit on line number 3. A second straight line is then drawn from the vapor pressure deficit through the nozzle diameter on line number 5. The intersection of this line with line number 4 is pivot point A.

A third straight line is drawn between the nozzle operating pressure on line number 7 and the wind velocity on line number 9. Where this line crosses line number 8 is pivot point B. The final step is to draw a straight line connecting pivot points A and B. This line intersects the evaporation and wind drift losses given in percent on line number 6. The procedure is demonstrated in the following example problem.

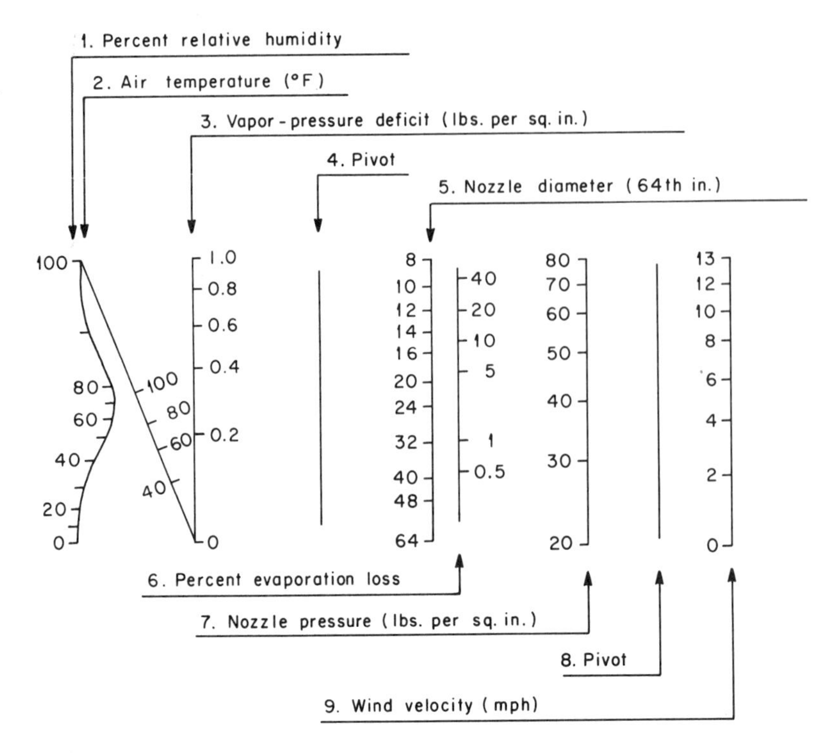

Figure 7-20 Nomograph for determination of evaporation and wind drift from sprinklers. (Adapted from Frost and Schwalen, 1955.)

Example Problem 7-2

Determine the expected evaporation and wind drift for a sprinkler system with the following operating characteristics:

$$RH = 10 \text{ percent} \qquad T = 90°F$$
$$d_{nozzle} = 12/64 \text{ inch} \qquad p = 40 \text{ psi}$$
$$U = 5 \text{ mph}$$

Solution Note that the units given for application of the Frost and Schwalen nomograph are English. Conversion of the data to SI units yields the following:

$$d_{nozzle} = 4.76 \text{ mm} \qquad T = 32.2°C$$
$$U = 8 \text{ km/h} \qquad p = 276 \text{ kPa}$$

The relative humidity, wind speed, and temperature data are for daytime operating conditions when evaporation and wind drift are maximum. The wind speed is assumed to correspond to that measured at approximately 2 m height.

A straight line drawn between the relative humidity and air temperature intersects the vapor pressure deficit line at 0.62 psi (4.27 kPa). The second line drawn between this point and a nozzle diameter of 12/64 inch establishes pivot point A. Pivot point B is determined by drawing a straight line between 40 psi nozzle operating pressure and 5 mph wind velocity. A straight line between the two pivot points results in an evaporation and wind drift of approximately 8.5 percent on line number 6.

Trimmer (1987) performed an analysis of the Frost and Schwalen nomograph to put it into terms more applicable to current computing methods. This work was done by using a computer to digitize the various nomograph axes and fitting the relationship between the related axes by regression equations. A final equation for evaporation and wind drift losses was developed based on the regression relationships and the geometry of the nomograph.

In his analysis, Trimmer (1987) bypassed drawing the first line for the vapor pressure deficit by employing an equation for this quantity given by Tetens (1930). This equation is given as

$$e_s - e_a = 0.61 \exp[17.27\, T/(T + 237.3)](1 - RH) \tag{7-5}$$

where

$$\begin{aligned} e_s - e_a &= \text{vapor pressure deficit, kPa} \\ T &= \text{air temperature, °C} \\ RH &= \text{relative humidity, fraction} \end{aligned}$$

The final equation for evaporation and wind drift is given as

$$L_s = [1.98(D)^{-0.72} + 0.22(e_s - e_a)^{0.63} + 3.6 \times 10^{-4}(h)^{1.16} + 0.14(U)^{0.7}]^{4.2} \tag{7-6}$$

where

$$\begin{aligned} L_s &= \text{evaporation and wind drift, percent} \\ D &= \text{nozzle diameter, mm} \\ h &= \text{nozzle operating pressure, kPa} \\ U &= \text{wind velocity, m/s} \end{aligned}$$

The advantages of Trimmer's equation is that it is given in SI units and is easily programmable. It does not, however, fit all sections of the nomograph perfectly well because it is a regression equation. It fits the middle range of the nomograph with differences of ± 10 percent. These differences are probably not significant because visual inspection of the field data shows that the nomograph solution fits the original data within approximately ± 20 percent. It is therefore important to note that solution of evaporation and wind drift by either the nomograph or Eq. (7-6) gives an approximate result. The realities of the degree of accuracy in this quantity are important to keep in mind as one proceeds with the design process. Trimmer (1987) recommends use of alternative coefficients for pressures below 207 kPa or nozzles larger than 12.7 mm in which cases differences of $+40$ percent exist between the Eq. (7-6) and the nomograph. Application of Eq. (7-6) is indicated in the following example problem.

Example Problem 7-3

Apply the Trimmer (1987) equation to solve for evaporation and wind drift for the case given in example problem 7-2. Compute the percent difference between the nomograph and equation estimates.

Solution Convert wind speed to m/s:

$$U = 8 \text{ km/h } (1000 \text{ m/km})(1 \text{ h}/3600 \text{ s})$$

$$U = 2.22 \text{ m/s}$$

Compute $e_s - e_a$ using Eq. (7-5):

$$e_s - e_a = 0.61 \exp[17.27(32.2°\text{C})/(32.2°\text{C} + 237.3)](1 - 0.10)$$

$$e_s - e_a = 4.32 \text{ kPa}$$

Compute L_s using Eq. (7-6):

$$L_s = [1.98(4.76 \text{ mm})^{-0.72} + 0.22(4.32 \text{ kPa})^{0.63} + 3.6 \times 10^{-4}(276 \text{ kPa})^{1.16} + 0.14(2.22 \text{ m/s})^{0.7}]^{4.2}$$

$$L_s = 9.0 \text{ percent}$$

Compute percent difference between solutions:

$$\Delta L_s = [(9.0 - 8.5)/8.5]100$$

$$\Delta L_s = 6 \text{ percent}$$

This difference is well within the expected error range of the estimate and points out the usefulness of applying the Trimmer equation.

General Guidelines

Table 7-2 indicates the potential combined application and distribution pattern efficiency of various types of sprinkler systems. The combined efficiency is the volume of water stored in the root zone compared to the volume delivered to the application devices. The application devices for sprinkler systems are the sprinkler nozzles. The combined efficiency must account for both deep percolation and evaporation and wind drift. It is given as

$$E_c = (1 - L_d)(1 - L_s) \tag{7-7}$$

where

E_c = combined application and distribution pattern efficiency, fraction

L_d = deep percolation, fraction

L_s = evaporation and wind drift, fraction

Table 7-2 gives estimates of the potential combined efficiency for values of the uniformity coefficient found to be feasible for a properly designed system based on field tests. The same level of adequacy was selected for each system so as not to bias the amount of deep percolation for a particular system. Note that for a properly designed system the runoff loss is zero which is assumed in the development of Eq. (7-7).

Table 7-2 gives a general range of evaporation and wind drift losses which may be applied for arid to humid environments. Criteria for low or high wind design

TABLE 7-2 Potential combined application and distribution pattern efficiencies of sprinkler irrigation systems.

System	Uniformity Coefficient (percent)	Adequacy (percent)	Runoff (percent)	Deep Percolation (percent)	Evaporation and Wind Drift [Arid-Humid][1] (percent)	Combined Efficiency [Arid-Humid][1] (percent)
Hand-move and Side-roll (with offsets)	82	75	0	18	15– 8	70–75
Solid set (low wind design)	70	75	0	30	22–15	55–60
Solid set (high wind design)	82	75	0	18	22–15	64–70
Center pivot and Linear move	90	75	0	10	18–10	74–81
Big gun (low wind design)	65	75	0	35	5– 3	62–63
Big gun (high wind design)	82	75	0	18	5– 3	78–80
Surface irrigation						20–90
Drip systems						≥ 90

[1]Arid conditions were estimated for 5.6 kPa vapor pressure deficit, humid conditions were estimated for 2.8 kPa vapor pressure deficit.

are given later in this chapter as is an explanation for an offset system. Combined efficiencies of surface and trickle irrigation systems are given in the table for comparison purposes.

7.5 Components of System Design

Application Rates

Table 7-2 indicated that there was no runoff with a properly designed sprinkler irrigation system. The application rate to the soil surface must therefore be less than the intake rate of the soil. This is the upper bound. The lower bound of the application rate must take into account that there will be some evaporation and wind drift of water from the nozzle. The discharge rate of the nozzle should be high enough that adequate water remains after evaporation and wind drift to enable a reasonable amount of water to be infiltrated into the root zone.

The average gross application rate is the nozzle discharge converted into the amount of water applied to the field considering only lateral and mainline geometry. The gross application rate is given by

$$d_g = \{q/[s_l(s_m)]\}360 \qquad (7\text{-}8)$$

where

$$d_g = \text{gross application rate, cm/h}$$
$$q = \text{sprinkler nozzle discharge, L/s}$$
$$s_l = \text{lateral spacing, m}$$
$$s_m = \text{mainline spacing, m}$$

Part of the gross application will go to evaporation and wind drift, and the remainder will be applied to the soil surface. That portion applied to the soil surface is called the net application rate. The net application rate is given by

$$d_a = d_g(1 - L_s) \tag{7-9}$$

where

$$d_a = \text{net application rate, cm/h}$$
$$L_s = \text{evaporation and wind drift, fraction}$$

General guidelines for the minimum gross application rate as a function of climate are given in Table 7-3. This table was developed considering that, relative to climatic zone, extremely low values of the gross application rate result in net application rates which are not economically feasible. The values given in the table are basic guidelines against which design gross application rates should be checked.

Table 7-4 indicates guidelines for maximum net application rates as a function of soil texture and slope. This table gives general values which are useful for preliminary design. In many locations in the United States, the Soil Conservation Service has developed Irrigation Guides for different soil types within a state. These Irrigation Guides include intake rate recommendations for sprinkler systems. This type of specific information for net application rate should be applied instead of general values in Table 7-4 if it is available.

Sprinkler nozzle discharge is a function of nozzle diameter, model, and operating pressure. This relationship is given by the following equation:

$$q = K(p)^{0.5} \tag{7-10}$$

TABLE 7-3 Recommended minimum gross application rates. (Adapted from USDA-SCS, 1984.)

	Minimum Gross Application Rate			
	(cm/h)		(in/h)	
Climatic Zone	From	To	From	To
Cool Maritime	0.25	0.40	0.10	0.15
Warm Maritime	0.40	0.50	0.15	0.20
Cool Dry Continental	0.40	0.50	0.15	0.20
Warm Dry Continental	0.50	0.75	0.20	0.30
Cool Desert	0.75	1.25	0.30	0.50
Hot Desert	1.25	1.90	0.50	0.75

TABLE 7-4 Suggested maximum net application rates for sprinklers for average soil, slope, and tilth.

Soil Texture and Profile	0–5% Slope in/h	0–5% Slope cm/h	5–8% Slope in/h	5–8% Slope cm/h	8–12% Slope in/h	8–12% Slope cm/h	12–16% Slope in/h	12–16% Slope cm/h
Coarse sandy soil to 2 m (6 ft)	2.0	5.0	1.5	3.8	1.0	2.5	0.50	1.3
Coarse sandy soils over more compact soils	1.5	3.8	1.0	2.5	0.75	1.9	0.40	1.0
Light sandy loams to 2 m (6 ft)	1.0	2.5	0.80	2.0	0.60	1.5	0.40	1.0
Light sandy loams over more compact soils	0.75	1.9	0.50	1.3	0.40	1.0	0.30	0.8
Silt loams to 2 m (6 ft)	0.50	1.3	0.40	1.0	0.30	0.8	0.20	0.5
Silt loams over more compact soils	0.30	0.8	0.25	0.6	0.15	0.4	0.10	0.3
Heavy textured clays or clay loams	0.15	0.4	0.10	0.3	0.08	0.2	0.06	0.2

where

q = nozzle discharge, L/s

p = nozzle operating pressure, kPa

K = nonlinear proportionality constant dependent on nozzle model and diameter

The value of the proportionality constant K varies for different ranges of nozzle discharge. Manufacturers of sprinkler nozzles have contributed to proper design procedures by developing tables of nozzle characteristics for various nozzle models. These tables indicate the nozzle discharge and wetted diameter as a function of operating pressure for different sprinkler models and nozzle diameters. Table 7-5 is an example of data available from such a table. The minimum recommended operating pressure, indicated below the dashed line in Table 7-5, is specified by the manufacturer to promote correct water distribution patterns from the nozzle. Data of this type are usually the most reliable on which to base system design.

Sprinkler Spacing

Wind criteria

Figures 7-13 and 7-14 indicated the effects of wind velocity on sprinkler distribution patterns. Reasonable values of uniformity of application can be maintained by decreasing the spacing between sprinkler nozzles as the normal operating wind velocity increases. Daytime wind velocities are typically higher than nighttime velocities. Use of mean daytime velocities to determine sprinkler spacing results in a conservative design if the sprinkler nozzles are to be operated during both daytime

TABLE 7-5 Manufacturer's sprinkler specifications. (Adapted from Rainbird, 1982.)

Model 29JH

Diameter	Pressure (kPa)	D-wet (m)	Flow Rate (L/s)	Pressure (psi)	D-wet (ft)	Flow Rate (gpm)
	138	22.3	0.128	20	73	2.03
3.175 mm	172	22.6	0.143	25	74	2.27
	207	23.2	0.157	30	76	2.49
1/8 inch	241	23.5	0.170	35	77	2.69
Minimum	276	23.8	0.181	40	78	2.87
Recommended	310	24.1	0.192	45	79	3.05
Pressure	345	24.4	0.203	50	80	3.21
	379	24.7	0.213	55	81	3.37
	414	24.7	0.222	60	81	3.52
	448	25.0	0.231	65	82	3.66
	483	25.3	0.240	70	83	3.80
	517	25.3	0.248	75	83	3.93
	552	25.6	0.256	80	84	4.06
	138	23.2	0.162	20	76	2.57
3.572 mm	172	23.5	0.181	25	77	2.87
	207	24.1	0.199	30	79	3.15
9/64 inch	241	24.4	0.215	35	80	3.40
Minimum	276	25.0	0.230	40	82	3.64
Recommended	310	25.3	0.244	45	83	3.86
Pressure	345	25.6	0.257	50	84	4.07
	379	25.9	0.269	55	85	4.26
	414	25.9	0.281	60	85	4.45
	448	26.2	0.293	65	86	4.64
	483	26.5	0.303	70	87	4.81
	517	26.8	0.313	75	88	4.96
	552	26.8	0.324	80	88	5.14
	138	23.8	0.200	20	78	3.17
3.969 mm	172	24.4	0.224	25	80	3.55
	207	25.0	0.245	30	82	3.89
5/32 inch	241	25.6	0.265	35	84	4.20
	276	25.9	0.283	40	85	4.49
Minimum	310	26.2	0.300	45	86	4.76
Recommended	345	26.8	0.317	50	88	5.02
Pressure	379	27.1	0.332	55	89	5.26
	414	27.1	0.347	60	89	5.50
	448	27.4	0.361	65	90	5.72
	483	27.7	0.375	70	91	5.94
	517	28.0	0.388	75	92	6.15
	552	28.3	0.401	80	93	6.35

and nighttime. The use of a weighted daytime-nighttime velocity is possible. The grower and designer should agree on the weighting factor since it will have an impact on the risk the grower faces in system operation. The use of an average 24-hour wind velocity will generally result in a design which is balanced for wind effects over a total growing season.

Table 7-6 indicates recommended values for ratios of lateral and mainline spacing to wetted diameter as a function of prevailing wind speed. The value of wetted diameter available from a manufacturer's catalog is based on no-wind conditions. Table 7-6 is normally used to calculate the required wetted diameter for a sprinkler nozzle when some idea of the possible lateral and mainline spacings is known. Possible values for lateral and mainline spacing are often dictated by available lengths of pipeline from equipment distributors and spacings typically used in different geographic areas for production of various crops. Decreasing lateral and mainline spacing increases potential uniformity of application, but at a higher equipment and operating cost.

Figure 7-21 indicates the change in uniformity coefficient with wind speed, spacing, and operating pressure for a double nozzle sprinkler set on various rectangular spacings. The trends shown in the figure are typical. The uniformity coefficient decreases with increases in wind speed and spacing between sprinklers. There is a slight increase in uniformity with increase in operating pressure up to a point. The operating pressures indicated are all above the minimum operating pressure recommended by the manufacturer.

The spacing configurations designated in Fig. 7-21 are the lateral spacing followed by the mainline spacing. Equations (7-8) and (7-9) indicate that for a given nozzle at a given operating pressure—that is, a given nozzle discharge—the gross and net application rates vary with sprinkler spacing. Any change in sprinkler spacing must therefore be evaluated to determine that the resulting application rates are within the maximum and minimum constraints.

Mainline spacing is typically larger than lateral spacing to reduce the cost of hookup devices between the lateral and mainline. The guidelines previously given for reasonable target uniformity levels may be applied for selection of sprinkler spacing. For the uniformities demonstrated in Fig. 7-21, high value specialty crops would warrant the tightest spacing of 9 m by 12 m (30 ft by 40 ft), 12 m by 12 m (40 ft by 40 ft) or 9 m by 15 m (30 ft by 50 ft) depending on wind speed. Spacing for moderate value field crops could be increased to 12 m by 15 m (40 ft by 50 ft),

TABLE 7-6 Recommended sprinkler spacings along laterals (s_l) and mains (s_m) as a fraction of the wetted diameter of the sprinkler under no-wind conditions (D_w).

Wind speed km/h	Sprinkler spacing, decimal s_l/D_w	s_m/D_w
0 to 8	0.60	0.65
8 to 16	0.50	0.50
16 and above	0.35	0.50

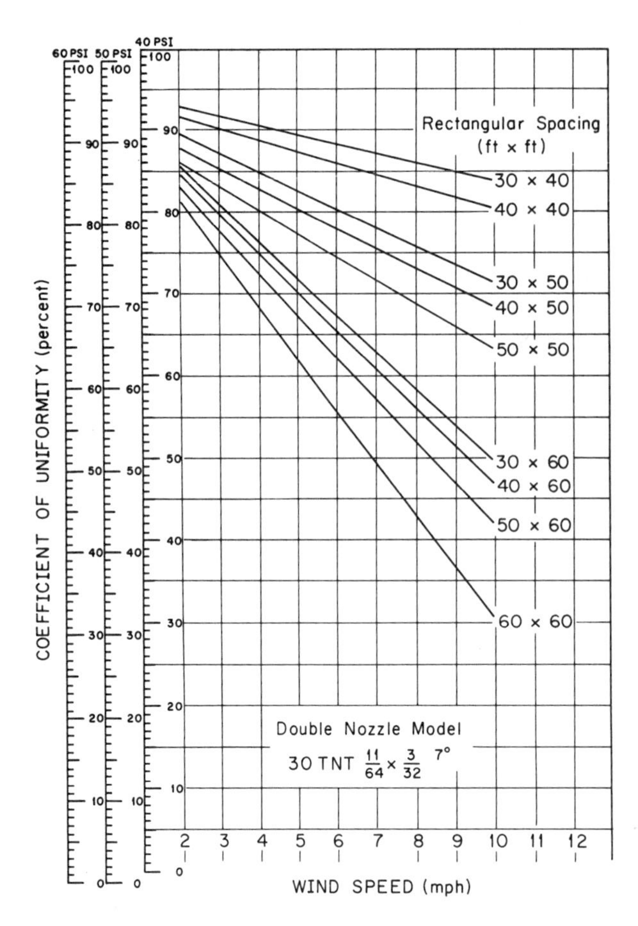

Figure 7-21 Change in uniformity coefficient with wind speed, spacing, and operating pressure for a particular double nozzle sprinkler model.

15 m by 15 m (50 ft by 50 ft) or 12 m by 18 m (40 ft by 60 ft) depending on wind speed. Wider spacings could be used for orchards. The use of offsets, discussed in the following section, can increase the uniformity coefficient in the range of from 5 to 15 percent. This enables use of more economical, wider spacings without loss of uniformity. Application of wind criteria for sprinkler spacing is indicated in the following example problem.

Example Problem 7-4

Side-roll sprinkler equipment and mainline pipe is available from a manufacturer with a lateral spacing of 12 m (40 ft) and mainline spacing of 18 m (60 ft). A system is to be designed for a region with high average wind velocities of 19 km/h. Determine the design wetted diameter for this system.

Solution Wind speed is in high range referring to Table 7-6. Rearrange the ratios in the table to solve for the wetted diameter. Based on s_l,

$$D_w = \frac{s_l}{0.35} = \frac{12\ \text{m}}{0.35} = 34.3\ \text{m}$$

Based on s_m,

$$D_w = \frac{s_m}{0.50} = \frac{18 \text{ m}}{0.50} = 36.0 \text{ m}$$

Choose the most restrictive case—that is, largest D_w—for the best uniformity. Sprinkler nozzle selection must therefore be made using the criteria that

$$D_w = 36 \text{ m}$$

Use of Offsets

Use of offsets refers to the practice of not placing the lateral in exactly the same position in the field each time a particular section of the field is irrigated. This type of operation is applicable in systems in which the position of the lateral is determined by the operator and the laterals are moved over the total area to be irrigated to conserve equipment costs. Hand-move and side-roll systems fall into this category. The offset procedure cannot be used on solid set, center pivot, or linear move systems in which the operator does not control the location of the sprinkler lateral.

Figures 7-11 and 7-12 demonstrated the uneven pattern of applied water which is the result of the lateral and mainline spacing between nozzles. If a lateral is always placed in the same position in a field, the parts of the field over-irrigated in previous irrigations continue to be over-irrigated and those under-irrigated continue to have higher deficits. The principal of using offsets is to change the position of the lateral so the high and low water application points tend to balance out over a growing season.

Figure 7-22 is a schematic which indicates application of the offset principle. The valve spacing for lateral hookups along the mainline is given by s_m. The laterals are positioned directly opposite these valves for irrigation number n. For the subse-

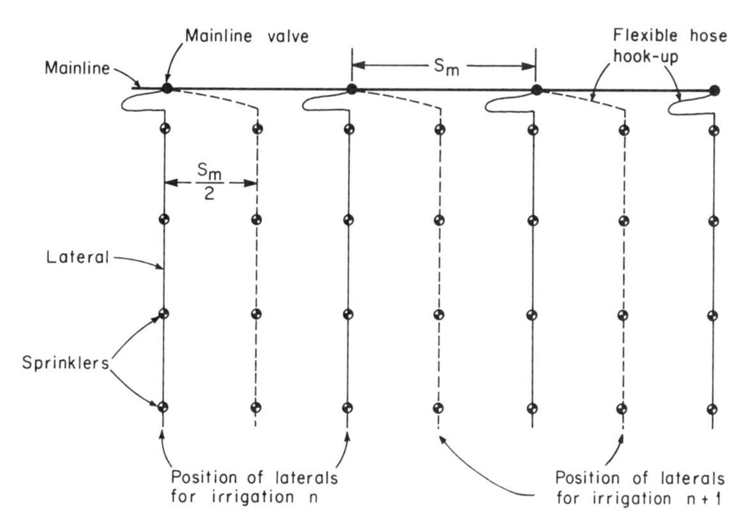

Figure 7-22 Schematic diagram indicating procedure for use of offsets in sprinkler systems for which position of lateral is set by the operator.

quent irrigation number n + 1, the laterals are placed midway between two valves on the mainline—that is, at distance $s_m/2$ from their position in the previous irrigation. The connection is made to the mainline by a flexible hose whose length is slightly greater than $s_m/2$. The flexible hose is used to connect the lateral in all positions. Using offsets, the field sees a different water application pattern with each irrigation interval.

This variation in the application serves to balance out the maximum and minimum applied water locations and increase uniformity over the growing season. The increase in the seasonal uniformity coefficient is greater for those systems which have a somewhat low uniformity coefficient under standard design conditions. Standard design conditions refer to repeated placement of laterals opposite the mainline valves. For example, if a system has a uniformity coefficient of 65 percent under standard design conditions, this coefficient can be expected to increase to 80 percent, a gain of 15 percent, using offset operations. If the standard design uniformity coefficient is a relatively high 80 percent, the use of offsets may be expected to increase the uniformity of application to 85 percent for a gain of 5 percent. Note that uniformity tests using offsets require that the lateral be operated in the standard position and in the offset position. The final catch for any can is the summation of the catches from the lateral in the two positions.

The use of offsets may also be used to gain flexibility in the manufacturer's specifications for nozzle operation. This is especially the case for the wetted diameter if the mainline spacing constrains the wetted diameter—that is, the largest required D_w is due to mainline and not lateral spacing. Under standard operating conditions, the minimum recommended overlapping distance of sprinkler wetted diameters parallel to the mainline is 6 meters. The wetted diameter of one line of sprinklers should overlap the wetted diameter of the lateral in the adjacent position by at least 6 meters. With the use of offsets, this minimum overlap of wetted diameters can be reduced to 3 meters. This is equivalent to enhancing the manufacturer's wetted diameter dimension by 3 meters. Such an enhancement can be useful in the design process when for some particular reason a nozzle with a slightly inadequate wetted diameter is preferred for the system. The use of the offset procedure therefore increases uniformity of application and allows for more flexibility in the design process.

Irrigation Interval

The relationship between soil-water holding capacity, rooting depth, allowable depletion, and net irrigation depth was discussed in the chapter on Crop Water Requirements. The major points are reviewed here since the irrigation interval derived from these relationships is a major aspect of sprinkler system design.

The key to the design irrigation interval is the total allowable depletion, TAD. The equation for the total allowable depletion is

$$TAD = TAM(MAD)D_r \qquad (7\text{-}11)$$

where

$$TAD = \text{total allowable depletion, mm}$$
$$TAM = \text{total available moisture, mm/m}$$
$$MAD = \text{management allowed depletion, fraction}$$
$$D_r = \text{depth of active root zone, m}$$

Total available moisture is defined as the difference between the field capacity and the limit of crop extractable water. It is given by

$$TAM = FC - CEW \tag{7-12}$$

where

$$FC = \text{field capacity, mm/m}$$
$$CEW = \text{limit of crop extractable water, mm/m}$$

Values of total available moisture for different soil textures were indicated in tabular form in the chapter on Soil Physics. Following are some very general guidelines for total available moisture. They should only be applied when more specific data are not available for a particular site. Listed are examples of available moisture for three soil textures. Values interpolated between those given may be used for soils with intermediate textures.

Light sandy soil	AM = 80 mm/m
Medium loam soil	AM = 140 mm/m
Heavy clay soil	AM = 200 mm/m

Guidelines for management allowed depletion were discussed in the Crop Water Requirements chapter. The following general guidelines are a function of crop value and rooting depth.

High value—Shallow rooted crop	MAD = 33 percent
Medium value—Medium rooted crop	MAD = 50 percent
Low value—Deep rooted crop	MAD = 67 percent

A table of depth of active root systems for various crops was also given in the Crop Water Requirements chapter. That table may be applied for sprinkler system design.

The irrigation interval is the time between successive irrigations in the same section of a field or between overlapping sections in the case of an offset operation. The interval is equal to the total allowable depletion divided by the peak period evapotranspiration rate:

$$T_i = \frac{TAD}{ET_{cp}} \tag{7-13}$$

where

$$T_i = \text{irrigation interval, d}$$

$$ET_{cp} = \text{peak period crop water requirement, mm/d}$$

Application of the irrigation interval concept is demonstrated in the following example problem.

Example Problem 7-5

A sprinkler system is to be designed to irrigate an alfalfa crop. Based on climatic data, the peak period crop water requirement is 11 mm/d. The soil is a light sand with a total depth of 1.7 m and bedrock with no available moisture content below that depth. The maximum crop rooting depth is 3 m. Compute the irrigation interval.

Solution Without more specific information on the available moisture content, the general value of 80 mm/m will be applied. Due to the limited depth of soil, the D_r variable will be constrained to 1.7 m instead of the maximum rooting depth. Alfalfa is generally a low value crop. Due to the limited depth of soil, it cannot be considered to be deep-rooted in this case. Therefore a compromise value of 60 percent will be applied for the management allowed depletion.

Compute the total allowable depletion:

$$TAD = (80 \text{ mm/m})(0.60)(1.7 \text{ m}) = 82 \text{ mm}$$

Compute the irrigation interval:

$$T_i = \frac{82 \text{ mm}}{11 \text{ mm/d}} = 7 \text{ to } 8 \text{ days}$$

The final decision of whether to use an interval of 7 or 8 days can be made by the designer and grower considering equipment costs and convenience of operations.

Nozzle Selection Criteria

Equation 7-10 indicated that there is a relationship between nozzle operating pressure and discharge. Table 7-5 showed a relationship between discharge, operating pressure, and wetted diameter for a given sprinkler model. There are a number of alternatives to consider in selection of a sprinkler nozzle. A nozzle operating at a low pressure will have reduced operating costs compared to one at higher pressure. However the lower pressure nozzle will have a smaller wetted diameter and lower discharge rate. These may be inadequate for the required application. Single nozzle sprinklers tend to perform better in high wind conditions than double nozzle sprinklers. In other than high wind conditions, double nozzle sprinklers generally have a higher uniformity coefficient than single nozzle systems. However, the increase in uniformity for a double nozzle sprinkler is normally accompanied by an increase in the discharge rate of a double nozzle sprinkler compared to a single nozzle model at the same operating pressure.

The nozzle selection process is one of balancing the operating characteristics of the nozzle with the physical requirements of the irrigation system. The nozzle discharge has to be high enough to meet the required gross application rate. The net application rate must be less than the intake rate of the soil. The net application rate has to be sufficient to apply the net irrigation requirement to the soil within a reasonable operating schedule. The wetted diameter must be compatible with lateral and mainline spacing.

Nozzle selection is always a trial-and-error procedure in which previous experience is helpful in making decisions regarding nozzle diameter, operating pressure, and sprinkler spacing. The process is not straightforward because so many factors must be brought into balance at the same time. Working out solutions for different design situations is the only way to gain experience in the selection process.

Tables 7-7 through 7-10 have been developed to aid in the nozzle selection process. These tables link the most important operating characteristics of the nozzle which are diameter, pressure, discharge rate, and wetted diameter. The tables should not be considered as replacing manufacturer's catalog data for specific sprinkler

TABLE 7-7 General recommendations for sprinkler operating pressure. (Adapted from Shearer et al., 1965.)

Nozzle Diameter		Recommended Pressure Range (kPa)		Recommended Pressure Range (psi)	
From	To	From	To	From	To
2.38 mm 3/32 in	4.76 mm 3/16 in	240	345	35	50
4.76 mm 3/16 in	6.35 mm 1/4 in	310	410	45	60
6.35 mm 1/4 in	9.53 mm 3/8 in	345	480	50	70

TABLE 7-8 Wetted diameters of moderate size nozzles. (Adapted from Shearer et al., 1965.)

Pressure (kPa)	D-wet (m), Nozzle Diameter (mm): 2.381	3.175	3.572	3.969	4.366	4.763	5.159	5.556
240	21.3	22.9	25.0	26.2	27.4	28.7	29.6	30.5
275	21.3	23.2	25.3	26.8	28.0	29.3	30.2	31.1
310	21.6	23.5	25.3	27.1	28.3	29.9	30.8	31.7
345	21.6	23.5	25.9	27.4	29.0	30.5	31.4	32.3

Pressure (psi)	D-wet (ft), Nozzle Diameter (in): 3/32	1/8	9/64	5/32	11/64	3/16	13/64	7/32
35	70	75	82	86	90	94	97	100
40	70	76	83	88	92	96	99	102
45	71	77	83	88	93	98	101	104
50	71	77	85	90	95	100	103	106

TABLE 7-9 Discharge rates of moderate size nozzles. (Adapted from Shearer et al., 1965.)

Pressure (kPa)	q (L/s) Nozzle Diameter (mm) 2.381	3.175	3.572	3.969	4.366	4.763	5.159	5.556
240	0.09	0.17	0.215	0.262	0.317	0.377	0.447	0.521
275	0.10	0.18	0.229	0.281	0.339	0.404	0.479	0.560
310	0.11	0.20	0.242	0.298	0.360	0.430	0.509	0.594
345	0.11	0.20	0.255	0.314	0.379	0.453	0.536	0.623
380	0.12	0.21	0.266	0.329	0.397	0.474	0.560	0.650

Pressure (psi)	q (gpm) Nozzle Diameter (in) 3/32	1/8	9/64	5/32	11/64	3/16	13/64	7/32
35	1.5	2.7	3.40	4.16	5.02	5.97	7.08	8.26
40	1.6	2.9	3.63	4.45	5.37	6.41	7.60	8.87
45	1.7	3.2	3.84	4.72	5.70	6.81	8.07	9.41
50	1.8	3.1	4.04	4.98	6.01	7.18	8.49	9.88
55	1.9	3.3	4.22	5.22	6.30	7.51	8.87	10.30

TABLE 7-10 Average gross application rates for various sprinkler spacings and nozzle discharges. (Adapted from Shearer et al., 1965.)

Sprinkler Spacing s-main (m)	s-lat (m)	d-gross (cm/h) Discharge per Nozzle (L/s) 0.126	0.189	0.252	0.315	0.379	0.505	0.631
6.1	6.1	1.22	1.83	2.44	3.06	3.67	4.89	6.11
6.1	12.2	0.61	0.92	1.22	1.53	1.83	2.44	3.06
9.1	9.1	0.54	0.81	1.09	1.36	1.63	2.17	2.72
9.1	12.2	0.41	0.61	0.81	1.02	1.22	1.63	2.04
9.1	15.2	0.33	0.49	0.65	0.81	0.98	1.30	1.63
12.2	12.2	0.31	0.46	0.61	0.76	0.92	1.22	1.53
12.2	15.2			0.49	0.61	0.73	0.98	1.22
12.2	18.3					0.61	0.81	1.02

Sprinkler Spacing s-main (ft)	s-lat (ft)	d-gross (in/h) Discharge per Nozzle (gpm) 2	3	4	5	6	8	10
20	20	0.48	0.72	0.96	1.20	1.44	1.93	2.41
20	40	0.24	0.36	0.48	0.60	0.72	0.96	1.20
30	30	0.21	0.32	0.43	0.54	0.64	0.86	1.07
30	40	0.16	0.24	0.32	0.40	0.48	0.64	0.80
30	50	0.13	0.19	0.26	0.32	0.39	0.51	0.64
40	40	0.12	0.18	0.24	0.30	0.36	0.48	0.60
40	50			0.19	0.24	0.29	0.39	0.48
40	60					0.24	0.32	0.40

models. However they can be used as a guide to narrow the nozzle parameters to a limited range of feasible options. Application of tables for nozzle operating characteristics to a specific design situation is indicated in the following example problem.

Example Problem 7-6

A side-roll sprinkler system is to be designed to irrigate bush beans in the Willamette Valley in Oregon. This region has a warm maritime climate during the period of peak crop water use. The peak period water requirement has been calculated at

$$ET_{cp} = 7 \text{ mm/d}$$

Meteorological data for the peak period rate

$$T_{min} = 10.5°C \qquad RH_{max} = 93 \text{ percent}$$

$$T_{max} = 27.8°C \qquad RH_{min} = 39 \text{ percent}$$

$$U_2 = 6.4 \text{ km/h}$$

The following information for total allowable depletion and maximum application rate is available from the Oregon Irrigation Guide for this location:

$$TAD = 82 \text{ mm}$$

$$I_{max} = 0.76 \text{ cm/h}$$

The design uniformity and adequacy criteria for a specialty crop will be applied, resulting in

$$UC_C = 85 \text{ percent}$$

$$a = 90 \text{ percent}$$

Side-roll equipment is available from the manufacturer with a lateral spacing of 12 m. Mainline sections are available in lengths of 15 m or 18 m. Previous tests under the expected low wind conditions and using offsets have indicated that uniformities of 85 percent are attainable with 18 m mainline spacing and 12 m lateral spacing. If the attainable uniformity coefficient is in question, a test should be run on a trial system for verification.

The operator wants to run the system on a schedule of 9 out of 10 days during the peak period with a total operating time of 20 out of 24 hours. The remaining four hours are required to drain and move the lateral. The operator wants to pressurize the system at the minimum feasible pressure to limit operating costs.

Assume the nozzle selection is constrained to those available in Table 7-5 for Model 29JH from the manufacturer's catalog. This will be a single nozzle system. Determine the design nozzle diameter, operating pressure, wetted diameter, nozzle discharge, gross application rate, and actual time of operation.

Solution Check feasibility of the irrigation interval.

$$T_{i-max} = TAD/ET_{cp} = 82 \text{ mm}/(7 \text{ mm/d})$$

$$T_{i-max} = 11.7 \text{ d}$$

The maximum allowable irrigation interval is therefore greater than the grower's preferred interval of 10 days. The grower's interval will not constrain the design.

Compute the expected evaporation and wind drift. Referring to Table 7-5, assume a nozzle diameter of 3.572 mm and minimum operating pressure of 276 kPa for the initial estimate of evaporation and wind drift. Do not recalculate this value unless the final nozzle diameter and pressure vary substantially from the initial estimates. The accuracy of the evaporation and wind drift estimate does not warrant recalculation for minor changes in nozzle diameter and operating pressure. Apply mean meteorological data to produce a result for average of day and night conditions. Applying Eq. (7-5),

$$e_s - e_a = 0.61 \exp[17.27(19.2°C)/(19.2°C + 237.3)](1 - 0.66)$$

$$e_s - e_a = 0.76 \text{ kPa}$$

Using Eq. (7-6),

$$L_s = [1.98(3.572 \text{ mm})^{-0.72} + 0.22(0.76 \text{ kPa})^{0.63} + 3.6 \times 10^{-4}(276 \text{ kPa})^{1.16} + 0.14(1.78 \text{ m/s})^{0.7}]^{4.2}$$

$$L_s = 4.5 \text{ percent} = 0.045$$

Compute the deep percolation and combined efficiency. Applying Table 7-1 for the given uniformity and adequacy criteria,

$$L_d = (1 - E) = (1 - 0.750) = 0.250$$

Applying Eq. (7-7) for combined application and distribution pattern efficiency,

$$E_c = (1 - L_d)(1 - L_s) = (1 - 0.250)(1 - 0.45)$$

$$E_c = 0.716$$

Compute required gross depth of irrigation:

$$i_g = i_n/E_c = (7 \text{ mm/d})(10 \text{ d})/0.716$$

$$i_g = 98 \text{ mm}$$

Develop an initial operating schedule. Compute a first approximation of net application rate to soil:

$$d_a = (1 - L_s)i_g/T_{set}$$

where

$$T_{set} = \text{set time, h}$$

The set time is the number of hours of operation for the lateral in one position.

$$d_a = (1 - 0.045)98 \text{ mm}/T_{set}$$

$$d_a = 93.6 \text{ mm}/T_{set}$$

Based on the grower's preferred operating schedule, two 10-hour sets or one 20-hour set per day are possible. For a set time of 10 hours,

$$d_a = 9.36 \text{ cm}/10 \text{ h} = 0.936 \text{ cm/h} > I_{max}$$

Therefore the maximum intake rate of the soil constrains T_{set} to 20 hours. Check minimum gross application rate against Table 7-3 for a warm maritime climate.

$$d_g = 9.8 \text{ cm}/20 \text{ h} = 0.49 \text{ cm/h}$$

This is towards the high end of the acceptable minimum rate. Assume the application rate is adequate.

Determine the design wetted diameter using an s_l of 12 m, s_m of 18 m, and low wind speed range in Table 7-6. Based on s_l,

$$D_w = s_l/0.60 = 12 \text{ m}/0.60 = 20.0 \text{ m}$$

Based on s_m,

$$D_w = s_m/0.65 = 18 \text{ m}/0.65 = 27.7 \text{ m}$$

Constraint is a wetted diameter equal to 27.7 m. Due to use of offsets, the wetted diameter given by the manufacturer can be extended approximately 3 m since the D_w constraint is based on mainline spacing. Therefore a manufacturer's D_w of approximately 25 m will be required.

Compute initial value of required discharge. Rearranging Eq. (7-8),

$$q = d_g(s_l)(s_m)/360$$

$$q = (0.49 \text{ cm/h})(12 \text{ m})(18 \text{ m})/360 = 0.294 \text{ L/s}$$

Required nozzle operating characteristics are a discharge of at least 0.294 L/s, D_w equal to approximately 25 m, and a reasonably low operating pressure. Based on choices available in Table 7-5, select Model 29 JH with

$$d = 3.969 \text{ mm}$$

$$p = 310 \text{ kPa}$$

$$D_w = 26.2 \text{ m}$$

$$q = 0.300 \text{ L/s}$$

Do not recompute evaporation and wind drift since final nozzle diameter and operating pressure are not substantially different from those used in the initial estimate. If a recomputed L_s is more than 10 percent different from that originally computed, the design should be reworked.

Calculate actual set time and operating schedule based on final q. Compute gross application rate using Eq. (7-8) and final q:

$$d_g = \{(0.300 \text{ L/s})/[12 \text{ m}(18 \text{ m})]\}360$$

$$d_g = 0.500 \text{ cm/h} = 5.00 \text{ mm/h}$$

$$T_{set} = i_g/d_g = 98 \text{ mm}/(5.00 \text{ mm/h}) = 19.6 \text{ h}$$

This result is sufficiently close to 20 hours to be an acceptable time of operation per day for the grower. The relatively low nozzle discharge will maintain d_n within the I_{max} constraint. The nozzle selected is therefore a feasible design solution.

System Capacity

The capacity of the system is the continuous flow rate required to irrigate the specified area within the selected operating schedule. It may be estimated as a function of the gross irrigation requirement, area, and operating schedule as follows:

$$Q = 2.778[i_g(A)]/[N_{op}(T_{op})] \qquad (7\text{-}14)$$

where

Q = continuous flow rate required, L/s

i_g = gross irrigation requirement, mm

A = total irrigated area, ha

N_{op} = number of days of operation per irrigation interval, d

T_{op} = hours of operation per day, h/d

The required number of sprinklers can be estimated by dividing the system capacity by the design discharge for the nozzle selected. This is given by

$$n = \frac{Q}{q} \tag{7-15}$$

where

n = number of sprinklers

q = design discharge per nozzle

The system capacity and number of nozzles required to be solved for using Eqs. (7-14) and (7-15) are approximate. The final solution depends on how the lateral and mainline spacing are accommodated within the actual dimensions of the field to be irrigated. The layout of the laterals and mainline will determine the actual number of sprinklers. The number of nozzles to be operated simultaneously times the design discharge per nozzle will determine the final system capacity. Normally the results given by Eqs. (7-14) and (7-15) will be very close to the final result, especially for fields without unusually shaped boundaries.

7.6 Distribution System Design and Layout

Overview

The water distribution system brings water from its source point in the area to be irrigated to the individual sprinkler nozzles. Mainline and lateral distribution systems are laid out to conform to the dimensions and topography of the fields to be irrigated and the hydraulic principles of flow in pipes. The design procedures for the majority of sprinkler systems including hand-move, side-roll, and solid set are all similar and are discussed in this section. Examples of the design of laterals, layout of mainlines and laterals, and the determination of the pressure distribution in the mainline will be presented. The material in this section will be referred to when related material in the chapters on Pipeline Systems and Pump Systems is presented.

Lateral System Design

Concept of lateral design

Equation (7-10) indicated that nozzle discharge was a function of the square root of the nozzle operating pressure. Previous relationships for uniformity, gross application rate, and net application rate all assumed that each nozzle was discharging at the same flow rate. In all but the rarest conditions, it is not possible to have the same operating pressure available for every nozzle on a lateral. The concept of lateral design is therefore based on limiting pressure differences along a lateral so the variation of nozzle discharge is within an acceptable range.

The usual criterion applied for the design of laterals is that the difference in nozzle discharge along a single lateral is less than ± 10 percent. To accomplish this goal, the difference in nozzle operating pressure is typically constrained to a variation of less than ±20 percent along the lateral. This pressure variation may be reduced to ±15 percent under circumstances in which the uniformity is of particular importance. The typical allowable variation of ±20 percent will be applied in this section.

The procedure for lateral design requires that a balance be developed between the length of the lateral, the headloss due to friction in the lateral, and the change in elevation head due to topographic effects. These factors are kept in balance so the pressure variation between the two critical sprinklers on a lateral is limited to ±20 percent. The two critical sprinklers are the ones with the maximum pressure difference. If the lateral is laid on level ground, on a constant slope, or on a constant moderate downslope relative to the friction headloss gradient, the two critical sprinklers are the first and last on the line. If there is a rise or depression in the middle of the line or the lateral is laid on a constant significant downslope, the change in elevation head may outweigh the head loss due to friction and a nozzle near the middle of the line may be one of the critical sprinklers.

The governing equation for the maximum allowable headloss due to friction between the two critical sprinklers is given by,

$$H_L = [\theta(H_a) - H_e]/l \qquad (7\text{-}16)$$

where

H_L = maximum allowable headloss due to friction, m/m

θ = maximum allowable pressure difference, fraction

H_a = nozzle design pressure expressed as head, m

H_e = increase in elevation in direction of water flow between the two critical sprinklers, m

l = distance between the two critical sprinklers, m

H_e is negative for downhill sloping laterals. The lateral will operate within the design constraints as long as the actual headloss due to friction along the lateral is less than the allowable headloss given in Eq. (7-16). Application of this equation is demonstrated in the following example problem.

Example Problem 7-7

A trial configuration of a hand-move sprinkler system has a lateral running downslope from a mainline along a constant grade of 0.005 m/m. The design operating pressure of the nozzle is 310 kPa. The trial length of the lateral results in a distance of 400 m between the first and last sprinkler. Compute the maximum allowable headloss due to friction as m/m.

Solution Since the lateral runs along a moderate constant grade line, the sprinklers at each end of the line are critical. Convert the nozzle operating pressure to head.

$$H_a = p/[\rho(g)] = 310{,}000 \text{ Pa}/[1000 \text{ kg/m}^3(9.807 \text{ m/s}^2]$$

$$H_a = 31.61 \text{ m}$$

Compute the increase in elevation between the two critical sprinklers. Since the elevation decreases along the lateral, the increase in elevation is negative and equal to the constant slope times the length of the lateral.

$$H_e = -S(l) = -0.005 \text{ m/m}(400 \text{ m}) = -2.0 \text{ m}$$

Set the allowable pressure differential between the critical nozzles to 20 percent. Compute the allowable headloss due to friction using Eq. (7-16).

$$H_L = [0.20(31.61\text{ m}) - (-2.0\text{ m})]/400\text{ m}$$
$$H_l = 0.021\text{ m/m}$$

Hydraulics of laterals

The allowable headloss due to friction computed using Eq. (7-16) must be compared with the actual headloss in the lateral. The actual headloss is a function of the volumetric flow rate and pipe diameter. The hydraulic analysis for friction in a sprinkler lateral is more complex than for a straight length of pipe because the analysis must account for the fact that water is removed at each nozzle and the volumetric flow rate is decreasing along the length of the lateral. This analysis is accomplished by application of the following assumptions:

(a) Sprinklers are evenly spaced at the lateral spacing s_1.
(b) Discharge is the same for each nozzle.
(c) Total flow into the lateral is discharged through the nozzles.

The actual headloss along a lateral is computed as a function of the headloss in an equivalent through-flow pipe. The through-flow pipe has the same diameter and friction coefficient as the lateral, but the volumetric flow rate is set constant at the inflow rate of the lateral. The actual headloss is given by

$$H_{L-ac} = F[H_{L-p}] \tag{7-17}$$

where

H_{L-ac} = actual headloss due to friction, m/m
F = friction factor to account for decrease in flow along lateral, fraction
H_{L-p} = equivalent headloss due to friction in through-flow pipe, m/m

Any convenient friction headloss formula may be applied to compute the equivalent headloss for a through-flow pipe. Examples are the Scobey equation, Hazen-Williams equation, and Darcy-Weisbach equation. The Hazen-Williams equation will be applied in this section. It is given as

$$h_f = kL\left[\frac{(Q/C)^{1.852}}{D^{4.87}}\right] \tag{7-18}$$

where

h_f = friction loss expressed as head
k = conversion constant
L = length of pipe
Q = volumetric flow rate
C = Hazen-Williams friction coefficient
D = pipe diameter

Tabulated values of different combinations of units for the Hazen-Williams equation and the corresponding conversion factor k are indicated in the chapter on Pipeline System Design. That chapter also contains tabulated values for the Hazen-Williams friction factor for different pipe materials. For problems in this section, friction headloss and pipe length will be expressed in m, volumetric flow rate in L/s, and pipe diameter in mm. The corresponding value of k is 1.22×10^{10}. The value which will be applied for the friction factor in this section will be for aluminum pipe with joints at every 12 m. The Hazen-Williams C factor for this condition is 135.

Equation (7-17) requires the friction headloss for through-flow pipe to be expressed on a per-unit length basis. This is accomplished by revising Eq. (7-18) to calculate H_{L-p} as

$$H_{L-p} = h_f/L \tag{7-19}$$

The F factor accounting for the decreasing flow rate along the sprinkler lateral in Eq. (7-17) is called the Christiansen friction factor. It is given by the following equation for the case in which the first sprinkler is at distance s_l from the mainline:

$$F = \frac{1}{m + 1} + \frac{1}{2N} + \frac{(m - 1)^{0.5}}{6N^2} \tag{7-20}$$

where

N = number of sprinklers along the lateral

m = exponent on velocity related term in friction headloss formula

m = 1.9 for Scobey equation

m = 1.852 for Hazen-Williams equation

m = 2.0 for Darcy-Weisbach equation

If the first sprinkler is at distance $s_l/2$ from the mainline, the Christiansen friction factor is given by

$$F = \frac{2N}{2N - 1}\left\{\frac{1}{m + 1} + \frac{(m - 1)^{0.5}}{6N^2}\right\} \tag{7-21}$$

The actual friction headloss will be a function of the pipe material, volumetric flow rate, pipe length, and diameter. The volumetric flow rate is controlled by the length of pipe for a sprinkler lateral since the length determines the number of nozzles. If convenience of operation and hydraulic efficiency require a specific orientation of the mainline and lateral in the field, only the pipe diameter can be adjusted to cause the actual headloss given by Eq. (7-17) to be less than the allowable headloss given by Eq. (7-16) for a given nozzle operating pressure. In some circumstances it may be possible to adjust the lateral length and therefore the volumetric flow rate to maintain the friction headloss within the allowable limit. The following example problem indicates a situation which requires adjustment of the pipe diameter.

Example Problem 7-8

Determine the required pipe diameter to maintain actual headloss within the allowable limit for conditions indicated in example problem 7-7. The sprinkler spacing s_l is 12 m and the first sprinkler is at distance s_l from the mainline. The design discharge per nozzle is 0.315 L/s.

Solution Compute the number of sprinklers on the lateral and volumetric flow rate.

$$N = L/s_l = 400 \text{ m}/12 \text{ m} = 33$$

$$Q = N(q) = 33(0.315 \text{ L/s}) = 10.395 \text{ L/s}$$

For a C factor of 135, set up the Hazen-Williams equation as a function of pipe diameter. This will facilitate evaluation of actual headloss as a function of pipe diameter. Expressing the through-flow pipe headloss on a per-unit length basis,

$$H_{L-p} = 1.22 \times 10^{10}\{[(10.395 \text{ L/s})/135]^{1.852}\}/D^{4.87}$$

$$H_{L-p} = (1.057 \times 10^{8})/D^{4.87}$$

Compute the F factor for the case where the first sprinkler is at distance s_l from the mainline:

$$F = 1/(1.852 + 1) + 1/[2(33)] + (1.852 - 1)^{0.5}/[6(33)^2]$$

$$F = 0.366$$

The results for the through-flow pipe friction headloss and actual lateral headloss as a function of a range of available inside pipe diameters is indicated in the following solution table.

Diameter (mm)	(in)	H_{L-p} (m/m)	H_{L-ac} (m/m)
50.8	2.0	0.521	0.1907
76.2	3.0	0.072	0.0264
101.6	4.0	0.018	0.0066
127.0	5.0	0.006	0.0022

The maximum allowable headloss from example problem 7-7 was

$$H_L = 0.021 \text{ m/m}$$

Therefore, based on constraining the pressure differential along the lateral to ≤20 percent, the minimum required pipe diameter is

$$D_{min} = 101.6 \text{ mm (4 in)}$$

The solution should be checked against a maximum velocity criterion to insure that the flow is reasonably uniform, including at points where there are changes in flow direction such as at the entrance to the lateral and at the sprinkler riser. The maximum velocity criterion which will be applied is

$$v_{max} \leq 1.5 \text{ m/s (5 ft/s)}$$

Check the flow velocity at the entrance to the lateral:

$$v = Q/A = (10.395 \text{ L/s})/[\pi(101.6 \text{ mm})^2/4](100 \text{ mm}^3/\text{L})$$
$$v = 1.282 \text{ m/s} < 1.5 \text{ m/s}$$

Therefore the velocity constraint is met.

The Christiansen F factors allow for convenient solution of the complex hydraulic problem of flow through a lateral. These factors are presented as a function of the exponent on the velocity term of the headloss equation and the number of nozzles in Tables 7-11 and 7-12 for the first nozzle at distance $s_1/2$ and s_1 from the mainline, respectively.

TABLE 7-11 Friction factor F to compute actual lateral friction loss. First nozzle at half lateral spacing from mainline.

Number of Nozzles N	Velocity Related Term Exponent m 1.852	1.900	2.000
1	1.009	1.006	1.000
2	0.519	0.512	0.500
3	0.441	0.435	0.422
4	0.412	0.405	0.393
5	0.396	0.390	0.378
6	0.387	0.381	0.369
7	0.381	0.375	0.363
8	0.377	0.370	0.358
9	0.373	0.367	0.355
10	0.371	0.365	0.353
11	0.369	0.363	0.351
12	0.367	0.361	0.349
13	0.366	0.360	0.348
14	0.364	0.358	0.347
15	0.363	0.357	0.346
16	0.363	0.357	0.345
17	0.362	0.356	0.344
18	0.361	0.355	0.343
19	0.361	0.355	0.343
20	0.360	0.354	0.342
22	0.359	0.353	0.341
24	0.358	0.352	0.341
26	0.358	0.352	0.340
28	0.357	0.351	0.340
30	0.357	0.351	0.339
35	0.356	0.350	0.338
40	0.355	0.349	0.338
50	0.354	0.348	0.337
100	0.352	0.347	0.335

TABLE 7-12 Friction factor F to compute actual lateral friction loss. First nozzle at full lateral spacing from mainline.

Number of Nozzles N	Velocity Related Term Exponent m 1.852	1.900	2.000
1	1.004	1.003	1.000
2	0.639	0.634	0.625
3	0.534	0.529	0.519
4	0.485	0.480	0.469
5	0.457	0.451	0.440
6	0.438	0.433	0.421
7	0.425	0.419	0.408
8	0.416	0.410	0.398
9	0.408	0.402	0.391
10	0.402	0.396	0.385
11	0.397	0.392	0.380
12	0.393	0.388	0.376
13	0.390	0.384	0.373
14	0.387	0.381	0.370
15	0.385	0.379	0.367
16	0.382	0.377	0.365
17	0.381	0.375	0.363
18	0.379	0.373	0.362
19	0.377	0.372	0.360
20	0.376	0.370	0.359
22	0.374	0.368	0.356
24	0.372	0.366	0.354
26	0.370	0.364	0.353
28	0.369	0.363	0.351
30	0.367	0.362	0.350
35	0.365	0.359	0.348
40	0.363	0.357	0.346
50	0.361	0.355	0.343
100	0.356	0.350	0.338

Lateral layout

The governing principle for positioning mainlines and laterals in a field is to minimize costs and if possible to use topographic effects to balance the pressure distribution along the lateral. Mainline costs are substantially higher than the cost of laterals and therefore the mainline length should be minimized. On moderate, uniform sloping fields, mainlines are normally placed at the center of the field running in the direction of the maximum slope. The laterals are run at right angles to the mainline in the direction of minimum slope to insure the highest pressure and distribution uniformity for the site. Figures 7-23 and 7-24 demonstrate a mainline and laterals installed in such a configuration.

On moderately sloping terrain, it may be possible to place the mainline along the upper edge of the field and run the laterals downslope at right angles to the main-

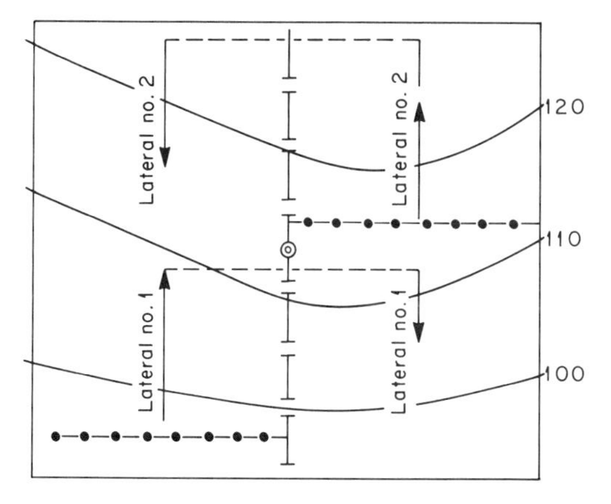

Figure 7-23 Layout on field with moderate, uniform slopes and water supply at the center.

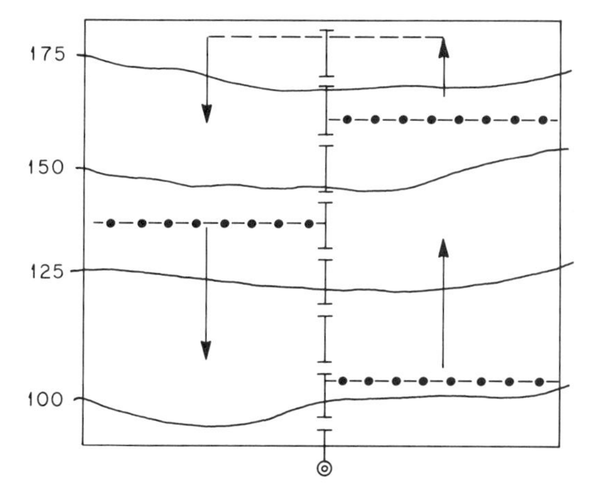

Figure 7-24 Layout on field with moderate, uniform slopes and water supply at one edge. Odd numbers of laterals are indicated to provide required number of sprinklers.

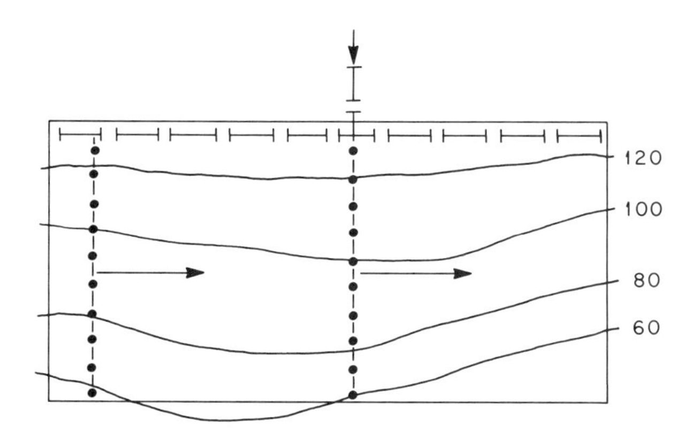

Figure 7-25 Layout with mainline along upper edge of field and lateral running downslope. Pressure losses in lateral can be compensated by gain in elevation head.

line. This type of installation aims at trying to balance the friction headloss along the lateral with the gain in elevation head to promote uniform pressures and discharge along the lateral. Figure 7-25 demonstrates an installation of this type. A similar situation is indicated in Fig. 7-26 except the orientation of the field relative to the topography does not provide for adequate mainline pressures at the upper end of the field. In this case a booster pump is required midway along the mainline.

On irregular terrain, mainlines are placed along the ridges and laterals are run downslope to maintain pressure and discharge uniformity. Figure 7-27 demonstrates a layout on irregular terrain in which mainlines are placed towards the center of the field. Figure 7-28 also demonstrates an installation on irregular terrain in which the topography dictates that a comparatively extensive mainline system be run on the edges of the irrigated area.

Mainline System Design

Detailed specifications, design procedures, and friction calculations for pipelines are indicated in the Pipeline System Design chapter. The material in this section will pertain to those analyses which are particular to systems comprised of sprinkler laterals and the required mainlines.

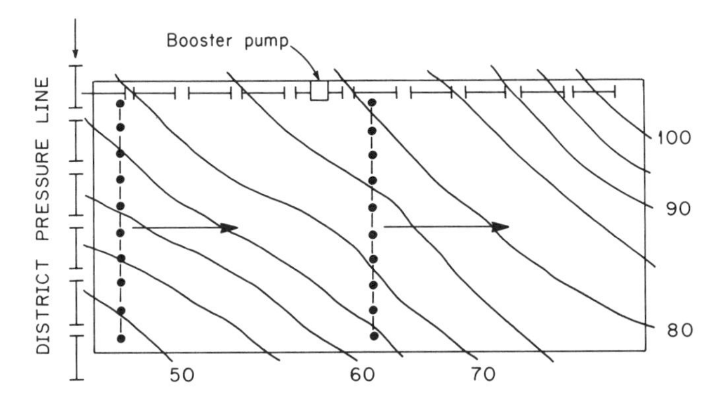

Figure 7-26 Layout with mainline along upper portion of field and laterals running downslope to balance friction and elevation head. Orientation of field requires booster pump to provide adequate pressures in upper half of mainline.

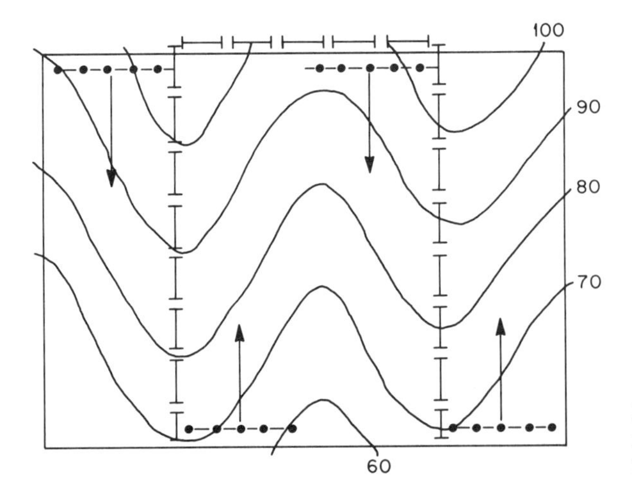

Figure 7-27 Layout on uneven terrain with mainlines run on ridges towards center of field and laterals run downslope.

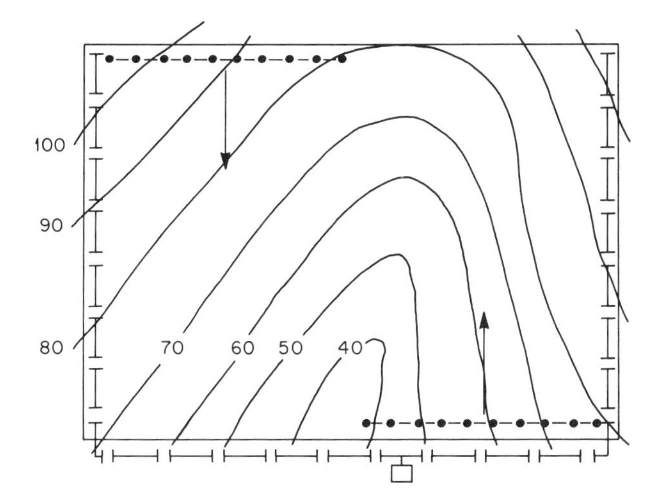

Figure 7-28 Layout on uneven terrain with extensive mainline system required to position laterals to run downslope.

Pressure required at mainline entrance to laterals

Adequate pressure must be available at the mainline take-out for the lateral to provide the correct operating pressure for the selected nozzle. The pressure required must also account for elevation changes along the lateral and the height of the connecting riser between the lateral line and the sprinkler nozzle. The design operating pressure of the nozzle is theoretically required at the nozzle itself.

The pressure requirement at the mainline entrance to the lateral is calculated by the following equation:

$$H_m = H_a + [0.75(H_f + H_e) + H_r]9.807 \text{ kPa/m} \tag{7-22}$$

where

H_m = required entrance pressure at mainline, kPa

H_a = design nozzle operating pressure, kPa

H_f = total friction headloss in the lateral, m

H_e = increase in elevation of lateral from inlet to position of critical sprinkler, m

0.75 = factor to produce the average operating pressure near the mid-point of the lateral

H_r = height of sprinkler riser, m

The critical sprinkler is that with the minimum operating pressure. It is normally the last sprinkler on a lateral unless there is a point of maximum elevation between the inlet and the end of the lateral or the downhill slope is significant relative to the friction headloss gradient. Minimum sprinkler riser heights are required to produce uniform flow conditions at the nozzle and to clear the crop canopy. Specifications for the minimum heights as a function of riser diameter are given in Table 7-13. Application of the equation to compute minimum required mainline pressure is demonstrated in the following example problem.

TABLE 7-13 Minimum acceptable sprinkler riser heights.

Riser Diameter		Minimum Height	
cm	in	cm	in
1.27	1/2	7.6	3
1.90	3/4	15	6
2.54	1	30	12
7.6	3 (big gun)	90	36

Example Problem 7-9

Determine the required entrance pressure at the mainline to serve the lateral described in example problems 7-7 and 7-8. Nozzle operating pressure is 310 kPa. The line is to be laid on the ground surface. Assume riser height is 1 m to clear the crop canopy.

Solution Compute actual friction headloss in the lateral using results of example problem 7-8 with the lateral diameter equal to 101.6 mm.

$$H_f = H_{l-ac}(L) = 0.0066 \text{ m/m } (400 \text{ m}) = 2.640 \text{ m}$$

Apply Eq. (7-22) using an increase in elevation head of −2.0 m.

$$H_m = 310 \text{ kPa} + [0.75(2.640 \text{ m} - 2.0 \text{ m}) + 1.0 \text{ m}](9.807 \text{ kPa/m})$$

$$H_m = 324 \text{ kPa}$$

The result is the required pressure in the mainline at the point of the lateral take-out.

Critical pressure requirement on mainline

The pressure required at any point of interest on the mainline is the sum of the following quantities:

(a) Pressure required at the next point on the mainline in direction of flow.
(b) Friction headloss between the point of interest and the next point on the mainline.
(c) Increase in elevation head between point of interest and the next point on the mainline.
(d) Increase in velocity head between point of interest and the next point on the mainline.

This relationship is expressed by the following equation in which the point of interest is designated as i and the next point in the direction of flow is designated as n:

$$H_i = H_n + h_{f-in} + H_{e-in} + H_{v-in} \tag{7-23}$$

where

$$H_i = \text{pressure head required at point i, m}$$
$$H_n = \text{pressure head required at point n, m}$$
$$h_{f-in} = \text{friction headloss from point i to point n, m}$$
$$H_{e-in} = \text{increase in elevation head from point i to point n, m}$$
$$H_{v-in} = \text{increase in velocity head from point i to point n, m}$$

The velocity head at point i is calculated as

$$H_{v-i} = (v_i)^2/2\ g = (v_i)^2/[2(9.807\ m/s^2)] \tag{7-24}$$

where

$$v_i = \text{flow velocity at point i, m/s}$$

The velocity heads are usually small and can be neglected without loss of accuracy. They are included in this discussion to indicate the complete procedure. The calculation of the head required at any point i requires the calculation of the pressure required at the next point n. Therefore pressure calculations are started at the end of the line and worked back towards the pump. The critical point on the mainline is the point with the highest pressure requirement taking into account all pressure losses starting from the pump. The pump must deliver adequate head at the critical point for proper operation of the system.

The pressure head required at the pump is equal to the sum of the following components:

(a) Pressure head required at the critical point in the mainline.
(b) Total friction headloss from the pump to the critical point in the mainline.
(c) Elevation head from the water source to the critical point in the mainline.
(d) Friction headloss from the pumping water level to the centerline of the pump.
(e) Velocity head at the critical point in the mainline.

The summation of these quantities equals the total dynamic head requirement of the pump. The total dynamic head required must theoretically be calculated for each point in the mainline, and the point with the highest requirement is the critical point. In equation form, the total dynamic head is given as

$$TDH_i = H_i + h_{f-Pi} + H_{e-si} + h_{f-s} + (v_i)^2/2g \tag{7-25}$$

where

$$TDH_i = \text{total dynamic head required for point i, m}$$
$$h_{f-Pi} = \text{friction headloss from the pump to point i, m}$$
$$H_{e-si} = \text{increase in elevation head from level of water source to point i, m}$$
$$h_{f-s} = \text{friction headloss on suction side of pump, m}$$

The pump must be able to produce the maximum calculated total dynamic head at the design flow rate to develop the required pressure distribution in the mainline. If the pressure required at the critical point in the mainline is adequate, all other points on the mainline with lower pressure requirements will have sufficient pressure.

In some cases development of adequate pressure at the critical point will produce exceedingly high pressures at other points in the distribution system. These high pressures could cause excessively high discharges in sprinkler laterals connected at these points. There are two correction procedures for this condition. One is to install pressure regulators at the lateral take-outs from the mainline or to adjust the opening of the hydrant valve at the inlet to achieve the same effect. The regulators or valve adjustment maintain the pressure at the inlet to the lateral at the level computed using Eq. (7-22). This is the normal method of correction for high pressures in parts of the mainline on small or moderate size irrigation systems.

In more extensive systems, it may be advisable to produce the required pressure at the critical point in stages. This is done by installing a booster pump at an intermediate location on the mainline to produce the proper pressure head at the critical point. Use of booster pumps avoids the cost of producing high pressure in certain sections of the pipeline only to have it controlled by a pressure regulator. An analysis can be run to compute the cost of developing high enough pressures throughout the mainline to meet the critical head and scaling the pressure down with regulating valves in the lower sections. This can be compared with the cost of developing moderate pressures at the lower sections of the mainline and adding a booster pump. These questions are discussed in more detail in the chapter on Pump Systems. The following example problem demonstrates calculation of the critical point on the mainline.

Example Problem 7-10

A distribution system is to be designed for a sprinkler system on which five laterals will be operated simultaneously. Each lateral is to have 30 nozzles with 0.360 L/s discharge per nozzle. The required pressure on the mainline at the inlet to each lateral is 393 kPa. The elevation of each point on the line and distance between points are given in the following table.

Point	Elevation (m)	Distance between Points (m)
Pump	48.41	
		36.58
A	47.55	
		18.29
B	48.46	
		18.29
C	48.77	
		18.29
D	47.85	
		18.29
E	47.24	

The mainline is made of 20.32 cm inside diameter aluminum pipe with a Hazen-Williams C factor of 144. The friction loss and elevation head from the water surface to the centerline of the pump is +0.333 m. Laterals are to operate on points A through E simultaneously. Determine the critical pressure point in the mainline and the total dynamic head requirement of the pump.

Solution Start at the end of the mainline and work towards the pump. Compute discharge required at each lateral,

$$Q_i = 30(0.360 \text{ L/s}) = 10.800 \text{ L/s}$$

Compute friction loss between any two points on the mainline as a function of flow rate in that section of the mainline and length of section using Eq. (7-18). Compute total flow rate and friction headloss from point D to point E as an example. Required flow rate from D to E is flow for one lateral.

$$Q_{t-DE} = 1(Q_i) = 10.800 \text{ L/s}$$
$$h_{f-DE} = 1.22 \times 10^{10}(18.29 \text{ m})\{[(10.800 \text{ L/s})/144]^{1.852}/(203.2 \text{ mm})^{4.87}\}$$
$$h_{f-DE} = 0.011 \text{ m}$$

Point E is the last point on the line. Therefore the only pressure required at E is the pressure at the entrance to the lateral. Convert pressure required at lateral take-out to head.

$$H_m = 393{,}000 \text{ Pa}/[1000 \text{ kg/m}^3(9.807 \text{ m/s}^2)]$$
$$H_m = 40.073 \text{ m}$$

The total head required at point E will be H_m plus a velocity head which is small because it is the last lateral on the line.
Compute increase in elevation head from D to E.

$$H_{e-DE} = 47.24 - 47.85 = -0.61 \text{ m}$$

Compute velocity heads at D and E using Eq. (7-24) and increase in velocity head from D to E.

$$H_{v-E} = \{(10.800 \text{ L/s})(1 \text{ m}^3/1000 \text{ L})[4/(\pi(0.2032 \text{ m})^2)]\}^2/[2(9.807 \text{ m/s}^2)]$$
$$H_{v-E} = (10.800 \text{ L/s})^2(0.0000485) = 0.00565 \text{ m}$$
$$H_{v-D} = (21.600 \text{ L/s})^2(0.0000485) = 0.02262 \text{ m}$$
$$H_{v-DE} = 0.00565 - 0.02262 = -0.01696 \text{ m}$$

Compute pressure head required at D using Eq. (7-23).

$$H_D = 40.073 \text{ m} + 0.011 \text{ m} - 0.61 \text{ m} - 0.0170 \text{ m}$$
$$H_D = 39.457 \text{ m}$$

The pressure head required at each point on the mainline back to the pump is computed in the same manner. Calculation is next made for the TDH at point E using Eq. (7-25). This equation requires calculation of all the friction headlosses from the pump to point E, h_{f-PE}, and the total increase in elevation from the pump to point E, H_{e-PE}. This type of analysis may be conveniently set up using a spreadsheet. Table 7-14 indicates the results of using a spreadsheet for analysis of the system treated in this example problem.

Table 7-14 is divided into three parts. Calculation for the pressure required at each point i in the mainline considering the requirements at the next point n are shown in part (a). This is the result of using Eq. (7-23). Part (b) indicates the results of computing the total dynamic head which must be produced at the pump to serve each point on the line. Equation (7-25) is used to compute these values.

The method of starting at the end of the line and working back toward the pump considering friction losses and changes in elevation and velocity heads insures that the TDH calculated for the pump will adequately serve all points on the mainline. This is the reason the TDH computed for all points in Table 7-14(b) are equal. The required TDH at the pump can be simply calculated by adding the required pressure head and velocity head at P to the friction loss and elevation head on the suction side of the pump.

Table 7-14(a) shows that the last point on the line, E, is not the critical point in this case. Although adequate pressure exists at E, all of the other points on the line have less pressure and point C is the critical point with the lowest pressure. It is customary to start an analysis such as this at the end point, as was done, in anticipation that it will be the critical point. The second place to inspect as a possible critical point is the point with the highest elevation, which is C in this problem. The pressure at point C must therefore be adjusted upward to the required 40.073 m, which will affect the pressure distribution in the rest of the line. This result is indicated in Table 7-14(c).

The required TDH at the pump can be computed using the results from Table 7-14(c) for the pressure head plus velocity head at P and adding the friction and elevation head on the suction side of the pump

$$TDH = 40.980 \text{ m} + 0.1414 \text{ m} + 0.333 \text{ m} = 41.454 \text{ m}$$

This TDH at the pump will produce the pressure distribution indicated in Table 7-14(c) in the mainline.

As previously mentioned, the velocity heads are usually negligible and are often not considered. They are included in these calculations to indicate the degree of significance and the complete procedure. Friction losses through the hydrant valve from the mainline to the lateral have been neglected in this example. They should not change the critical point in the line because the friction loss should be the same for each lateral take-out. However, hydrant losses will increase the required TDH a small amount. Often a safety factor of up to 3 m is added to the TDH to account for miscellaneous losses of this type, friction losses on the suction side of the pump, and unexpected operating conditions.

TABLE 7-14(a) Calculation of pressure head required and velocity head at each point on mainline.

Mainline Section	Flow Rate (L/s)	Friction Head (m)	Elevation Head (m)	Point	Velocity Head (m)	Required Head (m)
				E	0.0057	40.073
D − E	10.800	0.011	−0.610			
				D	0.0226	39.457
C − D	21.600	0.038	−0.920			
				C	0.0509	38.547
B − C	32.400	0.081	0.310			
				B	0.0905	38.898
A − B	43.200	0.138	0.910			
				A	0.1414	39.896
P − A	54.000	0.418	−0.860			
				P	0.1414	39.454

TABLE 7-14(b) Calculation of total dynamic head required for each point on the mainline starting from the pump.

Mainline Section	Friction Head (m)	Elevation Head (m)	Point	Total Dynamic Head* (m)
			P	39.928
P − A	0.418	−0.860		
			A	39.928
P − B	0.556	0.050		
			B	39.928
P − C	0.637	0.360		
			C	39.928
P − D	0.676	−0.560		
			D	39.928
P − E	0.686	−1.170		
			E	39.928

*Includes friction loss and elevation head on suction side of pump.

TABLE 7-14(c) Calculation of pressure head required and velocity head at each point on mainline.

Mainline Section	Flow Rate (L/s)	Friction Head (m)	Elevation Head (m)	Point	Velocity Head (m)	Required Head (m)
				E	0.0057	41.599
D − E	10.800	0.011	−0.610			
				D	0.0226	40.983
C − D	21.600	0.038	−0.920			
				C	0.0509	40.073
B − C	32.400	0.081	0.310			
				B	0.0905	40.424
A − B	43.200	0.138	0.910			
				A	0.1414	41.422
P − A	54.000	0.418	−0.860			
				P	0.1414	40.980

7.7 Center Pivot Systems

Types of Center Pivot Systems

Center pivot systems are differentiated from other types of sprinkler systems in that the water source is at the center of the irrigated field and the lateral pivots about the center point. The result is that nozzles towards the end of the pivot arm are moving at a much higher velocity than nozzles close to the pivot point. Special design considerations are required to produce an equal depth of applied water along the entire length of the pivot arm.

Various center pivot designs are possible to develop an even application rate. Systems are available with moderate to high pressure impact sprinklers or with relatively low pressure spray nozzles. What are called end-guns are sometimes placed at the end of the pivot arm to extend the area irrigated. These require high pressure at the end of the line due to the large nozzle diameter.

The two major categories of center pivot systems are constant and variable spacing. The distance between sprinklers is constant on a constant spacing system. The differential application rate is attained along the pivot arm by a combination of the pressure distribution in the lateral and selection of different diameter nozzles. Constant spacing designs use larger sprinkler nozzles with wider coverage than other types of center pivot systems. The pressure requirements are also higher to serve the larger nozzle diameters.

A variable spacing system incorporates nozzles with approximately the same discharge. Differences in application rate are attained by varying the distance between sprinklers. The sprinkler size and coverage are less than that for a constant spacing design. The operating pressure is also less because of the smaller nozzle diameters.

A spray nozzle system is one which replaces the impact sprinklers with low pressure nozzles which distribute water against a circular rotating plate or other device which breaks up the water stream. Spray nozzles are small in diameter and the area of application is narrow contrasted with the other designs. Figure 7-29 demonstrates the application patterns of the three systems discussed.

There are certain apparatuses required on center pivot systems which are generally not required on other types of sprinkler systems. Pressure regulators which serve individual sprinklers are sometimes required to maintain the proper pressure distribution along the lateral. Maintenance of the correct pressure distribution maximizes the uniformity of application but at an increased cost compared to more conventional sprinkler systems. A booster pump may be required at the end of the pivot arm to develop adequate pressure to operate a large diameter end-gun. End-guns are

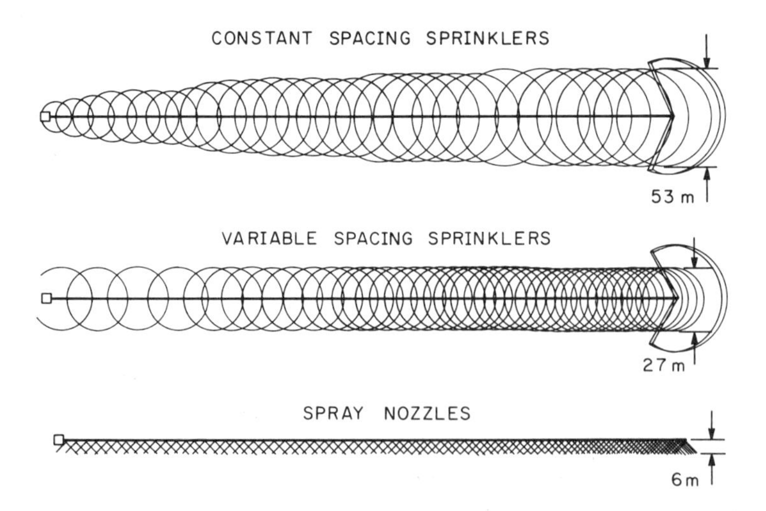

Figure 7-29 Various configurations of nozzle spacing and wetted area for center pivot systems.

popular in applications to expand the area irrigated to include part of the unirrigated corners which are otherwise lost to production (see Fig. 7-6). Systems which incorporate low-pressure spray nozzles have a narrow area of application as noted earlier. Such systems are sometimes mounted on booms holding four or five nozzles each to compensate for the narrow coverage. The booms are maintained in a position almost perpendicular to the pivot arm.

The vertical supports which elevate the pivot arm are called towers. These towers must be driven in unison so the pivot arm rotates about the pivot point at the design speed. Wheels on the towers are driven by hydraulic or electric motors mounted on the drive towers. On some systems, alignment of the pivot arm is maintained by short rods parallel to the arm which can activate circuit breakers. If part of the pivot arm gets too far out of alignment, the rod touches the circuit breaker completing the circuit and activating the drive wheels. All the drive towers along the pivot arm are thus positioned and keep the arm in a straight line.

Center pivot systems were originally developed in the American Midwest as a labor saving alternative to hand-move sprinkler lines in an area where labor was scarce. They have since evolved into systems with high uniformity of application because of the effort put into hydraulic analysis of such systems. The per-unit cost of center pivot systems tends to be high relative to other sprinkler systems. For this reason, they have proven most cost effective in regions where areas of 50 to 67 ha (125 to 165 ac) can be irrigated by a single pivot. These large areas reduce the cost per unit area to a minimum while keeping the size of the system at a level which is practical to operate and maintain.

Center Pivot Operation Parameters

The concepts of uniformity, adequacy, evaporation and wind drift, combined application and distribution pattern efficiency, and irrigation interval are generally the same for center pivot systems as previously presented for standard sprinkler systems. The major differences are that the pivot arm must make one complete revolution within the irrigation interval calculated using Eq. (7-13) for the peak period. The normal minimum time for one complete revolution of the pivot arm to avoid excessive wear on the equipment is 24 hours. It will be demonstrated later that the irrigation interval must sometimes be reduced towards this minimum to avoid runoff with this type of system.

The center pivot system applies the net irrigation requirement during each revolution. The seasonal operation time is therefore calculated as

$$T_{seas} = \left(\frac{ET_{seas}}{i_n}\right) T_r \tag{7-26}$$

where

T_{seas} = seasonal operation time, h

ET_{seas} = seasonal crop water requirement, mm

i_n = net depth of irrigation, mm

T_r = time per revolution of pivot arm, h

The most significant difference affecting the design between center pivot systems and other types of sprinkler systems is that the time of application varies along the length of the pivot arm. The towers at the end of the pivot move at a much faster rate than towers towards the center of the field. The time of application is therefore a minimum at the far end of the pivot arm. Since each revolution is to replace the net irrigation requirement, the application rate along the outer circumference of the field is the maximum.

The time of application at any radial distance r from the pivot point is calculated by dividing the wetted diameter of the sprinkler nozzle at distance r by the speed of the lateral. The lateral speed is computed as the circumference at distance r from the pivot point divided by the time per revolution. This relationship is expressed as

$$T_a = \frac{D_w T_r}{2\pi r} \tag{7-27}$$

where

T_a = time of application, h

D_w = wetted diameter of sprinkler at distance r from pivot point, m

r = radial distance from pivot point, m

At the outer edges of the irrigated circle, r is maximum and T_a is a minimum under normal conditions for the selection of wetted diameter. The application rate experienced by the soil at this point is the maximum since i_n is to be replaced along the total length of the lateral. Runoff will occur if the intake rate of the soil is less than the application rate. The application rate may be reduced by shortening the length of the pivot arm—that is, by reducing r_{max}. It may not be feasible to install a center pivot system if the soil intake rate severely limits the length of the pivot arm.

The rate of application of a center pivot system must take into account that the wetted diameter is circular but the center of this circle is moving along a circular path during the time of application, T_a. It has been shown that at a distance of $2.5D_w$ from the pivot point, this pattern can be described as a straight line over the time of application with negligible error (Heermann and Hein, 1968). The rate of application at any point on the field will reflect the discharge of the sprinkler nearest that point plus nearby sprinklers with overlapping patterns.

The application pattern of center pivot sprinkler systems is considered elliptical—that is, described by half an ellipse. This has proven a reasonable assumption when compared with field data (Dillon et al., 1972). It has also been demonstrated that the difference between assuming an elliptical or triangular application pattern from a single nozzle is negligible in calculation of the application rate (Herrmann and Hein, 1968). To proceed with calculation of the application rate, the concept of the net depth of application will be used. This value is the gross application depth minus evaporation and wind drift and is given as

$$i_a = i_g(1 - L_s) \tag{7-28}$$

where

$$i_a = \text{net depth of application, mm}$$

Taking into account that the pivot arm will pass through distance D_w in time T_a and integrating for the effects of overlapping sprinkler patterns, the maximum application rate experienced at any point along the pivot arm is computed as (USDA, 1983)

$$d_{a-max} = \left(\frac{4}{\pi}\right)\left(\frac{i_a}{T_a}\right) \tag{7-29}$$

where

$$d_{a-max} = \text{maximum application rate, mm/h}$$

Equation (7-27) demonstrated that the time of application was a function of the time of revolution. The minimum constraint for the time of revolution is the irrigation interval. Due to high application rates at the end of the pivot arm, the time of revolution for center pivot systems is normally constrained by the intake rate of the soil instead of the total allowable depletion given by Eq. (7-13). This constraint is exercised by reducing the net depth of irrigation so the center pivot system puts on frequent but light irrigations. Reducing the net irrigation depth in this manner reduces the required time of application.

This concept is demonstrated in Fig. 7-30. It was shown in the Soil Physics chapter that the infiltration rate for a particular soil decreased with time. Figure 7-30 indicates an infiltration rate curve superimposed on an elliptical pattern of water application over time T_a. Such a pattern would be experienced at any point in a field irrigated by a center pivot system. The application rate in the figure is seen to exceed the infiltration capacity of the soil at point T_{pr}, initiating the potential for runoff. The integral of the area under the elliptical curve is equal to the net depth of application. If the net depth of application is reduced, the time of application is decreased and the system provides a light but frequent irrigation. The peak application rate remains the same because i_a and T_a are reduced equivalently in Eq. (7-29). However, the application curve in Fig. 7-30 is shifted to the right, thereby minimizing the potential for runoff.

Specific intake rate data are required for design of center pivot systems due to the importance of correct infiltration information to limit runoff. The curves given in Fig. 7-31 may be used for preliminary estimates of infiltration rate as a function of soil texture. The figure indicates general intake rates superimposed over the USDA textural triangle. The adaptability of different soil textures to irrigation by center pivot system is indicated by the dashed lines which extend from the left side of the figure.

The final parameters which will be indicated are the discharge required at the pivot and the required sprinkler nozzle discharge. This discharge required at the pivot is calculated by

$$Q = K(Ai_g/T_r) \tag{7-30}$$

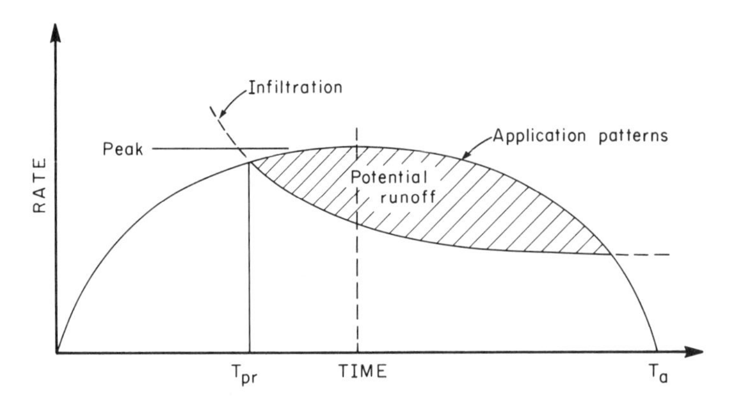

Figure 7-30 Intersection between infiltration rate curve and elliptical pattern of water application rate for center pivot systems.

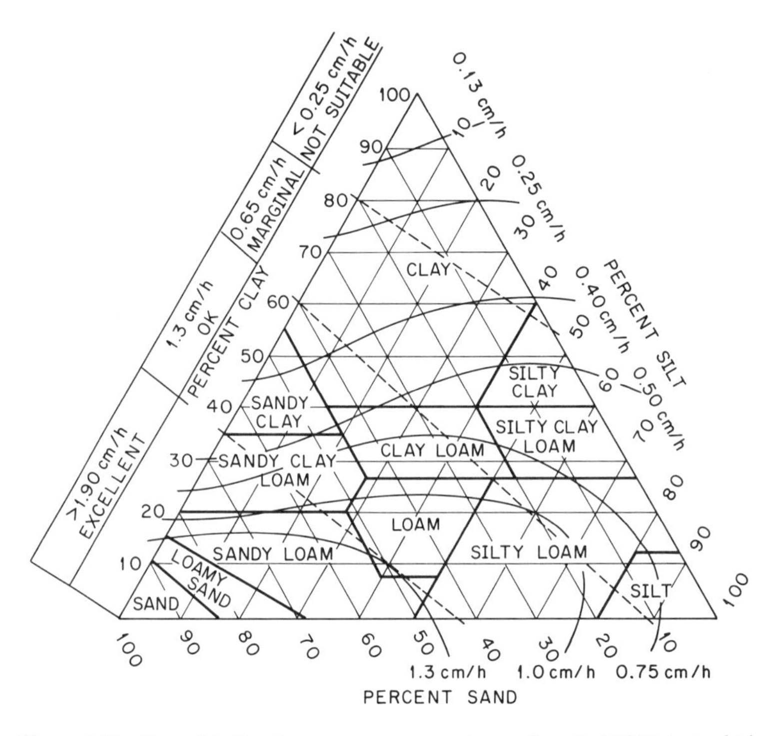

Figure 7-31 General infiltration rate contours superimposed on the USDA textural triangle. Adaptability of soil type to irrigation by center pivot indicated by dashed lines extending from the left of figure. (Adapted from USDA-SCS, 1983.)

where

Q = discharge required at the pivot, L/s

A = total area irrigated by the pivot system, ha

K = conversion constant

K is equal to 2.78 for the SI units indicated and to 453 for English units with Q in gpm, A in acres, and i_g in inches. The required sprinkler nozzle discharge is given as

$$q_r = rs_r[2Q/(r_{max})^2] \tag{7-31}$$

where

q_r = sprinkler discharge required at radial distance r from the pivot point, L/s

s_r = sprinkler spacing at distance r, m

The sprinkler spacing is equal to half the distance to the next upstream nozzle plus half the distance to the next downstream nozzle. Application of the operation parameters for a center pivot system are demonstrated in the following example problem.

Example Problem 7-11

A center pivot system is to be designed for a field with the maximum possible radius of the pivot arm equal to 400 m. Evaporation and wind drift is estimated as 9 percent and the potential combined application and distribution pattern efficiency considering uniformity, adequacy, and spray losses is 75 percent. The net irrigation requirement based on the depth of the root zone and water holding capacity of the soil is 29 mm and the peak period evaporation rate is 7 mm/d. The seasonal crop water requirement is 378 mm.

Compute the irrigation interval, seasonal operating time, required discharge at the pivot, and the maximum net application rate. Evaluate the maximum net application rate for a constant spacing system using a nozzle with a wetted diameter of 53 m and for a variable spacing system using a nozzle with a 27 m wetted diameter.

Solution Compute the gross depth of irrigation,

$$i_g = i_n/E_c = 29 \text{ mm}/0.75 = 39 \text{ mm}$$

Compute the net application depth using Eq. (7-28),

$$i_a = i_g(1 - L_s) = 39 \text{ mm}(1 - 0.09) = 35 \text{ mm}$$

Calculate the potential irrigation interval, T_i. The potential interval may have to be reduced if the soil intake characteristics cannot support the maximum net application rate.

$$T_i = i_n/ET_{cp} = 29 \text{ mm}/(7 \text{ mm/d}) = 4 \text{ d}$$

Set the initial estimate of the time of revolution to the calculated irrigation interval. This will have to be modified if the net application rate is judged to be too high. Calculate the seasonal operating time applying this estimate of the time of revolution using Eq. (7-26).

$$T_{seas} = (ET_{seas}/i_n)T_r$$

$$T_{seas} = (378 \text{ mm}/29 \text{ mm})4 \text{ d}(24 \text{ h/d}) = 1251 \text{ h}$$

Compute the area irrigated and apply the result in Eq. (7-30) to compute the required discharge at the pivot using the initial estimate of the time of revolution.

$$A = \pi(r_{max})^2 = \pi(400 \text{ m})^2(1 \text{ ha}/10{,}000 \text{ m}^2)$$

$$A = 50.3 \text{ ha}$$

$$Q = 2.78A(i_g)/T_r$$

$$Q = 2.78(50.3 \text{ ha})(39 \text{ mm})/96 \text{ h} = 56.8 \text{ L/s}$$

Compute time of application at the end of the pivot using Eq. (7-27). This is the area of the most critical application rates. For the constant spacing case,

$$T_a = (D_w T_r)/(2\pi r_{max})$$

$$T_{ac} = 53 \text{ m}(96 \text{ h})/2\pi 400 \text{ m}) = 2.02 \text{ h}$$

For the variable spacing case,

$$T_{av} = 27 \text{ m}(96 \text{ h})/(2\pi 400 \text{ m}) = 1.03 \text{ h}$$

Compute the maximum net application rate using Eq. (7-29). For the constant spacing case,

$$d_{a-max} = (4/\pi)(i_a/T_a)$$

$$d_{ac-max} = (4/\pi)(35 \text{ mm}/2.02 \text{ h}) = 2.2 \text{ cm/h}$$

For the variable spacing case,

$$d_{av-max} = (4/\pi)(35 \text{ mm}/1.03 \text{ h}) = 4.3 \text{ cm/h}$$

Reference to the recommended maximum net application rates in Table 7-4 or to Fig. 7-31 suggests that the center pivot with variable spacing would only be applicable on a coarse sand soil, while the constant spacing system might be applicable on a soil as fine as a sandy loam. More detailed analysis of the particular soil intake rates would be necessary to check if the soil could support the maximum application rate for a period of up to $T_a/2$ when the rate is a maximum as indicated in Fig. 7-30.

7.8 Linear Move Systems

System Description

Linear move systems have similar structural supports and guidance mechanisms as center pivot systems but the hydraulics are equivalent to those along the lateral of a standard sprinkler system. Linear move systems operate on rectangular fields. The entire field is irrigated which is an advantage over a center pivot system. The water source for the system is a pipeline or open channel run along one side or down the center of the irrigated field. Use of a pipeline requires dependable coupling devices. If an open channel is used for the water supply, the change in grade in the direction of the channel is limited.

The center pivot system is more convenient from the perspective of the water source since the water must only be delivered to a point in the center of the field. Another disadvantage of linear move systems is that following an irrigation the lateral arm should be moved back to the starting position before beginning the next irrigation. A center pivot system is automatically ready to begin the next irrigation at the correct position due to its circular motion.

A linear move can be conceptualized as a single lateral which is continuously moving down the field. This continuous movement in a direction perpendicular to the lateral allows for very high uniformity of application when proper spacing and pressure are maintained at the nozzles. Linear move systems have been measured to have the highest uniformity of application of any system under high wind conditions (USDA, 1983).

Linear Move Operation Parameters

The hydraulic analysis of linear move systems is similar to that for a standard sprinkler lateral because the sprinkler spacing and nozzle discharge are constant within reasonable constraints along the lateral arm. The speed of movement of the linear move system is normally chosen so the gross depth of application is similar to that of a standard sprinkler system. As with the center pivot system, application rates tend to be high which increases the amount of terrain which can be irrigated by the same lateral arm. This reduces the cost of the system per unit area irrigated. Precisely balancing the maximum rate of application at time $T_a/2$ with the specific soil intake rate, as demonstrated in Fig. 7-30, would allow for the maximum irrigated area.

The nozzle configuration of a linear move system is similar to the middle section of a center pivot arm where the discharge rate is between the extremes at the end of the arm or towards the center of the field. The maximum application rate assuming an elliptical application pattern as was done for the center pivot is the same as that given by Eq. (7-29). Revising this equation to account for sprinkler spacing along the lateral and the wetted diameter of the application pattern produces

$$d_{a-max} = (4/\pi)3600[Q_a/(D_w L)] \tag{7-32}$$

where

d_{a-max} = maximum application rate, mm/h
Q_a = discharge rate for water applied to soil surface for total lateral arm, L/s
D_w = wetted diameter of sprinkler nozzle, m
L = total length of lateral arm, m

Rewriting Eq. (7-32) relative to the spacing between nozzles along the lateral,

$$d_{a-max} = (4/\pi)3600[q_a/(D_w s_1)] \tag{7-33}$$

where

q_a = discharge rate for water applied to soil surface for a single nozzle, L/s
s_1 = lateral spacing, m

Q_a and q_a are equal to the total lateral arm or nozzle discharge, respectively, times one minus the fraction of evaporation and wind drift.

The concept of net depth of irrigation and maximum application rate for the linear move is identical to that discussed for the center pivot and described by Fig. 7-30. The depth of water applied to the field is equal to the integral of the area under the elliptical curve in the figure. Increasing the rate of travel of the lateral arm decreases T_a and therefore decreases the net depth of irrigation. However, the maximum application rate remains constant because it is governed by Eq. (7-29) in which i_a and T_a are reduced equivalently. The net result is that the elliptical application curve in Fig. 7-30 is shifted to the left thereby reducing the potential for runoff.

7.9 Big Gun and Boom Sprinkler Systems

System Description

Big gun sprinklers have large diameter tapered or orifice-ring nozzles and discharge a high volumetric flow rate of water in a circular pattern. These sprinklers require the highest operating pressure of the systems reviewed in this chapter to maintain proper spray formation for the large orifice nozzles and to overcome friction losses in typically long supply hoses.

The most common big gun sprinklers in the United States are drawn along tow paths and are sometimes referred to as traveling sprinklers or travelers. Frequently the nozzle does not make a full rotation but leaves a portion of the circle of about 25° dry in front of the vehicle which holds the sprinkler so it is always traveling over dry ground. Mounted on the vehicle is a water piston or turbine winch which wheels in a cable anchored at the opposite end of the tow path. This type of system is demonstrated in Fig. 7-32. The nozzle is connected by a flexible hose to the mainline. The mainline is normally positioned in the center of the field so the hose can be drawn in either direction. The total length of run is often on the order of 400 m.

Stationary boom sprinklers have a lateral arm at about the same height as a center pivot arm which makes a full rotation every one to five minutes. The lateral or boom arm is 35 to 75 m in length and irrigates a circular area up to 90 m in diameter. It is periodically towed to a different position on the field. Big gun sprinklers normally require less maintainance than boom sprinklers and are more stable when transported.

Some traveler vehicles are equipped with boom arms 20 to 60 m in length. The boom arm is rotated by the force of the directed nozzles and the vehicle is winched in on a cable as with the big gun system. The application pattern is like that of a small-scale center pivot.

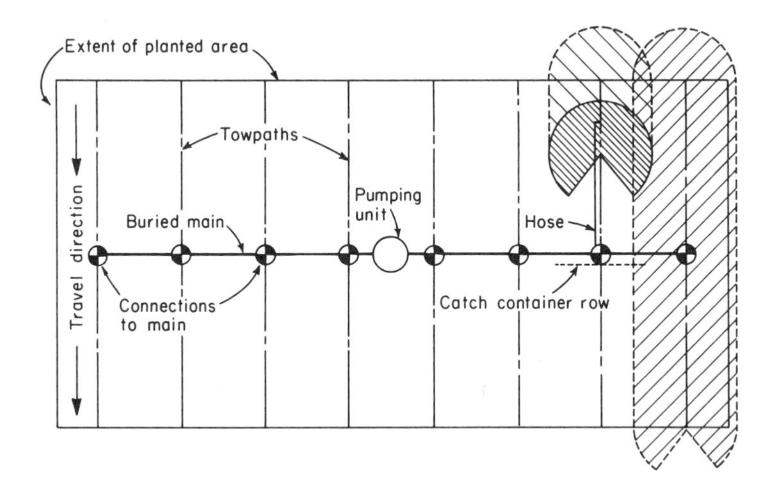

Figure 7-32 Layout for traveling big gun showing towpaths, mainline, catch-cans used to evaluate uniformity of application, and dry segment in direction of vehicle travel. (Adapted from USDA-SCS, 1983.)

Big Gun Operating Parameters

Application pattern and rate

Big gun sprinklers normally wet paths in the range of 120 m in width. Limited overlap between application patterns on adjacent tow paths produces uniformities which are not very high unless compensation is made for wind conditions by limiting the distance between paths (see Table 7-2). The concepts of uniformity and adequacy, combined application and distribution pattern efficiency, and total allowable depletion are the same as for a standard sprinkler system.

Big gun systems are fitted with tapered or orifice-ring nozzles with trajectory angles between 18° and 32°. The higher trajectory angles give maximum coverage in low wind conditions and have minimal drop impact on the soil and crops. The small drop impact reduces the potential for surface crusting and sealing on fine soils. Low trajectory angle nozzles give more uniform coverage in moderate to high winds but have a higher drop impact which may damage crops and cause crusting on soils which are not coarse. Nozzles with moderate trajectory angles of 23° to 25° are most satisfactory for general conditions.

Tapered nozzles produce a compact water jet which is less susceptible to distortion by wind. Orifice-ring nozzles produce a narrower wetted path than tapered nozzles but have a better stream break-up at lower pressures. Discharges for various diameter tapered nozzles as a function of operating pressure are indicated in Table 7-15. The discharges can be conveniently expressed as a function of the

TABLE 7-15 Typical discharges for big gun sprinklers with 24 degree trajectory angle and tapered nozzles.

Nozzle Pressure (kPa)	Nozzle Diameter (mm)				
	20	25	30	36	41
	Nozzle Discharge (L/s)				
414	9.0	14.2	20.8	—	—
483	9.8	15.5	22.4	30.3	—
552	10.4	16.4	24.0	32.5	42.6
621	11.0	17.3	25.6	34.4	45.1
689	11.7	18.3	26.8	36.3	47.6
758	12.3	19.2	28.1	38.2	49.8
827	12.9	20.2	29.3	39.7	52.0

Nozzle Pressure (psi)	Nozzle Diameter (inch)				
	0.8	1.0	1.2	1.4	1.6
	Nozzle Discharge (gpm)				
60	143	225	330	—	—
70	155	245	355	480	—
80	165	260	380	515	675
90	175	275	405	545	715
100	185	290	425	575	755
110	195	305	445	605	790
120	205	320	465	630	825

square root of operating pressure as given by Eq. (7-10) for different nozzle diameters. Table 7-16 indicates values for the K factors which will result in negligible errors in calculated discharge when substituted into Eq. (7-10).

Table 7-17 indicates the wetted diameters for a range of tapered nozzle sizes at 24° trajectory angle and at various operating pressures. The wetted diameter would increase or decrease about 1 percent for each degree increase or decrease in trajectory angle. Orifice-ring nozzles which are sized to give similar discharges at the same operating pressures would produce wetted diameters about 5 percent smaller than those indicated for tapered nozzles.

TABLE 7-16 K coefficient to calculate discharge as function of square root of nozzle operating pressure.

Nozzle Diameter (mm)	K (q in L/s) (p in kPa)	Nozzle Diameter (inch)	K (q in gpm) (p in psi)
20	0.4451	0.8	18.53
25	0.6991	1.0	29.10
30	1.021	1.2	42.51
36	1.382	1.4	57.52
41	1.812	1.6	75.39

TABLE 7-17 Typical wetted diameters for big gun sprinklers with 24 degree trajectory angle and tapered nozzles operating under no-wind conditions.

Nozzle Pressure (kPa)	Nozzle Diameter (mm)				
	20	25	30	36	41
	Wetted Diameter (m)				
414	87	99	111	—	—
483	91	104	116	133	—
552	94	108	120	139	146
621	98	111	125	143	151
689	101	114	128	146	155
758	104	117	131	149	158
827	107	120	134	152	163

Nozzle Pressure (psi)	Nozzle Diameter (inch)				
	0.8	1.0	1.2	1.4	1.6
	Wetted Diameter (ft)				
60	285	325	365	—	—
70	300	340	380	435	—
80	310	355	395	455	480
90	320	365	410	470	495
100	330	375	420	480	510
110	340	385	430	490	520
120	350	395	440	500	535

Big gun sprinklers at design operating pressures produce a modified elliptical pattern of application which dips in the center (USDA-SCS, 1983). The actual application rate which is to be matched to the infiltration rate of the soil is given approximately by

$$d_a = 360\{4q_a/[\pi(0.9\ D_w)^2]\}(360/\Phi) \qquad (7\text{-}34)$$

where

d_a = application rate, cm/h

q_a = nozzle discharge rate for water applied to the soil, L/s

Φ = portion of circle wet by sprinkler, degrees

The wetted area in this application rate equation is based on 90 percent of the wetted diameter. This results in a calculated application rate which corresponds to that experienced by the majority of the pattern rather than the average rate over the entire wetted area. Application of Eq. (7-34) and Tables 7-15 through 7-17 is demonstrated in the following example problem.

Example Problem 7-12

Compute the application rate to be matched to soil intake characteristics for a big gun sprinkler with a 24° trajectory angle and 30 mm diameter tapered nozzle. Operating pressure at the nozzle is 650 kPa. Assume a 25° arc in front of the traveler is kept dry and evaporation and wind drift equals 7 percent of nozzle discharge. Compute the application rate for the same set-up if the tapered nozzle is replaced by an orifice-ring nozzle.

Solution Use Eq. (7-10) to solve for discharge with K coefficient from Table 7-16. For the nozzle diameter given

$$K = 1.021$$

$$q = 1.021(650\ \text{kPa})^{0.5} = 26.03\ \text{L/s}$$

$$q_a = q(1 - L_s) = (26.03\ \text{L/s})(1 - 0.07)$$

$$q_a = 24.21\ \text{L/s}$$

Use linear interpolation in Table 7-17 to solve for wetted diameter for the tapered nozzle at a pressure of 650 kPa.

$$D_w = 125.0 + [(650 - 621)/(689 - 621)](128.0 - 125.0)$$

$$D_w = 126.3\ \text{m}$$

Degree of circle irrigated:

$$\Phi = 360° - 25° = 335°$$

Calculate application rate for tapered nozzle using Eq. (7-34).

$$d_a = 360\{(4)(24.21\ \text{L/s})/\pi[0.9(126.3\ \text{m})]^2\}(360/335)$$

$$d_a = 0.92\ \text{cm/h}$$

If the tapered nozzle is replaced with an orifice-ring nozzle, D_w is reduced by 5 percent.

$$D_{w-or} = 0.95(126.3\ \text{m}) = 120.0\ \text{m}$$

Substituting into Eq. (7-34),

$$d_a = 360\{(4)(24.21 \text{ L/s})/\pi[0.9(120.0 \text{ m})]^2\}(360/335)$$

$$d_a = 1.02 \text{ cm/h}$$

Towpath spacing

The uniformity of application for traveling big guns is affected by wind velocity, type of nozzle, jet trajectory, and operating pressure. Figure 7-32 demonstrates that the center sections of a field receive an overlapping water pattern on both sides and therefore get the maximum coverage. The application uniformity of a traveling big gun is better than that for a periodic move big gun. The application pattern of the continuous move system is like that of a periodic move system with very tight spacing. As with standard sprinkler systems, tighter spacing is required between towpaths for increasing wind conditions to maintain uniformity. With average windspeeds of approximately 15 km/h, uniformity coefficients of 70 to 75 percent have been measured in the central portion of a field with towpaths spaced at 70 to 60 percent of the wetted diameter (USDA-SCS, 1983).

Table 7-18 indicates recommended towpath spacing for sprinklers with a 23° to 25° trajectory angle as a function of nozzle type, wetted diameter, and anticipated average windspeed. Nozzles with trajectory angles of 20° to 21° are recommended for applications where average winds are expected to exceed 15 km/h. Nozzles with angles of 26° to 28° are recommended where winds are light.

Application depth

As we have seen for center pivot and linear move systems, the rate of application is unaffected by the speed of the application mechanism which is the traveler vehicle in the case of the big gun. However the depth of application is a function of the traveler speed. The gross depth of application as a function of nozzle discharge, towpath spacing and traveler speed is given by

$$i_g = \frac{6.0q}{s_t v_t} \qquad (7\text{-}35)$$

TABLE 7-18 Recommended towpath spacing for traveling big gun sprinklers as function of nozzle type and average windspeed. (Adapted from USDA-SCS, 1983).

	Towpath Spacing Percent of Wetted Diameter	
Wind Condition	Tapered Nozzle	Orifice-Ring Nozzle
No wind	80	80
Wind up to 8 km/h	75	70
Wind up to 15 km/h	65	60
Wind over 15 km/h	55	50

where

$$i_g = \text{gross depth of application, cm}$$
$$q = \text{nozzle discharge, L/s}$$
$$s_t = \text{towpath spacing, m}$$
$$v_t = \text{traveler speed, m/min}$$

Example Problem 7-13 demonstrates application of this equation and Table 7-18.

Example Problem 7-13

Calculate the gross depth of irrigation for a big gun system with a 36 mm orifice-ring nozzle operating at 690 kPa under average wind conditions of 6 km/h. The speed of the traveler vehicle is 0.30 m/min.

Solution Take nozzle discharge and wetted diameter from Tables 7-15 and 7-17.

$$q = 36.28 \text{ L/s}$$
$$D_w = 146.3 \text{ m}$$

Calculate the towpath spacing using Table 7-18 for orifice-ring nozzle and given wind conditions.

$$s_t = 0.70(D_w)$$
$$s_t = 0.70(146.3 \text{ m}) = 102.4 \text{ m}$$

Calculate gross depth of irrigation using Eq. (7-35).

$$i_g = \frac{6.0(36.28 \text{ L/s})}{102.4 \text{ m}(0.30 \text{ m/min})}$$
$$i_g = 7.1 \text{ cm}$$

REFERENCES

American Society of Agricultural Engineers, Standards—ASAE S330.1, "Procedure for Sprinkler Distribution Testing for Research Purposes," 1985, pp. 462–464.

Cuenca, R. H., G. H. Daskalakis, and J. A. Bondurant, "Energy Use Coefficients for Irrigation System Materials," Project Completion Report, USDA-SAE/AR: Cooperative Agreement No. 48-9AHZ-9-417, June, 1981.

Dillon, R. C., Jr., E. A. Hiler, and G. Vittetoe, "Center-Pivot Sprinkler Design Based on Intake Characteristics," Transactions of the American Society of Agricultural Engineers, vol. 15, no. 5, 1972, pp. 996–1001.

Frost, K. R., and H. C. Schwalen, "Sprinkler Evaporation Losses," *Agricultural Engineering,* vol. 36, no. 8, August, 1955, pp. 526–528.

Hart, W. E., and W. N. Reynolds, "Analytical Design of Sprinkler Systems," Transactions of the American Society of Agricultural Engineers, vol. 8, no. 1, 1965, pp. 83–85, 89.

HERRMANN, D. F., and P. R. HEIN, "Performance Characteristics of Self-Propelled Center Pivot Sprinkler Irrigation System," Transactions of the American Society of Agricultural Engineers, vol. 11, no. 1, 1968, pp. 11–15.

PETTYGROVE, G. S., and T. ASANO, eds., *Irrigation with Reclaimed Municipal Wastewater—A Guidance Manual*. Lewis Publishers, Inc., Chelsea, Michigan, 1986.

Rain Bird, "Agriculture Irrigation Equipment—Catalog," Rain Bird Sprinkler Manufacturing Corporation, Glendora, California, 1982.

SHEARER, M. N., "Comparative Efficiency of Irrigation Systems," Irrigation Association Technical Conference, Cincinnati, Ohio, February, 1978.

SHEARER, M., M. HAGOOD, D. LARSEN, and J. WOLFE, "Sprinkler Irrigation in the Pacific Northwest—A Troubleshooter's Guide," Pacific Northwest Cooperative Extension Publication, PNW 63, 1965.

TETENS, O., "Uber einige meterologische Begritte," Z. Geophys, vol. 6, 1930, pp. 297–309.

TRIMMER, W. L., "Sprinkler Evaporation Loss Equation," American Society of Civil Engineers, *Journal of Irrigation and Drainage Engineering,* vol. 113, no. 4, 1987, pp. 616–620.

USDA-SCS, "National Engineering Handbook, Section 15: Irrigation—Chapter 11: Sprinkle Irrigation," U.S. Government Printing Office, 1983, 389:097.

PROBLEMS

7-1. A sprinkler system is to be designed to irrigate a sandy loam soil planted to alfalfa in an arid environment. Soil tests and management decisions result in the following design parameters:

(a) Available moisture = 11 cm/m.
(b) Management allowed depletion = 55%.
(c) Depth of roots at peak period = 3.35 m.
(d) Peak period ET_c = 12.4 mm/d, where c is for alfalfa.

What irrigation frequency in days is theoretically possible?

7-2. A field to be sprinkler irrigated has a soil with an available water holding capacity of 120 mm per meter depth. The crop is to be irrigated at 30 percent depletion of available water and the root zone is 1.5 meters deep at mid-season. The mid-season peak water use rate is 11 mm/d. The sprinkler system is to be operated allowing for one day of down time during each irrigation interval. Compute the number of days of operation, to the nearest day, during the irrigation interval.

7-3. A hand-move sprinkler system is to be designed to irrigate a field for the conditions given. Determine the required time of operation per set as an integer (i.e., no fractions of an hour).

L_s = 10% (average over 24 hours) $ET_{c\,peak}$ = 6 mm/d
Adequacy = 75% Area = 8 ha
I_{max} = 10 mm/h s_l = 12.2 m
UC = 80% s_m = 18.3 m

Management allowed depletion = 99 mm

Operating schedule constraints: Run 12 out of 14 days

Minimum sets = 2 per day

Operating hours per set: Minimum = 8

Maximum = 11

7-4. A sprinkler system is to be designed for the following climatic and operating conditions:

(a) Evaporation loss equals 8%.
(b) Application efficiency equals 65%.
(c) Adequacy equals 75%.

What is the required design uniformity to operate this system?

7-5. A grower's field has 22 cm of available water. The management allowed depletion is 40 percent. During the peak period the crop ET is 8 mm/d. The grower wants to allow one day without operations for each irrigation interval during peak period and operate two sets per day. The hand-move system has 1180 m of mainline with valves at both ends and every 20 m. How many laterals are required to irrigate this system?

7-6. A sprinkler system with a normally distributed application pattern has a measured uniformity coefficient of 70 percent and a deep percolation loss of 31 percent.

(a) What percent of the total area is adequately irrigated by the system?
(b) The estimated evaporation and spray losses for this system are 6 percent. Compute the combined application and distribution pattern efficiency in percent.
(c) What percent of the total area would have to be adequately irrigated by the system to reduce deep percolation losses to 20 percent?

7-7. A solid set sprinkler system is designed to operate during the day only when the average wind speed is 12 mph, with minimum relative humidity at 45 percent and maximum temperature of 75°F. The nozzle chosen has a diameter of 11/64 inch, operating pressure of 60 psi, wetted diameter of 80 ft. The limiting criteria for design is soil intake rate which is 0.15 in/h.

(a) What is the recommended spacing between sprinklers along the lateral in feet?
(b) What is the recommended spacing between laterals in feet?
(c) What is the recommended nozzle discharge in gpm based on the foregoing information?

7-8. A sprinkler lateral is designed for the following conditions:

(a) Slope of land from inlet along lateral is constant at +0.012 m/m.
(b) Maximum allowable difference in pressure between two sprinklers along lateral is 20%.
(c) Sprinkler spacing along lateral is 12.2 m with first sprinkler 6.1 m from mainline.
(d) Lateral has 18 sprinkler nozzles.
(e) Equivalent headloss in equal diameter pipe of same length as lateral with all flow going through the pipe computed by Hazen-Williams formula is 0.064 m/m.

Compute the operating pressure of the nozzles in kPa if the system is operated at the design limit.

7-9. A sprinkler lateral is designed for the following conditions:

(a) Operating pressure of nozzles is 380 kPa.
(b) Slope of land from inlet along lateral is constant at +0.012 m/m.
(c) Maximum allowable difference in pressure between two sprinklers along lateral is 20%.
(d) Sprinkler spacing along lateral is 12.2 m with first sprinkler 6.1 m from mainline.

(e) Equivalent headloss in equal diameter pipe of same length as lateral with all flow going through the pipe computed by Hazen-Williams formula is 0.064 m/m.

What is the length of the lateral in m?

7-10. The following conditions are required for a sideroll system to sprinkler irrigate peppermint:

(a) Nozzle operating pressure is 448 kPa.
(b) Length of lateral is 183 m.
(c) Friction headloss in lateral is 0.023 m/m.
(d) Slope of land from mainline is constant at −0.004 m/m.
(e) Height of sprinkler nozzle above ground is 0.9 m.

What is the required entrance pressure to the lateral at the mainline in kPa?

7-11. A sprinkler system is laid out on a 20 m by 10 m rectangular spacing using a nozzle with a discharge of 0.59 L/s. Evaporation and spray losses are estimated at 5 percent. The field to be irrigated has an average slope of 9 percent. What type of soil(s) would such a system not be recommended for and why?

7-12. A lateral runs downslope on a ground surface with constant slope of 0.010 m/m. The actual friction loss in the lateral is 0.0085 m/m and the lateral length is 274 m. A 1.2 m riser is required for the crop. Compute the design operating pressure for the sprinkler nozzle if the pressure head required at the mainline is 448 kPa.

7-13. A center pivot system with variable spacing has a lateral length of 415 m and is designed to make one revolution in 24 hours. The wetted diameter of the sprinkler at the end of the lateral is 29.0 m. What is the rate of water application to the soil at the end of the lateral in mm/h if the gross depth of application per revolution is 9.4 mm and spray and drift losses are estimated at 12 percent.

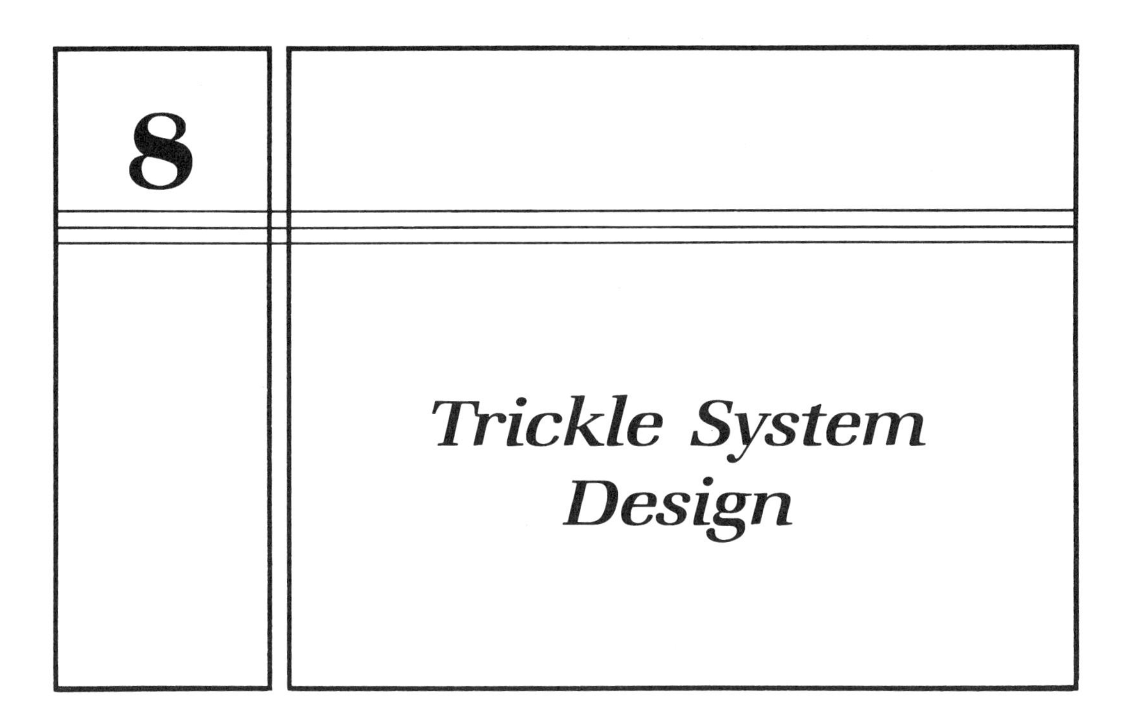

8.1 Concepts of Trickle Systems

Principles

Trickle irrigation, also referred to as drip irrigation, differs from the application systems previously discussed in that the water is applied at a point or over a very limited fraction of the total surface area of a field. Micro irrigation is another term which generally refers to application of water using miniature sprinklers called micro-jets placed close to the soil surface. Although the design principles are similar for micro and trickle systems, this chapter will concentrate on trickle system design.

Trickle systems apply water directly in the vicinity of the root zone, wetting a limited amount of surface area and depth of soil. Such systems were originally designed for applications in which there were relatively large distances between plants, such as orchards or vineyards. Figure 8-1 shows a close-up view of a trickle system in the same young peach orchard indicated in the first chapter. Subsequent to the successful use of trickle systems with widely spaced plants, special tubing was developed which made drip systems equally applicable to row crops.

A major difference between trickle systems and most other water application systems is that the balance between crop evapotranspiration and applied water is maintained over limited periods of 24 to 72 hours. The limited capacity of trickle systems operating over this short time interval requires that particular attention be

Figure 8-1 Trickle irrigation system installed on a peach orchard in the Willamette Valley, Oregon.

Figure 8-2 Tensiometers installed at three depths adjacent to a trickle irrigation line to monitor soil moisture status. (Photo courtesy of Marvin Shearer.)

applied to estimation of the crop water requirement or measurement of the soil-water status in the root zone. Application of tensiometers at different depths adjacent to a drip line to monitor soil-water content is indicated in Fig. 8-2.

Advantages

As for all types of application systems, there are advantages and disadvantages of trickle systems. A major advantage of trickle systems is that the close balance between applied water and crop evapotranspiration reduces surface runoff and deep percolation to a minimum. The fact that limited portions of the field surface are moistened also tends to reduce weed growth which would otherwise consume irrigation water.

There is evidence that trickle systems generally produce a higher ratio of yield per unit area and yield per unit volume of water than typical surface or sprinkler irrigation systems. There could be a number of reasons for this apparent increase in water use efficiency. One is that the frequent application of water in the vicinity of the root zone produces a continuously high soil-water content at the roots. The plants therefore come under very little stress if the trickle system is properly operated during the growing season. A second reason is that the limited moistened surface area reduces weed growth and the weeds are not competing with the crop for water and nutrients. Not all studies point to increased levels of yield per unit of water compared to other types of application systems. Pruitt et al. (1980) refer to a detailed study in which no appreciable increase in yield per unit of water was found when high frequency trickle irrigation was compared to furrow irrigation in tomatoes.

Trickle systems have been found applicable in marginal lands which could not be irrigated by other methods of application. Figure 8-3 demonstrates this principle in the irrigation of a high value crop in land which could not otherwise be irrigated. Trickle systems can be the most cost-effective method of irrigation in areas not well suited to surface or sprinkler irrigation.

Figure 8-3 Trickle irrigation system installed to irrigate a high value crop in difficult terrain not normally thought of as agriculturally productive. (Photo courtesy of Marvin Shearer.)

Disadvantages

There are certain disadvantages to trickle irrigation that the design engineer should be aware of in evaluating alternative systems. In general, the capital costs required for trickle distribution systems are higher than those required for surface or sprinkler systems. This is especially the case in comparison with surface irrigation systems unless excessive land leveling is necessary.

Trickle systems operate on the principle of applying a very precise amount of water directly in the vicinity of the root zone. This application normally requires very limited fluctuations of pressure from the design head at the point of water application. These constraints of accurate crop water accounting and tight pressure tolerances along the distribution line require a higher level of technology and sophisticated equipment for trickle systems than for surface or sprinkler applications. The design operating conditions of trickle systems must be carefully maintained during the growing season for the application system to be effective. The pressure constraints require increased maintenance throughout the season compared to surface or sprinkler systems.

Trickle systems have a limited buffering capacity within the soil profile in case of equipment malfunction compared to surface or sprinkler systems. This is a direct result of applying limited quantities of water to the root zone which is associated with the high overall irrigation efficiency of trickle systems. This limited buffering capacity puts further emphasis on a thorough maintenance program.

Impacts on Water Resources

The overall impact on water resources of using trickle irrigation systems compared to surface or sprinkler systems is uncertain due to the potential importance of the individual advantages and disadvantages just listed on any particular project. Considering the overall objective of irrigation project development to be long-term increases in crop production by the most cost-effective means, trickle systems may or may not be optimum depending on the specific soil, field topography, climate, expertise of irrigators, and market conditions for a location. In some cases where there is government support for development of irrigation projects, the improvement in crop production due to trickle systems may not be correctly evaluated against the true production costs which are highly subsidized within the project. Development of trickle irrigation projects should be analyzed applying the same economic and sociological concepts as for other application systems.

System Components

The components required for trickle systems are generally more involved than those for other application systems due to the need to filter the water supply and to maintain a specific pressure distribution throughout the system. The terminology used to describe the trickle system components is slightly different from that we have seen for surface and sprinkler systems.

The components of a typical trickle system are shown in Fig. 8-4. These can be divided into the mainline, submain, and lateral. The mainline has a pump to pres-

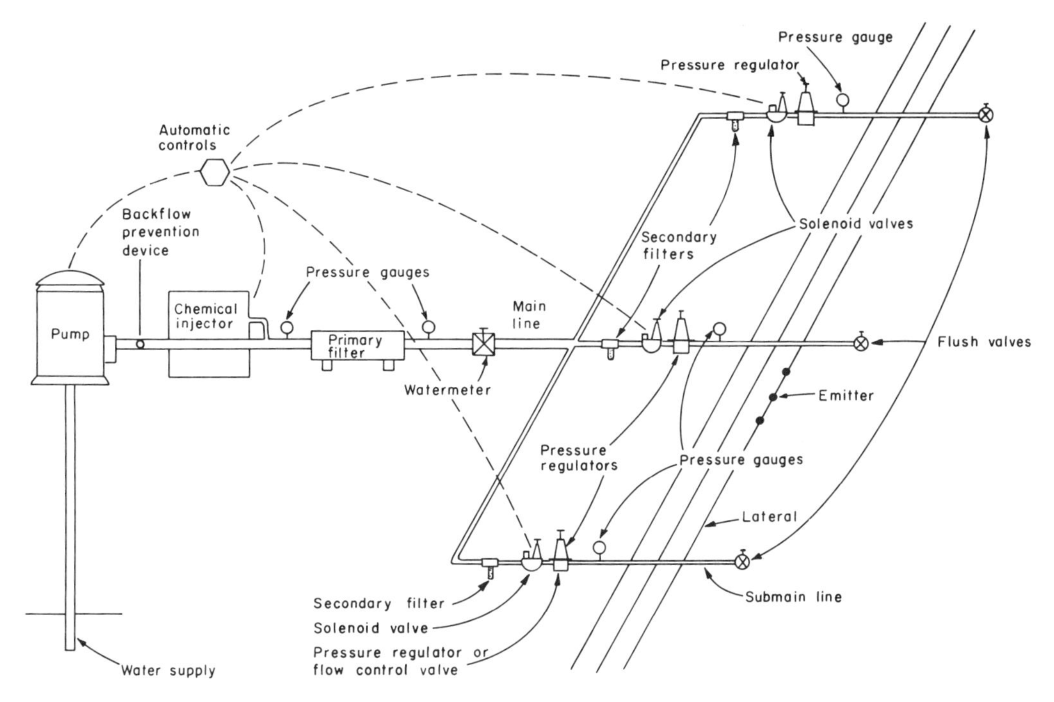

Figure 8-4 Components of a trickle irrigation system installation. (Adapted from Bucks et al., 1983.)

surize the system and possibly a chemical injector to conveniently apply nutrients through the distribution system. A primary filter is used to screen the largest particles out of the system. Primary pressure gauges on either side of the filter are used to evaluate when the pressure drop across the filter is high enough to require backflushing. The final components on the mainline are a discharge control valve and flow meter.

The submain line has a secondary filter for finer particles and a solenoid valve to aid in system automation. A pressure regulator is required on this line to keep the system operating within the close tolerances of discharge necessary for the water balance. Secondary pressure gauges are used to verify the operating pressure. Flush valves are shown at the end of the submain line to periodically clear accumulated debris from the line.

Lateral lines are shown coming off the submains. The laterals distribute water to the emitters which deliver water directly to the root zone. The various types of emitters possible are discussed in the next section. Table 8-1 indicates some general guidelines for equipment needs in trickle irrigation systems as a function of type of crop. The information in this table can be used for initial estimates of equipment requirements and cost evaluation. Application of the table is demonstrated in the following example problem.

TABLE 8-1 General estimates of trickle system equipment requirements. (Adapted from Hanks and Keller, 1972.)

Type of Crop	Row Spacing (m)	Plants per Hectare	Emitters per Hectare	Lateral Length (m/ha)
Ordinary orchards	6	250	500–1500	1,900
Dwarf orchards and vineyards	3.7	1,000	2,000	3.040
Berries and wide-spaced row crops	1.5	15,000	7,500	6,840
Greenhouse and close-spaced row crops	1	25,000	10,000	10,640

Example Problem 8-1

A typical orchard is to be developed on a field with dimensions of 253 m by 439 m. The orchard will be irrigated using a trickle system laid out so that each tree is served by four emitters. The following design conditions are based on peak period requirements at full tree maturity:

(a) Operating pressure head at the emitter = 10 m.
(b) Peak period crop water requirement = 5 mm/d.
(c) Distribution pattern efficiency = 92 percent.
(d) Operating time = 18 h/d.

Estimate the following design parameters:

(a) Number of emitters required.
(b) Required emitter discharge, L/h.
(c) Length of lateral, m.

Solution From Table 8-1 for an ordinary orchard, plant density is 250 trees per hectare. Calculate number of emitters required rounding to the nearest number of integer trees.

$$N = 253\ \text{m}(439\ \text{m})\,(1\ \text{ha}/10{,}000\ \text{m}^2)\,(250\ \text{trees/ha})(4\ \text{emitters/tree})$$

$$N = 11{,}108\ \text{emitters}$$

Calculate the required emitter discharge taking into account the crop water requirement, distribution pattern efficiency, and operating time. Begin by calculating the equivalent application rate.

$$d_a = (5\text{mm/d})\,(1/0.92)\,(1\ \text{d}/18\ \text{h}) = 0.302\ \text{mm/h}$$

$$q = (0.302\ \text{mm/h})\,(253\ \text{m})\,(439\ \text{m})\,(1\ \text{m}/1000\ \text{mm})\,(1000\ \text{L/m}^3)\,(1/11{,}108\ \text{emitters})$$

$$q = 3.0\ \text{L/h}$$

Calculate required lateral length using 1,900 m/ha from Table 8-1.

$$L = (1{,}900\ \text{m/ha})\,(253\ \text{m})\,(439\ \text{m})\,(1\ \text{ha}/10{,}000\ \text{m}^2)$$

$$L = 21{,}103\ \text{m}$$

8.2 Emitters

Types of Emitters

An emitter is a device which applies water to the soil from the distribution system. The two major categories of emitters are point source and line source. Both categories have been successfully used in various cropping situations.

There are numerous attributes sought in an optimum emitter. It should be available in small increments of discharge—that is, on the order of 1 L/h. The flow should be controlled within narrow limits as a function of operating pressure to be able to properly balance applied water with crop water use. A large flow area will be more resistant to clogging by particles which pass through the screening and filtration system or bacterial slime than a small flow area. The emitter should resist degradation due to temperature fluctuations and solar radiation, both of which are expected in typical installations. Finally the manufacturer should specify a useful life for the emitter during which it will operate according to design specifications. This will allow the designer to make more accurate cost projections for system operations and will be useful in development of a maintenance and replacement schedule.

Examples of different point and line source emitters are indicated in Fig. 8-5. Detailed section drawings for specific emitter types are shown in Figs. 8-6, 8-8, 8-9, and 8-11. Lateral line pressure is dissipated to design levels by headloss along an ex-

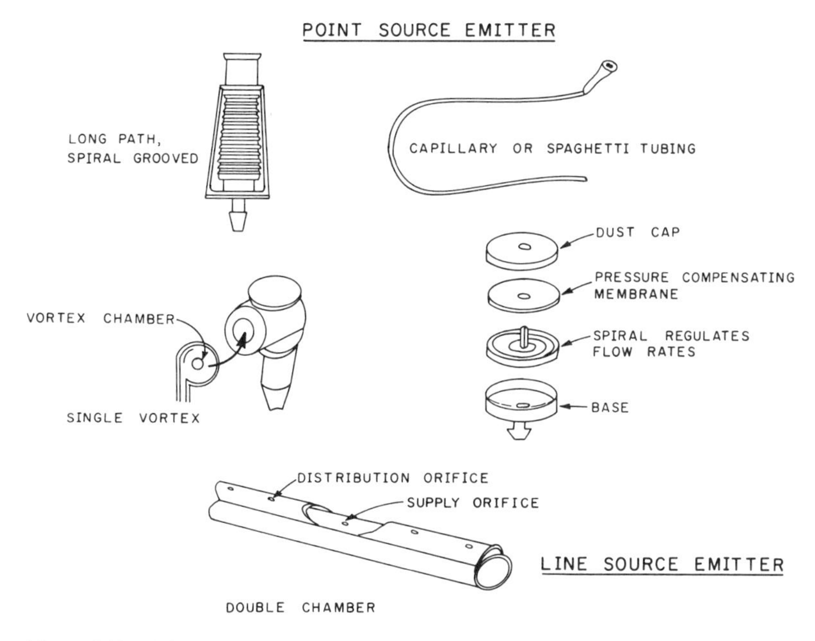

Figure 8-5 Different models of point and line source emitters. (Adapted from Pair et al., 1983.)

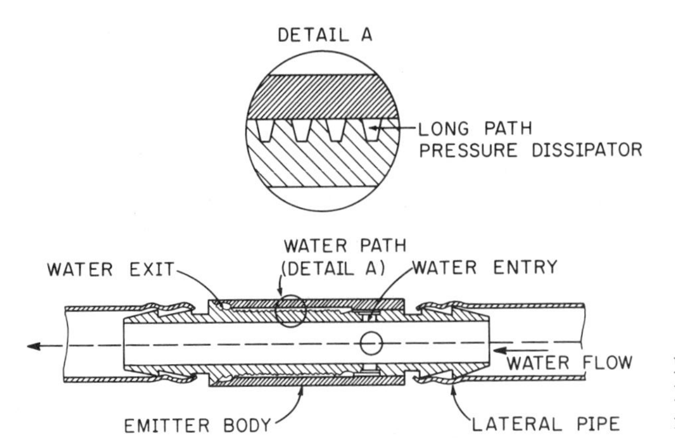

Figure 8-6a Section view of single-exit long-path emitter installed in-line. (Adapted from Keller and Karmeli, 1975.)

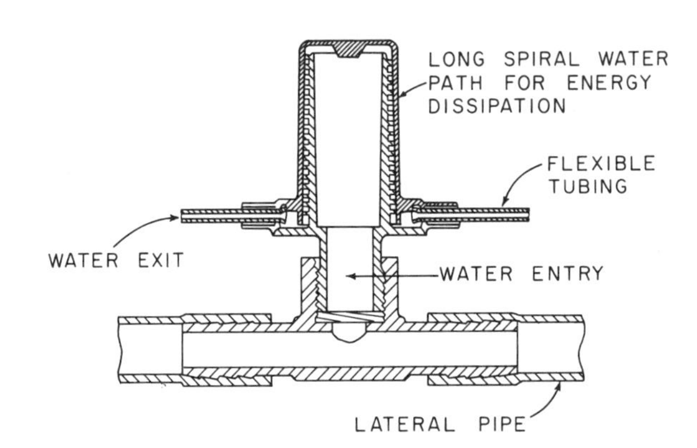

Figure 8-6b Section view of multiple-exit long-path emitter installed in-line. (Adapted from Keller and Karmeli, 1975.)

tended flow path in long-path or micro-tubing emitters (see Fig. 8-6). Figure 8-7(a) indicates a long-path emitter installed in a lateral line. Figure 8-7(b) indicates a long-path emitter which has been disassembled to demonstrate the spiral flow path.

Pressure is dissipated at the discharge orifice in the short orifice emitter shown in Fig. 8-8(a). A turbulent flow vortex emitter has increased pressure loss through the orifice compared to that operating in a laminar flow regime [see Fig. 8-8(b)]. A pressure compensating emitter, such as that shown in Fig. 8-9, aims at maintaining a constant discharge despite pressure fluctuations in the distribution system. The flexible membrane or diaphragm responds to pressure changes and keeps discharge constant within the design specifications. Figure 8-10 indicates a pressure compensating emitter installed on a lateral.

A porous pipe or double wall device is a line source emitter with closely spaced discharge points in the outer bore. These are supplied by less frequently spaced discharge points in the inner bore. This system is depicted in Fig. 8-11.

Typically, point source emitters are applied on widely spaced crops such as orchards and vineyards, and line source systems are placed on more closely spaced

Figure 8-7a Long-path emitter in operational position installed on a lateral.

Figure 8-7b Disassembled long-path emitter showing spiral flow path.

row crops. However Fig. 8-12 demonstrates that point and line source emitters may be applied in a mixture of cropping situations. The key design principle is to insure that water is applied with high uniformity to the root zone.

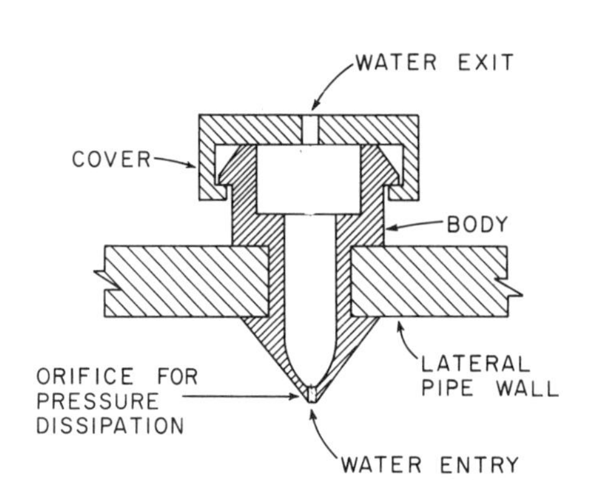

Figure 8-8a Section view of single-exit orifice type emitter installed on-line. (Adapted from Keller and Karmeli, 1975.)

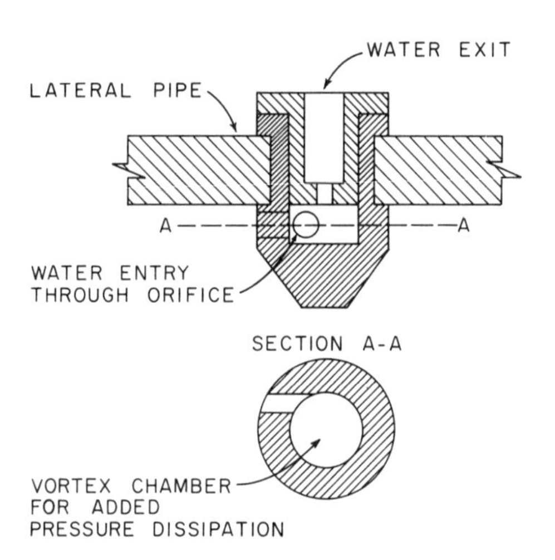

Figure 8-8b Section view of orifice-vortex type emitter installed on-line. (Adapted from Keller and Karmeli, 1975.)

Emitter Hydraulics

The relationship between emitter discharge and operating pressure is dependent upon flow regime. The flow regime is determined by the Reynold's number which is computed as

$$R_N = vD/(1000\nu) \tag{8-1}$$

where

R_N = Reynold's number, dimensionless
v = flow velocity, m/s
D = emitter diameter, mm
ν = kinematic viscosity of water, m^2/s

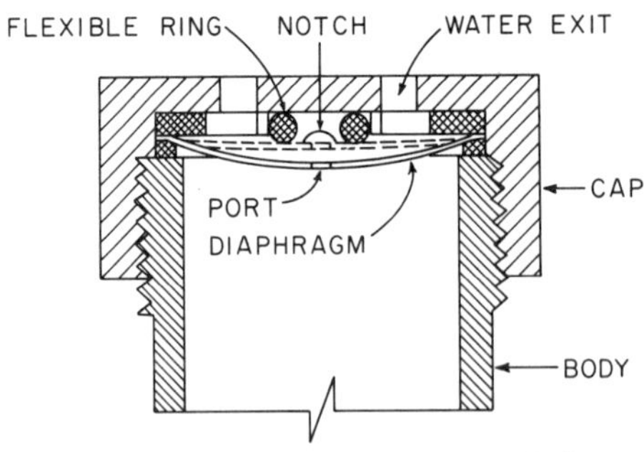

Figure 8-9 Section view of pressure compensating emitter. (Adapted from Keller and Karmeli, 1975.)

Figure 8-10 Pressure compensating emitter in operational position on lateral.

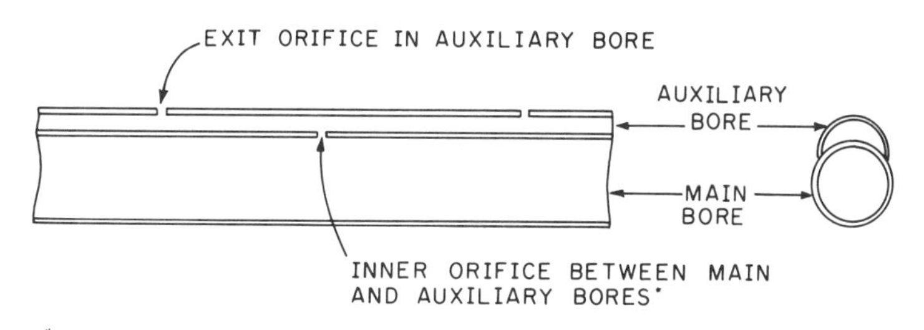

Figure 8-11 Schematic of double-wall emitter lateral. (Adapted from Keller and Karmeli, 1975.)

The standard value for the kinematic viscosity of water is 1.0×10^{-6} m^2/s at 20°C. The following categories are used to describe the flow regime as a function of Reynold's number:

(a) Laminar $R_N \leq 2000$.
(b) Unstable $2000 < R_N \leq 4000$.
(c) Partially turbulent $4000 < R_N \leq 10{,}000$.
(d) Fully turbulent $R_N > 10{,}000$.

The chapter on Pipeline System Design contains a Moody diagram which indicates the relationship between Reynold's number and the dimensionless friction factor f given by

$$f = \frac{h_f}{\left(\frac{L}{D}\right)\left(\frac{v^2}{2g}\right)} \tag{8-2}$$

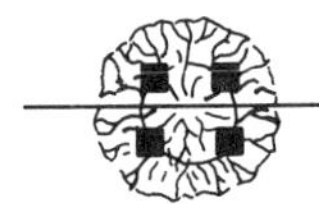

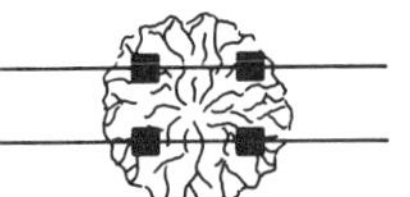

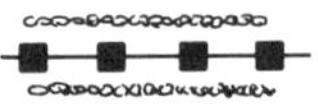

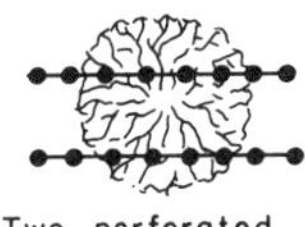

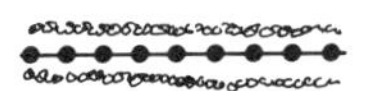

Figure 8-12 Installation of point and line source emitters in various cropping situations. (Adapted from Pair et al., 1983.)

where

h_f = headloss due to friction, m

L = length of pipe or tubing over which headloss is evaluated, m

D = diameter of pipe or tubing, m

$v^2/2g$ = velocity head of flow, m

From the Moody diagram it is evident that the friction factor is a linear function of Reynold's number in laminar flow and a nonlinear function of Reynold's number in the partially turbulent or transition zone. In fully turbulent flow, the friction factor is constant regardless of the value of the Reynold's number. A flow regime cannot be maintained in the unstable or critical zone in a field installation. This zone of flow regime is not important in design situations other than the fact that it is to be avoided because the flow is unstable.

Different equations relating emitter discharge and operating pressure are required depending on the type of emitter and flow regime. The discharge equation for an orifice emitter in fully turbulent flow is given as

$$q = 3.6(A)C_o(2gH)^{0.5} \tag{8-3}$$

where

q = emitter discharge, L/h
A = emitter cross-sectional flow area, mm^2
C_o = orifice coefficient, dimensionless
g = acceleration of gravity = 9.807 m/s^2
H = orifice operating pressure head, m

The orifice coefficient is typically assumed equal to 0.6 for application in Eq. (8-3).

The governing equation for a long-path emitter operating in a laminar flow regime is given as

$$q = 0.11384(A)\left[2g\left(\frac{HD}{fL}\right)\right]^{0.5} \tag{8-4}$$

where

f = friction factor, dimensionless
L = emitter length, m

Under turbulent flow, the governing equation for discharge from a long-path emitter is given by

$$q = 0.11384(A)\left[2g\left(\frac{\sqrt{H}\,D}{fL}\right)\right]^{0.5} \tag{8-5}$$

Different expressions are required for the friction factor f dependent upon flow regime. The linear expression for laminar flow is given as

$$f = \frac{64}{R_N} \tag{8-6}$$

For fully turbulent flow, f is said to be a function of the relative roughness of the pipe or tubing material. The expression for the friction factor in the fully turbulent regime is given by

$$\frac{1}{\sqrt{f}} = 2\log\left(\frac{D}{\epsilon}\right) + 1.14 \tag{8-7}$$

where

ϵ = absolute roughness of pipe or tubing, mm

The relative roughness is equal to ϵ/D and requires that the absolute roughness and diameter be expressed in the same units. The logarithmic expression in Eq. (8-7) is

TABLE 8-2 Values of the absolute roughness for various pipe and tubing materials. (Adapted from Albertson et al., 1960.)

Material	Absolute Roughness (mm) Minimum	Maximum
Plastic	0.003	0.03
Commercial steel and wrought iron	0.03	0.09
Galvanized iron	0.06	0.02
Aluminum	0.1	0.3
Concrete	0.3	3.0
Riveted steel	0.9	9.0
Corrugated metal pipe	30.0	60.0

base 10. Table 8-2 indicates absolute roughness values for different pipe and tubing materials. Application of Eqs. (8-3) through (8-7) are demonstrated in the following example problems.

Example Problem 8-2

Determine the required diameter for an orifice emitter in a turbulent flow regime with a design discharge of 10 L/h and operating pressure head of 10.0 m. Assume a value of 0.6 for the orifice coefficient.

Solution Applying Eq. (8-3),

$$q = 3.6(A)(0.6)[2(9.807\ \text{m/s}^2)(10.0\ \text{m})]^{0.5}$$

Substituting the design discharge and solving for the cross-sectional area and diameter,

$$A = \frac{10.0\ \text{L/h}}{30.26} = 0.3305\ \text{mm}^2 = \frac{\pi D^2}{4}$$

$$D = 0.65\ \text{mm}$$

Example Problem 8-3

Compute the required length of a long-path emitter for a system with a design discharge of 4.0 L/h and operating pressure head of 10.0 m using smooth plastic micro-tubing with an inside diameter of 1.0 mm. Assume the standard value of 1.0×10^{-6} m²/s for the kinematic viscosity of water.

Solution Compute the Reynold's number using Eq. (8-1) to determine the flow regime.

$$v = \frac{q}{A} = \frac{4.0\ \text{L/h}}{\frac{\pi(1.0\ \text{mm})^2}{4}} = 1.4147\ \text{m/s}$$

$$R_N = \frac{1.4147\ \text{m/s}(1.0\ \text{mm})}{1000\ \text{mm/m}(1.0 \times 10^{-6}\ \text{m}^2/\text{s})}$$

$$R_N = 1415 < 2000$$

Therefore the flow is laminar. Compute the friction factor using Eq. (8-6):

$$f = \frac{64}{1415} = 0.0452$$

Substituting into Eq. (8-4), the only unknown is the length of micro-tubing.

$$4.0 \text{ L/h} = 0.11384\left(\frac{\pi(1.0 \text{ mm})^2}{4}\right)\left\{2(9.807 \text{ m/s}^2)\left[\frac{10.0 \text{ m}(1.0 \text{ mm})}{0.0452 \text{ L}}\right]\right\}^{0.5}$$

$$\sqrt{L} = 1.472$$

$$L = 2.167 \text{ m}$$

Example Problem 8-4

Compute the required length of a long-path emitter for the same system as that given in example problem 8-3 except that the design discharge is 28 L/h.

Solution Compute the Reynold's number for this flow regime. Since the Reynold's number is a linear function of flow velocity and the velocity is directly proportional to the discharge, we can set up the Reynold's number as a function of that computed in example problem 8-3.

$$R_N = 1415\left(\frac{28 \text{ L/h}}{4.0 \text{ L/h}}\right) = 9905$$

Assume a flow regime will be fully turbulent at this high a Reynold's number. Apply Eq. (8-7) for the friction factor. Use the value of ϵ from Table 8-2 for plastic equal to 0.003 mm for the absolute roughness.

$$\frac{1}{\sqrt{f}} = 2 \log(1.0 \text{ mm}/0.003 \text{ mm}) + 1.14$$

$$f = 0.0261$$

Substituting into Eq. (8-5) and solving for emitter length,

$$28.0 \text{ L/h} = 0.11384(0.7854 \text{ mm}^2)\left(\frac{2(9.807 \text{ m/s}^2)\sqrt{10.0 \text{ m}}(1.0 \text{ mm})}{0.0261 \text{ L}}\right)^{0.5}$$

$$L = 0.0242 \text{ m}$$

The length requirements have therefore been substantially reduced by increasing the discharge and going from a laminar to turbulent flow regime.

Emitter Uniformity

A key element in the design of trickle irrigation systems is the close balance between the crop water requirement and the emitter discharge. To properly maintain this balance, it is important that the discharge along a lateral have a high degree of uniformity. Quantification of the uniformity is given by the emission uniformity,

$$U_e = 100[1.0 - (1.27/n)C_v]\frac{q_{min}}{q_{avg}} \quad (8\text{-}8)$$

where

U_e = emission uniformity, percent

n = number of emitters per plant for point source emitters on a permanent crop

n = whichever is the greater of (a) or (b) for a line source emitter on an annual row crop

(a) spacing between plants divided by same unit length of lateral line used by manufacturer to compute C_v

(b) 1

C_v = manufacturer's specified coefficient of variation

q_{min} = minimum emitter discharge, L/h

q_{avg} = average or design emitter discharge, L/h

The minimum emitter discharge corresponds to the point of minimum pressure in the distribution system. The manufacturer's coefficient of variation is a function of the type of emitter and the quality control exercised during the manufacturing process. Examples of values of the coefficient of variation for different types of available emitters are indicated in Table 8-3. This table also indicates values of the emitter discharge exponent which is explained later in this section. Standards have been developed by the American Society of Agricultural Engineers (ASAE) to classify emitters based on the coefficient of variation. These classifications are indicated in Table 8-4.

Standards have also been developed by ASAE for recommended ranges of design emission uniformity as a function of type of emitter, crop spacing, and field to-

TABLE 8-3 Manufacturer's coefficient of variation and emitter exponent for various types of emitters. (Adapted from Solomon, 1979.)

Type of Device	Coefficient of Variation	Emitter Discharge Exponent
Single vortex	0.07	0.42
Multiple-flexible orifice	0.05	0.70
Multiple-flexible orifice	0.07	0.70
Ball and slotted seat—non-compensating	0.27	0.50
Ball and slotted seat—pressure compensating	0.35	0.15
Ball and slotted seat—pressure compensating	0.09	0.25
Small tube	0.05	0.70
Small tube	0.05	0.80
Spiral long-path—nonflushing	0.02	0.65
Spiral long-path—manual flushing	0.06	0.75
Long-path—pressure compensating	0.05	0.40
Long-path—pressure compensating	0.06	0.20
Tortuous long-path	0.02	0.65
Groove and flap, short-path	0.02	0.33
Slot and disk, short-path	0.10	0.11
Slot and disk, short-path	0.08	0.11
Porous pipe	0.40	1.00
Twin-wall lateral	0.17	0.61

TABLE 8-4 Recommended emitter classification based on manufacturer's coefficient of variation. (Adapted from ASAE Standard EP405, 1985.)

Emitter Type	Coefficient of Variation	Classification
Point source	< 0.05	Good
	0.05–0.10	Average
	0.10–0.15	Marginal
	> 0.15	Unacceptable
Line source	< 0.10	Good
	0.10–0.20	Average
	> 0.20	Marginal to unacceptable

TABLE 8-5 Design standards for emission uniformity and uniformity coefficient for arid areas. (Adapted from ASAE Standard EP405, 1985.)

Type of Emitter	Crop Spacing	Field Topography	Emission Uniformity (percent)
Point source	Wide[a]	Uniform[c]	90–95
		Steep[d] or undulating	85–90
	Close[b]	Uniform	85–90
		Steep or undulating	80–90
Line source	Close	Uniform	80–90
		Steep or undulating	75–85

[a]Spaced greater than 4 m apart
[b]Spaced less than 2 m apart
[c]Slope less than 2 percent
[d]Slope greater than 2 percent

pography. These design standards are given in Table 8-5. Application of Eq. (8-8) and Table 8-5 are indicated in the following example problem.

Example Problem 8-5

A trickle system is to be designed for an established orchard in which the field slope is greater than 2 percent and spacing between trees is greater than 4 m. Four point source emitters corresponding to long-path–type C in Fig. 8-13 are to be used per tree. Design emitter discharge is 8.0 L/h and the manufacturer's coefficient of variation is 0.08.

Compute the minimum emitter discharge and corresponding minimum emitter pressure.

Solution From ASAE Standards in Table 8-5 the minimum design emission uniformity is

$$U_{e-min} = 85 \text{ percent}$$

Substituting problem variables into Eq. (8-8) for emission uniformity,

$$85 = 100[1.0 - (1.27/4)0.08][q_{min}/(8.0 \text{ L/h})]$$

Solving for q_{min},

$$q_{min} = 8.0 \text{ L/h}(0.85)/0.9746$$
$$q_{min} = 6.977 \text{ L/h} = 7 \text{ L/h}$$

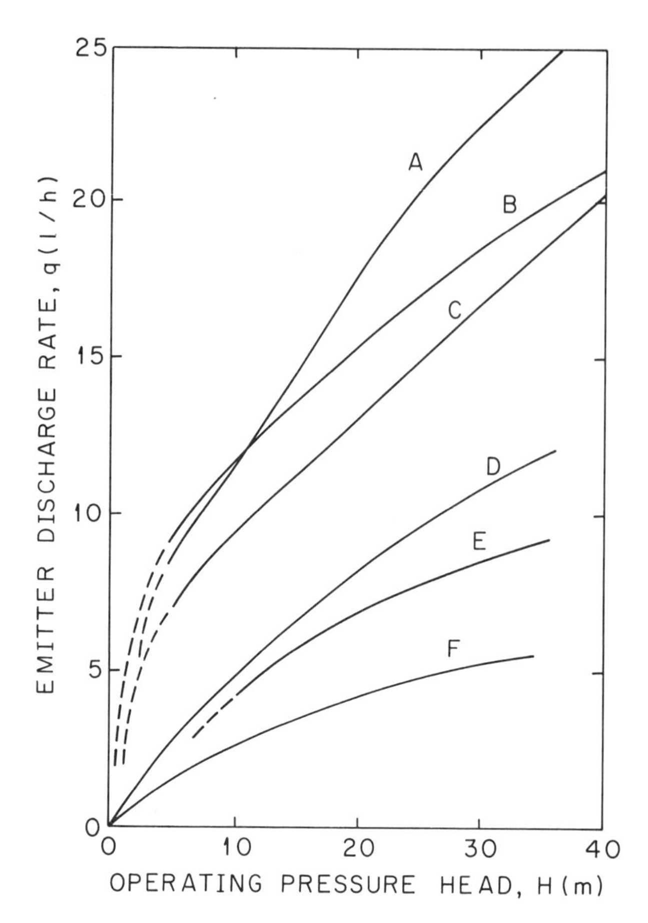

Figure 8-13 Discharge rates for various emitter designs as a function of operating head. Corresponding values of discharge at 10 m operating head tabulated in Table 8-6. (Adapted from Keller and Karmeli, 1975.)

Referring to Fig. 8-13 for long-path–type C emitter,

$$H_{min} = 5 \text{ m}$$

This is the minimum pressure head required on the lateral to maintain the design standards for emission uniformity.

If measured values of C_v, q_{min} and q_{avg} are substituted into Eq. (8-8), the result is the field emission uniformity. Specific procedures have been established so results from different fields are comparable (Merriam and Keller, 1978). These procedures are based on making measurements of emitter discharge along four lateral lines on a submain: one at the inlet, one at the far end, and two in the middle at the one-third and two-thirds positions. Four positions are tested on each lateral: one at the inlet, one at the far end, and two in the middle at the one-third and two-thirds positions. This gives a total of 16 measurement positions.

Measurements of emitter discharge are made at each measurement position for two adjacent emitters. This is done by measuring the flow volume collected in a graduated cylinder over a one-minute period. The average discharge, minimum discharge and coefficient of variation are calculated using data from the 16 positions and the results substituted into Eq. (8-8). General criteria for the field emission uniformity are (a) 90 percent or greater—excellent; (b) 80 to 90 percent—good; (c) 70 to 80 percent—fair; and (d) less than 70 percent—poor (Bralts, 1986).

A simplified way to quickly determine a measure of uniformity is to calculate the emitter flow variation using the following equation:

$$q_{var} = 100\left[1 - \frac{q_{min}}{q_{max}}\right] \tag{8-9}$$

where

q_{var} = emitter flow variation, percent

q_{max} = maximum emitter discharge, L/h

General criteria for the emitter flow variation are (a) 10 percent or less—desirable; (b) 10 to 20 percent—acceptable; and (c) greater than 20 percent—not acceptable (Bralts, 1986). This method does not give as much information about the system as using the procedure to compute the field emission uniformity, but it is simple to apply.

The emitter discharge exponent given in Table 8-3 is used to relate emitter discharge to operating pressure by the following equation:

$$q = k(H)^x \tag{8-10}$$

where

k = empirical factor

x = emitter discharge exponent

The empirical factor k also includes conversion constants to balance both sides of Eq. (8-10). The discharge exponent is a measure of the change in emitter discharge with variation in operating pressure. For discharges measured at two operating pressures, Eq. (8-10) can be rearranged as

$$x = \frac{\ln(q_1/q_2)}{\ln(H_1/H_2)} \tag{8-11}$$

Values of x should range from 0 for a pressure compensating emitter to 1.0 for an emitter in a laminar flow regime. The discharge exponent should equal about 0.5 for emitters operating in a turbulent flow regime. The higher the value of the emitter discharge exponent, the greater degree of care is required to maintain the proper pressure distribution along the lateral for the same uniformity of application.

8.3 Lateral Hydraulics

Governing Relationships

The hydraulics of trickle irrigation distribution systems up to the lateral are virtually the same as those for other distribution systems. The head losses in lines and across control devices are computed using the same procedures described in the chapters on

Sprinkler Systems and Pipelines. The pump must deliver adequate pressure head to overcome friction losses in the distribution line, filters, and control devices, compensate for changes in elevation head, and still deliver the design pressure to the emitters.

Computation of the friction headloss along a lateral will be handled similarly in this chapter as the loss along a sprinkler lateral. The hydraulic principles are identical in that the actual friction headloss must account for the fact that the flow rate is reduced along the lateral with the discharge at each emitter. The friction headloss along a lateral with no discharge through the emitters will be calculated using the following form of the Darcy-Weisbach equation:

$$h_f = 6.377fL\frac{Q^2}{D^5} \tag{8-12}$$

where

h_f = friction loss along the lateral, m
L = lateral length, m
Q = total lateral flow rate, L/h
D = lateral diameter, mm

The friction factor is computed using Eqs. (8-6) or (8-7) for laminar or turbulent conditions, respectively.

The actual friction headloss along a lateral accounting for discharge through the emitters is given as

$$h_{ac} = Fh_f \tag{8-13}$$

where

h_{ac} = actual friction headloss, m
F = Christiansen's friction factor

The Christiansen friction factor in Eq. (8-13) is the same as that introduced in the Sprinkler System Design chapter. As was demonstrated for sprinkler systems, the increase in elevation head would have to be added to the actual friction headloss in Eq. (8-13) to determine the total head required at the inlet to provide the required head at the emitter. Application of Eqs. (8-12) and (8-13) will be demonstrated in the following example problem.

Example Problem 8-6

A lateral is to be designed for a vineyard with 30 single orifice emitters along the line. The lateral is 10 mm in diameter. The emitters are spaced 1.0 m apart with the first emitter at one full space from the lateral entrance from the submain. The field surface is level. A secondary pressure regulator maintains a pressure head of 10.0 m at the entrance to the lateral which corresponds to a design emitter discharge of 12.0 L/h. Calculate the discharge of the final emitter on the lateral using Fig. 8-13 for the discharge to pressure relationship.

Solution Compute the total discharge on the lateral.

$$Q = N(q) = 30(12.0 \text{ L/h}) = 360 \text{ L/h}$$

Compute the flow velocity and Reynold's number to verify the flow regime assuming the standard value of dynamic viscosity.

$$v = \frac{Q}{A} = \frac{360\ \text{L/h}(1\ \text{m}^3/1000\ \text{L})(1\ \text{h}/3600\ \text{s})}{\dfrac{\pi(10\ \text{mm})^2}{4(1\ \text{m}/1000\ \text{mm})^2}}$$

$$v = 1.273\ \text{m/s}$$

Substituting into Eq. (8-1) for the Reynold's number,

$$R_N = \frac{1.273\ \text{m/s}(10\ \text{mm})}{1000(1.0 \times 10^{-6}\ \text{m}^2/\text{s})}$$

$$R_N = 12{,}732$$

Flow is therefore fully turbulent. Compute the friction factor for smooth plastic using Eq. (8-7).

$$\frac{1}{\sqrt{f}} = 2\log(10\ \text{mm}/0.003\ \text{mm}) + 1.14$$

$$f = 0.0149$$

Compute the friction headloss in the lateral for all flow passing through the lateral using Eq. (8-12).

$$h_f = 6.377(0.0149)(30\ \text{m})(360\ \text{L/h})^2/(10\ \text{mm})^5$$

$$h_f = 3.70\ \text{m}$$

Compute the actual friction headloss accounting for discharge through the emitters. Apply the Christiansen friction factor from Table 7-12 for first discharge point one full spacing from the entrance with $m = 2$ and $N = 30$.

$$F = 0.350$$

$$h_{ac} = 0.350(3.700\ \text{m}) = 1.30\ \text{m}$$

Since the field is level, the actual friction headloss equals the total headloss on the lateral. Calculate the operating pressure head at the last emitter in the lateral.

$$H_{final} = 10.0\ \text{m} - 1.30\ \text{m} = 8.7\ \text{m}$$

Referring to Fig. 8-13 for an orifice emitter and pressure head of 8.7 m,

$$q = 11\ \text{L/h}$$

The total variation of discharge in percent along the lateral as a ratio of the design discharge is therefore

$$\Delta q = \left(\frac{12\ \text{L/h} - 11\ \text{L/h}}{12\ \text{L/h}}\right)100$$

$$\Delta q = 8.3\ \text{percent}$$

Equation 8-13 indicates the actual friction headloss in a lateral not including losses through emitters and connectors. For some installations, especially where the emitters are of the on-line type indicated in Fig. 8-8, these losses can be considered minor. Solomon and Keller (1978) indicated that the manufacturer's coefficient of variation often has a more important effect on discharge variability than hydraulic losses. Keller and Karmeli (1975) indirectly show by variation of the Hazen-Williams C factor that friction losses due to an in-line emitter as shown in Fig. 8-6 can be five times larger than in a lateral with no emitters. They give an alternate procedure for computing the effects of friction losses in emitters and connectors by increasing the length L in Eq. (8-12) as a function of the type of emitter. The increase in length to account for emitter and connector friction effects, L_e, is given as follows (Keller and Karmeli, 1975):

(a) L_e equals from 1.0 to 3.0 m for in-line emitters as shown in Fig. 8-6.
(b) L_e equals 0.1 to 0.6 m for on-line emitters as shown in Fig. 8-8.
(c) L_e equals 0.3 to 1.0 m for a "T" connector solvent welded onto a buried plastic lateral line.

Wu et al. (1986) indicate an alternative expression for computing L_e. It is given by

$$L_e = 3.43 H_e \frac{(D)^{4.871}}{(Q)^{1.852}} \tag{8-14}$$

where

L_e = equivalent length to account for friction loss in emitters and connectors, m
H_e = emitter and connector friction loss, m
D = lateral diameter, mm
Q = lateral flow rate, L/h

Wu et al. (1986) indicate that H_e must be evaluated in laboratory experiments. This requirement will make application of Eq. (8-14) limited until data for H_e are developed or provided by emitter manufacturers. Nonetheless, the concept of adding an equivalent length of lateral for emitter and connector friction losses is convenient and potentially useful.

8.4 Filtration and Water Treatment Systems

Emitter Clogging

The most serious problem in proper operation and maintenance of trickle irrigation systems is emitter clogging. The orifices of trickle system emitters are extremely small in comparison to sprinkler nozzles and the potential for emitter clogging by particulates, chemical precipitates, or bacteria is very high. Emitter clogging can have a severe impact on crop development because of the careful balance between crop water requirements and emitter discharge built into the design of trickle irrigation systems.

Materials which can clog emitters can be divided into physical components in the form of suspended solids, chemical precipitates, and biological material produced by bacteria or algae. Table 8-6 indicates different materials which have contributed to clogging of emitters. Some of these materials are easily identifiable in the water source, such as suspended sand, silt, or clay, and the need for removal by filtration is obvious. Other components such as chemical precipitates and bacteria are not visible and can become apparent only after initiation of system operation.

Criteria to judge the susceptability of water supplies to emitter clogging have been developed to enable the design engineer to make decisions about the utility of water supplies for trickle irrigation. Table 8-7 indicates physical, chemical, and biological parameters which have been ranked as to their potential impact on operation of trickle irrigation systems. These criteria can aid the design engineer in deciding if trickle irrigation is the most adaptable system for a particular water supply and what type of filtration and water treatment devices may be required depending on the severity of the potential clogging problem.

TABLE 8-6 Principal physical, chemical, and biological components contributing to clogging of trickle system emitters. (Adapted from Bucks et al., 1979.)

Physical (suspended solids)	Chemical (precipitation)	Biological (bacteria and algae)
Inorganic particles	Calcium or magnesium carbonate	Filaments
Sand	Calcium sulfate	Slimes
Silt	Heavy metal hydroxides, carbonates, silicates, and sulfides	Microbial depositions
Clay	Oil or other lubricants	Iron
Plastic	Fertilizers	Sulfur
Organic particles	Phosphate	Manganese
Aquatic plants (phytoplankton/algae)	Aqueous ammonia	
Aquatic animals (zooplankton)	Iron, copper, zinc, manganese	
Bacteria		

TABLE 8-7 Criteria for using irrigation water used in trickle irrigation systems. (Adapted from Bucks and Nakayama, 1980.)

Type of Problem	Minor	Moderate	Severe
Physical			
Suspended solids[a]	50	50–100	100
Chemical			
pH	7.0	7.0–8.0	8.0
Dissolved solids[a]	500	500–2000	2000
Manganese[a]	0.1	0.1–1.5	1.5
Iron[a]	0.1	0.1–1.5	1.5
Hydrogen sulfide[a]	0.5	0.5–2.0	2.0
Biological			
Bacteria population[b]	10,000	10,000–50,000	50,000

[a]Maximum measured concentration from representative number of water samples using standard procedures for analysis (mg/L).

[b]Maximum number of bacteria per milliliter obtained from portable field samplers and laboratory analysis.

Filters

Filters are used to remove undesirable material from the water supply before it enters the distribution system and creates the potential for emitter clogging. Some type of filtration system is required on virtually any trickle system to insure efficient system operation. Different types of filtration devices are effective in removing different size material from the water source. When possible, the manufacturer's recommendation should be used to size the required filter media. If such a recommendation is not available, the filter should be sized based on the emitter opening and the type of material to be removed.

The most common filtration systems used for trickle irrigation systems are discussed in the following subsections. Figure 8-14 indicates sketches of sand, screen, and centrifugal filtration systems. Figure 8-15 is a photograph of a double sand filter.

Screen filters

Screen filters are probably the most common filters applied in trickle irrigation systems. They are sometimes used as the first step in the filtration process being operated in series with media filters discussed in the next subsection. As shown in the sketch in Fig. 8-14, water enters the interior of the typical screen filter and must pass through the screen before it enters the distribution line. Screens are sized according to the diameter of the particulates they are expected to remove from the flow. Table 8-8 indicates the standard mesh classification for screens and the equivalent diameter of the screen opening. Many commercially available emitters are sized to require screening of material of the 150 or 75 micron (100 or 200 mesh) size.

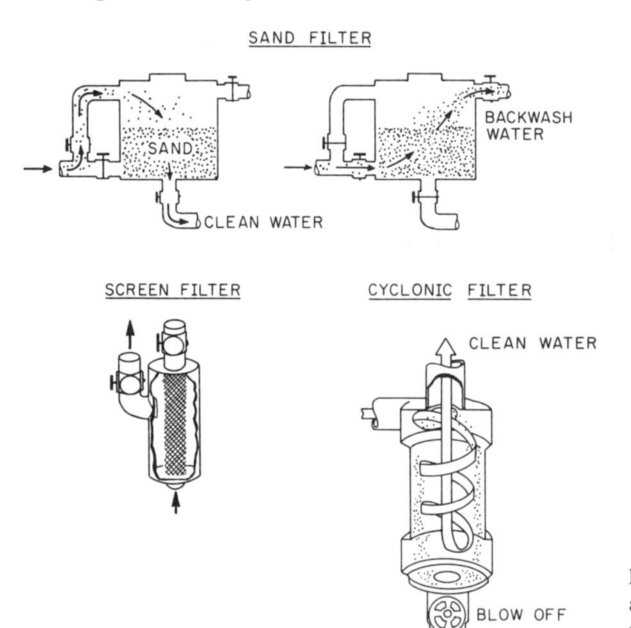

Figure 8-14 Examples of screen, media, and centrifugal filtration systems. (Adapted from Nakayama, 1986.)

Figure 8-15 Double sand filtration system installed on typical trickle system.

Media filters

Media filters are made up of graded fine gravel and sand within pressurized tanks. They can remove relatively large amounts of suspended solids before requiring back-flushing. They are available to remove particulate matter in the 25 to 100 micron size. The ASAE standard for flow rates through this type of filter is a maximum of 14 L/s per square meter of filtration surface. The thickness of the filtration media should be at least 50 cm.

Centrifugal filters

The operating principal of a centrifugal filter is shown in Fig. 8-14. They are used to remove particles which are heavier than water and larger than 75 microns and which can settle out within the resident time of the fluid passing through the filter. This type of filter is not effective in removing organic compounds. Sometimes a centrifugal filter is placed on the suction side of a pump as a primary sand removal

TABLE 8-8 Classification of screens and particle sizes. (Adapted from Nakayama, 1986.)

Screen Mesh Number	Equivalent Diameter (microns)	Particle Designation	Equivalent Diameter (microns)
16	1180	Coarse sand	>1000
20	850	Medium sand	250–500
30	600	Very fine sand	50–250
40	425	Silt	2–50
100	150	Clay	<2
140	106	Bacteria	0.4–2
170	90	Virus	<0.4
200	75		
270	53		
400	38		

system. This reduces abrasion on the pump impeller due to sand as well as removing sand particles from the trickle system distribution line.

Back-flushing

Back-flushing is the operation whereby the flow is reversed from its normal direction through the filter and accumulated particulate matter is removed from the filtration system. The method of back-flushing a media filter is indicated in Fig. 8-14. The dual sand filter in Fig. 8-15 is designed so one chamber can be back-flushed while the other chamber continues normal operation. The centrifugal filter in Fig. 8-14 is flushed by opening the blow-off valve and closing the normal discharge valve.

The capacity of the filtration system should be large enough so it can operate without frequent back-flushing or cleaning. Systems which require manual operations for cleaning should be able to operate for 24 hours without maintenance. Pressure drops in excess of 70 kPa across the filter media indicate that back-flushing or cleaning are required.

Bacterial Treatment

Trickle distribution systems can be clogged in a very short time—on the order of one week—by bacterial slime which renders the system inoperative. Figure 8-16 indicates a screen filter which was completely plugged by iron bacteria buildup within a week.

High bacteria concentrations require treatment by bactericides containing chlorine or other compounds which must remain in contact with the water source long enough to kill off the bacteria. Ten to 30 minutes of contact in a solution with a free residual chlorine concentration of 1 mg/L is normally an adequate time for this purpose. Contact is made in small reservoirs which are sized dependent on the design

Figure 8-16 Screen filter which has become completely clogged and inoperable after use during a single week due to bacterial slime. Water source was high in iron bacteria. (Photo courtesy of Marvin Shearer.)

TABLE 8-9 Chlorine equivalents of commercial sources and quantities required to treat 1,233 cubic meters of water to obtain 1 mg/L chlorine. (Adapted from Nakayama, 1986.)

Chemical	Quantity Equivalent to 454 g Cl_2	Quantity to Treat 1,233 m^3 to 1 mg/L Cl_2
Chlorine gas	454 g	1226 g
Calcium hypochlorite		
65–70% available chlorine	681 g	1816 g
Sodium hypochlorite		
15% available chlorine	2.54 L	6.81 L
10% available chlorine	3.78 L	10.22 L
5% available chlorine	7.57 L	20.44 L

flow rate of the system. Table 8-9 indicates the quantity of commercially available sources of chlorine required to treat 1,233 m^3 of water to obtain a concentration of 1 mg/L.

The standard criterion for bacterial control using chlorine is that enough chlorine should be applied in the contact reservior to have at least 0.1 mg/L free chlorine available at the end of the lateral lines under normal operating conditions. Chlorine concentrations may be tested by convenient portable kits such as those used for monitoring chlorine levels in swimming pools.

Chemical Treatment

Chemical treatment is required to reduce the potential for precipitation of insoluble salts at the orifice and internal parts of emitters which causes clogging. These precipitates are commonly caused by the reaction of soluble calcium and carbonates in the irrigation water producing calcium carbonate. The normal procedure to inhibit this reaction is to control the solution pH by addition of acid. The tendency for calcium carbonate precipitate to occur in irrigation water was described in the Soil Chemistry chapter. It will be briefly reviewed here.

The tendency for $CaCO_3$ formation is based on the Langelier Saturation Index (LSI). A calculated pH, pH_c, is compared to the measured pH of the water source, pH_m. The pH is defined as the negative base 10 logarithm of the hydrogen ion activity and p will be used to indicate a negative log transformation in general. The following relationship for the calculated pH_c involves the solubility product of $CaCO_3$, K_s, and the dissociation constant for HCO_3^-, K_d. These are defined by the following (Nakayama, 1986):

$$K_s = (Ca^{2+})(CO_3^{2-}) \tag{8-15}$$

$$K_d = \frac{(H^+)(CO_3^{2-})}{(HCO_3^-)} \tag{8-16}$$

where the terms in parenthesis are concentrations in meq/L. The equation for the calculated pH is,

$$pH_c = (pK_d - pK_s) + p[Ca^{2+}] + p[HCO_3^-] + p(ACF) \tag{8-17}$$

where

$$ACF = \text{activity coefficient factor for } Ca^{2+} \text{ and } HCO_3^-$$

The terms in brackets in Eq. (8-17) are concentrations in moles/L. HCO_3^- is used in Eq. (8-17) instead of total alkalinity because it is the dominant species (Nakayama, 1986). K_d and K_s are temperature dependent and an expression applicable to Eq. (8-17) is

$$pK_d - pK_s = 2.586 - 2.621 \times 10^{-2}T + 1.01 \times 10^{-4}T^2 \tag{8-18}$$

where

$$T = \text{solution temperature, °C}$$

The activity coefficient factor is dependent upon solution concentration and is calculated by

$$p(ACF) = 7.790 \times 10^{-2} + 2.610 \times 10^{-2}TDS - 5.477 \times 10^{-4}TDS^2 + 5.323 \times 10^{-6}TDS^3 \tag{8-19}$$

where

$$TDS = \text{total dissolved ion concentration, meq/L}$$

If the measured pH of the irrigation water is greater than the pH calculated using Eq. (8-17), there is potential for $CaCO_3$ precipitate to form. The following example problem demonstrates application of Eqs. (8-17) through (8-19).

Example Problem 8-7

A water source is measured to have the following concentrations, pH, and temperature:

$$Ca^{2+} = 1.8 \text{ meq/L}$$
$$HCO_3^- = 4.6 \text{ meq/L}$$
$$TDS = 8.6 \text{ meq/L}$$
$$T = 27°C$$
$$pH = 7.8$$

Determine if there is a potential $CaCO_3$ precipitation problem.

Solution Compute the $pK_d - pK_s$ and p(ACF) terms using Eqs. (8-18) and (8-19).

$$pK_d - pK_s = 2.586 - 2.621 \times 10^{-2}(27°C) + 1.01 \times 10^{-4}(27°C)^2$$
$$= 1.952$$
$$p(ACF) = 7.790 \times 10^{-2} + 2.160 \times 10^{-2}(8.6 \text{ meq/L}) - 5.477 \times 10^{-4}(8.6 \text{ meq/L})^2 + 5.323 \times 10^{-6}(8.6 \text{ meq/L})^3$$
$$= 0.227$$

Calculate the molar concentration of Ca^{2+} and HCO_3^- for substitution into Eq. (8-17). By the procedure indicated in the Soil Chemistry chapter, the molecular weight of HCO_3^- is

61.018 g and the equivalent weight is 61.018 g, the molecular weight of Ca^{2+} is 40.03 g and the equivalent weight is 20.015 g. Calculate concentrations in mg/L:

$$HCO_3^- = 4.6 \text{ meq/L}(61.018 \text{ g/eq}) = 280.683 \text{ mg/L}$$

$$Ca^{2+} = 1.8 \text{ meq/L}(20.015 \text{ g/eq}) = 36.027 \text{ mg/L}$$

Calculate the concentrations in mole/L and p factors:

$$HCO_3^- = \frac{280.683 \text{ mg/L}}{61.018 \text{ g/mole}} = 0.00460 \text{ mole/L}$$

$$p[HCO_3^-] = 2.337$$

$$Ca^{2+} = \frac{36.027 \text{ mg/L}}{40.03 \text{ g/mole}} = 0.00090 \text{ mole/L}$$

$$p[Ca^{2+}] = 3.046$$

Substitute values into Eq. (8-17):

$$pH_c = 1.952 + 3.046 + 2.337 + 0.227 = 7.562$$

Compute the Langelier index:

$$LSI = pH_m - pH_c = 7.8 - 7.56 = 0.24$$

The positive difference indicates a potential for $CaCO_3$ precipitation.

Decreasing the pH of the irrigation water, pH_m, by the addition of acid will cause the saturation index to become negative indicating that precipitation will not occur if all other factors remain equal. However, in Eq. (8-18) a portion of the saturation index value is temperature dependent to the extent that the sign of the LSI can be reversed by temperature effects. Thus if a water source has a negative saturation index at a temperature of 25°C at the system intake, this index could become positive as the water temperature is elevated along the submains and laterals. Once precipitates are formed, redissolution takes place at only very slow rates.

Source waters respond differently to acid treatment due to differing buffering capacities. The only sure means of determining the effect of acid treatment on pH is by the standard water quality laboratory method of titration. In general, most waters with an initial pH of approximately 8 decrease 1 unit with the addition of 0.5 meq/L acid. The use of excessive acid is uneconomical and may also cause corrosion problems on metallic parts. Methods of injecting chemicals into trickle system distribution lines are discussed in the next section under fertilizer injection.

Although the most common precipitation problem in trickle irrigation is calcium carbonate, manganese and ferric sulfides also cause problems in some locations. The reactions for these precipitates are

$$Mn^{2+} + S^- = MnS \tag{8-20}$$

$$Fe^{2+} + S^- = FeS \tag{8-21}$$

These precipitates have a black color and are difficult to remove chemically. The pH levels used to control calcium carbonate are not the same as those required to control the sulfide precipitates.

The addition of acid to a solution causes an increase in total dissolved solids as well as adjusting the pH. This increase may have some effect on the overall irrigation water salinity. The effects of the final concentrations should be evaluated using the principles described in the Soil Chemistry chapter.

8.5 Fertilizer Injection Systems

Advantages and Disadvantages

Trickle irrigation systems are particularly adaptable to fertilizer injection into the distribution system because the water is applied directly to the plant root zone. This can lead to increased efficiency of fertilizer use, reduced energy and labor costs since the same system is used for distribution of water and fertilizer, and flexibility in field management since tractors need not be scheduled for use in fertilizer application.

At the same time, care must be taken to see that the fertilizer and concentrations used are not corrosive to distribution system parts. Table 8-10 indicates the potential degree of corrosion problems on different types of metal from various sources of fertilizers. Due to the addition of nutrients to the water supply, algal growth, and bacterial slime, problems may also occur. These can cause clogging of emitters and screen and filter blockage.

Injection Equipment

Similar equipment may be used for chemical injection to control bacteria and pH and for fertilizers. This is with the exception of chlorine gas which is normally bled into the line from pressurized tanks. The equipment available includes venturi injectors, differential pressure systems with pressurized supply tanks, and diaphragm or piston driven positive displacement injectors (Nakayama, 1986). Fertilizers can be directly fed into the suction side of a pump, but this causes the corrosive fertilizer solution to flow through the expensive metal pump.

Probably the most common type of injection system is an auxiliary positive displacement pump which feeds the fertilizer solution into the distribution line.

TABLE 8-10 Severity of corrosion damage to common metals caused by fertilizers. (Adapted from Martin, 1955.)

Type of Metal	Calcium Nitrate	Ammonium Nitrate	Ammonium Sulfate	Urea	Phosphoric Acid	Diammonium Phosphate	Complete Fertilizer 17N-17P-10K
Galvanized Iron	M	SV	C	N	SV	N	M
Sheet Aluminum	N	SL	SL	N	M	M	SL
Stainless Steel	N	N	N	N	SL	N	N
Phospho-Bronze	SL	C	C	N	M	SV	SV
Yellow Brass	SL	C	M	N	M	SV	SV
pH of Fertilizer Solution	5.6	5.9	5.0	7.6	0.4	8.0	7.3

Note: N = None SL = Slight M = Moderate C = Considerable SV = Severe

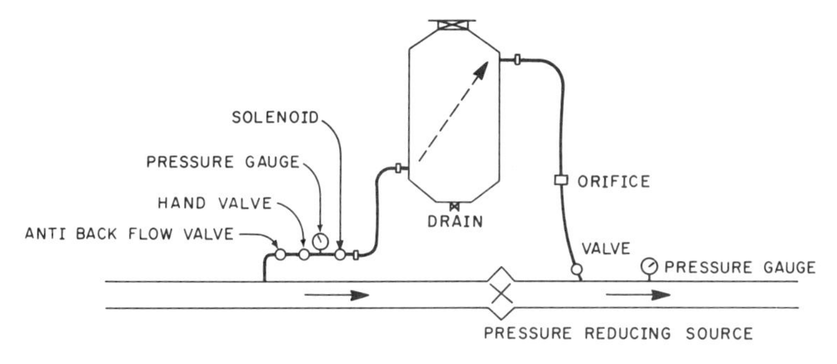

Figure 8-17 Schematic of a pressure-differential fertilizer injection system.

Another common system for fertilizer injection is the pressure differential type. A schematic of this type of system is shown in Fig. 8-17. The flow rate through the pressurized fertilizer holding tank is controlled by valves on either side of the tank. A pressure reducing obstruction or device in the line causes adequate difference in pressure to drive fluid through the tank. The advantage of this type of system is that there are no moving parts. A disadvantage is that the nutrient concentration is not constant but continuously changing with time.

The function for the concentration of material remaining in the tank as a ratio of the original concentration is given as

$$C = 100 \exp\left[\frac{-q_t(t)}{100}\right] \tag{8-22}$$

where

C = concentration of material remaining in the tank, percent

q_t = flow rate through the tank per unit time as ratio of tank capacity, percent

t = time with units corresponding to q_t

Flow rates through the tank can be controlled by valves and meters. Disk orifices installed in the tank discharge line can also control flow. The governing equation for the orifice diameter is given by

$$D = \left[\frac{15.13Q_t}{(C_o\sqrt{P})}\right]^{0.5} \tag{8-23}$$

where

D = orifice diameter, mm

Q_t = flow rate through the tank, L/min

C_0 = orifice coefficient

P = pressure differential across the orifice, kPa

The standard value for the orifice coefficient is 0.62. Application of Eqs. (8-22) and (8-23) are demonstrated in the following example problem.

Example Problem 8-8

A fertilizer injection tank with a capacity of 350 L has a 2 mm diameter disk orifice installed on the discharge side. Under operating conditions, the pressure differential across the orifice is maintained at 20 kPa. Compute the time required in hours for the concentration of fertilizer in the tank to be 50 percent of the original concentration.

Solution Rearrange Eq.(8-23) to solve for the discharge through the tank.

$$Q_t = \frac{D^2}{15.13} C_o \sqrt{P} = \frac{(2 \text{ mm})^2}{15.13} [0.62(20 \text{ kPa})^{0.5}] = 0.733 \text{ L/min} = 43.98 \text{ L/h}$$

Calculate q_t and solve for the time in hours to reach 50 percent concentration by rearranging Eq. (8-22).

$$q_t = \left(\frac{43.98 \text{ L/h}}{350 \text{ L}}\right) 100 = 12.6\%/\text{h}$$

$$t = \frac{1}{q_t}\left[-100 \ln\left(\frac{C}{100}\right)\right] = \frac{1}{12.6\%/\text{h}}\left[-100 \ln\left(\frac{50}{100}\right)\right] = 5.50 \text{ h}$$

For the conditions given, the fertilizer concentration should equal 50 percent of the original concentration after approximately 6 hours of operation.

Operating Procedures

The basic principle of operating fertilizer and chemical injection systems is that the material should not be allowed to set in the lines when the system is not operating. This is done to avoid potential corrosion problems and excessive bacterial growth. Material should not be injected into the system until all lines are filled and emitters are discharging. A standard practice to accomplish the objectives is not to begin injection until one hour after flow has begun and to terminate injection one hour before shutting down the system. This time period should insure adequate flushing of potentially problem chemicals from the line.

REFERENCES

ALBERTSON, M. L., J. R. BARTON, and D. B. SIMMONS, *Fluid Mechanics for Engineers*. New Jersey: Prentice-Hall, 1960.

American Society of Agricultural Engineers, "Design, Installation and Performance of Trickle Irrigation Systems," ASAE Standard EP405, ASAE, St. Joseph, Michigan, 1985, pp. 507–510.

BRALTS, V. F., "Chapter 3—Operational Principles—Field Performance and Evaluation," in *Trickle Irrigation for Crop Production,* eds. F. S. Nakayama and D. A. Bucks. Amsterdam: Elsevier, 1986, pp. 216–240.

BUCKS, D. A., and F. S. NAKAYAMA, "Injection of Fertilizer and Other Chemicals for Drip Irrigation," Proceedings of Agri-Turf Irrigation Conference, Houston, Texas, Irrigation Association, Silver Springs, Maryland, 1980, pp. 166–180.

BUCKS, D. A., F. S. NAKAYAMA, and R. G. GILBERT, "Trickle Irrigation Water Quality and Preventive Maintenance," *Agricultural Water Management,* vol. 2, 1979, pp. 149–162.

BUCKS, D. A., F. S. NAKAYAMA, and A. W. WARRICK, "Principles, Practices and Potentialities of Trickle (Drip) Irrigation," in *Advances in Irrigation,* vol. 1, ed. D. Hillel. New York: Academic Press, 1982, pp. 219–298.

HANKS, R. J., and J. KELLER, "New irrigation method saves water but is expensive." Utah Science, 1972. vol. 33, pp. 75–82.

KELLER, J., and D. KARMELI, *Trickle Irrigation Design.* Glendora, California: Rain Bird Sprinkler Manufacturing Corp., 1975.

MERRIAM, J. L., and J. KELLER, *Farm Irrigation System Evaluation: A Guide for Management.* Logan, Utah: Agricultural and Irrigation Engineering Department, Utah State University, 1978.

MARTIN, W. E., "Do Fertilizers Ruin Sprinkler Systems?" Proceedings Irrigation Sprinkler Conference, University of California, Davis, California, 1953.

MOODY, L. F., "Friction Factors for Pipe Flow," Transactions of the American Society of Mechanical Engineers, vol. 66, 1944, p. 671.

NAKAYAMA, F. S., "Chapter 3—Operational Principles—Water Treatment," in *Trickle Irrigation for Crop Production,* eds. F. S. Nakayama and D. A. Bucks. Amsterdam: Elsevier, 1986, pp. 164–187.

PAIR, C. H., W. W. HINZ, K. R. FROST, R. E. SNEED, and T. J. SCHLITZ, *Irrigation,* fifth edition. Silver Springs, Maryland: Irrigation Association, 1983.

PRUITT, W. O., A. A. KAMGAR, D. W. HENDERSON, and R. M. HAGAN, "Production Functions for Tomatoes as Affected by Irrigation and Row Spacing," Proceedings American Society of Civil Engineers, Irrigation and Drainage Division Specialty Conference, Boise, Idaho, 1980, pp. 346–364.

SOLOMON, K., "Manufacturing Variation of Trickle Emitters," Transactions American Society of Agricultural Engineers, vol. 22, no. 5, 1979, pp. 1034–1038, 1043.

SOLOMON, K., and J. KELLER, "Trickle Irrigation Uniformity and Efficiency," *Journal of the Irrigation and Drainage Division,* American Society of Civil Engineers, vol. 104, no. IR3, 1978, pp. 293–306.

WU, I. P., H. M. GITLIN, K. H. SOLOMON, and C. A. SARUWATARI, "Chapter 2—Design Principles—System Design," in *Trickle Irrigation for Crop Production,* eds. F. S. Nakayama and D. A. Bucks, Amsterdam: Elsevier, 1986, pp. 53–92.

PROBLEMS

8-1. A drip system is to be designed for a vineyard in an arid area with vines spaced 1.5 m apart. The soil has a slope of less than 2 percent. One emitter will serve each vine. The design emitter discharge is 8.0 L/h and the minimum discharge is 7.5 L/h. Compute the maximum allowable manufacturer's coefficient of variation for this system.

8-2. A fully turbulent orifice emitter with a diameter of 1.5 m operates in a drip system with a head of 8.0 m. Assuming the orifice coefficient equals 0.6 and the kinematic viscosity of water equals 1.0×10^{-6} m^2/s, what is the Reynolds number of flow through the emitter?

8-3. A long-path trickle system emitter is to be operated under laminar flow conditions. The design discharge is 8.5 L/h, the emitter length is 4.5 m, and the emitter diameter is 2.0 mm. Compute the required operating pressure head in meters for a circular conduit flow area. Assume the kinematic viscosity is 1.00×10^{-6} m^2/s.

8-4. A new model emitter is designed which has a discharge relation given by

$$q = KAC(2gH)^{1/2}$$

where

q = emitter discharge, L/h

A = orifice area, cm^2

C = dimensionless orifice constant

g = acceleration of gravity, 9.81 m/s^2

H = operating head, cm

K = conversion constant

Determine the magnitude and units for K.

8-5. A long-path emitter with a diameter of 2.0 mm has a design discharge of 8.5 L/h and an equivalent length of 4.5 m. Assume the water temperature is 20°C. Compute the design operating pressure head in meters.

8-6. Compute the limits of emitter discharge to irrigate an area of 1.1 ha with a flow of 78 m^3/h if laterals are spaced every 2 m and emitter spacing along the laterals is 75 cm. The irrigation requirement is 7 mm/d and the distribution pattern efficiency is 85 percent. The total time of irrigation per 8-day period is 15 hours.

8-7. A multiple outlet lateral has 50 emitters, an inlet pressure of 8 m, and a pressure head loss equal to 15 percent of average pressure. Assume that the lateral is level and that the emitter discharge is given by

$$q = 0.95H^{0.695}$$

where q is in L/h and H in meters. Compute the pressure at the last emitter on the line, the average pressure and its location, and the average, maximum and minimum discharge.

8-8. A lateral line laid on a 1 percent uniform downslope has a length of 90 m and an inside diameter of 15 mm. Emitters with a design discharge of 4.2 L/h at a pressure head of 10 m are placed along the line every 0.70 m. The discharge of the emitter selected is a function of the square-root of the operating pressure head. Determine the pressure and discharge distribution of emitters along the lateral and the percent variation of discharge.

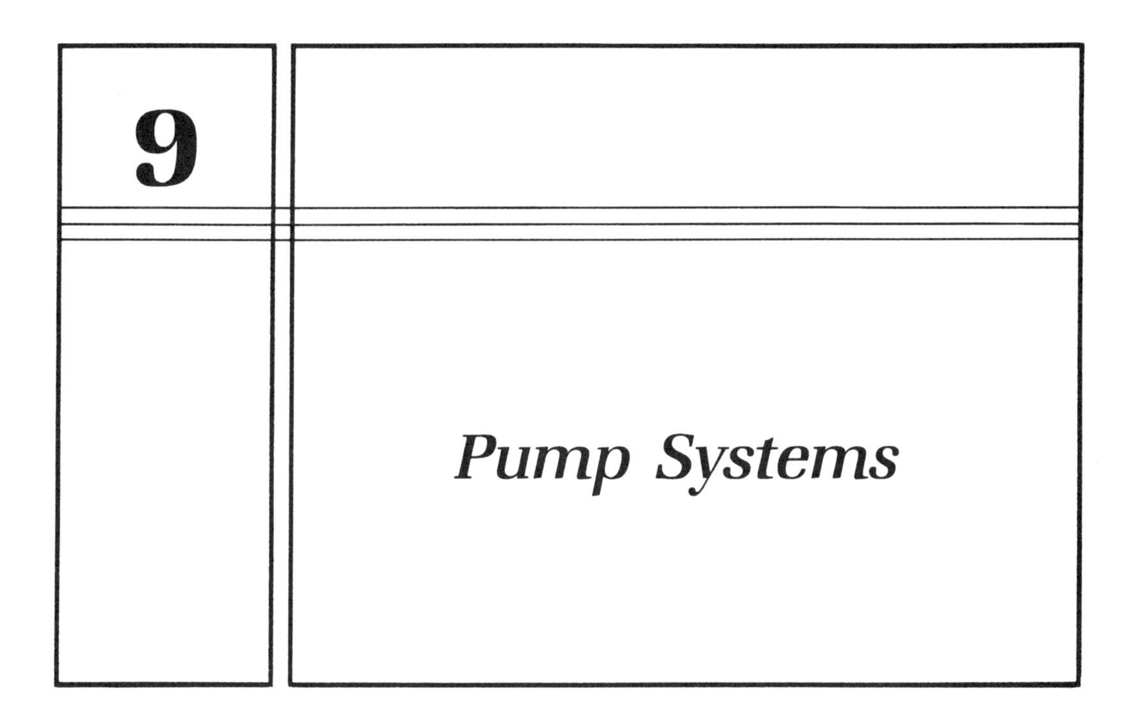

9 Pump Systems

Pumps are used in irrigation systems to impart a head to the water so it may be distributed to different locations on the farm and used effectively in application systems. Figures 9-1 and 9-2 show hand- and animal-driven pumps not usually seen in many parts of the world. Both the tambour in Fig. 9-1 and sakia in Fig. 9-2 are used in low lift situations in the Nile River Valley. In this application they lift water from a supply ditch to a distribution ditch for use in surface irrigation.

The key requirement in pump selection and design of pump systems for typical irrigation installations is that there is a correspondence between the requirements of the irrigation system and the maximum operating efficiency of the pump. The requirements of the irrigation system are the flow rates and pressure output necessary to operate the system as it was designed. This chapter will look at the performance characteristics of different types of pumps and describe procedures to match pump performance characteristics with system requirements. Procedures for proper installation of pumps to insure that they operate as expected will also be demonstrated.

9.1 Types of Pumps

Pumps used in irrigation systems are available in a wide variety of pressure and discharge configurations. Pressure and discharge are inversely related in pump design so pumps which produce high pressure have a relatively small discharge and pumps

Figure 9-1 Archemedes screw type pump (locally called a tambour) used to produce low lift to transport water from a supply ditch in the Nile River Valley.

Figure 9-2 Animal driven water wheel (locally called a sakia) used to lift water from the supply ditch on the left of the photo to the distribution ditch on the right. Wheel rotates in a clockwise direction. Water from the supply ditch is scooped up at the bottom of the wheel and channeled into the center of the wheel for distribution to the delivery ditch. Photo taken in the Nile River Valley.

which produce a large discharge are capable of relatively low pressures. This relationship is shown schematically in Fig. 9-3 which indicates the characteristics of centrifugal, turbine, and propeller pumps.

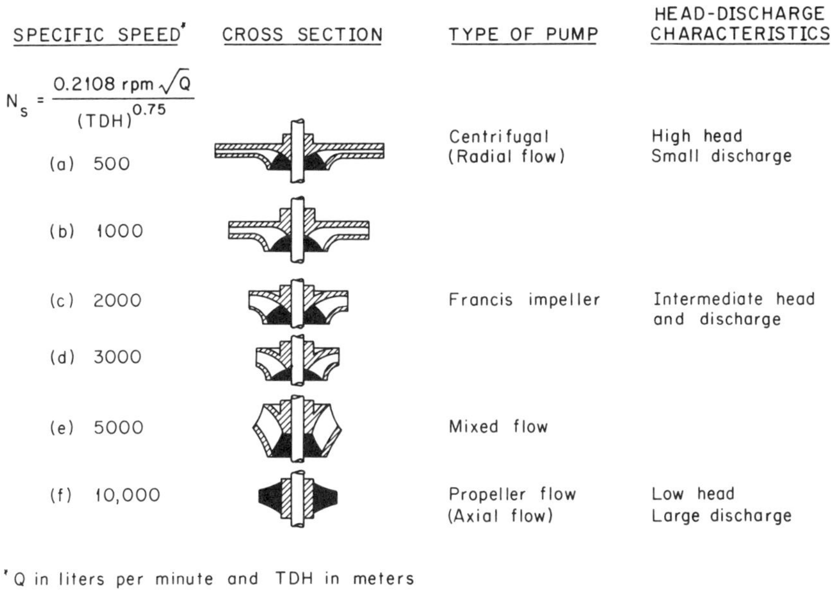

SPECIFIC SPEED*	CROSS SECTION	TYPE OF PUMP	HEAD-DISCHARGE CHARACTERISTICS
$N_s = \frac{0.2108 \text{ rpm} \sqrt{Q}}{(TDH)^{0.75}}$			
(a) 500		Centrifugal (Radial flow)	High head Small discharge
(b) 1000			
(c) 2000		Francis impeller	Intermediate head and discharge
(d) 3000			
(e) 5000		Mixed flow	
(f) 10,000		Propeller flow (Axial flow)	Low head Large discharge

*Q in liters per minute and TDH in meters

Figure 9-3 Variation in impeller design and pump characteristics as function of specific speed. (Adapted from Hansen et al., 1980.)

For the pump impellers demonstrated in Fig. 9-3, the flow enters the pump from the bottom of the diagram. The figure demonstrates that in a centrifugal pump, for example, energy is imparted to the flow by the impeller which directs the flow radially outward. Centrifugal pumps are used in applications requiring high heads but limited discharge. Proceeding down the diagram, the impellers have increased cross-sectional flow area but reduced diameter. The result is that pumps such as those using a Francis impeller may deliver intermediate flow rates but there is less energy available to pressurize the fluid. Single stage turbine type pumps which use such impellers are capable of delivering moderate rates of flow but at pressures lower than those produced by a centrifugal pump.

Proceeding towards the bottom of Fig. 9-3, the discharge flow direction goes from radially outward, through what is termed mixed flow, and finally to axial flow in a propeller type pump. A propeller type system is able to deliver large flow volumes but is capable of imparting a very small pressure differential to the fluid.

One means of quantitatively categorizing the operating characteristics of a pump is the specific speed. This quantity is given by

$$N_s = 0.2108N\left(\frac{Q^{0.5}}{H^{0.75}}\right) \tag{9-1}$$

where

N_s = specific speed, dimensionless

N = revolutionary speed of pump, rev/min (rpm)

Q = pump discharge, L/min

H = discharge pressure head, m

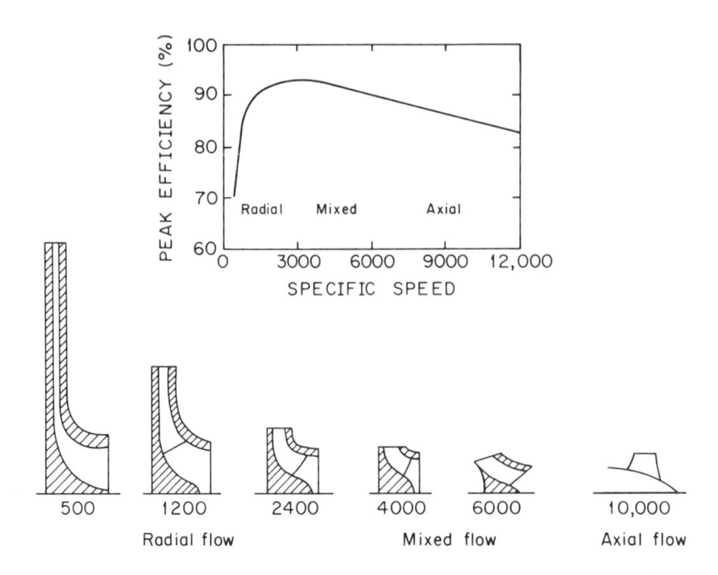

Figure 9-4 Impeller shape and maximum efficiency as function of specific speed. (Adapted from Linsley and Franzini, 1972.)

The specific speed is seen in Fig. 9-3 to range from approximately 500 for a centrifugal pump to 10,000 for a propeller pump.

The specific speed of a pump is closely related to the maximum operating efficiency of the pump. This relationship is demonstrated in Fig. 9-4. The operating efficiency is the ratio of the power imparted by the impeller to the water compared to the power supplied to the pump. The power is supplied by a motor which rotates the pump shaft. In the beginning of this chapter we refer to pump efficiency which considers only the power imparted to the water and the power coming in the drive shaft. Later on in this chapter, we discuss *pumping plant efficiency* which considers the pump and the motor driving the pump as a single unit. The pumping plant efficiency therefore includes the efficiency of the motor.

The performance curve in Fig. 9-4 indicates that careful attention must be given to the discharge requirements of the pump, which determine the specific speed, so the most suitable pump may be selected. This point is demonstrated in the following example problem.

Example Problem 9-1

An irrigation system is designed which has a required discharge of 1140 L/min. The water must be lifted from a surface source to a field 30.0 m higher in elevation for application by sprinklers. The total friction loss from the water source to the sprinkler nozzle is estimated at 5.0 m and the required operating pressure of the nozzle is equivalent to 35.0 m head. The pump is to be driven by a motor which has a shaft rotational speed of 1750 rpm.

(a) Determine the appropriate type of pump for this application.

(b) Calculate the head which would be produced by a propeller pump at peak efficiency with the same flow rate and shaft rotational speed.

Solution

(a) Apply Eq. (9-1) for specific speed.

$$N_s = \frac{0.2108N(Q)^{0.5}}{H^{0.75}}$$

Compute total head.

$$H = 30.0 \text{ m} + 5.0 \text{ m} + 35.0 \text{ m} = 70.0 \text{ m}$$

$$N_s = 0.2108(1750 \text{ rpm}) \frac{(1140 \text{ L/min})^{0.5}}{70.0 \text{ m}^{0.75}}$$

$$N_s = 515$$

Referring to Fig. 9-4, the pump with the maximum efficiency corresponding to this N_s is centrifugal.

(b) From Fig. 9-3 for a propeller pump.

$$N_s = 10{,}000$$

Rearranging Eq. (9-1),

$$(H)^{0.75} = \frac{.02108N(Q)^{0.5}}{N_s}$$

$$(H)^{0.75} = \frac{0.2108(1750 \text{ rpm})(1140 \text{ L/min})^{0.5}}{10{,}000}$$

$$(H)^{0.75} = 1.2455$$

$$H = 1.34 \text{ m}$$

9.2 Hydraulic Principles

Affinity Laws

The volumetric flow rate, discharge head, and power required to drive a pump are related to the impeller diameter and the shaft rotational speed. The relationships between these parameters are called the affinity laws for pump performance. The governing relations which describe the result of variation of rotational speed and impeller diameter on the pump operating characteristics are given by the following:

(a) Discharge capacity, Q, versus rotational speed:

$$\frac{Q_1}{Q_2} = \frac{N_1}{N_2} \qquad (9\text{-}2)$$

(b) Discharge capacity versus impeller diameter, D:

$$\frac{Q_1}{Q_2} = \frac{D_1}{D_2} \qquad (9\text{-}3)$$

(c) Discharge head versus rotational speed:

$$\frac{H_1}{H_2} = \left[\frac{N_1}{N_2}\right]^2 \tag{9-4}$$

(d) Discharge head versus impeller diameter:

$$\frac{H_1}{H_2} = \left[\frac{D_1}{D_2}\right]^2 \tag{9-5}$$

(e) Power, P, versus rotational speed:

$$\frac{P_1}{P_2} = \left[\frac{N_1}{N_2}\right]^3 \tag{9-6}$$

(f) Power versus impeller diameter:

$$\frac{P_1}{P_2} = \left[\frac{D_1}{D_2}\right]^3 \tag{9-7}$$

Application of the affinity laws is demonstrated in the following example problem.

Example Problem 9-2

An existing irrigation system has a pump which delivers 3000.0 L/min flow at a total head of 62.0 m. The impeller diameter is 25.0 cm and it is rotated at 1750 rpm. A motor with an output shaft power of 54 metric horsepower is required to drive the pump.

The existing irrigation system is modified to incorporate low pressure nozzles and make other reductions in friction head loss. The required flow rate is unchanged but the discharge pressure requirement is reduced to 50.0 m head. It is decided to keep the existing pump but to pull and trim the impeller to match the new system requirements.

(a) Determine the impeller diameter required for the new discharge conditions and the metric shaft horsepower necessary to drive the modified pump.

(b) Compute the flow rate which will have to be produced by an additional pump to meet the original discharge requirements when the modified pump is re-installed.

Solution

(a) Apply Eq. (9-5) for modified head conditions.

$$\frac{H_1}{H_2} = \left[\frac{D_1}{D_2}\right]^2$$

$$D_2 = D_1\left[\frac{H_2}{H_1}\right]^{0.5} = 25.0\text{ cm}\left[\frac{50.0\text{ m}}{62.0\text{ m}}\right]^{0.5}$$

$$D_2 = 22.45\text{ cm}$$

or approximately a 10 percent trim of the impeller.

Apply Eq. (9-7) for the power relationship.

$$\frac{P_1}{P_2} = \left[\frac{D_1}{D_2}\right]^3$$

$$P_2 = P_1\left[\frac{D_2}{D_1}\right]^3 = 54\text{ mhp}\left[\frac{22.45\text{ cm}}{25.0\text{ cm}}\right]^3$$

$$P_2 = 39.1\text{ mhp}$$

(b) Compute the flow rate for the modified system using Eq. (9-3).

$$\frac{Q_1}{Q_2} = \frac{D_1}{D_2}$$

$$Q_2 = Q_1\left[\frac{D_2}{D_1}\right] = 3000.0\ \text{L/min}\left[\frac{22.45\ \text{cm}}{25.0\ \text{cm}}\right]$$

$$Q_2 = 2694.0\ \text{L/min}$$

Since the required flow rate remains 3000.0 L/min, the flow rate which must be produced by an additional pump is

$$Q_{add} = Q_1 - Q_2 = 3000.0\ \text{L/min} - 2694.0\ \text{L/min}$$

$$Q_{add} = 306\ \text{L/min}$$

Power

The power required to operate a pump is directly proportional to the flow rate, discharge pressure head, and specific gravity of the fluid, and is inversely proportional to the pump efficiency. This power must be supplied by a motor to the pump drive shaft so the pump impeller can impart the power to the water at the relevant pump efficiency. The following formulas are expressions for the power which vary dependent upon the input units.

$$P = \frac{Q(H)(Sg)}{4634(E)} \tag{9-8}$$

where

P = power, metric horsepower (mhp)
Q = pump discharge, L/min
H = discharge pressure head, m
Sg = specific gravity of the fluid, dimensionless
E = pump efficiency, fraction

$$P = \frac{Q(H)(Sg)}{278.04(E)} \tag{9-9}$$

where

Q = pump discharge, m^3/hr

and all other units remain unchanged.

$$P = \frac{Q(H)(Sg)}{0.102(E)} \tag{9-10}$$

where

P = power, kW
Q = pump discharge, m^3/s

and all other units remain unchanged. In the English system the formula for power is given by

$$P = \frac{Q(H)(Sg)}{3960(E)} \tag{9-11}$$

where

P = power, brake horsepower (bhp)

Q = pump discharge, gallons per minute (gpm)

H = discharge pressure head, ft

and other units remain unchanged. Note that the metric horsepower and brake horsepower are nearly identical in magnitude:

$$1.0000 \text{ mhp} = 1.01422 \text{ bhp} \tag{9-12}$$

Since this book is concerned with the design of irrigation systems, the specific gravity required in the preceding power formulas will be that of water. It may be observed from Table B-1 in the Appendix that the specific gravity of water is slightly dependent upon temperature. For the purposes of irrigation system design, the specific gravity can be assumed to equal 1.0 with no loss in accuracy. When systems involving fluids with viscosities much different than water are designed, the specific gravity values relevant to the particular fluid will have to be applied in the foregoing equations.

Net Positive Suction Head

The net positive suction head principle will be described for the case in which the source of water is exposed to the atmosphere and the intake of the pump is above the level of the water source. This is the most common condition for evaluation of net positive suction head in irrigation systems. The analysis indicated applies equally to sources of water under pressure or cases in which the pump intake is below the level of the water source. Analysis of such conditions requires only proper definition of terms and sign.

The net positive suction head relates the atmospheric pressure on water on the intake side of the pump to the vapor pressure of the liquid. By the time the water reaches the intake of the pump, the original head due to atmospheric pressure has been reduced by the height of the pump intake above the water source, the velocity head developed, and friction losses in the intake pipe. If the pressure head remaining is less than the vapor pressure of the fluid at the operating temperature, the fluid will vaporize as it is accelerated by the impeller of the pump. By Bernoulli's principle there is an inverse relationship between pressure head and velocity head. Along the impeller where pressure once again increases to levels greater than vapor pressure, the vapor bubbles collapse causing a condition termed *cavitation*. The forces involved in cavitation are extremely destructive to metal impellers and cause the impellers to operate inefficiently and far below their design conditions after a very short time.

To insure that pumps operate according to their specified performance characteristics, pump manufacturers specify a required net positive suction head. The required net positive suction head must be available if cavitation is not to occur at the impeller tip at the design rotational speed. This concept is extremely important in the installation of centrifugal pumps which are often placed at some height above the source of water. To insure that the pump is properly installed, the engineer must verify that the net positive suction head available is greater than that required by the manufacturer. This condition is expressed as

$$NPSH_a > NPSH_r \tag{9-13}$$

where

$NPSH_a$ = net positive suction head available, m

$NPSH_r$ = net positive suction head required, m

The $NPSH_a$ is computed by

$$NPSH_a = p_{atmos} - z_s - h_{fs} - p_v \tag{9-14}$$

where

p_{atmos} = atmospheric pressure, m

z_s = static suction lift, m

h_{fs} = friction head loss on suction side of pump, m

p_v = vapor pressure of the liquid at operating temperature, m

The *static suction lift* is equal to the difference in height between the level of the water source and the centerline of the pump intake. It is defined as positive if the pump is above the water source and negative if the pump is below the water source. The friction head loss is computed for the straight length of pipe and any fittings on the suction side of the pump. The friction head loss is normally quite small because it is proportional to the fluid velocity squared which is less than 1 m/s on the suction side for a properly installed pump.

Atmospheric pressure is commonly based on what is referred to as the Standard U.S. Atmosphere which quantifies the pressure as a decreasing function of altitude. Values for atmospheric pressure based on the U.S. standard are listed in Table 9-1. Formulas for atmospheric pressure were given in the Crop Water Requirements chapter. The equivalent atmospheric pressure available for a given problem is computed by dividing the value given in Table 9-1 by the specific gravity of the fluid involved. As previously indicated, the specific gravity of water for the design of irrigation systems can be set equal to 1.0 with no loss in accuracy. This allows the atmospheric pressure to be read directly off Table 9-1 as long as the fluid to be pumped is water.

If the water source is not exposed to the atmosphere, as would be the case if the water supply was a pressurized tank, the atmospheric pressure term in Eq. (9-14) must be replaced by the absolute pressure at the water surface. The vapor pressure of water is a function of temperature. Values for the vapor pressure of water are given in Table B-1 of the Appendices. The application of the vapor pressure concept is demonstrated in the following example problem.

TABLE 9-1 Standard U.S. atmospheric pressures at various altitudes.

Altitude m	mm Hg*	Atmospheric Pressure kPa	m H_2O[†]
−300	788	105.0	10.7
−200	778	103.7	10.6
0	760	101.3	10.3
200	742	98.9	10.1
400	725	96.6	9.9
600	707	94.3	9.6
800	690	92.0	9.4
1000	674	89.8	9.2
1200	657	87.6	8.9
1400	641	85.5	8.7
1600	626	83.4	8.5
1800	611	81.4	8.3
2000	596	79.4	8.1
2200	581	77.5	7.9
2400	567	75.6	7.7
2600	553	73.7	7.5
2800	539	71.8	7.3
3000	526	70.1	7.1
3200	512	68.3	7.0
3400	500	66.6	6.8
3600	487	64.9	6.6
3800	474	63.2	6.4
4000	462	61.6	6.3
4200	450	60.0	6.1
4400	439	58.5	6.0
4600	427	56.9	5.8

*Mercury at 0°C
[†]Water at 4°C
Source: PSEUDO-ADIABATIC CHART: lower levels, Chart WEA D-11. U.S. Department of Commerce, National Oceanic and Atmospheric Administration, National Weather Service.

Example Problem 9-3

A centrifugal pump is to be installed at a site with an elevation of 400 m where it will be required to pump water at 30°C. The water source is exposed to the atmosphere and the friction losses on the suction side of the pump are estimated at 0.6 m. The net positive suction head required from the manufacturer's specifications is 5.2 m. Compute the maximum height that the centerline of the pump intake may be placed above the level of the water source.

Solution From Table 9-1 of Standard U.S. Atmosphere for 400 m elevation,

$$p_{atmos} = 9.9 \text{ m}$$

From Table B-1 for properties of water at 30°C,

$$p_v = 0.43 \text{ m}$$

From the problem description,

$$NPSH_r = 5.2 \text{ m}$$

$$h_{fs} = 0.6 \text{ m}$$

Rearranging Eq. (9-14) and setting

$$NPSH_a = NPSH_r$$

as the minimum condition for $NPSH_a$, we can calculate the maximum height as

$$z_s = p_{atmos} - p_v - NPSH_a - h_{fs}$$

$$z_s = 9.9 \text{ m} - 0.43 \text{ m} - 5.2 \text{ m} - 0.6 \text{ m}$$

$$z_s = 3.7 \text{ m}$$

This solution is demonstrated schematically in Fig. 9-5. Note that in practice a factor of safety of 0.6 m is generally subtracted from the calculated z_s for unanticipated conditions on the suction side of the pump. This reduces the maximum static suction lift to

$$z_{s-max} = 3.7 \text{ m} - 0.6 \text{ m}$$

$$z_{s-max} = 3.1 \text{ m}$$

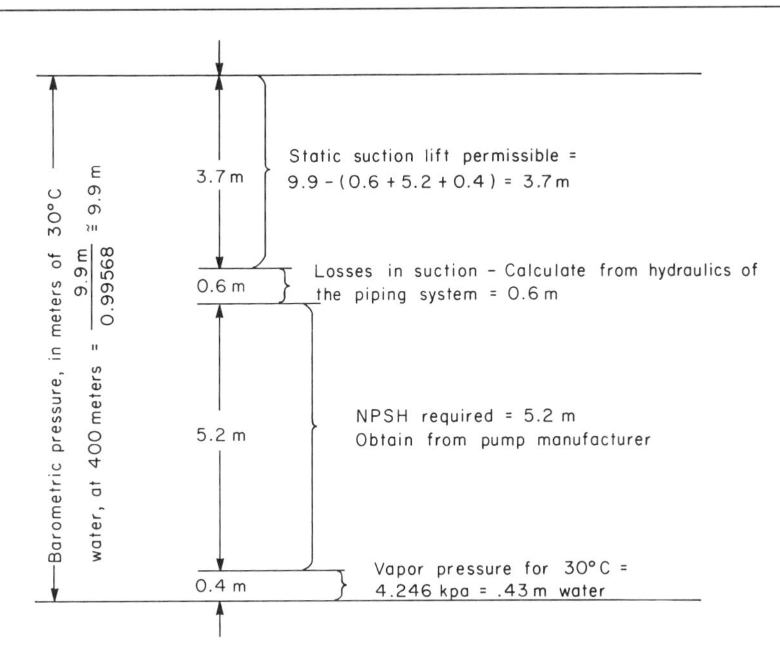

Figure 9-5 Graphical representation of application of net positive suction head.

9.3 Pump Selection

Pump Performance Curves

For a particular pump, manufacturers produce pump performance curves which graphically relate the head-discharge relation, pump efficiency, and shaft input requirements of the pump. An example of such a curve for a single impeller diameter is indicated in Fig. 9-6. Considerable useful information is available to the design engineer from such a curve.

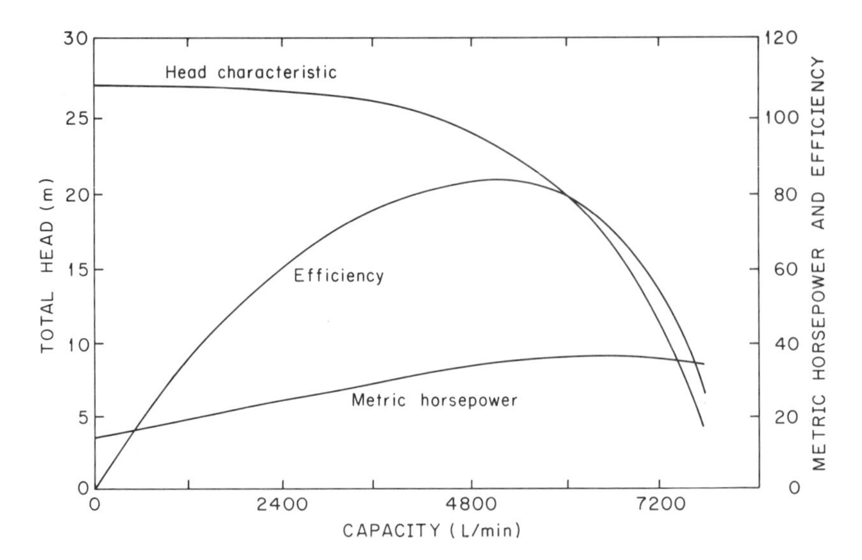

Figure 9-6 Components of a pump performance curve.

Let us first examine the head-discharge relationship which is described by what is called the head characteristic curve. The shaft power delivered to the pump is partitioned into producing head or into causing flow according to the hydraulics of the particular pump. Figure 9-6 indicates that at zero discharge—that is, with the pump running but the discharge valve closed—the pump produces the maximum head. This maximum head corresponding to zero discharge is called the *dead head* or *shut-off head*. It is recommended that the pump be operated in this manner only long enough to determine the dead head reading in a pump test due to the fact that the cooling and lubrication of the pump bearings are dependent on flow through the pump.

The efficiency is a ratio of the power output by the pump divided by the power delivered to the pump through the shaft. The power output by the pump is termed the *water horsepower* and is equivalent to the power given in Eqs. (9-8) to (9-11) but without the pump efficiency term. The water horsepower is given by

$$P_w = \frac{Q(H)}{4634} \tag{9-15}$$

where

P_w = water horsepower, mhp

Q = pump discharge, L/min

H = discharge pressure head, m

The units in Eq. (9-15) correspond to the units in Eq. (9-8) and equations corresponding to the other equations given for power [i.e., Equations (9-9), (9-10), and (9-11)] could be similarly written. Note that Eq. (9-15) has been written for water as the fluid with the specific gravity assumed equal to 1.0. If other fluids were used, the specific gravity of the fluid would have to be input into the equation as was done in Eq. (9-8).

Both the shape and magnitude of the pump efficiency curve vary depending on the pump design and diameter of the impeller. This efficiency approaches a maximum at a particular combination of discharge and head and then drops below the maximum at increasing values of discharge. The shaft input power required is a function of the pump efficiency and is also seen to go through a maximum point. The task of the design engineer is to select the pump which has the highest efficiency at the flow rate and pressure conditions required for operation of the irrigation system.

Figure 9-7 is a reproduction of a pump performance curve developed by a manufacturer from pump tests. Additional information is available in this curve to aid the design engineer. The most obvious addition is that the curve has been drawn for various possible impeller diameters which can all be used within the same pump housing. As expected from application of the affinity laws, the potential maximum head decreases with decrease in diameter. Contours of pump efficiency curves have been drawn to cover the full range of diameters presented in the figure. The efficiencies indicated for this model range from approximately 40 to 83 percent. These efficiencies are relative to the shaft power input to the pump and do not include the efficiency of the motor used to drive the shaft.

Figure 9-7 also indicates contour lines of net positive suction head required to avoid cavitation on the impeller. These contours cover the full operating range of the impeller and can be applied using linear interpolation for points between lines. The power requirements for the full range of pump operation are also indicated on the

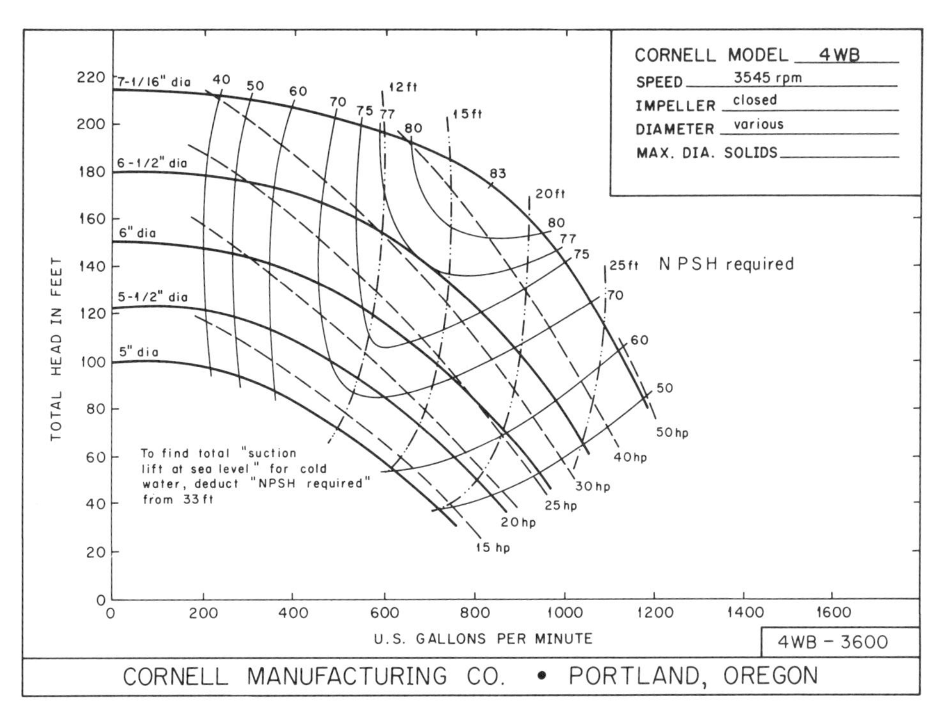

Figure 9-7 Pump performance curve for a Cornell pump model 4WB. (Courtesy of Cornell Manufacturing Company, Portland, Oregon.)

graph. The power ratings indicated refer to the nominal shaft *output* of motor sizes available in the United States. For any point falling between two contours of motor size, the next larger size motor must be applied. The percentage of the motor output actually required under pump operating conditions is referred to as the motor *loading*. Typical applications of manufacturer's pump curves for design purposes are demonstrated in the following example problem.

Example Problem 9-4

The Cornell Model 4WB pump operated at 3545 rpm is to be applied in an irrigation system which requires a discharge of 2650 L/min (700 gpm) at a total head of 42.67 m (140 ft). Determine the following responses based on the pump characteristic curve in Fig. 9-7.

(a) Select the impeller diameter to the nearest 2 mm (1/16 inch) for the required discharge.
(b) Determine the pump efficiency in percent for the operating conditions.
(c) Determine the nominal power requirement for a motor to power the pump.
(d) Compute the actual power required by the pump and the percent load on the motor.
(e) What will the pressure be on a gate valve on the discharge side of the pump if the valve is closed and the pump is operating?
(f) Compute the maximum distance in feet the center line of the pump may be placed above a source of water in an installation at 1600 m altitude pumping water at 10°C.

Solution

(a) Applying the pump characteristic curve for the required operating conditions,

$$D = 6.5 \text{ inch} = 165 \text{ mm}$$

(b) Pump efficiency:

$$E = 77 \text{ percent}$$

(c) From the pump characteristic curve, the nominal motor size required to power the pump is 40 bhp.

(d) Applying Eq. (9-8) to compute the actual power required,

$$P = \frac{Q(H)}{4634(E)}$$

$$P = \frac{2650 \text{ L/min}(42.67 \text{ m})}{4634(0.77)}$$

$$P = 31.7 \text{ mhp} = 32.2 \text{ bhp}$$

The next larger manufactured pump motor is rated at 40 bhp. The percent load is then

$$\text{Motor loading} = \frac{32.2 \text{ bhp}}{40 \text{ bhp}} = 80.4 \text{ percent load}$$

(e) Pressure on the discharge side of the pump against the closed valve is the maximum head from the pump performance curve for a 6.5 inch (165 mm) diameter impeller:

$$h_{max} = 180 \text{ ft} = 54.9 \text{ m}$$

(f) To compute maximum installation height from Table 9-1 for 1600 m elevation,

$$p_{atmos} = 8.5 \text{ m}$$

From Table B-1 for 10°C,

$$p_v = 0.13 \text{ m}$$

Assume 0.90 m losses for suction side friction and miscellaneous losses. Allow 0.67 m factor of safety. Total losses,

$$h_{fs} = 1.57 \text{ m}$$

From pump performance curve,

$$NPSH_r = 14 \text{ ft} = 4.27 \text{ m}$$

Set $NPSH_a = NPSH_r$ to compute maximum height above water source. Rearranging Eq. (9-14)

$$z_s = p_{atmos} - h_{fs} - p_v - NPSH_r$$

$$z_s = 8.5 \text{ m} - 1.57 \text{ m} - 0.13 \text{ m} - 4.27 \text{ m}$$

$$z_s = 2.5 \text{ m}$$

This is the maximum height the centerline of the pump may be placed above the water source to operate within the manufacturer's specifications.

Discharge and Pressure Requirements

Pumps are chosen to match the required performance characteristics of the irrigation system at a high level of efficiency. The first step is to develop an irrigation system performance curve which relates the total system head to discharge. The total system head is divided into two components: fixed head and variable head.

The fixed system head does not vary with discharge. It is made up of the difference between the static water level and elevation to the discharge point. Referring to Fig. 9-8 as an example, the fixed system head is given by

$$h_{fix} = h_{zs} + h_{zd} \qquad (9\text{-}16)$$

where

h_{fix} = fixed system head, m

h_{zs} = static head on suction side of pump, m

h_{zd} = static head on discharge side of pump, m

The variable system head increases with increase in discharge. It is made up of well drawdown if the water source is a well, friction loss in pipeline and fittings, pressure at the outlet, plus velocity head. If the discharge is into an open ditch or surface irrigation system at atmospheric pressure, the outlet pressure is zero and constant. For a sprinkler system such as shown in Fig. 9-8, the discharge pressure is

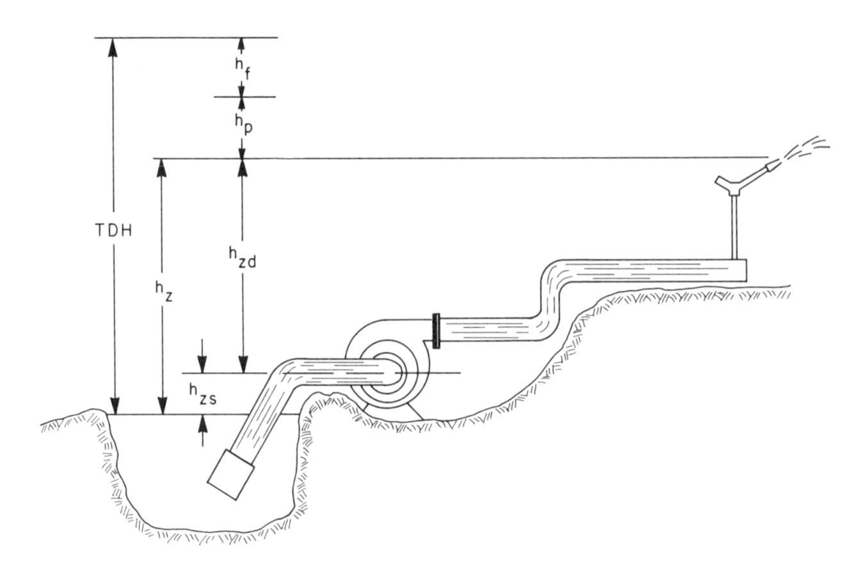

Figure 9-8 Distribution of elevation, pressure, and friction head in a centrifugal pump installation. The total dynamic head, TDH, is the sum of the fixed and variable heads.

not constant. In this case, the pressure at the outlet is variable because the sprinkler nozzle is an orifice for which nozzle discharge is a function of nozzle pressure. The variable system head is given by

$$h_{var} = s_{well} + h_f + h_p + v^2/(2g) \tag{9-17}$$

where

h_{var} = variable system head, m
s_{well} = well drawdown, m
h_f = total friction headloss in mainline and fittings, m
h_p = pressure head at critical discharge point in distribution system, m
v = velocity at critical discharge point in distribution system, m/s
g = acceleration of gravity, m/s^2

The sum of the fixed and variable system operating heads is the total dynamic head, TDH, given by

$$\text{TDH} = h_{fix} + h_{var} \tag{9-18}$$

It is the total dynamic head which is substituted for head in the power calculations in Eqs. (9-8) through (9-11). The velocity head given by $v^2/2g$ is normally a small component of the total dynamic head in properly designed irrigation distribution systems [less than approximately 0.1 m (0.4 ft) for typical mainline velocity constraints]. It is often neglected in calculations of the TDH without loss of accuracy. Computation of the fixed plus variable system head is demonstrated in the following example problem.

Example Problem 9-5

A centrifugal pump is to be installed at a height 2.74 m (9.0 ft) above the water level in an irrigation pond which is the water source for a sprinkler system. The elevation difference from the center of the pump to the point in the distribution pipeline with the critical (i.e., highest) pressure requirement is 4.57 m (15.0 ft). The critical pressure requirement at that point is equivalent to 35.20 m (50.0 psi) at a system design discharge of 47.3 L/s (750.0 gpm). The following fittings are on the suction side of the pump:

(a) 305 mm (12.0 in.) foot valve with strainer and poppet disk with L/D = 420.
(b) 2.74 m (9.0 ft) length of steel pipe 305 mm (12.0 in.) in diameter.
(c) 90 degree flanged elbow 305 mm (12.0 in.) in diameter with r/D = 10 (r/D is ratio of radius of curvature to pipe diameter).
(d) Eccentric reducer from 305 mm (12.0 in.) to 254 mm (10.0 in.) with 28 degree interior angle.
(e) 1.22 m (4.0 ft) length of steel pipe 254 mm (10.0 in.) in diameter.

The diameter of the pump discharge is 102 mm (4.0 in.). The following fittings are between the pump discharge and point of critical pressure requirement in the distribution line:

(f) Concentric expansion from 102 cm (4.0 in.) to 203 mm (8.0 in.) with 12 degree interior angle.
(g) 203 mm (8.0 in.) stop-check valve with $\frac{L}{D} = 400$.
(h) 203 mm (8.0 in.) globe valve with $\frac{L}{D} = 340$.
(i) 90 degree flanged elbow 203 mm (8.0 in.) in diameter with $\frac{r}{D} = 10$.
(j) 91.44 m (300.0 ft) length of steel pipe 203 mm (8.0 in.) in diameter.

Perform a fixed and variable head analysis for pump discharge from 31.6 L/s (500.0 gpm) to 47.3 L/s (750.0 gpm) in increments of 3.16 L/s (50.0 gpm). Check that the maximum intake velocity at the pump is less than 1 m/s (3 ft/s) and maximum flow velocity in the mainline is less than 1.5 m/s (5 ft/s).

Solution Apply Eq. (9-16) to compute fixed system head:

$$h_{fix} = h_{zs} + h_{zd}$$
$$h_{fix} = 2.74 \text{ m} + 4.57 \text{ m}$$
$$h_{fix} = 7.31 \text{ m}$$

Apply Eq. (9-17) to compute variable system head:

$$h_{var} = s_{well} + h_f + h_p + \frac{v^2}{2g}$$

We can set up the following functional relationships for the variables in Eq. (9-17):

$$s_{well} = f(Q)$$
$$v = f(Q^2, D)$$
$$h_f = f(Q^2, D, \text{type of fitting})$$
$$h_p = f(Q^2)$$

Therefore the variable system head is a function of the discharge (Q), diameter of the flow section (D), and the fittings used.

Apply tables from Cameron Hydraulic Data (Westaway and Loomis, 1979) in Appendix C for headloss through fittings. Assume that the formula for a concentric reducer is applicable for the eccentric reducer. [Note: L/D ratio is used to calculate the length L of straight pipe of diameter D with equivalent friction headloss as through the fitting. In this example problem it is used to specify the resistance coefficient from the hydraulic data table.]

Apply the Hazen-Williams equation with C equal to 100 for headloss in steel pipe. As indicated in Appendix C, the form of the Hazen-Williams equation applicable to this problem is

$$h_f = 1.22 \times 10^{10}(L)\left[\frac{(Q/C)^{1.852}}{D^{4.87}}\right]$$

where

$$h_f = \text{friction loss, m}$$
$$L = \text{pipe length, m}$$
$$Q = \text{discharge, L/s}$$
$$D = \text{pipe diameter, mm}$$

The headloss through fittings is computed as

$$h_f = K(v^2/2\ g)$$

where

$$K = \text{resistance coefficient from Appendix C}$$
$$v = \text{flow velocity through fitting unless noted otherwise, m/s}$$

The headloss through the fitting is in the same units as the velocity head since K is dimensionless.

Assuming that the sprinklers can be treated as smooth bore nozzles, the relationship between discharge pressure and flow rate is given as,

$$h_p = K_b(Q^2)$$

where

$$K_b = \text{smooth bore discharge-pressure coefficient}$$

For the design conditions given, namely a discharge pressure head of 35.20 m at the design discharge of 47.3 L/s, the value of K_b is calculated as,

$$K_b = 0.01573$$

Table 9-2 indicates the results of the analysis for variable system head. It indicates the flow velocities and headlosses through each section of the flow problem as described in this example. The bottom of the table is a summary of the results for variable, fixed, and total system head requirements.

The results of the analysis conducted in example problem 9-5 are conveniently displayed in tabular form as indicated in Table 9-2. The velocities printed out in the table allow a quick check that flow rates are within the maximum velocity constraints. With proper definition of the data locations in Table 9-2, fixed and variable system head analysis can be efficiently done by use of computerized spreadsheets.

Such an analysis has the advantage that once the table is properly set up, variables such as flow rates, pipe diameters, and lengths can be modified and the impact on the operation of the system ascertained immediately. A certain amount of time and effort is required to establish the original spreadsheet, but once it is developed subsequent analyses can be made with tremendous efficiency. The results indicated in Table 9-2 are referred to as the system performance curve when plotted as indicated in Fig. 9-9.

System performance curves are applied in conjunction with the pump performance curve investigated earlier. An example is demonstrated in Fig. 9-10. A pump in good working order always operates on its own performance curve. The affinity laws constrain the pump from moving off that curve. The system performance curve indicates the range of values over which the system may operate. However, in observing Fig. 9-10, there is only one point on the system performance curve which intersects the pump performance curve. Since the pump is constrained to operate on its

TABLE 9-2(a) Analysis of fixed and variable heads to develop system head curve (SI units).

CENTRIFUGAL PUMP INSTALLATION
FIXED/VARIABLE HEAD ANALYSIS

Suction line:		Discharge line:	
Diameter (mm)	305	Discharge diameter (mm)	102
Length (m)	2.74		
Pump intake:			
Diameter (mm)	254	Mainline diameter (mm)	203
Length (m)	1.22	Mainline length (m)	91.44
Hazen-Williams C factor	100		

Flow Section	Flow Velocity (m/s), Headloss (m) and Pressure Head (m) Discharge (L/s) 31.55	34.70	37.85	41.01	44.16	47.32
v-suction	0.432	0.475	0.518	0.561	0.604	0.648
Foot Valve	0.052	0.063	0.075	0.088	0.102	0.118
Pipe	0.003	0.004	0.005	0.005	0.006	0.007
90 Elbow	0.004	0.004	0.005	0.006	0.007	0.008
Eccen Red-28	0.001	0.001	0.001	0.001	0.001	0.001
v-intake	0.623	0.685	0.747	0.809	0.872	0.934
Pipe	0.003	0.004	0.005	0.006	0.007	0.007
v-discharge	3.860	4.247	4.633	5.019	5.405	5.791
Concn Exp-12	0.116	0.141	0.167	0.196	0.228	0.261
v-main	0.975	1.072	1.170	1.267	1.365	1.462
Vel. Head	0.048	0.059	0.070	0.082	0.095	0.109
Check Valve	0.271	0.328	0.390	0.458	0.531	0.610
Globe Valve	0.232	0.281	0.335	0.393	0.456	0.523
90 Elbow	0.019	0.023	0.027	0.032	0.037	0.042
Pipe	0.781	0.931	1.094	1.269	1.456	1.654
Pressure Head	15.653	18.940	22.540	26.453	30.679	35.219
h-variable	17.184	20.779	24.714	28.990	33.605	38.560
h-fixed	7.310	7.310	7.310	7.310	7.310	7.310
H-total	24.494	28.089	32.024	36.300	40.915	45.870

TABLE 9-2(b) Analysis of fixed and variable heads to develop system head curve (English units).

CENTRIFUGAL PUMP INSTALLATION
FIXED/VARIABLE HEAD ANALYSIS

Suction line:		Discharge line:	
Diameter (in)	12	Discharge diameter (in)	4
Length (ft)	9		
Pump intake:			
Diameter (in)	10	Mainline diameter (in)	8
Length (ft)	4	Mainline length (ft)	300
Hazen-Williams C factor	100		

Flow Section	Flow Velocity (fps), Headloss (ft) and Pressure Head (ft) Discharge (gpm)					
	500	550	600	650	700	750
v-suction	1.418	1.560	1.702	1.844	1.986	2.128
Foot Valve	0.172	0.208	0.247	0.290	0.337	0.387
Pipe	0.010	0.012	0.015	0.017	0.019	0.022
90 Elbow	0.012	0.015	0.018	0.021	0.024	0.027
Eccen Red-28	0.002	0.002	0.003	0.003	0.004	0.004
v-intake	2.042	2.247	2.451	2.655	2.859	3.064
Pipe	0.011	0.013	0.016	0.018	0.021	0.024
v-discharge	12.765	14.042	15.319	16.595	17.872	19.148
Concn Exp-12	0.387	0.468	0.557	0.654	0.759	0.871
v-main	3.191	3.511	3.830	4.149	4.468	4.787
Vel. Head	0.158	0.191	0.228	0.267	0.310	0.356
Check Valve	0.886	1.072	1.275	1.497	1.736	1.993
Globe Valve	0.759	0.919	1.093	1.283	1.488	1.708
90 Elbow	0.062	0.075	0.089	0.104	0.121	0.139
Pipe	2.496	2.977	3.498	4.057	4.654	5.288
Pressure Head	51.334	62.114	73.921	86.754	100.615	115.501
h-variable	56.289	68.067	80.959	94.966	110.086	126.320
h-fixed	23.984	23.984	23.984	23.984	23.984	23.984
H-total	80.273	92.051	104.943	118.950	134.070	150.304

curve, it will produce the head and discharge indicated for the intersection of the system and pump curves. In Fig. 9-10 this intersection is seen to occur at a discharge of 3785 L/min (1000 gpm) with a total dynamic head of approximately 48.8 m (160 ft).

The responsibility of the design engineer is to match the system and pump performance curves so the system can be operated at its designed capacity and pressure. The performance curve for the pump chosen must then be inspected to check that the pump is able to operate at a high level of efficiency relative to the design requirements of the irrigation system. If this is not the case, another pump must be chosen, the match between the system and pump performance curves verified, and the pump efficiency evaluated to see if it is satisfactory. Taking the time to insure that the pump will be operating at a high level of efficiency at this stage in the design will insure a cost-effective operation over the life of the system.

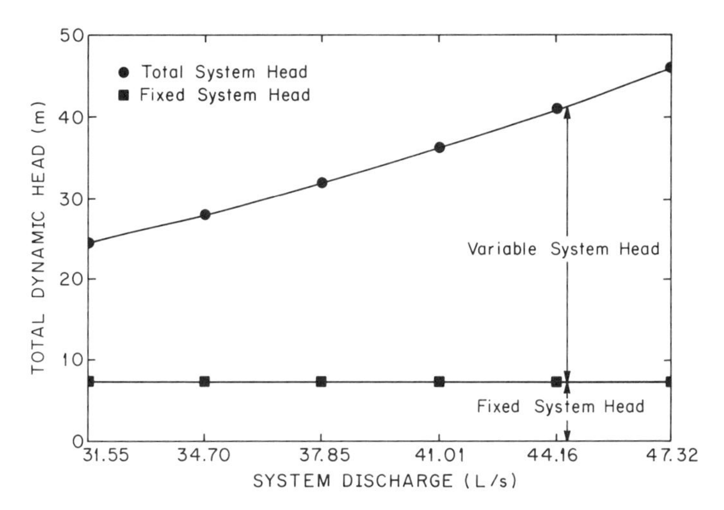

Figure 9-9 System head-capacity curve shown as summation of fixed and variable heads.

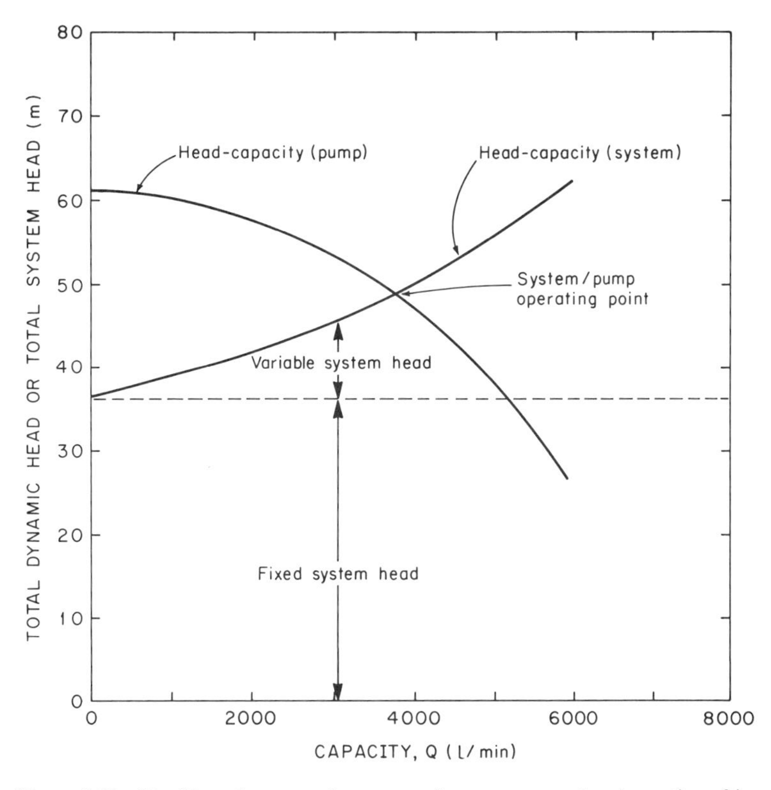

Figure 9-10 Matching of pump and system performance curves by observation of intersection point.

Figure 9-11 shows another common application of matching system and pump performance curves. In the case depicted, two system performance curves are plotted representing two projected types of operation for a sprinkler system. In one case a single sprinkler lateral will be operated and in the second case two laterals will be operated simultaneously. The head-capacity curve of a trial pump for this operation is indicated as is the pump efficiency curve. The design engineer has to check that the discharge and pressure at the intersections of the pump and system curves is adequate for the demands of the irrigation system. If they are satisfactory, the efficiency of the pump at the two points of operation must be investigated to see that the maximum point falls about midway between the two curves. In this way the pump is not operating far from maximum efficiency in either case. This is assuming that the irrigation system will be operated for approximately the same amount of time on both system curves. The pump performance indicated in Fig. 9-11 is not particularly well suited to the system performance curves since the peak efficiency occurs very close to the operating point for the two lateral system. If the normal operation was to be with two laterals and the single lateral system used only rarely, the pump indicated in Fig. 9-11 would probably be adequate in terms of efficiency.

Application of Performance Curves for Pump Selection

Earlier in this section the type of information presented by a single pump performance curve was investigated. Manufacturers of pumps have many types of pumps available with various operating characteristics. Now that the development of a system curve has been demonstrated, we are in position to choose the pump which is best suited to the irrigation system.

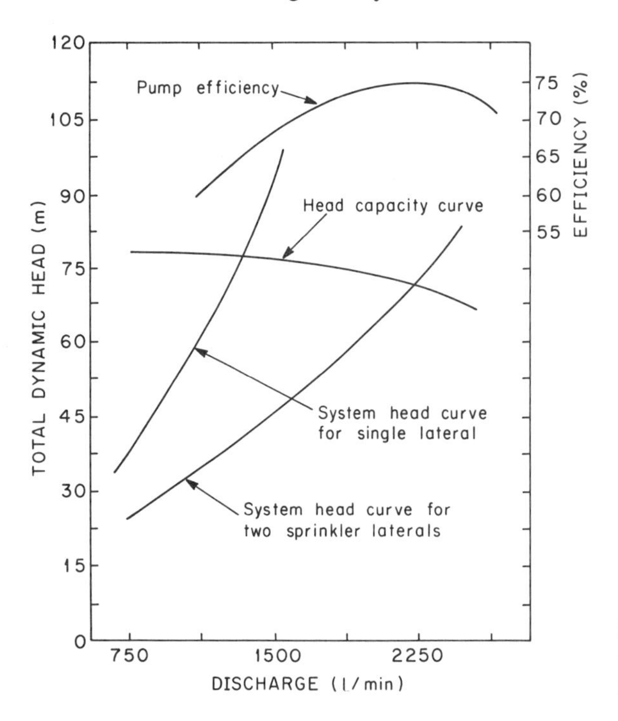

Figure 9-11 Variation of system head curve with change in sprinkler system layout and intersection of pump performance curve.

The specific speed calculation indicated earlier gives an idea of what type of pump will have the highest efficiency for the system requirements. There are some additional guidelines which help direct the proper pump selection. We have already seen how propeller type pumps can move large quantities of water but have little head producing capabilities. Therefore any system with moderate or high head requirements will have to employ a turbine or centrifugal type pump.

A single turbine pump has moderate discharge and moderate pressure producing capabilities. However multiple turbine sections can be constructed in series to produce very high heads so pressure requirements are not generally a constraint in the use of turbine pumps. More explanation of pumps in series and in parallel is given in the following section. Turbine pumps can also be placed directly into the water source within a well casing due to their limited radius. The motors are connected to the pump by a drive shaft from the surface or directly coupled to the pump in the well by the use of a submersible electric motor.

Centrifugal pumps produce high pressure but lower flow rates than turbine pumps. However, centrifugal pumps can be used in parallel to overcome the limitation of discharge. A limitation of centrifugal pumps that cannot be overcome is that their construction and radius prohibits them from being conveniently placed in a well of normal diameter. The net positive suction head requirements restrict the position of a centrifugal pump to a maximum of approximately 4.6 m above the water source. Centrifugal pumps are therefore useful only with shallow wells if the source of water is groundwater. When the water source is a reservior, stream, surface channel, or shallow well, centrifugal pumps can be placed on the ground surface where they can be conveniently operated and maintained. A wide range of centrifugal pumps with different operating capabilities is available for application to irrigation systems.

The selection of a pump based on models available from a manufacturer will be demonstrated for a centrifugal pump. The general performance characteristics for various models are demonstrated in Fig. 9-12. Pumps with electric motors are driven at either 1750 or 3500 rpm. Models driven at 1750 rpm are capable of lower head but higher discharge than those driven at 3500 rpm. The envelope curves in the figure refer to the model number which has the highest efficiency for the operating conditions which fall within the envelope. The first one or two digits of the model number indicate the nominal impeller diameter in inches and the last two digits the discharge diameter in tenths of an inch. Once the correct model is selected, the pump performance curve for that model is used to determine the specific operating characteristics for the system as previously demonstrated. A typical sample page of performance curves from a manufacturer's catalog is given in Fig. 9-13. Application of these charts is indicated in the following example problem.

Example Problem 9-6

An irrigation system has a discharge requirement of 2270 L/min (600 gpm) at a total head of 18.29 m (60.0 ft). Select the most efficient pump model for this application from those demonstrated in Fig. 9-12. Indicate the correct impeller diameter to the nearest millimeter and the pump efficiency at the design operating conditions from Fig. 9-13.

Solution From Fig. 9-12, select model 1030A. Referring to the pump performance curve for model 1030A in Fig. 9-13,

$$D = 9.5 \text{ inch} = 241 \text{ mm}$$

$$E = 82 \text{ percent}$$

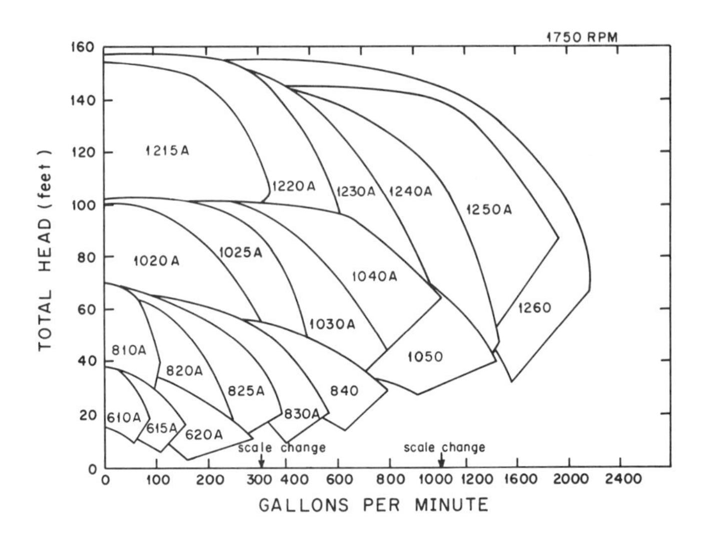

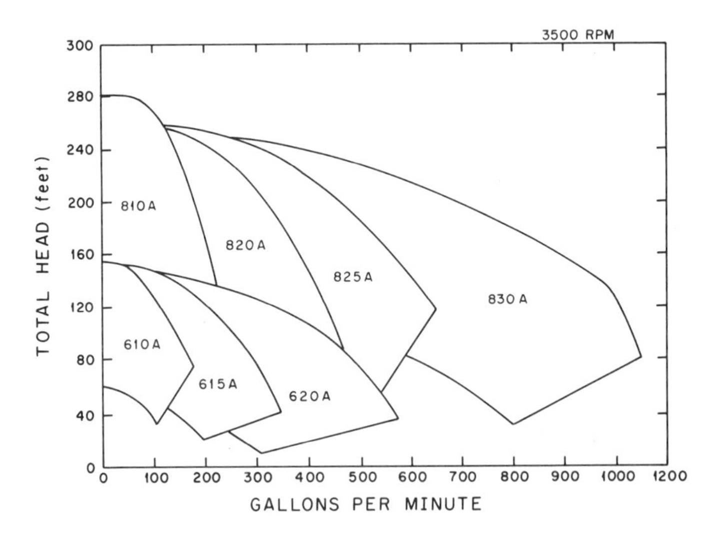

Figure 9-12 Envelop curves for pump selection based on system operating requirements. (Courtesy of Peerless Pump, Montebello, California.)

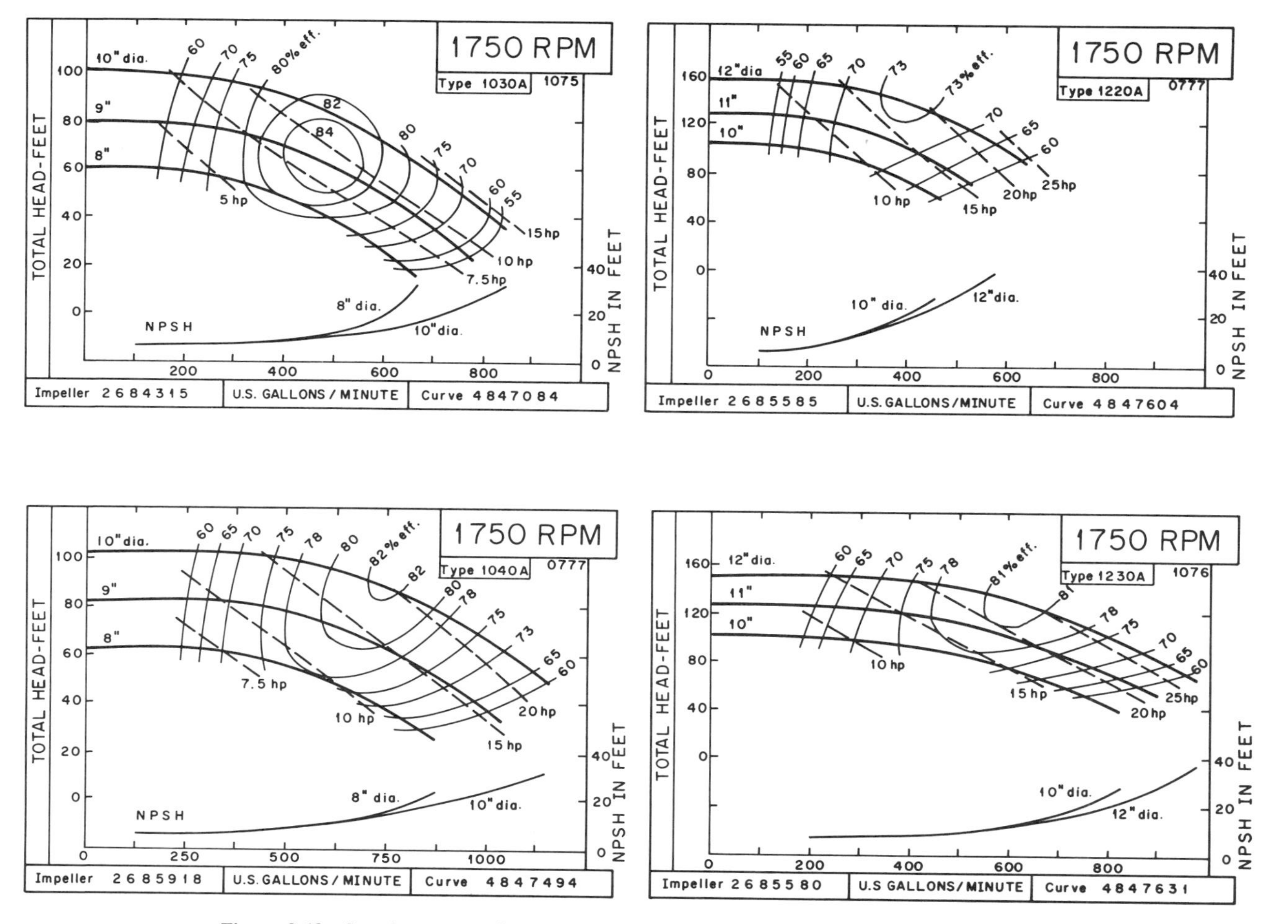

Figure 9-13 Sample pump performance curves corresponding to models indicated in Fig. 9-12. (Courtesy of Peerless Pump, Montebello, California.)

9.4 Pump System Configuration

Discharge requirements of various types of irrigation systems sometimes necessitate the combination of different pumps to achieve design flow rates and pressures. Pumps are placed in series—that is, sequentially on the same line—to produce increased pressure. Pumps are placed in parallel when the flow rate of a single pump is inadequate. Booster pumps are normally considered a special case of pumps in series in which additional pressure is developed for limited portions of the distribution system.

Pumps in Series

Pumps are in series when the downstream pump discharges directly into the intake of the following pump. The distance between the two pumps is normally limited to a few meters. This type of installation is used when the discharge flow rate from a single pump is adequate, but a larger head is required than can be produced from the single pump. Consider two pumps A and B in series with discharge Q_A and Q_B, output heads H_A and H_B, and power requirement P_A and P_B. The following equations govern the series system:

$$Q_{series} = Q_A = Q_B \tag{9-19}$$

$$H_{series} = H_A + H_B \tag{9-20}$$

$$P_{series} = P_A + P_B \tag{9-21}$$

From Eq. (9-19) we see that the discharge of the two pumps is the same. The efficiency of the pump system is therefore evaluated by considering the H_{series} and P_{series} just given in the following equation,

$$E_{series} = \frac{Q_{series}(H_A + H_B)}{0.102(P_A + P_B)} \tag{9-22}$$

where

$$Q_{series} = \text{discharge, m}^3/\text{s}$$

$$H_A \text{ and } H_B = \text{discharge pressure head, m}$$

$$P_A \text{ and } P_B = \text{power, kW}$$

Equation (9-22) may be evaluated for other units of discharge and head by reference to the units given in the power relations in Eqs. (9-8), (9-9), and (9-11).

Example Problem 9-7

Compute the discharge pressure head in meters, total power required in kW, and overall pumping system efficiency for model 1040A with a 25.4 cm (10.0 inch) impeller and model 1230A with a 30.48 cm (12.0 inch) impeller given in Fig. 9-13 when the pumps are placed in series. The pumps are to be operated at a rotational speed of 1750 rpm and at a design discharge of 0.03155 m^3/s (500.0 gpm).

Solution From Fig. 9-13,

Model 1040A	Model 1230A
$H_A = 99 \text{ ft} = 30.18 \text{ m}$	$H_B = 144 \text{ ft} = 43.89 \text{ m}$
$E_A = 0.765$	$E_B = 0.79$

Compute power required:

$$P_A = \frac{0.03155\ m^3/s(30.18\ m)}{0.102(0.765)}$$

$$P_A = 12.20\ kW$$

$$P_B = \frac{0.03155\ m^3/s(43.89\ m)}{0.102(0.79)}$$

$$P_B = 17.18\ kW$$

For the pumps installed in series,

$$Q_{series} = 0.03155\ m^3/s$$

$$H_{series} = 30.18\ m + 43.89\ m = 74.1\ m$$

$$P_{series} = 12.20\ kW + 17.18\ kW = 29.38\ kW$$

where

P_{series} = *total* power required for both pumps

$$E_{series} = \frac{0.03155\ m^3/s(74.1\ m)}{0.102(29.38\ kW)}$$

$$E_{series} = 0.78 = 78\ \text{percent}$$

Multistage turbines

Multistage turbine pumps may be considered a special case of pumps in series. Multistage turbines are specifically manufactured so one pump or stage bolts directly onto the next. Since the discharge from one pump enters directly into the intake of the following pump, they are in series. However, there is no pipe connecting the discharge and intake as there would be in connecting centrifugal pumps in series.

Multistage turbines are conveniently used in wells where they are normally placed below the water level. The limited diameter of turbine pumps is efficient for installing the pump within the well casing. The fact that successive *bowls* made up of pump housing and impeller can be bolted together allows adequate head to be produced for almost all irrigation system requirements. In the case of a motor at the surface, the pump is connected to the motor by a drive shaft which passes through the center of the well casing. Motors which are insulated so they may be submersed in water are also used in well installations. In this case the electric submersible motor is coupled directly to the multistage pump with a short shaft.

Figure 9-14 indicates a pump performance curve for a multistage vertical turbine. There are some distinct differences between this pump performance curve and others previously demonstrated. The basic difference is that the curve has been developed for one, two, three, or four stages placed in series. The increase in discharge pressure head per stage at a given discharge is not constant but decreases with the number of stages. The efficiency of all stages considered together is also seen to be nonlinearly related to the number of stages. The performance curve is employed by taking information from the curve corresponding to the number of stages in series. The efficiency is read directly from the curve. The discharge head and power

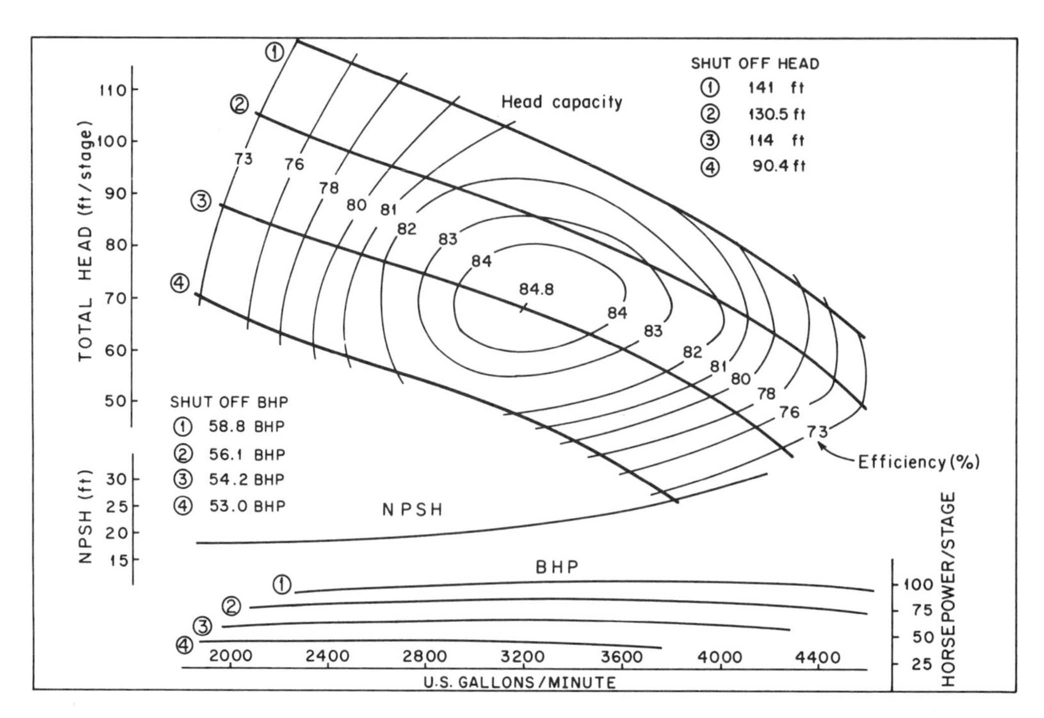

Figure 9-14 Pump performance curve for a multistage vertical turbine. (Courtesy of Peerless Pump, Montebello, California.)

required are indicated on a per-stage basis and multiplied by the number of stages. Application of such pump performance curves is indicated in the following example problem.

Example Problem 9-8

An irrigation system has a design flow rate of 0.1514 m^3/s (2400 gpm). Determine the total head produced in meters, total power required in kW, net positive suction head required in meters, and overall pump efficiency if the vertical turbine depicted in Fig. 9-14 with three stages is selected for this installation.

Solution From Fig. 9-14,

$$H/\text{stage} = 80.0 \text{ ft} = 24.384 \text{ m}$$

$$H_T = 24.384 \text{ m/stage (3 stages)} = 73.152 \text{ m}$$

$$P/\text{stage} = 65 \text{ bhp } (0.7457 \text{ kW/bhp}) = 48.47 \text{ kW}$$

$$P_T = 48.47 \text{ kW/stage (3 stages)} = 145.41 \text{ kW}$$

$$NPSH_r = 18.5 \text{ ft} = 5.64 \text{ m}$$

$$E = 0.794 = 79.4 \text{ percent}$$

Booster pumps

Booster pumps may be considered a special case of pumps in series. They are normally placed on a distribution line some distance from the water source. Booster pumps serve to increase the pressure to downstream points on the line. Although a booster pump may be placed in a mainline to increase pressure for the total flow, they are more commonly used on a submain through which only part of the flow passes. The discharge of the booster pump is selected to match the flow in that portion of the distribution line where it is to be placed. In contrast to bringing the total flow to a higher pressure, a smaller pump with less power required may be selected.

An example case would be one in which there were a number of center pivots on a distribution system including one or more towards the end of the line far from the water source. If there is inadequate pressure to operate the pivots towards the end of the line, a booster pump may be installed which would increase the pressure for only that portion of the flow going towards the more distant pivots. This may be more economical than developing high pressure for the total flow so enough head is available to operate the far pivots.

Booster pumps can also be cost-effective solutions to required increases in pressure due to changes in topography along the distribution line. In general, if capital and operating costs of an irrigation system with moderate to extensive distribution lines are to be minimized for the economic life of the system, booster pumps will be required. The smaller booster pump which pressurizes only a portion of the total flow will have a lower capital and operating cost than alternative systems.

Pumps in Parallel

Two or more pumps which discharge directly into the same distribution line are considered in parallel. An example of a parallel pump installation with three pumps is indicated in Fig. 9-15. Parallel pump installations allow pumps which produce adequate pressure head but insufficient discharge to be used together effectively. The operator is given increased flexibility in terms of the amount of flow in the distribution line. Irrigation systems which require different flow rates during the season due to a mix of crop water requirements or area to be irrigated may benefit from a parallel pump installation.

Figure 9-15 Parallel pump installation with three centrifugal pumps in background feeding into distribution system.

A design discharge pressure head is selected and the discharge from each pump in parallel is taken from its respective pump performance curve at that pressure. The power required is the sum of the power required for the individual pumps at the same discharge pressure. The governing equations for two pumps A and B in parallel are given by

$$H_{para} = H_A = H_B \tag{9-23}$$

$$Q_{para} = Q_A + Q_B \tag{9-24}$$

$$P_{para} = P_A + P_B \tag{9-25}$$

The efficiency is computed considering that both pumps operate at the same discharge pressure head:

$$E_{para} = \frac{(Q_A + Q_B)H_{para}}{0.102(P_A + P_B)} \tag{9-26}$$

where the units are as described for Eq. (9-22). Efficiency calculations for other units may be made by referring to the units in Eqs. (9-8), (9-9), and (9-11). The following example problem demonstrates calculations for a parallel pump installation.

Example Problem 9-9

Models 1030A with a 25.4 cm (10.0 inch) impeller and 1040A with a 25.4 cm (10.0 inch) impeller from Fig. 9-13 are to be installed in parallel. The pumps will be operated at a rotational speed of 1750 rpm and a discharge pressure head of 24.384 m (80.0 ft). Compute the discharge in cubic meters per second, power required in kW, and resulting efficiency for this pump installation.

Solution From Fig. 9-13,

Model 1030A	Model 1040A
$Q_A = 0.03312\ m^3/s$	$Q_B = 0.05489\ m^3/s$
$E_A = 0.83$	$E_B = 0.80$

Compute power requirements:

$$P_A = \frac{0.03312\ m^3/s(24.383\ m)}{0.102(0.83)}$$

$$P_A = 9.54\ kW$$

$$P_B = \frac{0.05489\ m^3/s(24.384\ m)}{0.102(0.80)}$$

$$P_B = 16.40\ kW$$

For parallel installation,

$$Q_{para} = 0.03312\ m^3/s + 0.05489\ m^3/s$$

$$Q_{para} = 0.08801\ m^3/s$$

$$P_{para} = 9.54\ kW + 16.40\ kW$$

$$P_{para} = 25.94\ kW$$

where

$$P_{para} = \textit{total} \text{ power required for both pumps}$$

$$E_{para} = \frac{0.08801 \text{ m}^3/\text{s}(24.384 \text{ m})}{0.102(25.94 \text{ kW})}$$

$$E_{para} = 0.81 = 81 \text{ percent}$$

Pumps operating in parallel must be balanced in terms of discharge pressure head so they are not acting against each other on the same distribution line. For example, if the discharge pressure head from pump A minus the friction and elevation headloss from A to B is different from the discharge pressure at B, the pumps will not be balanced. The result is that the heads produced by the pumps fluctuate which causes subsequent fluctuations in total discharge. This type of situation can occur in distribution systems in which widely spaced pumps feed into the same distribution line.

If the discharge pressure head from pump A minus headlosses from A to B is greater than the dead head or shut-off head of pump B, the flow from pump B may in fact be completely blocked. In extreme cases of poorly designed systems it is even possible that a portion of the flow from pump A is forced backwards through pump B. The balancing of the discharge pressure heads is accomplished by proper valving and gauging to control the flow. Often the discharge from two parallel pumps is fed into a large diameter baffle. A properly designed baffle insures that the flow comes to a uniform pressure before it enters the main distribution line.

9.5 Pump Installation

Proper installation of pumps is required to insure that (a) cavitation does not occur on the pump impeller, (b) the pump is operating on that portion of the pump characteristic curve for which it was selected, and (c) friction loss on the discharge side is minimized. Increased energy prices during the latter half of the 1970s caused increased awareness of energy usage in irrigation systems in general (Whittlesey, 1986) and pumping systems in particular. Tests of pump systems were conducted in many of the western states in the United States on hundreds of farms. These tests revealed that in many areas the majority of pump systems were operating below a reasonable field efficiency and that many pumps were not producing the expected discharge and pressure head due to improper installation. Material for this chapter was developed from recommendations developed during the pump test program.

Minimizing Air Entrapment and Cavitation

As discussed in the section on net positive suction head, air bubbles are formed when the pressure on the fluid drops below its vapor pressure at the operating temperature. As the fluid is repressurized across the impeller, these bubbles implode causing cavitation. Flow velocity is near its minimum and fluid pressure therefore maximized at the impeller tip. This is the location where damage due to cavitation can first be noted. Cavitation is normally avoided by insuring that the net positive suction head available is greater than the net positive suction head required as specified by the manufacturer [see Eq. (9-13)]. This is the first requirement of any properly designed pump installation.

Certain conditions can promote cavitation and air entrapment in the pump even when net positive suction head requirements are met. These conditions are related to entrainment of air and excessive turbulence on the intake side of the pump. The entrainment of air can be caused by aeration of the flow before it enters the suction line. An aeration condition may be caused by a falling stream of water which occurs close to a pump intake. Small bubbles generated by the falling stream pass through the suction line and into the pump where they expand at points of low pressure along the impeller. This results in reduced discharge in comparison to that indicated by the pump characteristic curve. Flows with high concentrations of air in solution may also have the air come out of solution at low pressure points along the impeller.

Three methods are available to reduce the potential of air entrainment and cavitation. The first two deal with allowing the flow to enter the pump with a minimum of turbulence. This is achieved by maintaining velocities in the suction line equal to or less than 1 m/s (3 ft/s). If the flow on the intake side must change direction by 90 degrees, this is most efficiently done through two 45 degree elbows. Using D for the diameter of the suction line before necking down to the pump intake, the nearest obstruction to flow should be no less than 4D from the pump intake. These recommendations are indicated schematically in Fig. 9-16.

The final recommendation regards the relative position of the pump intake and intake fittings. The pump intake should be at the highest position on the intake line so no air is allowed to accumulate and form an air pocket which reduces the flow area. It is recommended that an eccentric reducer be used on the pump intake with the straight portion being on the upper part of the intake line. Such a reducer is indicated in Fig. 9-16. Eccentric reducers with an interior angle of approximately 28 degrees have been found to give an optimum balance between reducer length and minimum friction loss.

It is also generally recommended that flanged fittings, which are bolted externally, be used instead of plumbing fittings which are threaded. The flanged fittings do not have the friction losses and turbulence associated with the abrupt edges of the threaded fittings. Air release valves which bleed air out of the fitting or line may

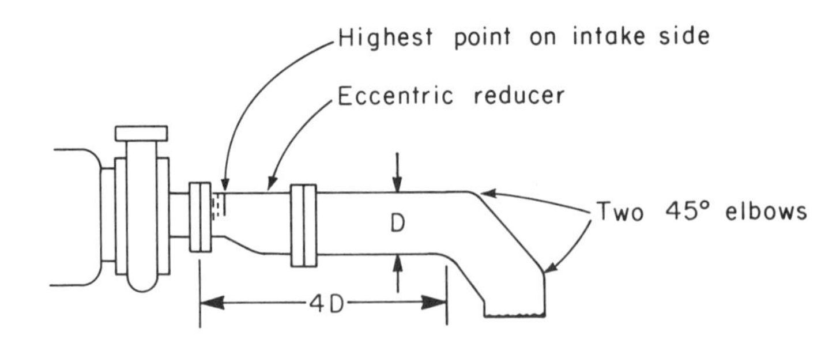

Figure 9-16 Recommended fittings on intake side of centrifugal pump installation.

also be required in a pump installation. Their recommended placement is on top of the pump housing or at the high point in the high velocity discharge. Air release valves preclude the formation of an air pocket which may partially block the flow area. Air release valves are further described in the chapter on Pipeline System Design.

Minimizing Headloss on the Discharge Side

Pumps are required to pressurize the fluid at the design discharge. Needless headloss which must be paid for by the operator in energy costs but which serves no benefit to the system is to be avoided. The rules to minimize friction loss on the discharge side of pumps are relatively straightforward. Yet pump tests conducted throughout many of the western states indicated that they were often neglected. Excessive headloss results in systems operating on different portions of the pump characteristic curve than for which they were selected—that is, at a higher head and lower discharge—and at a generally lower efficiency. It also results in higher energy costs per volume of water delivered to the soil surface through the application device.

Figure 9-17 is an example of a centrifugal pump installation with excessive headloss on the discharge side. Moving along the pipeline from right to left, the figure indicates (a) intake line with flexible Dresser coupling, (b) centrifugal pump with side-throw discharge, (c) primer pump, gate valve, and pressure gauge all connected using threaded fittings, and (d) abrupt expansion to mainline diameter. The placement of a series of devices on the discharge side in the high velocity—small diameter flow stream using threaded fittings causes excessive headloss in this installation.

Energy costs are not automatically reduced when discharge fittings which create excessive headloss are replaced by better suited fittings. This is due to the tradeoff between discharge and pressure on the pump performance curve. The system with a higher headloss, and therefore higher head produced by the pump, will produce less discharge than the same pump operating at a lower head. Referring to a power equation such as Eq. (9-8), the final power required may be less than, more

Figure 9-17 Centrifugal pump installation showing high headloss system on discharge side. Excessive headloss can be expected due to the following factors: (a) threaded fittings on discharge line, (b) primer pump and gate valve located in high velocity (small diameter) flow stream, and (c) abrupt expansion to discharge line.

than, or greater than that required for an installation with poor discharge fittings depending on the change in operating efficiency.

The power lost through poorly selected discharge fittings does nothing to improve the irrigation system application efficiency. The cost per unit volume of water applied to the soil surface at the design pressure is therefore minimized by application of proper installation criteria. This cost efficiency is realized every time the pump is turned on. The cost associated with proper pump installation and correct fittings is therefore recuperated very quickly, normally within the first season of operation.

Figure 9-18 contrasts two discharge systems used on the same pump installation. The fittings shown in part (a) are all threaded, placed in the high velocity flow

Figure 9-18a High headloss pump discharge fittings due to use of threaded fittings, gate valve in high velocity flow stream, and abrupt expansion to mainline diameter. (Photo courtesy of Marvin Shearer.)

Figure 9-18b Replacement discharge fittings with reduced headloss for same system shown in Fig. 9-18a. Headloss is reduced due to concentric expansion to mainline diameter, large diameter gate valve, and use of flanged fittings. (Photo courtesy of Marvin Shearer.)

stream and include an abrupt expansion. They allowed the sprinkler system pressurized by the pump to adequately discharge to three out of an intended four laterals. Part (b) indicates the replacement fittings installed after a pump test. The recommendations for the replacement fittings are described in the following paragraph. Because of the reduced headloss with the replacement fittings, the pump was able to produce adequate pressure and discharge to operate all four laterals.

Fluid velocity at the pump discharge normally ranges from 6 m/s (20 ft/s) to 9 m/s (30 ft/s). The maximum design velocity for flow in mainlines is usually 1.5 m/s (5 ft/s). Since friction loss through fittings is proportional to $v^2/2g$, the headloss in the high velocity flow stream is significantly higher than the loss through the same type of fittings in the lower velocity portion of the distribution system. The key requirement to reduce excessive headloss is to place the required discharge fittings in the lower velocity, larger diameter portion of the flow stream. This requires larger diameter fittings which have a higher capital cost. But reduced operating costs per unit of water applied to the field at design conditions allow for rapid recovery of the increased capital costs.

There are two major criteria for reducing the friction loss on the discharge side of the pump to the lowest possible amount: (a) water does not like to turn corners, and (b) fittings should be placed in the low velocity flow stream. These recommendations are accomplished by minimizing the number of elbows on the discharge and by expanding up to the mainline diameter before installing flow control fittings.

The number of elbows may be minimized by inspecting the pump location and the direction required for the discharge (i.e., direction of the mainline). Mainlines are normally laid out in a horizontal or near-horizontal direction. If a pump is installed with a vertical-throw discharge, the flow must first be turned through a 90 degree elbow before entering the mainline. The energy loss through the elbow could be completely avoided by specifying that the pump be installed with a horizontal-throw discharge. Adjustment of the discharge direction can often be accomplished by a simple unbolting and rotation of the pump discharge housing which can be performed in the field.

The recommended means of expanding to the mainline diameter is by installation of a concentric expansion which goes directly from the pump discharge to the mainline diameter. The absolute minimum friction loss is accomplished by a fitting which has a concentric expansion up to approximately 80 percent of the downstream diameter followed by an abrupt expansion to the full diameter (Lalibertie et al., 1983). Fittings of this type are not generally manufactured. The typical low-headloss fitting employed is a direct expansion with an interior angle of between 10 and 16 degrees. Interior angles of this magnitude strike a balance between reduced friction loss and required length of the fitting. Friction loss through such fittings is considerably reduced compared to that through an abrupt expansion. As was the case for the intake side, flanged fittings reduce turbulence and friction compared to threaded fittings. Inspection of the resistance coefficient tables in Appendix C demonstrate that it is probably impossible to conceive of situations when branch flow through a tee would be cost-effective. The relative magnitude of headloss for a proper pump installation compared to an obviously poor one employing branch flow through a tee and an abrupt expansion is indicated in the following example problem.

Example Problem 9-10

A pump with an electric motor is to be operated for 2200 h per year. Assume the cost of electricity is \$0.04/kWh and the electric motor efficiency is 85 percent. The discharge is required for flow in a horizontal mainline 20.0 cm in diameter. The required flow rate is 0.0378 m^3/s (2270 L/min), the pump discharge diameter is 7.5 cm, and the pump efficiency will be assumed constant in order to analyze the effects of the discharge fittings alone. Two alternative pump installations are to be evaluated:

(1) Vertical-throw discharge from pump
- **(a)** Flow turned horizontal by branch flow through a 7.5 cm tee with threaded fittings and L/D ratio equal to 60.
- **(b)** 7.5 cm diameter check valve with L/D ratio equal to 400.
- **(c)** 7.5 cm diameter gate valve with L/D ratio equal to 8.
- **(d)** Abrupt expansion in diameter from 7.5 cm to 20.0 cm.

(2) Horizontal-throw discharge from pump
- **(a)** Concentric expansion in diameter from 7.5 cm to 20.0 cm with 12 degree interior angle.
- **(b)** 20.0 cm diameter check valve with L/D ratio equal to 400.
- **(c)** 20.0 cm diameter gate valve with L/D ratio equal to 8.

Compute the difference in annual operating costs to overcome the headloss experienced by the discharge through the two systems.

Solution The solution is obtained by applying the appropriate resistance coefficients from Appendix C to compute the headloss through each fitting. For all fittings other than the abrupt and concentric expansions, the headloss is computed as

$$h_f = \frac{Kv^2}{2g}$$

where

h_f = friction headloss, m

K = resistance coefficient

v = flow velocity, m/s

g = acceleration of gravity = 9.81 m/s^2

For the abrupt expansion, K is calculated from

$$K = [1 - (d_1)^2/(d_2)^2]^2$$

where

d_1 = upstream pipe diameter, cm

d_2 = downstream pipe diameter, cm

For the concentric expansion, K is calculated using

$$K = 2.6 \sin(\theta/2)[1 - (d_1)^2/(d_2)^2]^2$$

where

θ = interior angle, degrees

The power required to overcome the headloss to push the water through the fittings is calculated using Eq. (9-10),

$$P = \frac{Q(H)}{0.102E}$$

where

$$P = \text{power, kW}$$
$$Q = \text{discharge, m}^3\text{/s}$$
$$H = \text{total friction headloss through fittings, m}$$
$$E = \text{efficiency, fraction}$$

In this application, E will only refer to the motor efficiency since pumps for both installations are assumed to have the same efficiency. In reality, E should be the pumping plant efficiency which is the product of the motor and pump efficiencies. The annual operating cost is computed by

$$C = P(T)C_{elec}$$

where

$$C = \text{annual operating costs, \$/yr}$$
$$T = \text{annual time of operation, h}$$
$$C_{elec} = \text{cost of electricity, \$/kWh}$$

The results of this problem are most conveniently set up in a computerized spreadsheet such as that shown in Table 9-3. Establishing the analysis in this manner allows for evaluation of effects of changes in the system parameters by merely changing the value of the parameter in its proper position in the spreadsheet. The headlosses and costs are automatically updated for the new conditions using the automatic calculation capabilities of the spreadsheet.

Table 9-3 indicates that for the two systems compared, the difference in annual operating costs is

$$\Delta C = C_1 - C_2$$
$$\Delta C = \$1195/\text{yr} - \$41/\text{yr}$$
$$\Delta C = \$1154/\text{yr}$$

This is the additional cost per year to pressurize the water to get it through the high headloss fittings. This cost can be completely avoided by selection of the correct installation configuration given by alternative #2.

Pump Support

Pump support refers to the support of the intake and discharge lines connected to the pump. Pump housings are not constructed to support the weight of the intake and discharge lines, especially when they are full of water. The pump and motor should be securely anchored in their permanent position. The intake and discharge pipes should then be brought to the position of the pump and supported so they can be directly bolted to the pump using flanged fittings. Proper intake and discharge line

TABLE 9-3 Comparison of pump installation annual energy costs.

Flow rate (L/min) = 2270
Pump discharge diam. (cm) = 7.5
Mainline diameter (cm) = 20.0
Operating time (h/yr) = 2000
Electrical cost ($/kWh) = 0.04
Motor efficiency (%) = 85

Installation #1 Vertical-Throw				Installation #2 Horizontal-Throw			
Fitting	Inlet Diameter (cm)	K-factor	Headloss (m)	Fitting	Inlet Diameter (cm)	K-factor	Headloss (m)
Tee Branch— L/D = 60 Threaded	7.5	1.08	4.04	Concentric Expansion 12 degrees	7.5	0.20	0.75
Check Valve L/D = 400	7.5	7.2	26.91	Check Valve L/D = 400	20.0	5.6	0.41
Gate Valve L/D = 8	7.5	0.14	0.52	Gate Valve L/D = 8	20.0	0.11	0.01
Abrupt Expansion	7.5	0.74	2.77				
TOTAL (m)			34.24	TOTAL (m)			1.17
POWER REQD (kW)			14.94	POWER REQD (kW)			0.51
ANNUAL COST ($)			1195	ANNUAL COST ($)			41

support is demonstrated in Fig. 9-19. Proper support of the pipelines will insure that power is transmitted to the pump from the motor as efficiently as possible.

It is also common to fit the intake and discharge lines with Dresser couplings. This type of flexible coupling basically absorbs elongation and shrinkage of the pipeline which may occur due to temperature changes and during pump operations. Dresser couplings are indicated in the sketch in Fig. 9-19 and in the photograph in Fig. 9-17.

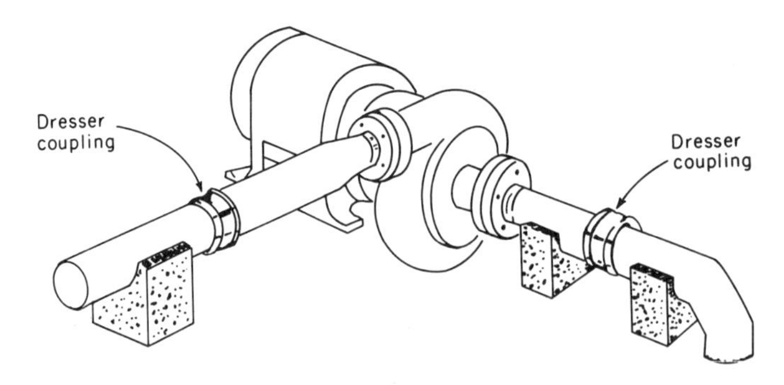

Figure 9-19 Schematic of installation with correct support of intake and discharge lines to pump.

REFERENCES

HANSEN, V. E., O. W. ISRAELSEN, and G. E. STRINGHAM, *Irrigation Principles and Practices*, fourth edition. New York: John Wiley and Sons, 1980.

LALIBERTE, G. E., M. N. SHEARER, and M. J. ENGLISH, "Design of Energy-Efficient Pipe-Size Expansion," American Society of Civil Engineers, *Journal of Irrigation and Drainage Engineering*, vol. 109, no. 1, 1983, pp. 13–28.

LINSLEY, R. K. and J. B. FRANZINI, *Water-Resources Engineering*, second edition. New York: McGraw-Hill, 1972.

WHITTLESEY, N. K., *Energy and Water Management in Western Irrigated Agriculture*. Boulder, Colorado: Westview Press, 1986.

PROBLEMS

9-1. A single-stage radial flow pump with a rotational speed of 3500 rpm is to be installed in a system which has a total head requirement of 74.0 m. Estimate the maximum pump discharge in liters per minute.

9-2. A model 1250 centrifugal pump with a 12-inch (30.48 cm) diameter impeller for which the performance curve is given in Fig. P9-1 will be operated at 1750 rpm and at a discharge of 1600 gpm (8722 m^3/d). Compute the percent loading on a nominal 60 bhp (59.16 mhp) motor at the required discharge.

9-3. A sprinkler system has a gross depth of irrigation required equal to 131 mm. The operating pressure at the sprinkler nozzle is 380 kPa. The area to be irrigated is 2.0 ha with a time of operation of 20 hours. The overall pump efficiency is 70 percent. At full operation, the pump is taking water from a water table 23 m below the height of the sprinkler nozzle. What metric horsepower pump is required to meet this demand if the head losses up to the sprinkler nozzle are equivalent to 7.6 m of head?

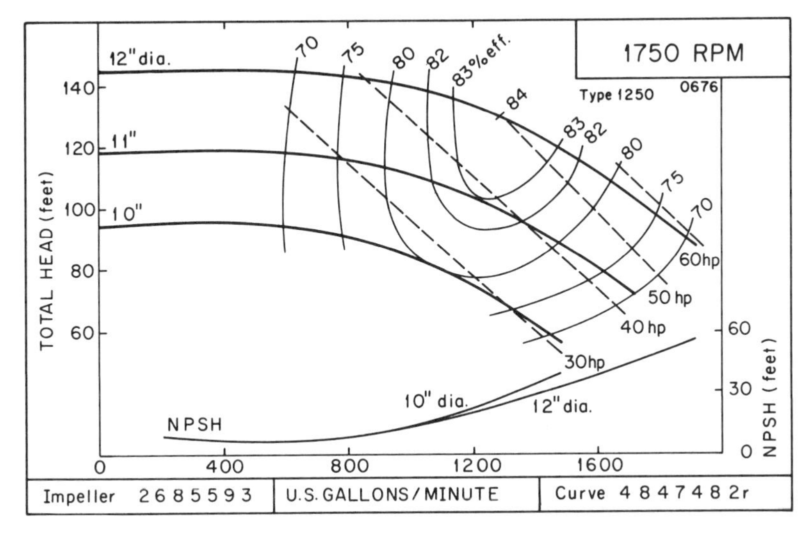

Figure P9-1 Sketch for problem 9-2.

9-4. A pumping plant is to be selected for a surface irrigation system. The pump will be required to produce 9 m of lift including friction losses. Each field to be irrigated has an area of 10 ha and will require 4 hours of pumping time during each irrigation interval. During the peak period, the fields will be irrigated every 5 days and the crop ET is 11 mm/d. The conveyance efficiency of the system is 85 percent and the distribution pattern efficiency is 75 percent. Other efficiencies may be assumed to be 100 percent. Compute the metric horsepower pump required if the pump efficiency is 73 percent.

9-5. A sprinkler system is tested which uses water pumped from a well. The pressure at a point in the mainline is measured at 427.5 kPa and all friction losses including the well column, pump, fittings, and mainline from the pump to the point where the pressure is measured are equal to 1.6 m of head. At a discharge of 984 L/min, the electric horsepower into the pumping system (including the motor) is measured at 22.6 mhp. Compute the dynamic water level in meters if the pumping plant efficiency is 68 percent, velocity head is assumed negligible, and there is no elevation difference between the pump and the point in the mainline where the pressure is measured.

9-6. A pump is to be selected for a sprinkler system which will be required to produce 9.0 m of head including friction losses. Plots of 10 ha in area are to be irrigated by the system at one time and 8.0 hours pumping will be required per 10 ha plot during each irrigation interval. The peak period crop water requirement is 11 mm/d and the irrigation frequency is 9 days. Spray/drift losses are estimated to be 10 percent and the design deep percolation losses are 15 percent. Compute the metric horsepower pump required if the pump efficiency is 75 percent.

9-7. A pump produces 65 L/s through a 100 mm discharge at a pressure of 1030 kPa. The flow then passes through a globe valve (L/D = 340), lift check valve (L/D = 600), and through the branch of a standard tee. All these fittings have a 100 mm diameter. After the tee, flow passes through a concentric expansion with a 30 degree interior angle to a 250 mm diameter pipe. Compute the pressure available just after the expansion to 250 mm. Assume the pump discharge and all the fittings are on the same horizontal plane.

9-8. A pump discharges 58 L/s horizontally through a 100 mm diameter pipe. The flow then enters a flanged 90 degree elbow of the same diameter with L/D equal to 42 and its direction is changed to vertical. After 1.5 m vertical travel, the flow passes through an abrupt expansion into a 250 m diameter pipe. The pressure just after the abrupt expansion is measured at 630 kPa. Compute the pressure in kPa at the discharge of the pump. Neglect friction losses in straight pipe sections.

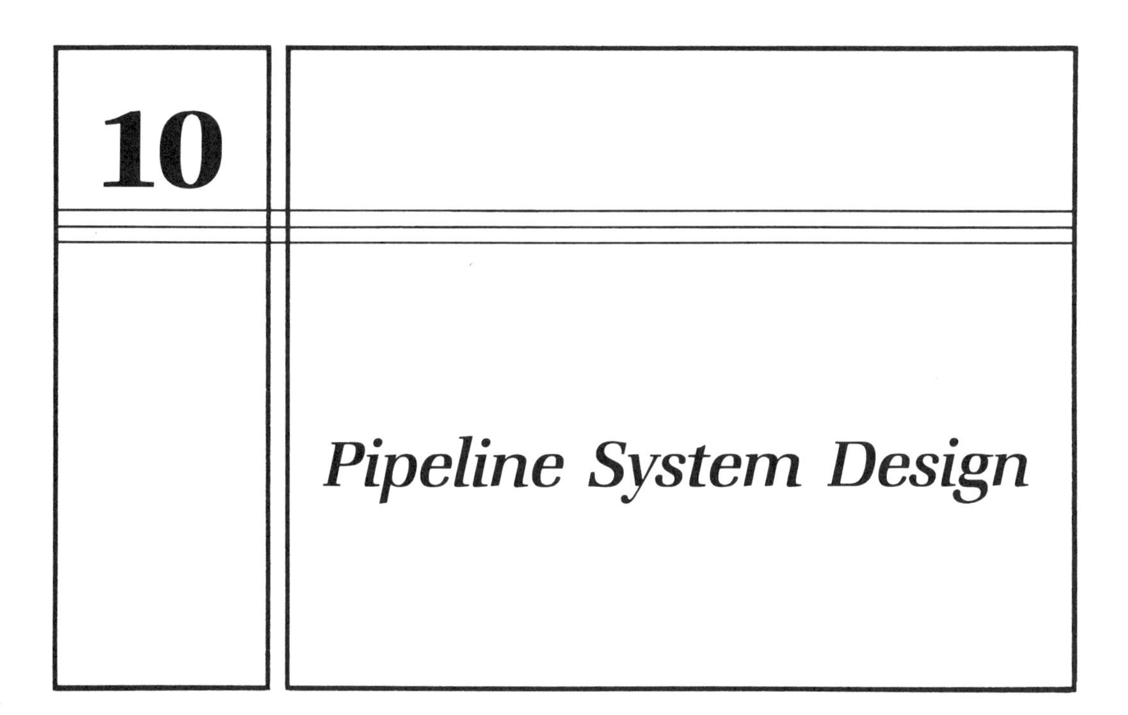

10.1 Pipeline Standards

Standards are established for pipeline materials so engineers can select the proper material for the required application and have confidence in the ability of the pipeline to perform as expected. Standards are established which relate the pipeline dimensions and type of material to its strength and ability to resist loads. The standards described in this section are taken from those developed by the American Society of Agricultural Engineers (ASAE) for thermoplastic pipe (Standard ASAE S376.1). Common thermoplastic pipe materials are polyvinyl chloride (PVC) and polyethylene (PE). These materials are used in a wide variety of applications in irrigation systems in the United States and other countries. Because of ease of fabrication, locally produced PVC pipe is employed in many irrigation development projects worldwide. The standards described in this section apply specifically to thermoplastic pipe. Other types of pipe materials have similar standards which are found in other references (e.g., ASAE Standards, 1987; Stephenson, 1981). Standards for aluminum tubing are found in Appendix D.

Pressure Category

Thermoplastic pipes are divided into low and high pressure categories. The pressure categories are based on both the pipe diameter and the design operating pressure.

- Low Pressure
  Nominal diameter—114 to 630 mm (4 to 24 inch)
  Internal pressure—545 kPa (79 psi) or less
- High Pressure
  Nominal diameter—21 to 710 mm (0.5 to 27 inch)
  Internal pressure—550 to 2170 kPa (80 to 315 psi) including surge pressure

Surge pressure occurs in unsteady flow regimes and is associated with rapid changes in flow velocity and resulting rapid changes in pressure. It is commonly referred to as water hammer and is covered later in this chapter.

Pressure Rating and Hydrostatic Design Stress

The pressure rating and hydrostatic design stress both refer to parameters related to long-term operation of the pipeline. The pressure rating is the maximum pressure that water in the pipe can exert continuously with a high degree of certainty that failure will not occur. The hydrostatic design stress is the maximum tensile stress due to an internal hydrostatic pressure that can be applied continuously with a high degree of certainty that failure will not occur.

These two parameters are related to the dimension ratio, DR, which is given by

$$DR = \frac{D}{t} \tag{10-1}$$

where

D = outside or inside pipe diameter depending on how the pipe size is controlled, mm

t = wall thickness, mm

The dimension ratio is dimensionless. The minimum wall thickness for thermoplastic pipe is specified as 1.52 mm (0.060 inch). Certain dimension ratios have been selected as standard and are designated in tabulated data as Standard Dimension Ratios.

For outside diameter based pipe, the pressure rating is given as

$$PR = \frac{2S}{DR - 1} \tag{10-2}$$

or

$$PR = \frac{2S}{\frac{D_0}{t} - 1} \tag{10-3}$$

where

PR = pressure rating, kPa (psi)

S = hydrostatic design stress, kPa (psi)

D_0 = average outside diameter, mm (in.)

t = minimum wall thickness, mm (in.)

For inside diameter based pipe,

$$PR = \frac{2S}{DR + 1} \qquad (10\text{-}4)$$

or

$$PR = \frac{2S}{\frac{D_i}{t} + 1} \qquad (10\text{-}5)$$

where

D_i = average inside diameter, mm (in.)

Hydrostatic design stresses for different thermoplastic pipe material and different strengths of pipe are given in Table 10-1. ABS refers to pipes fabricated from acrylonitrile-butadiene-styrene. The hydrostatic design stress is derived from the long-term hydrostatic strength which is determined by standardized tests to failure of the pipe material. The relationship is given by

$$S = \frac{S_{lt}}{2.0} \qquad (10\text{-}6)$$

where

S_{lt} = long-term hydrostatic strength

and the 2.0 represents a factor of safety. Table 10-2 indicates pressure ratings for different strengths of PVC, PE, and ABS materials as a function of Standard Dimension Ratios.

TABLE 10-1 Maximum hydrostatic design stress for thermoplastic pipe. (Adapted from ASAE Standard S376.1.)

Compound	Standard Code Designation	Hydrostatic Design Stress (MPa)	Hydrostatic Design Stress (psi)
PVC	PVC 1120	13.8	2000
PVC	PVC 1220	13.8	2000
PVC	PVC 2120	13.8	2000
PVC	PVC 2116	11.0	1600
PVC	PVC 2116	8.6	1250
PVC	PVC 2110	6.9	1000
PE	PE 3408	5.5	800
PE	PE 3406	4.3	630
PE	PE 3306	4.3	630
PE	PE 2306	4.3	630
PE	PE 2305	3.4	500
ABS	ABS 1316	11.0	1600
ABS	ABS 2112	8.6	1250
ABS	ABS 1210	6.9	1000

TABLE 10-2 Pressure ratings (PR) for nonthreaded thermoplastic pipe.* (Taken from ASAE Standard S376.1.)

SDR[‡]		PVC materials (all pipes OD based)								PE materials (pipes made to both OD & ID basis)						ABS materials (all pipes OD based)					
		PVC 1120 PVC 1220 PVC 2120		PVC 2116		PVC 2112		PVC 2110		PE 3408		PE 3406 PE 3306 PE 2306		PE 2305		ABS 1316		ABS 2112		ABS 1210	
OD based pipe	ID based pipe	psi	kPa[§]	psi	kPa	psi	kPa	psi	kPa	psi	kPa	psi	kPa	psi	kPa	psi	kPa	psi	kPa	psi	kPa
	5.3									250	1725	200	1380	160	1105						
	7.0									200	1380	160	1105	125	860						
11.0	9.0									160	1105	125	860	100	690						
13.5	11.5	315	2170	250	1725	200	1380	160	1105							250	1725	200	1380	160	1105
17.0	15.0	250	1725	200	1380	160	1105	125	860	100	690	80	550	63	435	200	1380	160	1105	125	860
21.0		200	1380	160	1105	125	860	100	690	80	550	64	440			160	1105	125	860	100	690
26.0		160	1105	125	860	100	690	80	550	64	440	50	345			125	860	100	690	80	550
32.5		125	860	100	690	80	550	63	435	50	345	40	275			100	690	80	550	64	440
41.0		100	690	80	550	63	435	50	345	40	275	31	215			80	550	64	440	50	345
51.0		80	550	63	435	50	345	40	275							64	440	50	345	40	275
64.0		63	435	50	345	40	275	30	205												
81.0		50	345	40	275	30	205	25	170							40	275	30	205	25	170
93.5[∥]		43	295																		
50 ft head		22	150																		

*For water at 23°C (73.4°F).

‡SDR = Standard Dimension Ratio

§kPa = kilopascals, kN/m^2

∥The dimension ratio 93.5 is nonstandard and is referred to as DR (Dimension Ratio)

Example Problem 10-1

A PVC pipe is to be manufactured from PVC 2116 compound. The pipe size is based on its inside diameter which is 25 cm. Two categories of the pipe will be manufactured: (a) low pressure with a wall thickness of 1.8 mm and (b) high pressure with a wall thickness of 9.6 mm. Compute the pressure rating for each category of pipe.

Solution For ID based pipe,

$$PR = \frac{2S}{\frac{D_i}{t} + 1}$$

Case (a) low strength:

$$\frac{D_i}{t} = \frac{250 \text{ mm}}{1.8 \text{ mm}} = 139$$

Case (b) high strength:

$$\frac{D_i}{t} = \frac{250 \text{ mm}}{9.6 \text{ mm}} = 26$$

From Table 10-1 for PVC 2116,

$$S = 11.0 \text{ MPa} = 11{,}000 \text{ kPa}$$

Pressure rating:

Case (a) low strength:

$$PR = \frac{2(11{,}000 \text{ kPa})}{139 + 1} = 157 \text{ kPa } (\cong 23 \text{ psi})$$

Case (b) high strength:

$$PR = \frac{2(11{,}000 \text{ kPa})}{26 + 1} = 815 \text{ kPa } (\cong 117 \text{ psi})$$

10.2 Pressure Distribution in Pipelines

By application of Bernoulli's law, the total head available at any point in a pipeline is equal to the pressure head plus elevation head plus velocity head. The difference in total head between any two points 1 and 2 on a pipeline under steady-state flow conditions is equal to the friction headloss between the two points so that

$$H_{T2} = \left(\frac{p}{\gamma}\right)_1 + z_1 + \left(\frac{v^2}{2\,g}\right)_1 - (h_f)_{1-2} \quad (10\text{-}7)$$

where

$$H_T = \text{total head, m}$$
$$\frac{p}{\gamma} = \text{pressure head, m}$$
$$\gamma = \text{specific weight of the fluid, kN/m}^3$$
$$z = \text{elevation head, m}$$
$$h_f = \text{friction headloss, m}$$

The specific weight of the fluid is given by

$$\gamma = Sg(\gamma_w) \tag{10-8}$$

where

$$Sg = \text{specific gravity of the fluid, dimensionless}$$
$$\gamma_w = \text{specific weight of water, kN/m}^3$$

The specific weight of water under standard conditions applicable to design of irrigation systems is 9.81 kN/m^3. The units required for pressure in Eq. (10-7) are kPa (equal to kN/m^2) to enable the pressure head to be expressed in meters.

Figure 10-1 is a schematic of the distribution of different forms of energy given as head in a pipeline. The static head line refers to the distribution of head if

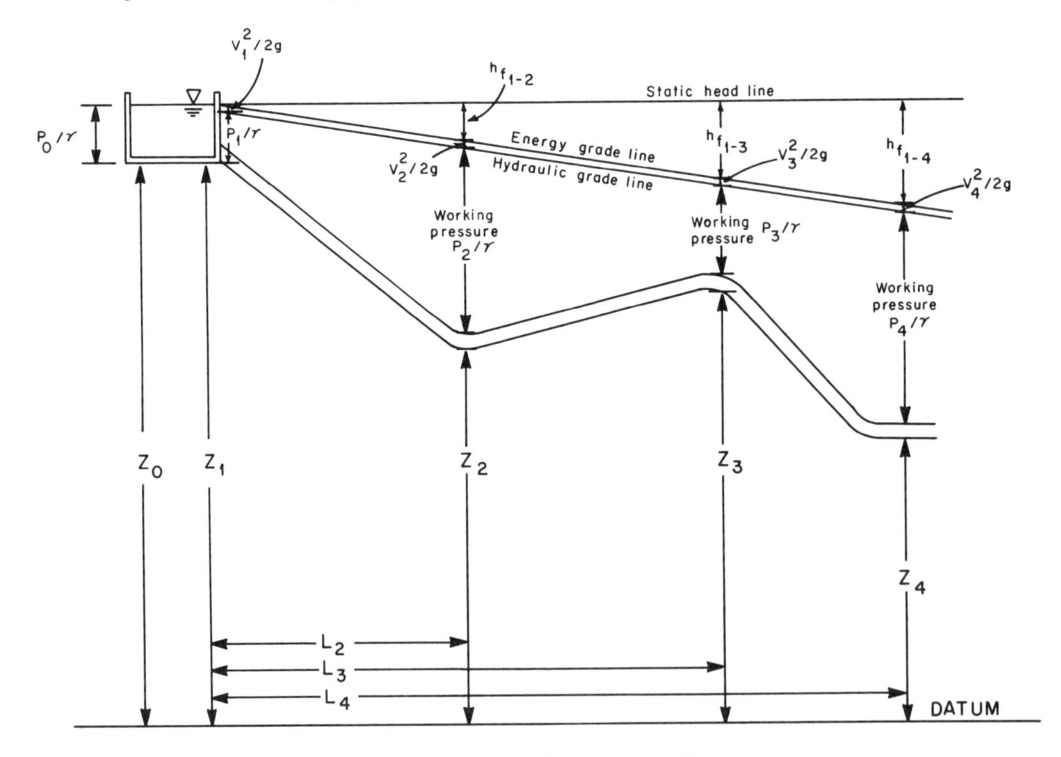

Figure 10-1 Distribution of heads in a pipeline system.

there is no flow in the system. The energy grade line describes the variation of total head along the pipeline, including velocity head, when flow occurs. The hydraulic grade line is defined by the energy grade line minus the velocity head. It indicates the height to which water would rise in a piezometer inserted at that position in the pipeline. The hydraulic grade line defines the working pressure available at any point along the pipeline. The pressure distribution in a pipeline is determined by the relative positions of water sources along the line and friction headloss. The following sections consider two methods of computing friction headloss.

Hazen-Williams Friction Headloss Equation

There are numerous methods for computing friction headloss in pipelines. One of the most common and convenient methods applicable to pumping water through irrigation systems is the Hazen-Williams equation given as

$$h_f = k_1 L \frac{\left(\frac{Q}{C}\right)^{1.852}}{D^{4.87}} \tag{10-9}$$

where

k_1 = conversion constant

L = length of pipe, L

Q = volumetric flow rate, L^3/T

C = Hazen-Williams coefficient

D = pipe diameter, L

Table 10-3 indicates common units associated with flow in pipes in the SI and English systems and the required conversion constant k_1 for Eq. (10-9). Table 10-4 lists the Hazen-Williams C coefficient for various types of pipe materials. The Hazen-Williams equation is only applicable to water at standard operating temperature (i.e., 20°C), or more specifically to fluids with a specific gravity of 1.0. Such an assumption is almost always valid for analysis of flow in irrigation systems. Example problem 10-2 indicates application of the Hazen-Williams equation to calculation of pressure distribution in a pipeline. For fluids with different specific gravities, and therefore different viscosities, a friction factor which accounts for the effect of fluid viscosity must be applied. Such a method is described following the example problem.

TABLE 10-3 Conversion constants for the Hazen-Williams equation given different combinations of units.

h_f	L	Q	D	k_1
m	m	L/s	mm	1.22×10^{10}
m	m	L/h	mm	3163
m	m	m^3/d	mm	3.162×10^6
ft	ft	ft^3/s	ft	4.73
ft	ft	gpm	in	10.46

TABLE 10-4 Friction factor C for Hazen-Williams equation.

Pipe Material	Values of C: Design	New Pipe	Corroded Pipe
Polyethylene (PE) and polyvinyl chloride (PVC)	140	150	130
Cement–Asbestos	140	150	140
Fiber	140	150	—
Bitumastic-enamel-lined iron or steel centrifugally applied	140	148	130
Cement-lined iron or steel centrifugally applied	140	150	—
Copper, brass, lead, tin, or glass pipe and tubing	130	140	120
Wood-stave	110	120	110
Welded and seamless steel	100	130	80
Interior riveted steel (no projecting rivets)	100	139	—
Wrought-iron, cast-iron	100	130	80
Tar-coated cast iron	100	130	50
Girth-riveted steel (projecting rivets in girth seams only)	100	130	—
Concrete	100	120	85
Full-riveted steel (projecting rivets in girth and horizontal seams)	100	115	—
Vitrified, spiral-riveted steel (flow with lap)	100	110	—
Spiral-riveted steel (flow against lap)	90	100	—
Corrugated steel	60	60	—

Example Problem 10-2

A pipeline with 200 mm inside diameter and 340 m in length made of new PVC is laid along a horizontal grade. The required flow rate in the pipeline at steady state is 40.0 L/s and the total head available at the inlet is 330 kPa. Compute the working pressure head at the discharge 340 m from the inlet by application of the Hazen-Williams friction headloss equation.

Solution From Eq. (10-9),

$$h_f = k_1 L \frac{\left(\frac{Q}{C}\right)^{1.852}}{D^{4.87}}$$

From Table 10-3 for L in m, Q in L/s, and D in mm,

$$k_1 = 1.22 \times 10^{10}$$

From Table 10-4 for new PVC,

$$C = 150$$

$$h_f = 1.22 \times 10^{10}(340 \text{ m})\frac{[(40.0 \text{ L/s})/150]^{1.852}}{(200 \text{ mm})^{4.87}}$$

$$h_f = 2.232 \text{ m}$$

To convert to kPa,

$$h_f' = h_f\gamma = 2.232 \text{ m}(9.81 \text{ kN/m}^3)$$

$$h_f' = 21.897 \text{ kN/m}^2 = 21.897 \text{ kPa}$$

$$H_{T\text{-}340} = 330 \text{ kPa} - 22 \text{ kPa}$$

$$= 308 \text{ kPa}$$

Darcy-Weisbach Friction Headloss Equation

When the fluid to be pumped is other than water or the specific weight is substantially different from water due to additions of chemicals to the flow or temperature effects, an equation which accounts for differences in viscosity must be applied to calculate the friction headloss. A common equation applied is that of Darcy-Weisbach which is given in its most basic form as

$$h_f = f\frac{L}{D}\left(\frac{v^2}{2g}\right) \tag{10-10}$$

where

$$f = \text{Darcy-Weisbach friction factor, dimensionless}$$

and the units for the friction headloss are that of the velocity head if L and D have the same units. Revising this equation to a form conveniently applicable to pipe flow

$$h_f = k_2 f L \frac{Q^2}{D^5} \tag{10-11}$$

where

$$k_2 = \text{conversion constant}$$

The conversion constant k_2 is given for various combinations of SI and English units common to pipe flow in Table 10-5.

The friction factor is a function of the flow regime—that is, whether the flow is laminar, turbulent, or in transition between the two, and the roughness of the pipe material. The flow regime is designated as a function of the dimensionless Reynold's number, R_N, which is defined by the following equation:

$$R_N = \frac{v\,D}{1000\,\nu} \tag{10-12}$$

where

$$v = \text{flow velocity, m/s}$$
$$D = \text{pipe diameter, mm}$$
$$\nu = \text{kinematic viscosity of the fluid, m}^2\text{/s}$$

TABLE 10-5 Conversion constants for the Darcy-Weisbach equation given different combinations of units.

h_f	L	Q	D	k_2
m	m	L/s	mm	8.2627×10^7
m	m	L/h	mm	6.3755
m	m	m^3/d	mm	1.10686×10^4
ft	ft	ft^3/s	ft	0.02517
ft	ft	gpm	in	0.03107

The kinematic viscosity of water under standard operating conditions for irrigation systems (i.e., 20°C) is 1.0×10^{-6} m^2/s.

For $R_N \leq 2000$, flow is laminar and the Darcy-Weisbach friction factor is given as

$$f = \frac{64}{R_N} \qquad (10\text{-}13)$$

In the range $2000 < R_N < 4000$, the flow is in the critical state and it cannot be said precisely whether it is laminar or turbulent. In typical engineering applications, flow tends to be either laminar or fully turbulent so the critical range is not of concern (Mott, 1979). For $R_N > 4000$ flow is said to be turbulent. However there are two zones of turbulence as indicated in the Moody diagram for the Darcy-Weisbach f versus R_N shown in Fig. 10-2. To the right of the transition zone line indicated in the figure, the flow is fully turbulent and the friction factor is independent of R_N and depends only on the roughness of the pipe material. This roughness is described by the relative roughness which is expressed as a dimensionless ratio of wall roughness to pipe diameter, ε/D. For flow in the completely turbulent zone, the Darcy-Weisbach f is given by the expression

$$\frac{1}{(f)^{1/2}} = 2 \log\left(3.7\frac{D}{\varepsilon}\right) \qquad (10\text{-}14)$$

From the left of the transition line in Fig. 10-2 to the lower limit of $R_N = 4000$, the flow is in transition and not yet fully turbulent. In this zone the friction factor is a function of both the Reynold's number and the relative roughness. The expression for the Darcy-Weisbach f in this zone developed by Colebrook (1939) is given as

$$\frac{1}{(f)^{1/2}} = -2 \log\left\{\frac{(\varepsilon/D)}{3.7} + \frac{2.51}{R_N(f)^{1/2}}\right\} \qquad (10\text{-}15)$$

Equation (10-15) requires an iterative solution since f appears on both sides of the equation. However a method has been developed to simplify this procedure which gives an initial estimate of f as a function of the relative roughness and R_N (Murdock, 1976). Labelling the first estimate as f_1, the function is

$$f_1 = 0.0055\left\{1 + \left[\frac{20000}{(D/\varepsilon)} + \frac{10^6}{R_N}\right]^{1/3}\right\} \qquad (10\text{-}16)$$

This initial estimate is substituted into the right-hand side of Eq. (10-15) to solve for the first iteration of f. The value of f_1 computed in Eq. (10-16) is close enough to the final value of f that a second iteration is normally not required (Mott, 1979). Table 10-6 indicates values of surface roughness for various types of pipeline materials.

The lower boundary of the turbulent zone indicated in Fig. 10-2 is that for smooth pipes. Pipes made of plastic and glass normally fall into this range as do

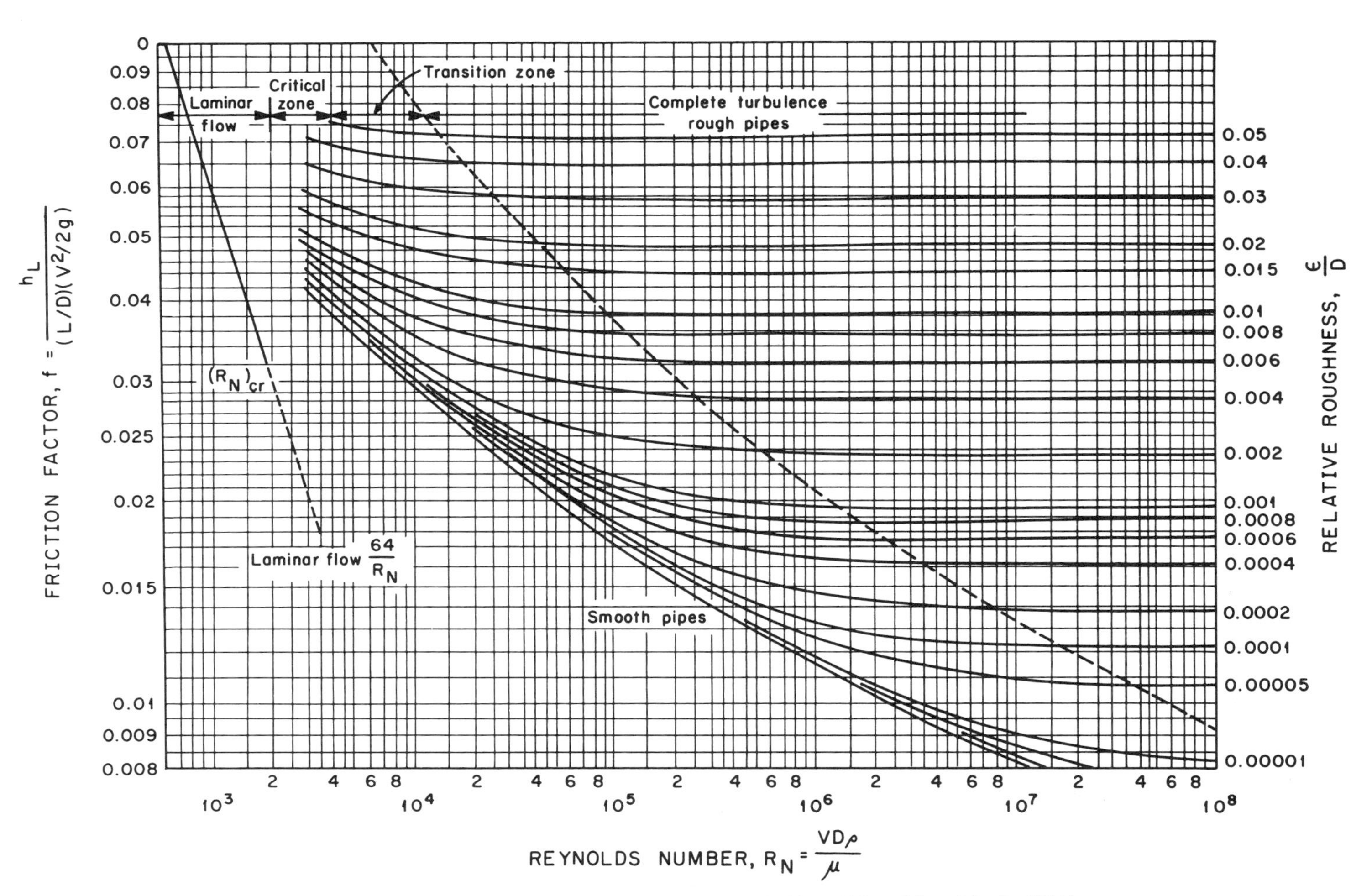

Figure 10-2 Moody diagram for friction loss in pipes and tubing. (Adapted from Moody, 1944.)

TABLE 10-6 Surface roughness of pipe materials. (Adapted from Stephenson, 1981.)

Pipe Material	ε(mm)		
	Smooth	Average	Rough
Glass, drawn metals	Smooth	0.003	0.006
Steel or polyvinyl chloride	0.015	0.03	0.06
Coated steel	0.03	0.06	0.15
Galvanized, vitrified clay	0.06	0.15	0.3
Cast iron or cement lined	0.15	0.3	0.6
Spun concrete or wood stave	0.3	0.6	1.5
Riveted steel	1.5	3	6
Foul sewers	6	15	30
Unlined rock, earth	60	150	300

large diameter pipes made of drawn brass or copper. The equation for the Darcy-Weisbach f as defined by the smooth pipe line is given by

$$\frac{1}{(f)^{1/2}} = 2 \log\left[\frac{R_N(f)^{1/2}}{2.51}\right] \tag{10-17}$$

This equation also requires an iterative solution. The following example problem indicates the Darcy-Weisbach solution to the same situation given in example problem 10-3.

Example Problem 10-3

A pipeline with 200 mm inside diameter and 340 m in length made of new PVC is laid along a horizontal grade. The required flow rate in the pipeline at steady state is 40.0 L/s and the total head available at the inlet is 330 kPa. Compute the working pressure head at the discharge 340 m from the inlet by application of the Darcy-Weisbach headloss equation.

Solution From Eq. (10-11),

$$h_f = k_2 f L \frac{Q^2}{D^5}$$

From Table 10-5 for L in m, Q in L/s and D in mm,

$$k_2 = 8.263 \times 10^7$$

From Table 10-6 for smooth PVC,

$$\varepsilon = 0.015 \text{ mm}$$

Solving for flow velocity,

$$v = \frac{Q}{A} = 40.0 \text{ L/s}\left(\frac{1 \text{ m}^3}{1000 \text{ L}}\right) \Big/ \left\{\pi\left[\frac{(200 \text{ mm})^2}{4}\right]\left[\frac{1 \text{ m}}{1000 \text{ mm}}\right]^2\right\}$$

$$v = 1.273 \text{ m/s}$$

Solving for Reynold's number,

$$R_N = \frac{v\,D}{1000\,\nu}$$

$$R_N = 1.273\ \text{m/s}\left[\frac{200\ \text{mm}}{1000(1 \times 10^{-6}\ \text{m}^2/\text{s})}\right]$$

$$R_N = 2.55 \times 10^5$$

Flow is fully turbulent and f is function of ε/D only. Using Eq. (10-14).

$$\frac{1}{\sqrt{f}} = 2\log\left[3.7\left(\frac{200\ \text{mm}}{0.015\ \text{mm}}\right)\right]$$

$$\frac{1}{\sqrt{f}} = 9.386$$

$$f = 0.0114$$

Substituting into Eq. (10-11),

$$h_f = 8.263 \times 10^7(0.0114)(340\ \text{m})\frac{(40.0\ \text{L/s})^2}{(200\ \text{mm})^5}$$

$$h_f = 1.594\ \text{m}$$

Converting to kPa,

$$h_f' = h_f\gamma = 1.594\ \text{m}(9.81\ \text{kN/m}^3)$$

$$h_f' = 15.64\ \text{kPa}$$

$$H_{T-340} = 330\ \text{kPa} - 16\ \text{kPa}$$

$$= 314\ \text{kPa}$$

It should be noted that the Darcy-Weisbach equation for headloss can be applied for fluids of different viscosities in contrast to the Hazen-Williams equation which is only valid for water. This is because fluid viscosity is required for the Reynold's number which is in turn required to compute the Darcy-Weisbach friction factor for all cases other than fully turbulent. However, for the wide majority of applications in irrigation system design, the Hazen-Williams equation is sufficient and will normally be applied in this book.

10.3 Unsteady Flow in Pipelines

Concept of Water Hammer

Water hammer, or a pressure surge, is caused in pipelines as a result of changes in fluid flow velocities. Changes in the flow velocity essentially cause the kinetic energy associated with velocity to be converted to pressure. This condition in a pipeline is termed one of hydraulic transients. The more rapid the change in flow velocity, the greater will be the magnitude of the resulting pressure surge. Pressure surges occur when valves are opened or closed, when pumps are started or stopped, or by sudden releases of entrapped air.

The water hammer phenomenon will be described by reference to Fig. 10-3. The figure illustrates a pipeline fed by a reservoir with constant head H. A valve is located along the pipeline at distance L from the reservoir. The energy grade line and hydraulic grade line are indicated in the figure and the difference between them is the velocity head. Initial conditions are a steady-state flow at velocity v_0.

At time t = 0 the valve is suddenly closed and the fluid behind the valve becomes compressed and causes the pipe to expand. The high pressure in the compressed fluid layer adjacent to the valve is transferred to the next fluid layer upstream resulting in a pressure wave which moves up the pipe. On the reservoir side of the pressure wave, the velocity is the initial v_0. On the valve side of the pressure wave, the velocity is zero. This is the case illustrated in part (b) of Fig. 10-3. The change in head caused by the stopped flow is indicated as ΔH. The pressure wave moves up the pipeline at velocity a. At t = L/a the pressure wave has moved all the way upstream to the reservoir.

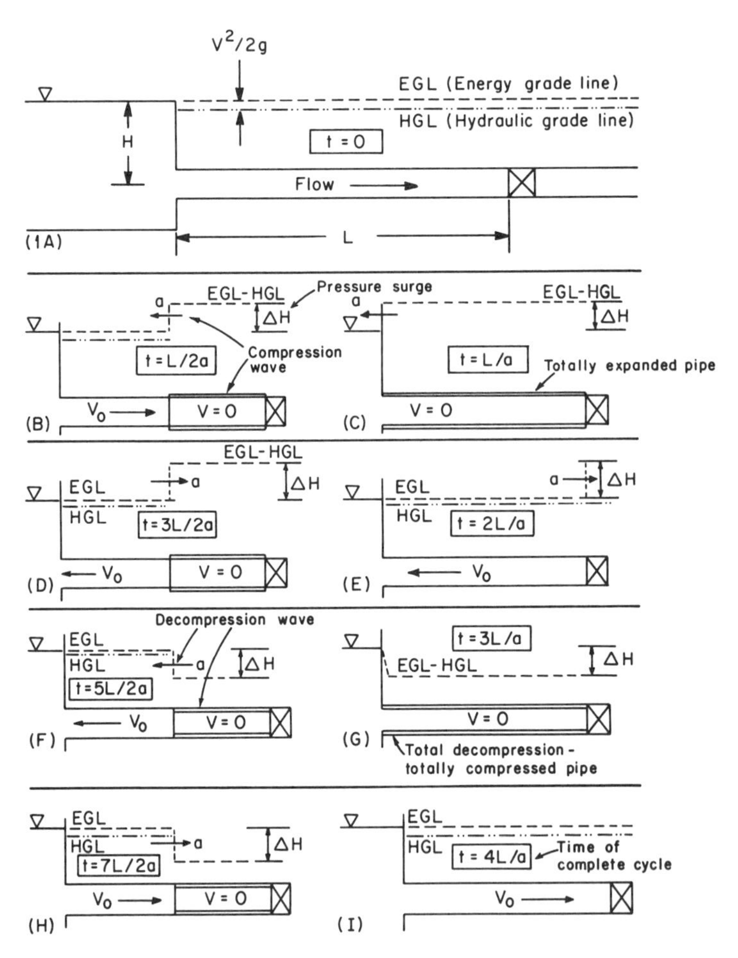

Figure 10-3 Schematic of the water hammer phenomenon.

At this time [refer to part (c) of Fig. 10-3] all the fluid in the pipeline has been brought to rest and the pipeline is subject to a pressure increase of ΔH over its entire length. As the pressure wave is reflected in the reservoir, the static pressure H is re-established at the reservoir end of the pipeline. The resulting pressure differential causes the fluid to move out of the pipeline towards the reservoir at velocity v_0 as indicated in part (d) of the figure. At $t = 2\ L/a$ the pressure wave has returned to the valve and the pressure in the pipeline is back to the original distribution that existed at $t = 0$ [as shown in part (e)].

Since the valve is still closed, the velocity at the valve is zero and the velocity differential causes a decrease in pressure. Now a reduced pressure wave moves back up the pipeline towards the reservoir at speed a [see Fig. 10-3(f)]. This continues until the pressure wave returns to the reservoir at time $t = 3\ L/a$ when the entire pipeline is in compression. This condition is shown in part (g) of the figure. At this point pressure is re-established in the pipeline at head H and the pressure wave again moves towards the valve at velocity a. The wave arrives back at the valve at $t = 4\ L/a$ at which time the pressure distribution is returned to the original condition depicted in Fig. 10-3(a). This process continues every 4 L/a seconds until the fluctuations dampen out due to fluid friction and elasticity of the pipeline. When the fluctuations completely dampen out, the fluid comes to rest.

Hydraulics of Water Hammer

The magnitude of pressure surges associated with water hammer is a function of

(a) System geometry.
(b) Magnitude of velocity change.
(c) Velocity of the pressure wave for a particular system.

The governing equation for this relationship is given by

$$\Delta H = \left(\frac{a}{g}\right)\Delta v \tag{10-18}$$

where

ΔH = surge pressure, L
a = pressure wave velocity, L/T
g = acceleration of gravity, L/T^2
Δv = change in fluid velocity, L/T

The velocity, or celerity, of the pressure wave is a function of both pipe material properties and fluid properties. The pipe material properties which affect the celerity are (a) modulus of elasticity, (b) diameter, and (c) wall thickness. The fluid properties of major importance are (a) modulus of elasticity, (b) fluid density, and (c) amount of air entrained in the fluid. Applying the impulse-momentum equation and the principle of conservation of mass, the following equation for pressure wave velocity can be derived (Watters, 1984):

$$a = \frac{k_1\left[\frac{K}{\rho}\right]^{0.5}}{\left[1 + \left(\frac{K}{E}\right)\left(\frac{D}{t}\right)C_1\right]^{0.5}} \tag{10-19}$$

where

a = pressure wave velocity, L/T
k_1 = conversion constant dependent upon units
K = bulk modulus of elasticity of water, F/L^2
ρ = density of water, M/L^3
D = inside pipe diameter, L
t = pipe wall thickness, L
E = pipe modulus of elasticity, F/L^2
C_1 = pipe support coefficient

The pipe support coefficient is a function of the method of pipe constraint and the relationship between longitudinal and circumferential stress given by the Poisson ratio. The function is given by the following equation (Watters, 1984):

$$C_1 = 1.25 - \mu, \text{ pipes anchored on one end only} \tag{10-20a}$$

$$C_1 = 1 - \mu^2, \text{ pipes anchored at both ends only} \tag{10-20b}$$

$$C_1 = 1.0, \text{ pipes with expansion joints along the length} \tag{10-20c}$$

where

μ = Poisson ratio for pipe material

The value of k_1 depends on the system of units employed. Consider that the same units are used for K and E in Eq. (10-19) and the same units for D and t. For SI units with K in Pa and ρ in kg/m^3, k_1 is equal to 1.0. For the English system with K given in psi and ρ in $slugs/ft^3$, k_1 is equal to 12.0. The relative properties of water and pipe materials required for Eq. (10-19) are given in Tables 10-7 and 10-8. Equation (10-19) does not account for entrainment of air in the pipeline which serves to reduce K and a. The result of Eq. (10-19) is therefore a conservative value of a which predicts the most severe water hammer pressure surge. This value is reasonable for design purposes since it represents the upper limit of operating conditions.

Application of Eq. (10-19) for the analysis of pressure surges in unsteady flow regimes is demonstrated in the following example problems. The analyses demonstrated are simplified in that they do not account for the nonlinearity of the pressure surge with time. An analysis considering nonlinearity requires solution of two simultaneous partial differential equations by the method of characteristics or more commonly by computerized numerical methods (Watters, 1984; Stephenson, 1981).

TABLE 10-7 Various properties of water as a function of temperature.

Temperature		Specific Weight		Density		Modulus of Elasticity	
(C)	(F)	(kN/m^3)	(lb/ft^3)	(kg/m^3)	(slug/ft^3)	(MPa)	(psi)
0	32	9.81	62.4	1000	1.94	1979	287,000
5	41	9.81	62.4	1000	1.94	2048	297,000
10	50	9.81	62.4	1000	1.94	2103	305,000
15	59	9.81	62.4	1000	1.94	2151	312,000
20	68	9.79	62.3	998	1.94	2193	318,000
25	77	9.78	62.3	997	1.93	2224	322,500
30	86	9.77	62.2	996	1.93	2251	326,500
35	95	9.75	62.1	994	1.93	2272	329,500
40	104	9.73	61.9	992	1.92	2286	331,500
45	113	9.71	61.8	990	1.92	2289	332,000
50	122	9.69	61.7	988	1.92	2289	332,000
55	131	9.67	61.6	986	1.91	2282	331,000
60	140	9.65	61.4	984	1.91	2275	330,000
65	149	9.62	61.2	981	1.90	2261	328,000
70	158	9.59	61.0	978	1.90	2248	326,000
75	167	9.56	60.9	975	1.89	2227	323,000
80	176	9.53	60.7	971	1.88	2206	320,000
85	185	9.50	60.5	968	1.88	2172	315,000
90	194	9.47	60.3	965	1.87	2137	310,000
95	203	9.44	60.1	962	1.87	2103	305,000
100	212	9.40	59.8	958	1.86	2068	300,000

TABLE 10-8 Modulus of elasticity and Poisson's ratio for common pipe materials.

Pipe Material	Modulus of Elasticity (MPa)	Modulus of Elasticity (10^5 psi)	Poisson's Ratio
Asbestos-Cement	20,684	30	0.20
Cast Iron	103,421	150	0.29
Ductile Iron	165,474	240	0.29
Permastran	9,653	14	0.35
Polyvinyl Chloride	2,758	4	0.46
Polyethylene	689	1	0.40
Steel	206,843	300	0.30

Example Problem 10-4

A steel pipe 1500 m in length and 0.5 m in diameter has a wall thickness of 5.5 cm. The pipe carries water from a reservoir and discharges into the atmosphere 50 m below the water level of the reservoir. A valve installed at the downstream end of the pipe allows a flow rate of 0.4 m^3/s. If the valve is completely closed in 2.0 s, calculate the maximum water hammer pressure surge at the valve and the time for the pressure wave to return to the valve. Assume the pipe is anchored at one end.

Solution Assume standard temperature 20°C. From Tables 10-7 and 10-8,

$$K = 2.19 \times 10^9 \text{ N/m}^2$$
$$E = 2.07 \times 10^{11} \text{ N/m}^2$$
$$\rho = 998 \text{ kg/m}^3$$
$$\mu = 0.30$$

Based on units and pipe constraints,

$$k_1 = 1.0$$
$$C_1 = 1.25 - \mu = 0.95$$

Compute the celerity,

$$a = \frac{1.0\left[\dfrac{2.19 \times 10^9 \text{ N/m}^2}{998 \text{ kg/m}^3}\right]^{1/2}}{\left[1 + \left(\dfrac{2.19 \times 10^9}{2.07 \times 10^{11}}\right)\left(\dfrac{0.5 \text{ m}}{0.055 \text{ m}}\right)0.95\right]^{1/2}}$$

$$a = 1418 \text{ m/s}$$

Time required for pressure wave to return to valve:

$$t = \frac{2L}{a} = \frac{2(1500 \text{ m})}{1418 \text{ m/s}}$$

$$t = 2.12 \text{ s}$$

The valve is completely closed by this time. The change in velocity equals

$$\Delta v = \frac{0.4 \text{ m}^3/\text{s}}{\dfrac{\pi}{4}(0.5 \text{ m})^2}$$

$$\Delta v = 2.04 \text{ m/s}$$

Pressure surge:

$$\Delta H = \left(\frac{a}{g}\right)\Delta v$$

$$\Delta H = \left(\frac{1418 \text{ m/s}}{9.81 \text{ m/s}^2}\right)(2.04 \text{ m/s})$$

$$\Delta H = 295 \text{ m}$$

More gradual valve closure and water hammer protection devices are required on this pipeline to reduce the potential for damage due to such high pressure surges.

Example Problem 10-5

(a) Compute the pressure surge in psi in a 12-inch outside diameter class 150 PVC 1220 pipe carrying water at 60°F due to a sudden change in velocity of 2.0 ft/s. Assume that the pipe is fitted with expansion joints throughout its length.

(b) Compute the maximum allowable change in velocity for the given pipe if the design is based on a 2.8 to 1 safety factor.

Solution

(a) Pressure rating = PR = 150 psi

$$PR = \frac{2\,S}{\frac{D_0}{t} - 1}$$

From Table 10-1 for PVC 1220,

$$S = 2000\ \text{psi}$$

Substituting into PR equation,

$$150\ \text{psi} = \frac{4000\ \text{psi}}{(12\ \text{in.}/t) - 1}$$

$$t = 0.4337\ \text{in.}$$

From Tables 10-7 and 10-8,

$$K = 313{,}000\ \text{psi}$$
$$E = 400{,}000\ \text{psi}$$
$$\rho = 1.938\ \text{slugs/ft}^3$$
$$D_i = D_0 - 2\,t$$
$$= 12\ \text{in.} - 2(0.4337\ \text{in.})$$
$$= 11.133\ \text{in.}$$

Calculate celerity:

$$a = \frac{12\left[\dfrac{313{,}000\ \text{psi}}{(1.938\ \text{slugs/ft}^3)}\right]^{1/2}}{\left[1 + \left(\dfrac{313{,}000\ \text{psi}}{400{,}000\ \text{psi}}\right)\left(\dfrac{11.133\ \text{in.}}{0.4337\ \text{in.}}\right)1.0\right]^{1/2}}$$

$$a = 1052\ \text{ft/s}$$

Calculate the pressure surge,

$$\Delta H = \frac{a}{g}\Delta v$$

$$\Delta H = \left(\frac{1050\ \text{ft/s}}{32.17\ \text{ft/s}^2}\right)(2.0\ \text{ft/s})$$

$$\Delta H = 65.29\ \text{ft} = 28.3\ \text{psi}$$

(b) Maximum operating pressure:

$$P_{max} = \frac{\dfrac{2.0\,S_{lt}}{S_f}}{\dfrac{D_0}{t} - 1}$$

From Table 10-1,

$$S_{lt} = 2.0(S)$$
$$= 2.0(2000 \text{ psi})$$
$$S_{lt} = 4000 \text{ psi}$$

Applying 2.8 safety factor,

$$P_{max} = \frac{\dfrac{2.0(4000 \text{ psi})}{2.8}}{\left(\dfrac{12 \text{ in.}}{0.4337 \text{ in.}}\right) - 1}$$

$$P_{max} = 107 \text{ psi}$$

Maximum surge pressure magnitude:

$$\Delta H_{max} = PR - P_{max}$$
$$\Delta H_{max} = 150 \text{ psi} - 107 \text{ psi}$$
$$\Delta H_{max} = 43 \text{ psi}$$

Rearranging the surge pressure equation,

$$\Delta v_{max} = \Delta H_{max}(g/a)$$
$$= 43 \text{ psi}(2.31 \text{ ft/psi})\left[\frac{(32.17 \text{ ft/s}^2)}{(1052 \text{ ft/s})}\right]$$
$$\Delta v_{max} = 3.04 \text{ ft/s}$$

Table 10-9 indicates the maximum operating pressure (maximum working pressure) for thermoplastic pipe from the ASAE standards in the case when the surge pressure is unknown. As long as the operating pressure is less than the maximum indicated in Table 10-9, the pipeline should be able to sustain surge pressures developed in normal operations without failure and without additional pipeline protection devices. Even if pipeline protection devices for water hammer are installed, it is prudent to operate the pipe at pressures below the maximum indicated in case of failure of the protection devices.

Standard pipeline design requires specification of protection devices which alleviate the effects of water hammer and other damaging conditions on the pipeline. These devices are key elements to the safe and continuous operation of the pipeline. They are described in the following section.

10.4 Pipeline System Components

Various devices need to be specified in the plans for a pipeline to insure safe and efficient operation of the system. These devices are required to control entrapped air, to allow the pipeline to drain after operations are completed, and to protect the line from pressure surges caused by hydraulic transients. Without these devices, the pipeline may not only operate inefficiently, it is subject to failure. This section describes the type of devices required to overcome the most common pipeline operational problems and their placement on the line.

TABLE 10-9 Maximum allowable pressure for nonthreaded thermoplastic pipes when surge pressures are not known.*† (Taken from ASAE Standard S376.1.)

SDR		PVC materials (all pipes OD based)								PE materials (pipes made to both OD & ID basis)						ABS materials (all pipes OD based)					
		PVC 1120 PVC 1220 PVC 2120		PVC 2116		PVC 2112		PVC 2110		PE 3408		PE 3406 PE 3306 PE 2306		PE 2305		ABS 1316		ABS 1212		ABS 1210	
OD based pipe	ID based pipe	psi	kPa	psi	kPa	psi	kPa	psi	kPa	psi	kPa	psi	kPa	psi	kPa	psi	kPa	psi	kPa	psi	kPa
	5.3									180	1240	144	995	115	795						
	7.0									144	995	115	795	90	620						
11.0	9.0									115	795	90	620	72	495						
13.5	11.5	227	1565	180	1240	144	995	115	795							180	1240	144	995	115	795
17.0	15.0	180	1240	144	995	115	795	90	620	72	495	58	400	45	310	144	995	115	795	90	620
21.0		144	995	115	795	90	620	72	495	58	400	45	310			115	795	90	620	72	495
26.0		115	795	90	620	72	495	58	400	45	310	36	250			90	620	72	495	58	400
32.5		90	620	72	495	58	400	45	310	36	250	29	200			72	495	58	400	46	315
41.0		72	495	58	400	45	310	36	250	29	200	22	150			58	400	46	315	36	250
51.0		58	400	45	310	36	250	29	200							46	315	36	250	29	200
64.0		45	310	36	250	29	200	22	150												
81.0		36	250	29	200	22	150	18	125							29	200	22	150	18	125
93.5		31	215																		
50 ft head		21	145																		

*Maximum allowable working pressure = pressure rating (PR) × 0.782 for SDR and DR pipe.

†For water at 23°C (73.4°F).

Air in Pipelines

Occurence

Air in pipelines may occur due to absorption at free water surfaces or entrainment in turbulent flow at pipeline entrances. Air may be present in solution or in the form of pockets or bubbles. Air in solution does not generally cause a problem unless pressure is reduced sufficiently that the dissolved air forms bubbles. The formation of bubbles may lead to cavitation damage on pump impellers or the bubbles may coalesce to form air pockets along the pipeline. The occurence of air in pipelines causes a reduction in effective flow area resulting in increased flow velocity and increased headloss. Air intermittently drawn into pumps and pipelines can cause vibration due to unsteady flow.

Pump inlet arrangements which promote either intermittent air entry or bubbles in pipelines and pumps can be avoided by (a) providing adequate submergence of the pump intake, (b) providing for minimum flow turbulence at the pump intake—that is, by assuring that the intake velocity is less than 1 m/s (3 ft/s) and that the nearest obstruction to flow is at a distance greater than four pipe diameters from the pump inlet, and (c) minimizing the opportunity for entrapping air on the suction side of the pump—that is, by using an eccentric reducer and checking that the pump intake is the highest point in the suction line. These points are more fully discussed in the Pump Installation section of the Pump System chapter.

Air release—vacuum relief valves

Air which does enter pipelines requires venting to the atmosphere by air release valves. These valves vent air which may collect at high points in the pipeline due to entrainment at the pump intake and release air during pipe filling. Vacuum relief valves vent into the pipe and draw air from the atmosphere when the pipeline is draining.

Venting of entrapped air is accomplished by small orifice valves—that is, valves with orifices up to 3 mm (0.12 inch) in diameter. These are activated by the accumulated air forcing a ball to be released from the seat around an orifice or forcing a ball to activate a lever which vents air to the atmosphere. Figure 10-4 demonstrates the design of such a valve.

Relatively rapid purging of air from or induction of air into pipelines, as occurs during pipeline filling or draining, is accomplished by large orifice valves. These

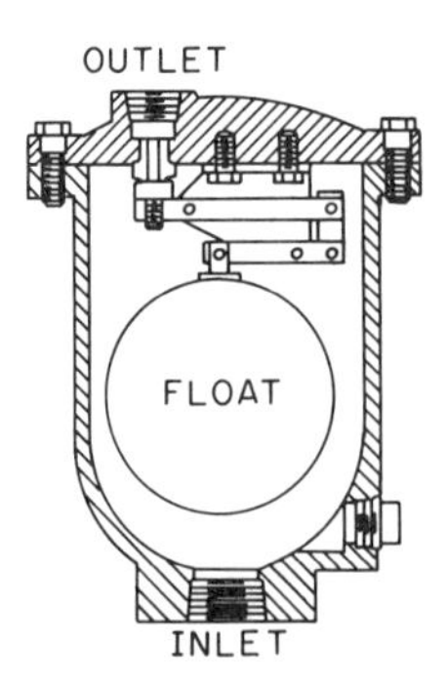

Figure 10-4 Graphic of air release valve.

valves are normally activated by releasing a ball from the orifice seat. Small and large orifice valves may be installed independently. However, because the specified location of such valves on the pipeline is often the same, dual or combination type valves are common. Such a valve is depicted in Fig. 10-5.

Referring to Fig. 10-5 we observe that during pipeline filling the large orifice ball is at the bottom of the valve—that is, unseated—and air is free to escape to the atmosphere. When the pipeline is filled with water, the large orifice ball becomes seated. If entrained air collects at the valve site, the small orifice ball is depressed and air is released to the atmosphere. When the pipeline is drained, atmospheric pressure exceeds the internal pressure in the pipeline and the large ball becomes unseated. This allows air at atmospheric pressure to freely enter the pipeline and promotes complete drainage of the line. Without a vacuum relief valve, drainage may not be complete and in fact the vacuum developed could collapse the pipeline.

Valve placement

Large orifice valves should be installed at high points in the physical layout of the pipeline and at high points along the hydraulic gradient. Together with small orifice valves, or more commonly as a combination valve, they should also be installed at the ends and intermediate points along the length of pipeline which is parallel to the hydraulic grade line. It is recommended that combination valves be placed every 0.5 to 1.0 km (0.3 to 0.6 mi) along descending pipeline sections, especially at points where the descending pipeline grade is steep.

Air may be released from solution if there is adequate drop in pressure. Such conditions may occur along a mainline which rises in elevation or where velocity is increased through a flow restriction such as a partially closed valve. This air should be bled off by appropriately placed small orifice air release valves to avoid blockage of the pipeline. Air release valves are therefore required along all ascending lengths of pipeline, particularly at points where there is a decrease in the upward gradient. Other reasonable points of air release valve placement are on the discharge side of pumps, at high points on large flow control valves, and upstream of orifice plates and reducing tapers if they are located in relatively high velocity flow streams.

Valve sizing

Air release/vacuum relief valves are sized by the diameter of the connection between the pipeline and the valve. The actual orifice diameter is generally smaller

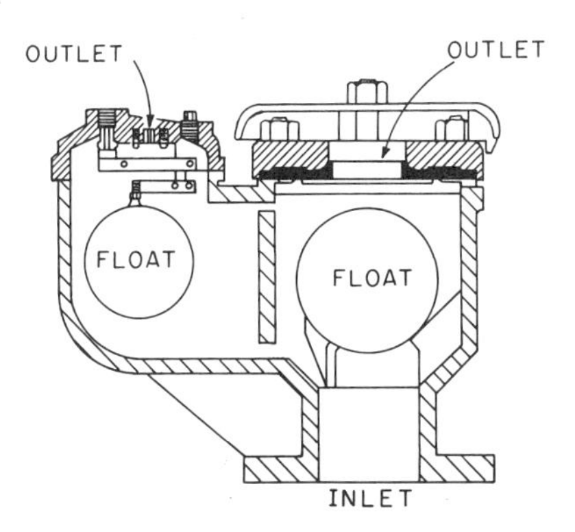

Figure 10-5 Graphic of combination air release/vacuum relief valve.

and can be obtained by referring to an equipment manufacturer's catalog. The ratio of valve diameter to pipeline diameter for high pressure systems should be 0.1 or greater—that is, valve diameter should be ≥10 percent of pipeline diameter. Table 10-10 indicates minimum recommended valve sizing for high and low pressure systems taken from the ASAE Standards (1987).

Pipeline Protection

The need to protect pipelines from surge pressure that develops during sudden valve closure has been previously described in the section on water hammer. An inverse condition exists if a pump is suddenly shut off. In this case the flow stops and the sudden deceleration of the water column causes a pressure drop at the point of pump discharge. This negative pressure surge moves along the pipeline in an analagous manner to the positive pressure surge demonstrated for a sudden valve closure in Fig. 10-3. The sudden pressure drop may cause vaporization, pipeline collapse, or induction of air at atmospheric pressure into the pipeline through combination air release/vacuum relief valves.

High pressure surges can be controlled through installation of pressure relief valves at those points in a pipeline where it is anticipated pressure waves will be generated or reflected. Relatively simple valves of this type suitable for on-farm distribution systems operate using a spring-loaded discharge valve. Under normal operating pressures the valve remains in the closed position. If water hammer causes a pressure surge in the line, the excess pressure depresses the spring thereby opening the valve and causing water to be discharged from the line. The result is that the magnitude of the pressure surge is considerably reduced and the reflected shock wave does not damage the pipeline. The level of pressure at which the valve will be activated can be preset by the manufacturer or set in the field to correspond to the operating pressure of a particular system.

The opposite problem is encountered when sudden downsurges in pressure are caused by the stopping of pumps. The objective of pipeline protection in this case is

TABLE 10-10 Air release and vacuum relief valve standards. (Adapted from ASAE Standard S376.1.)

High Pressure Systems			
Pipe Diameter		Minimum Valve Outlet Diameter	
(mm)	(in.)	(mm)	(in.)
102	4 or less	13	0.5
127–203	5–8	25	1.0
254–500	10–20	51	2.0
530	21 or larger	0.1 × pipe diameter	
Low Pressure Systems			
Pipe Diameter		Minimum Valve Outlet Diameter	
(mm)	(in.)	(mm)	(in.)
152	6 or less	51	2.0
178–254	7–10	76	3.0
305	12 or larger	102	4

to reduce the magnitude of the downsurge so the reflected high pressure wave is also reduced. The most common method of limiting downsurge is to feed water into the pipeline as soon as the pressure drops. Different methods of accomplishing this objective are applicable depending on the operating pressure of the system and anticipated magnitude of the pressure surge.

Pump bypass

Figure 10-6 indicates a pump bypass which is coupled with a check valve to allow water to enter the pipeline after the pump is shut off. This system is only operational if the pressure in the pipeline drops below the net positive suction head available to induce flow through the check valve. The minimum pressure the pipeline will experience with this type of installation is the net positive suction head available minus friction headloss through the bypass. This method is not applicable if the reflected surge pressure wave, equal in magnitude to the initial pressure drop, is damaging to the pipeline. This method is only recommended when the operational discharge pressure head is considerably less than the product involving pressure wave celerity and design flow velocity as given by the equation

$$H_O \ll a\left(\frac{v_O}{g}\right) \tag{10-21}$$

where

H_0 = operating pressure head, m
a = celerity of pressure wave, m/s
v_0 = design flow velocity, m/s
g = acceleration of gravity, m/s^2

The celerity of the pressure wave is computed using Eq. (10-19). Typical ranges of celerity are approximately 200 m/s for thermoplastic pipe and 850 m/s for thin wall steel pipes.

Surge tank

Surge tanks are connected to the pipeline and may be vented to the atmosphere. Tanks which are vented to the atmosphere are often called stand pipes. The height of water in the stand pipe is at the hydraulic grade line during normal operating conditions. Water in the tank dampens both decompression and compression waves generated by water hammer. The hydraulic transients between the pump and stand pipe are high frequency pressure waves which cause only slight changes in the

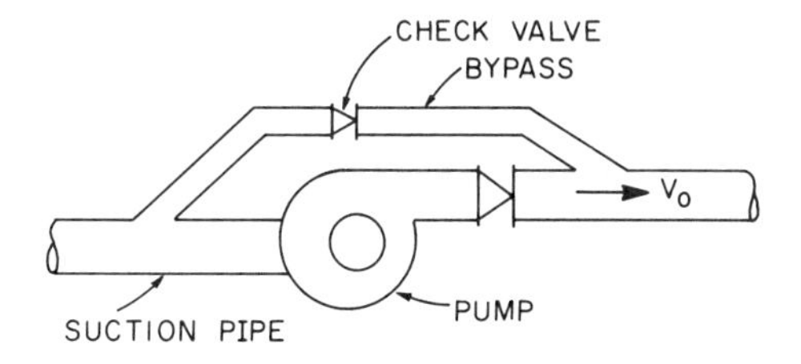

Figure 10-6 Schematic of pump bypass with check valve.

water level in the tank. However, the installation of the tank causes dampening of the pressure waves between the tank and the delivery end of the pipeline.

Closed or pneumatic surge tanks are used in high pressure systems where stand pipe heights are uneconomical. The tank contains air which serves as a pressure buffer. An air compressor is often necessary to replace air absorbed by the pressurized water. The discussion which follows is with reference to Fig. 10-7 which indicates possible locations for a number of water hammer protection devices along a pipeline.

The rate of deceleration of the flow stream between the stand pipe and end of the pipeline is given by

$$\frac{dv}{dt} = \frac{-g\Delta h}{l} \tag{10-22}$$

where

Δh = difference in height between hydraulic grade line and depressed water level in stand pipe following passage of pressure wave, m

l = length of pipeline between stand pipe and delivery end, m

By continuity,

$$A_p(v) = A_t\frac{dh}{dt} \tag{10-23}$$

where

A_p = cross-sectional area of pipe, m^2

A_t = cross-sectional area of stand pipe, m^2

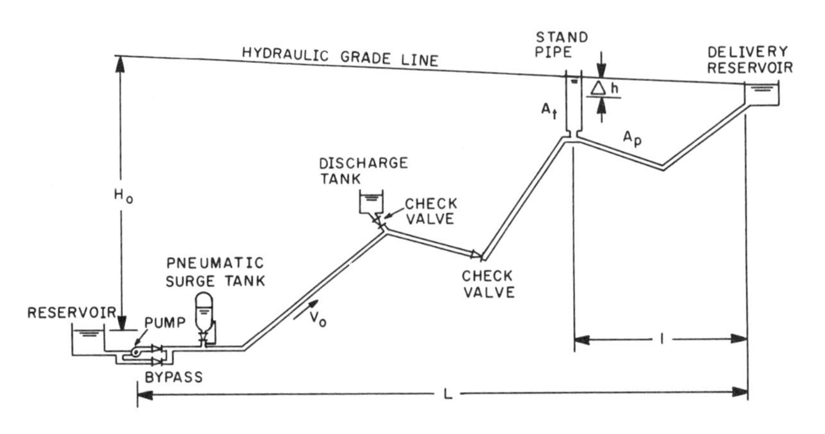

Figure 10-7 Pipeline profile with practical locations for various pipeline protection devices.

Combining Eqs. (10-22) and (10-23) and noting that at initial conditions $t = 0$, $\Delta h = 0$, and $v = v_0$, the following equation is derived for the change of water level in the tank (Stephenson, 1981):

$$\Delta h = v_0 \left[\frac{A_p(1)}{A_t g}\right]^{0.5} \sin\left\{t\left[\frac{A_p(g)}{A_t 1}\right]^{0.5}\right\} \qquad (10\text{-}24)$$

where

t = time required for flow stoppage following pump shut off, s

The term in brackets following the sin is in radians. Application of Eq. (10-24) to estimate the expected depression of water in a surge tank is indicated in the following example problem.

Example Problem 10-6

A stand pipe 20 cm in diameter is installed on a pipeline for water hammer protection. The pipeline diameter is 30 cm and the flow velocity is 1.5 m/s under normal operating conditions. Assume that 0.5 seconds is required for flow stoppage after the pump is shut off. The distance from the surge tank to the end of the pipeline is 300 m. Using a safety factor of 1.5 to avoid underestimation, compute the expected change in height of water in the tank.

Solution Let h equal the height of the tank calculated considering pipe hydraulics and h_{sf} the height adjusted for the safety factor.

$$h_{sf} = 1.5\,\Delta h$$

From Eq. (10-24),

$$\Delta h = v_0 \left[\frac{A_p(1)}{A_t g}\right]^{0.5} \sin\left\{t\left[\frac{A_p(g)}{A_t 1}\right]^{0.5}\right\}$$

Substituting the problem values,

$$\Delta h = 1.5\ \text{m/s}\left\{\frac{(0.30\ \text{m})^2\, 300\ \text{m}}{(0.20\ \text{m})^2\, 9.81\ \text{m/s}^2}\right\}^{0.5} \sin\left\{0.5\ \text{s}\left[\frac{(0.30\ \text{m})^2\,(9.81\ \text{m/s}^2)}{(0.20\ \text{m})^2\ 300\ \text{m}}\right]^{0.5}\right\}$$

$$\Delta h = 1.68\ \text{m}$$

$$h_{sf} = 1.5(\Delta h) = 2.5\ \text{m}$$

Discharge tank

Discharge tanks are applied where the pipeline profile is considerably lower than the hydraulic grade line. It is also recommended that the pipeline be concave downwards at the location of the discharge tank. The discharge tank must be separated from the pipeline by a check valve which allows water to flow one way from the tank to the pipeline. This valve isolates the tank from positive pressure surges in the pipeline. The discharge tank protects the pipeline from negative pressure surges following pump shut down. The analysis of surge protection for a discharge tank is similar to that indicated for the stand pipe.

Not all of the pipeline protection devices indicated in this section are required on irrigation system mainlines. The extent to which surge protection must be installed depends on the flow velocities and operating pressures of the system. For many moderate size irrigation systems without high operating pressures, the basic spring-loaded pressure relief valve previously discussed will adequately protect the pipeline.

10.5 Pipeline Installation

This section indicates specifications recommended for installation of underground pipelines. The recommendations are taken for the most part from the ASAE standards for underground irrigation lines and are considered adequate for a broad range of irrigation systems. Additional specifications may be required for very high capacity lines or lines expected to operate at very high pressures (Stephenson, 1981).

Trenching Requirements

Trench width

Trench widths above the top of the pipe should not be greater than 0.6 m (2 ft) wider than the pipe diameter. This standard may be exceeded in unstable soils where sloughing or caving may occur and it is required to slope the sidewalls above the pipe. Minimum and maximum trench widths for low pressure lines as a function of pipe diameter are indicated in Table 10-11.

Trench depth

The normal required minimum depth of cover as a function of pipe diameter is given in Table 10-12. If extra fill must be placed above the soil surface to provide the minimum depth of cover, the top width of the fill should be no less than 3 m (10 ft) and the side-slope ratio no less than 4 Horizontal:1 Vertical. The minimum depth of cover for locations in which vehicular wheel loads will be applied to the

TABLE 10-11 Minimum and maximum trench widths. (Adapted from ASAE Standard S376.1.)

Pipe Diameter		Approximate Trench Width			
		Minimum		Maximum	
(in.)	(mm)	(in.)	(mm)	(in.)	(mm)
4	102	16	400	30	760
6	152	18	450	30	760
8	203	20	510	30	760
10	254	22	560	30	760
12	305	24	610	30	760
14	356	26	660	30	760
15	381	27	690	30	760
18	457–475	30	760	36	910
20	508	32	810	36	910
24	610–630	36	910	42	1070
27	710	40	1020	46	1170

TABLE 10-12 Minimum depth of cover.

Pipe Diameter		Minimum Depth	
(mm)	(in.)	(mm)	(in.)
13–64	0.5–2.5	460	18
76–102	3–4	610	24
102	4	760	30

trench is 0.76 m (30 in.) above the top of the pipeline. This depth of cover is required before wheel loads are applied to the pipeline for the first time. This minimum depth of cover standard applies to both low and high pressure lines.

Thrust Blocking

Thrust blocking prevents the pipeline from shifting in the trench and is required to increase the capability of the line to resist deformation and failure at the joints. Thrust loading results from unequal forces developed by a change in direction of water in the pipeline. The thrust block serves to transfer the load from the pipe to a large load bearing surface.

Thrust blocks are required at the following locations:

(a) Where the pipeline changes the direction of water flow (i.e., at ties, elbows, crosses, wyes, and tees).
(b) Where pipe size changes (i.e., at reducers, reducing tees and crosses).
(c) At the end of a pipeline (i.e., at caps and plugs).
(d) Where there is an in-line flow control valve.

Thrust blocks are formed against a solid trench wall that has been excavated by hand. The pipe couplings themselves are not to be thrust blocked. This is indicated in Fig. 10-8 which demonstrates the proper location and placement of thrust-block material. The following steps to calculate the required load-bearing area are given with reference to Tables 10-13 and 10-14:

(a) Multiply the pipeline operating pressure by the appropriate value in Table 10-13 to obtain an estimate of the total thrust.
(b) Determine an estimation of the bearing strength of the soil from Table 10-14.
(c) Divide the total thrust determined in step (a) by the bearing strength to calculate the required load bearing area.

An alternative procedure applies the magnitude of the deflection of the flow stream in degrees. The amount of deflection is multiplied times the value tabulated in Table 10-15 as a function of pipe diameter. The result is multiplied by the operating pressure divided by 100 to obtain the total side thrust. To compute the total side thrust in newtons (N), the operating pressure is required in kPa. To obtain the side thrust in lb, the operating pressure is required in psi. The resulting total thrust is divided by the bearing strength of the soil from Table 10-14 to obtain the required

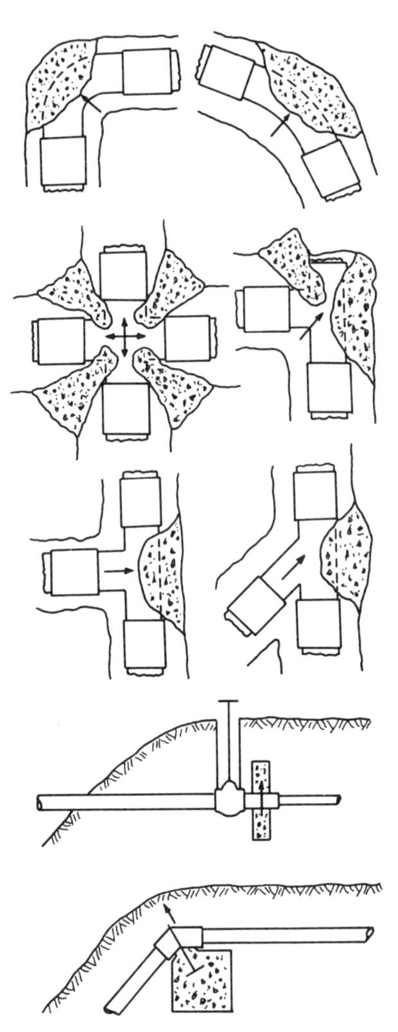

Figure 10-8 Examples of various configurations for installation of thrust blocks. (Adapted from ASAE Standard S376.1.)

load-bearing area as was indicated in step (c) of the preceding. The procedure to calculate the load-bearing area is demonstrated in the following example problem.

Example Problem 10-7

Compute the required load-bearing area for a thrust block at a 90 degree elbow in a 273 mm diameter pipeline which has an operating pressure of 586 kPa. The pipeline is installed in a trench made up of a coarse sand.

Solution From Table 10-13 for 90° elbow,

$$\text{total thrust} = T = 130.0(586 \text{ kPa})$$

$$T = 76{,}180 \text{ N}$$

From Table 10–14 for coarse sand,

$$\text{soil bearing strength} = s_b = 150 \text{ kPa}$$

Area required:

$$A_r = \frac{T}{s_b} = \frac{76{,}180 \text{ N}}{150 \text{ kPa}}$$

$$A_r = 0.51 \text{ m}^2$$

Thrust load is therefore to be transferred to approximately 0.51 m^2 bearing area of thrust block.

TABLE 10-13 Pipeline thrust factors. (Adapted from ASAE Standard S376.1.)

Pipe Diameter					
(in.)	(mm)	Dead End or Tee	90 Degree Elbow	45 Degree Elbow	22.5 Degree Elbow
1.5	48	2.94	4.16	2.25	1.15
2	60	4.56	6.45	3.50	1.78
2.5	73	6.65	9.40	5.10	2.60
3	89	9.80	13.9	7.51	3.82
3.5	102	12.8	18.1	9.81	4.99
4	114	16.2	23.0	12.4	6.31
5	141	24.7	35.0	18.9	9.63
6	168	34.8	49.2	26.7	13.6
8	219	59.0	83.5	45.2	23.0
10	273	91.5	130.0	70.0	35.8
12	324	129.0	182.0	98.5	50.3
14	363	160.0	226.5	122.6	62.6
15	389	183.9	260.0	140.7	71.9
16	406	201.1	284.4	153.8	78.6
18	475	274.7	388.4	210.1	107.4
20	518	326.9	462.2	250.1	127.8
21	560	381.8	539.9	292.1	149.3
24	630	483.2	683.2	369.6	188.9
27	710	613.7	867.8	469.5	239.9

TABLE 10-14 Bearing strength of soils. (Adapted from ASAE Standard S376.1.)

	Safe Bearing Load	
Soil Type	(kPa)	(lb/ft^2)
Sound shale	500	10,000
Cemented to gravel and sand difficult to pick	200	4,000
Coarse and fine compact sand	150	3,000
Medium clay—can be spaded	100	2,000
Soft clay	50	1,000
Muck	0	0

TABLE 10–15 Data for alternative side thrust procedure. (Adapted from ASAE Standard S376.1.)

Pipe Diameter		Side Thrust per Degree	
(in.)	(mm)	(lb)	(N)
1.5	48	5.1	22.7
2	60	7.9	35.1
2.5	73	11.6	51.6
3	89	17.1	76.1
3.5	102	22.4	99.6
4	114	28.3	125.9
5	141	43.1	191.7
6	168	60.8	270.5
8	219	103.0	458.2
10	273	160.0	711.7
12	324	225.0	1000.8
14	363	278.2	1237.4
15	389	319.6	1421.6
16	406	349.3	1553.7
18	475	477.3	2123.0
20	518	568.0	2526.5
21	560	663.6	2951.7
24	630	839.6	3734.5
27	710	1066.2	4742.5

Thrust block material

The recommended thrust block material is concrete which has a compressive strength of at least 13.8 MPa (2000 psi). The concrete mixture should be one part cement, two parts washed sand, and four parts gravel. The blocks should be constructed so the bearing surface is to the greatest extent possible in a direct line with the maximum force exerted by the pipeline. The direction of these forces is indicated by the arrows in Fig. 10-8.

REFERENCES

American Society of Agricultural Engineers, "Design, Installation and Performance of Underground Thermoplastic Irrigation Pipelines," ASAE Standard S376.1, ASAE, St. Joseph, Michigan, 1987, pp. 501–511.

Colebrook, C. F., "Turbulent Flows in Pipes with Particular Reference to the Transition Region between Smooth and Rough Pipe Laws," *Journal Institute of Civil Engineers,* London, 1939.

Moody, L. F., "Friction Factors for Pipe Flow," Transactions of the American Society of Mechanical Engineers, vol. 66, 1944, p. 671.

Mott, R. L., *Applied Fluid Mechanics,* second edition. Columbus, Ohio: Charles E. Merrill Publishing, 1979.

Murdock, J. W., *Fluid Mechanics and Its Application.* Boston: Houghton Mifflin, 1976.

STEPHENSON, D., *Pipeline Design for Water Engineers,* second edition. Amsterdam: Elsevier, 1981.

WATTERS, G. Z., *Analysis and Control of Unsteady Flow in Pipelines,* second edition. Boston: Butterworths, 1984.

PROBLEMS

10-1. The pressure rating of a PE pipe according to the ASAE standards is 160 kPa. The pipe is inside diameter based and has an inside diameter of 500 mm and a wall thickness of 12 mm. Determine the standard code designation for the pipe.

10-2. The annual power cost of a section of new PVC pipeline 1 km in length is $12,000. The pipeline has a diameter of 400 m and carries a flow of 180 L/s. Calculate the annual cost of energy for the same section of pipeline if it became badly corroded but everything else remained the same.

10-3. Compute the average annual total cost over a period of 50 years for a section of galvanized steel pipe 40.0 cm in diameter, 2 km in length, with a continuous flow rate of 440 L/s. Cost of new pipe is $1740 per 100 m including installation, power cost is 6 cents per kWh, and the interest rate is 9 percent. Power cost and interest rate will be assumed constant over the 50-year period. Assume the pipeline is horizontal and the pumping plant efficiency is 65 percent.

10-4. A 183 m long aluminum mainline with couplers of 15 cm diameter has a friction headloss of 1.93 m computed using the Hazen-Williams equation with C = 120. What is the volumetric flow rate through the pipe in m^3/s?

10-5. Water is discharged through a 90° flanged elbow which has a ratio of radius of curvature (r) to diameter (d) of 10. Discharge through the line is 12 L/s and friction loss is estimated as 0.0555 meters. What is the nominal diameter of the pipe in cm?

10-6. A 1600 ft long steel pipe discharges water from a reservoir at the upstream end and is fitted with an instantaneously closing valve on the downstream end. The pipe has an outside diameter of 12.750 inch and a wall thickness of 0.375 inch. Design discharge through the pipe is 3500 gpm with water at 50°F. The pipe is fitted with expansion joints throughout its length. Compute the time in seconds for a pressure wave to reach the reservoir if the valve is instantaneously closed.

10-7. A 203 mm inside diameter steel pipe set on the horizontal has a wall thickness of 8 mm and carries 65.56 degrees C water at a flow rate of 56 L/s. The pipeline is 457 m long, securely anchored at both ends only, and subject to an operating pressure of 3000 kPa. At t = 0 a gate valve is closed at the downstream end of the pipeline. Neglecting friction, compute the pressure in kPa at the middle of the pipeline at t = 1.0 s.

10-8. List the three main purposes of orifice type air release vacuum relief valves.

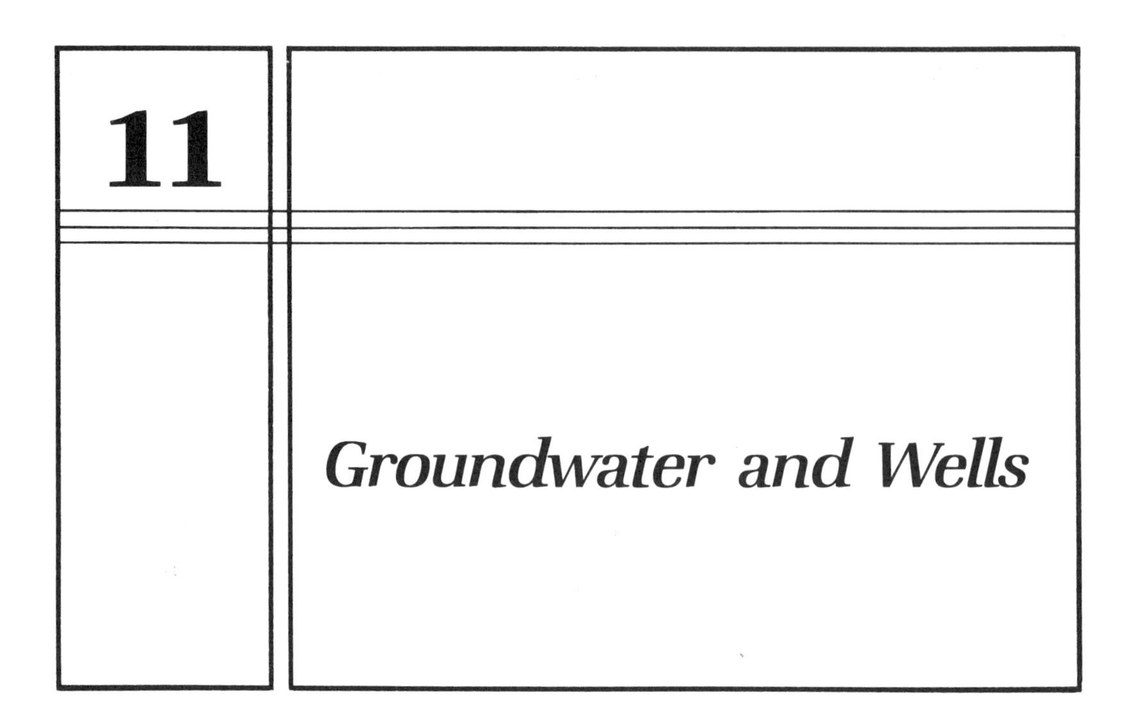

11.1 Definitions and Concepts

Groundwater resources often play a decisive role in the development of irrigation projects. This may be true of a single well which is the sole source of water at the farm level to the regional scale in which hundreds of wells must be managed for the long-term benefit of society. This chapter will concentrate on analysis of groundwater resources at the local as contrasted to the regional scale. Procedures for well design and installation will be described.

Aquifer Designation

An *aquifer* is a geologic formation which contains water and which allows for the movement of water under normal conditions. The description of various types of aquifer systems will be made with reference to Fig. 11-1. An *aquaclude* is a geologic formation which may contain water but which does not allow for the movement of water under normal conditions. Such a formation is said to be impervious. An *aquitard* is a geologic formation which transmits water at a very low rate compared to an aquifer. Such a formation is considered to be semipervious or leaky. An *aquifuge* is an impervious formation which neither contains nor transmits water.

A *confined aquifer* is one which is bounded above and below by impervious formations—that is, by aquicludes or aquifuges. An observation well penetrating

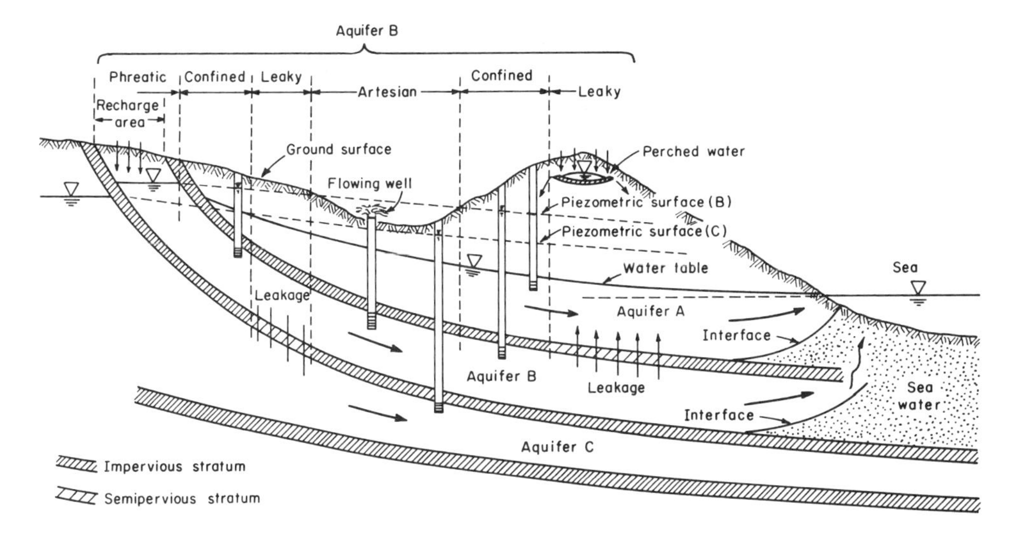

Figure 11-1 Schematic demonstrating different types of aquifer formations. (Adapted from Bear, 1979.)

such a formation will measure a piezometric head which is higher than the upper confining layer of the aquifer. An *artesian aquifer* is a special case of a confined aquifer in which the piezometric surface is above the ground surface. The result is a well which will flow without pumping and is called a flowing or artesian well. A leaky confined aquifer is one which is bounded on top and/or bottom by an aquitard.

An *unconfined aquifer* is bounded on top by the water table and not by a confining layer. It is bounded on the bottom by an impervious formation. The piezometric head measured in an observation well in an unconfined aquifer is at the level of the water table. A leaky unconfined aquifer is bounded on the bottom by an aquitard.

Porous Media Designation

Methods to designate the porous media will be made with reference to a variable which denotes some physical property of the porous media system. This property may be measured in the x-direction, y-direction, and the z-direction as well as at distinct points within the system.

Isotropic medium. At a point, properties do not change with respect to direction.

Anisotropic medium. At a point, properties vary depending on the direction in which they are measured.

Homogeneous medium. Properties of isotropy or anisotropy are constant throughout the medium.

Heterogeneous medium. Properties of isotropy or anisotropy vary throughout the medium.

Porous media systems need to be categorized with respect to both homogeneity and isotropy. It will be convenient to designate the medium with respect to intrinsic permeability, k, which is a physical property of the media only (and not also of the fluid as in hydraulic conductivity, K).

$$k = \frac{K\mu}{\rho g} \tag{11-1}$$

where

$$\mu = \text{dynamic viscosity}$$
$$\rho = \text{fluid density}$$

Homogeneous-isotropic. Permeability is the same in any direction from a point and constant throughout the medium.

Heterogeneous-isotropic. Permeability is the same in any direction from a point but varies spatially throughout the medium.

Homogeneous-anisotropic. Permeability varies dependent upon direction from a point but the variation is constant throughout the medium.

Heterogeneous-anisotropic. Permeability varies dependent upon direction from a point and this variation is not constant throughout the medium.

Storativity

The storativity of an aquifer is defined as the volume of water released from storage per unit horizontal area of aquifer per unit decline in piezometric head. For a confined aquifer, the relationship for the storativity is given by

$$S = \frac{\Delta V_w}{A\Delta h} \tag{11-2}$$

where

$$S = \text{storativity, dimensionless}$$
$$\Delta V_w = \text{volume of water released, m}^3$$
$$A = \text{horizontal area of aquifer, m}^2$$
$$\Delta h = \text{change in piezometric head, m}$$

The storativity is seen to be a dimensionless ratio which gives an indication of the potential yield of the aquifer under certain development conditions. The storativity in confined aquifers is normally within the ranges indicated by

$$10^{-6} \leq S \leq 10^{-4} \tag{11-3}$$

The storativity of a confined aquifer reflects the fact that water within the aquifer is under high enough pressure to be compressed and to expand the skeletal soil matrix of the porous media. Approximately 40 percent of the storativity is due to the expansion of water when the piezometric head is reduced in the aquifer due to pumping and 60 percent is due to the resulting compression of the porous media matrix. The compression of the porous media matrix cannot always be reversed even if the aquifer is brought back to the original piezometric head. Significant withdrawal of water from a confined aquifer may therefore result in permanent reduction of aquifer storage capacity.

The piezometric head in an unconfined aquifer is defined by the position of the water table. A unit decline in h in Eq. (11-2) is therefore represented by a unit decline in the level of the water table. Recognizing this fact, Eq. (11-2) can be also used to compute the storativity of an unconfined aquifer. The storativity of an unconfined aquifer is commonly referred to as the *specific yield,* S_y. General values of specific yield are given by the following range:

$$0.20 \leq S_y \leq 0.30 \qquad (11\text{-}4)$$

Representative values of specific yield for various aquifer materials are indicated in Table 11-1.

Water retained in the soil matrix after drainage of upper portions of the unconfined aquifer by lowering the water table is called the *specific retention,* S_r. (Specific retention in analysis of groundwater systems is analogous to field capacity in evaluating soil-moisture extraction by roots.) The sum of the specific yield and specific retention is the aquifer porosity as indicated by

$$S_y + S_r = N \qquad (11\text{-}5)$$

TABLE 11-1 Representative values of specific yield. (Johnson, 1967.)

Material	Specific Yield (percent)
Coarse Gravel	23
Medium Gravel	24
Fine Gravel	25
Coarse Sand	27
Medium Sand	28
Fine Sand	23
Silt	8
Clay	3
Fine-grained Sandstone	21
Medium-grained Sandstone	27
Limestone	14
Dune Sand	38
Loess	18
Peat	44
Schist	26
Siltstone	12
Predominantly Silt Till	6
Predominantly Sand Till	16
Predominantly Gravel Till	16
Tuff	21

where

$$N = \text{porosity, fraction}$$

Perennial Yield

Perennial yield is defined as the rate at which water may be withdrawn from a groundwater basin without producing an undesirable result. The following is a listing of possible undesirable results:

(a) Progressive reduction of the water resource.
(b) Development of uneconomic pumping conditions.
(c) Degradation of groundwater quality.
(d) Interference with prior water rights.
(e) Land subsidence caused by lowered groundwater levels.

As long as the amount of extraction from the aquifer is balanced by the recharge, whether natural or artificial, the withdrawal rate is within the perennial yield. An increase in perennial yield may be made possible by artificially recharging the aquifer with volumes of water greater than those available through natural recharge. If long-term discharge of an aquifer exceeds recharge, the aquifer is said to be in an *overdraft* or *mining* condition.

Example Problem 11-1

A volume of water equal to 40×10^6 m^3 is pumped from an unconfined aquifer through wells uniformly distributed over an area of the aquifer equal to 100 km^2. The specific yield of the aquifer as determined by pump tests is $S_y = 20$ percent.

Determine the average drawdown of the water table in meters assuming it is uniformly distributed over the area of the aquifer.

Solution Applying the equation for specific yield,

$$S_y = 0.20 = \frac{\Delta V_w}{[A\,\Delta h]}$$

which can be rearranged to

$$\Delta h = \frac{\Delta V_w}{[A(0.20)]}$$

$$\Delta h = \frac{40 \times 10^6 \text{ m}^3}{[100 \text{ km}^2(1000 \text{ m/km})^2(0.20)]}$$

$$\Delta h = 2.00 \text{ m}$$

11.2 Hydraulics of Groundwater Flow

Darcy's Law

The original work of Henry Darcy (Darcy, 1856) to model the hydraulics of flow in the sand beds used to filter the municipal water supply of Dijon, France, resulted in

Darcy's Law. It is given as

$$Q = -K(A)\left[\frac{h_2 - h_1}{L}\right] \tag{11-6}$$

where

Q = volumetric flow rate, L^3/T
A = cross-sectional area of flow, L^2
h_1 and h_2 = piezometric head at points 1 and 2, L

in which

$$h = \frac{p}{\gamma} + z \tag{11-7}$$

$\frac{p}{\gamma}$ = pressure head, L
p = pressure, F/L^2
γ = specific weight of fluid, F/L^3
z = elevation head above some arbitrary datum, L
L = distance between measurement points 1 and 2, L
K = proportionality factor, L/T

The term in brackets in Eq. (11-6), $[(h_2 - h_1)/L]$, is usually termed the hydraulic gradient. As we observed in the chapter on Soil Physics, flow always goes from the higher to the lower piezometric head. The minus sign preceding the K in Eq. (11-6) is to insure agreement between the direction of flow and the orientation of the coordinate system.

It is often convenient to apply the specific discharge, q, which is defined as the volume of water flowing per unit time through a unit cross-sectional area normal to the direction of flow. It is given by

$$q = \frac{Q}{A} = -K\left[\frac{h_2 - h_1}{L}\right] \tag{11-8}$$

The specific discharge can be interpreted as a descriptor of the velocity of flow through the porous media. This velocity is with respect to the total cross-sectional area of flow and not just the connected pore area through which the flow actually takes place. The specific discharge is often termed the Darcian velocity.

The actual velocity of fluid movement through interconnected pores is given by the pore velocity, v,

$$v = \frac{Q}{N(A)} = \frac{q}{N} \tag{11-9}$$

where

N = soil porosity as a fraction

The proportionality constant, K, between the hydraulic gradient for flow and the Darcian velocity is termed the hydraulic conductivity. It is clear from Eqs. (11-6) and (11-8) that the hydraulic conductivity may be considered a coefficient between the driving force for flow in a soil water system and the flow velocity. It is a key element in the quantification of groundwater flow volumes. The hydraulic conductivity in groundwater flow systems is a function of the composition of the porous media. Some typical values of hydraulic conductivity are given in Table 11-2.

The transmissivity of an aquifer is defined as the rate of flow per unit width through the entire thickness of the aquifer per unit hydraulic gradient. It is an important parameter in the characterization of flow in aquifers. The transmissivity, T, is a function of the hydraulic conductivity and is given as

$$T = K(b) \tag{11-10}$$

where

T = transmissivity, L^2/T

b = saturated thickness of aquifer, L

TABLE 11-2 Representative values of hydraulic conductivity. (Adapted from Todd, 1980.)

Material	Hydraulic Conductivity (m/d)	Type of Measurement
Coarse Gravel	150	R
Medium Gravel	270	R
Fine Gravel	450	R
Coarse Sand	45	R
Medium Sand	12	R
Fine Sand	2.5	R
Silt	0.08	H
Clay	0.0002	H
Fine-grained Sandstone	0.2	V
Medium-grained Sandstone	3.1	V
Limestone	0.94	V
Dolomite	0.001	V
Dune Sand	20	V
Loess	0.08	V
Peat	5.7	V
Schist	0.2	V
Slate	0.00008	V
Predominantly Sand Till	0.49	R
Predominantly Gravel Till	30	R
Tuff	0.2	V
Basalt	0.01	V
Weathered Gabbro	0.2	V
Weathered Granite	1.4	V

H = Horizontal hydraulic conductivity
R = Repacked sample
V = Vertical hydraulic conductivity

The hydraulic conductivity is conceptualized over a unit cross-sectional area of flow while the transmissivity considers flow over the total saturated thickness of the aquifer. Figure 11-2 demonstrates the difference in the affected depth from which water is removed in an aquifer by pumping from a well as a function of the transmissivity, all other factors being equal. It can be observed in this figure that the hydraulics of flow in wells requires that both the depth and radius of the affected aquifer vary with variation in the transmissivity.

The intrinsic permeability, k, is related to the hydraulic conductivity of an aquifer but is a function of only the properties of the porous media and not of the fluid. The relationship between the hydraulic conductivity and the intrinsic permeability is given as

$$K = \frac{k\gamma}{\mu} = \frac{kg}{\nu} \tag{11-11}$$

where

k = intrinsic permeability of porous media, L^2

μ = dynamic viscosity of fluid, $F(T)/L^2$

g = acceleration of gravity, L/T^2

ν = kinematic viscosity, L^2/T

Note that

$$\nu = \frac{\mu}{\rho} \tag{11-12}$$

where

ρ = fluid density, M/L^3

Hydraulic conductivity is therefore a function of both fluid and porous media properties. The porous media properties which affect the hydraulic conductivity and intrinsic permeability are porosity, grain-size distribution, and the shape of the grains. Table 11-3 indicates standard values for properties of water which are applicable in the equation for hydraulic conductivity.

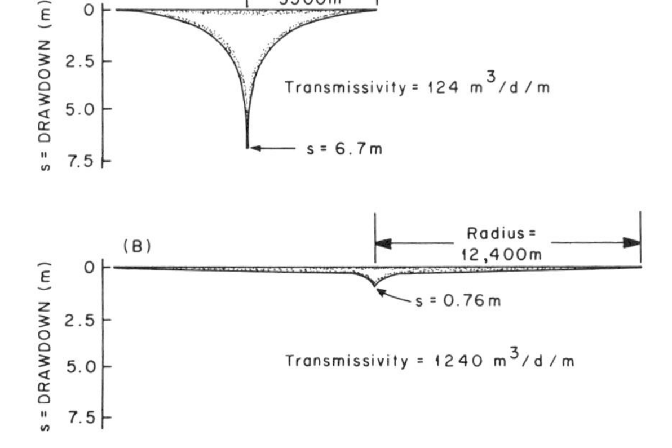

Figure 11-2 The effect of different values of transmissivity on the depth and extent of the piezometric response surface (cone of depression) due to pumping in a well. The amount of well discharge is the same for both parts of the figure.

TABLE 11-3 Standard values for properties of water.

Property	Standard Value SI Units	English Units
Temperature	15°C	60°F
Specific Weight	$9.81\ kN/m^3$	$62.4\ lb/ft^3$
Density	$1000\ kg/m^3$	$1.94\ slugs/ft^3$
Dynamic Viscosity	$1.15 \times 10^{-3}\ Ns/m^2$	$2.35 \times 10^{-5}\ lb(s)/ft^2$
Kinematic Viscosity	$1.15 \times 10^{-6}\ m^2/s$	$1.21 \times 10^{-5}\ ft^2/s$

Well Hydraulics

Darcy's Law will be applied to describe the flow towards a well. Recall that the specific discharge may be interpreted as a descriptor of the velocity of flow. Defining the specific discharge with respect to radial flow towards a well,

$$q = -K\left(\frac{dh}{dr}\right) \tag{11-13}$$

where

$\frac{dh}{dr}$ = derivative of piezometric head with respect to radial distance from the well

Confined Aquifer

Figure 11-3 indicates general flow conditions in a confined aquifer. Note that the flow velocity q is in the direction of declining piezometric head h. Considering that the positive x-direction is to the right, the hydraulic gradient dh/dx is negative. Figure 11-4 indicates the distribution of piezometric head measured by observation wells for radial flow towards a producing well in a confined aquifer. Defining the positive direction as radially outward, dh/dr is positive but the direction of flow is negative (i.e., radially inward) as indicated by the negative sign on the right-hand side of Eq. (11-13). Note that Fig. 11-4 shows that the confined aquifer is fully

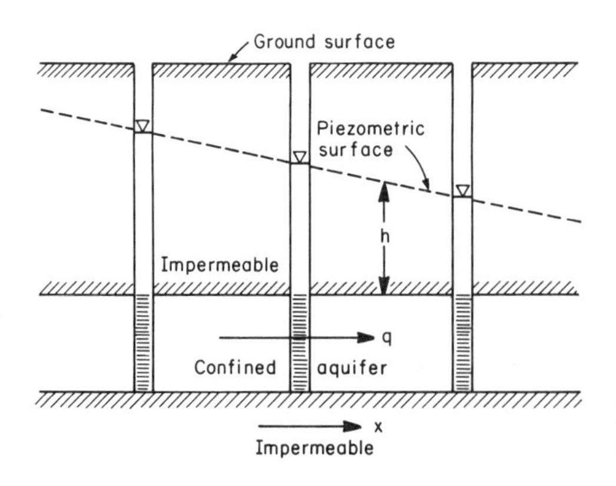

Figure 11-3 Steady unidirectional flow in a confined aquifer of uniform thickness.

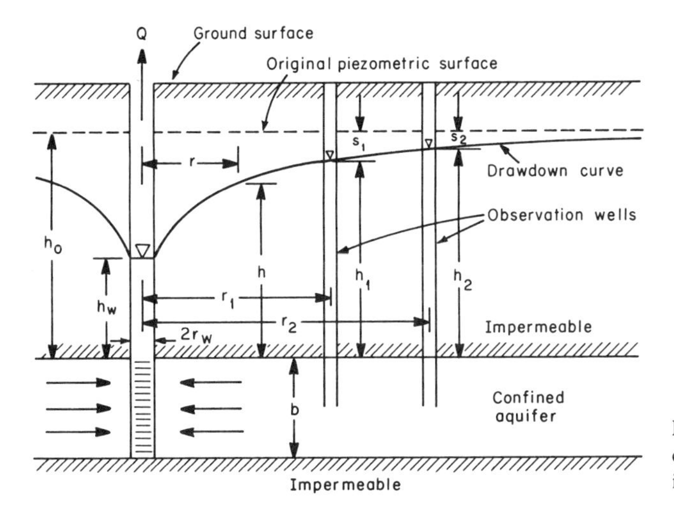

Figure 11-4 Radial flow to a well in a confined aquifer with constant thickness and infinite in areal extent.

screened throughout the thickness of the aquifer b. The original piezometric surface in the figure is the position of the piezometric surface before pumping begins.

Under steady-state conditions, the quantity of radial flow towards the well is equal to the well discharge Q given by

$$Q = A(q) \tag{11-14}$$

where

A = cross-sectional area of flow normal to direction of flow velocity

Applying Eq. (11-13) for the specific discharge and noting that flow is in the negative direction (i.e., q is negative because flow is radially inward), Eq. (11-14) may be rewritten as

$$Q = 2\pi rb\left[K\left(\frac{dh}{dr}\right)\right] \tag{11-15}$$

where dh/dr is the hydraulic gradient measured at radial distance r from the center of the producing well. Equation (11-15) will be integrated over the limits,

$$h = h_w \quad \text{at } r = r_w$$

and

$$h = h_0 \quad \text{at } r = r_0$$

in which r_w is the well radius and r_0 is a radial distance from the well at which there is no measurable decline in the piezometric head due to pumping. Rearranging Eq. (11-15) and integrating over the limits indicated,

$$Q\int_{r_w}^{r_0}\left(\frac{dr}{r}\right) = 2\pi bK\int_{h_w}^{h_0} dh \tag{11-16}$$

$$Q[\ln(r_0) - \ln(r_w)] = 2\pi bK(h_0 - h_w) \tag{11-17}$$

which can be rewritten as

$$Q = 2\pi bK\left[\frac{h_0 - h_w}{\ln\left(\frac{r_0}{r_w}\right)}\right] \tag{11-18}$$

r_0 is termed the radius of influence, since beyond this distance the piezometric surface is not affected by pumping in the well.

Equation (11-15) can be integrated with respect to other limits such as those defined by the observation wells at radial distances r_1 and r_2 in Fig. 11-4. The result of such an integration is

$$Q = 2\pi bK\left[\frac{h_2 - h_1}{\ln\left(\frac{r_2}{r_1}\right)}\right] \tag{11-19}$$

Equation (11-19) is a better representation of the governing equation for a confined aquifer since well losses caused by flow through the well screen and interior to the well introduce errors in the actual measurement of h_w.

Equation 11-19 can be rewritten to express the transmissivity of the aquifer, assuming the well is screened throughout the total thickness of the aquifer, as

$$T = Kb = \left\{\frac{Q}{2\pi(h_2 - h_1)}\right\}\ln\left(\frac{r_2}{r_1}\right) \tag{11-20}$$

In practice, the drawdown s is more easily measured than the piezometric head. As indicated in Fig. 11-4,

$$s = h_0 - h \tag{11-21}$$

Equation (11-20) can therefore be rewritten as

$$T = \left\{\frac{Q}{2\pi(s_1 - s_2)}\right\}\ln\left(\frac{r_2}{r_1}\right) \tag{11-22}$$

Example Problem 11-2

A well penetrates a confined aquifer of coarse gravel 7.0 m thick and is screened throughout the thickness of the aquifer. Hydraulic conductivity measurements made in a well in the same region yield a value of 180 m/d. Two observation wells are installed at radial distances of 20 m and 120 m from the pumped well. The well is to be tested by pumping at a constant discharge of 2725 m^3/d. If the drawdown in the observation well at 120 m distance is 1.26 m under steady-state conditions, compute the expected drawdown in the well at 20 m distance.

Solution Use the drawdown form of the discharge equation:

$$Q = 2\pi bK\left[\frac{s_1 - s_2}{\ln\left(\frac{r_2}{r_1}\right)}\right]$$

Upon substitution,

$$2725 \text{ m}^3/\text{d} = 2\pi(7.0 \text{ m})(180 \text{ m/d})\left[\frac{s_1 - 1.26 \text{ m}}{\ln\left(\dfrac{120 \text{ m}}{20 \text{ m}}\right)}\right]$$

Rearranging,

$$s_1 = \left[\frac{2725 \text{ m}^3/\text{d}}{4418 \text{ m}^2/\text{d}}\right] + 1.26 \text{ m}$$

$$s_1 = 1.88 \text{ m}$$

Unconfined Aquifer

The water table serves as the upper bound of the piezometric surface in an unconfined aquifer. Figure 11-5 indicates groundwater flow between two bodies of water through an unconfined aquifer. It can be observed that actual flow is not horizontal but converges towards the downstream vertical boundary of the flow system. This contrasts with the horizontal flow observed in the confined aquifer in Fig. 11-3.

The specific discharge will be defined as a function of the piezometric head and distance s along the surface of the water table as

$$q = -K\left[\frac{dh}{ds}\right] \tag{11-23}$$

The water table is defined as a surface at which the pressure is atmospheric—that is, $p = 0$. Substituting this result into Eq. (11-7) shows that the piezometric head is equal to the elevation head:

$$h = z \tag{11-24}$$

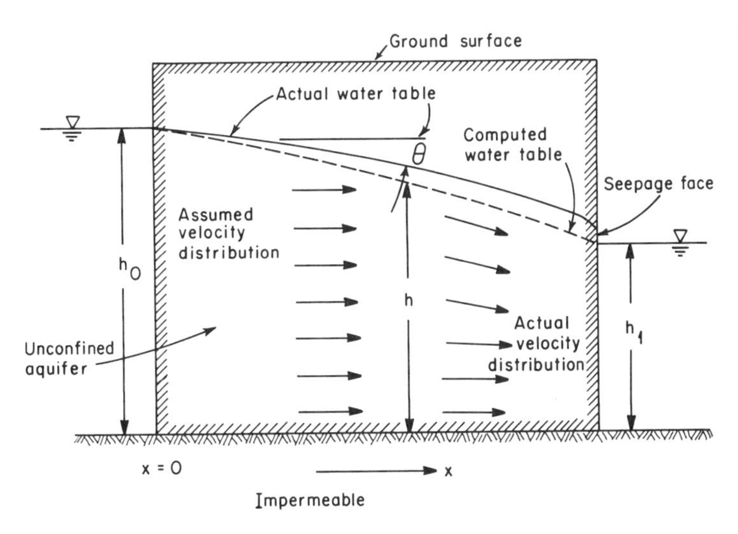

Figure 11-5 Steady flow in an unconfined aquifer with vertical boundaries between two bodies of water at different elevations.

Equation (11-23) can therefore be written as a function of the change in elevation head with distance along the water table:

$$q = -K\left[\frac{dz}{ds}\right] \tag{11-25}$$

If θ is the angle between the horizontal and the water table surface, Eq. (11-25) can be rewritten as

$$q = -K \sin \theta \tag{11-26}$$

where

$$\sin \theta = \frac{dz}{ds} \tag{11-27}$$

Dupuit-Forchheimer discharge formula

Dupuit (1863) observed that water table slopes in field situations are typically very small, often in the range of from 1/100 to 1/1000. He developed the following assumptions for flow in unconfined aquifers exhibiting small values of θ—that is, for water table slopes of less than 1/10 which corresponds to a horizontal angle of less than 6 degrees:

(a) $\sin \theta$ in Eq. (11-26) may be replaced by $\tan \theta$.

$$\tan \theta = \frac{dh}{dx} \tag{11-28}$$

(b) Flow in the aquifer is horizontal and of uniform velocity with depth.

These assumptions have the effect of setting the hydraulic gradient equal to the change in piezometric head in the direction of flow with change in *horizontal* distance. Equation (11-25) can therefore be rewritten as

$$q = -K\left[\frac{dh}{dx}\right] \tag{11-29}$$

Figure 11-6 demonstrates radial flow towards a well in an unconfined aquifer in which the well is screened throughout the depth of the aquifer. Substituting the derivative of radial distance dr for dx in Eq. (11-29) and recognizing, as was done for the confined aquifer case, that the hydraulic gradient is positive radially outward while the flow is radially inward and therefore negative in sign, we can rewrite Eq. (11-15) for the quantity Q pumped from the well as

$$Q = 2\pi rKh\left[\frac{dh}{dr}\right] \tag{11-30}$$

Rearranging Eq. (11-30) and integrating between the same limits defined for the confined aquifer case,

$$Q\int_{r_w}^{r_0}\left(\frac{dr}{r}\right) = 2\pi K\int_{h_w}^{h_0} h\, dh \tag{11-31}$$

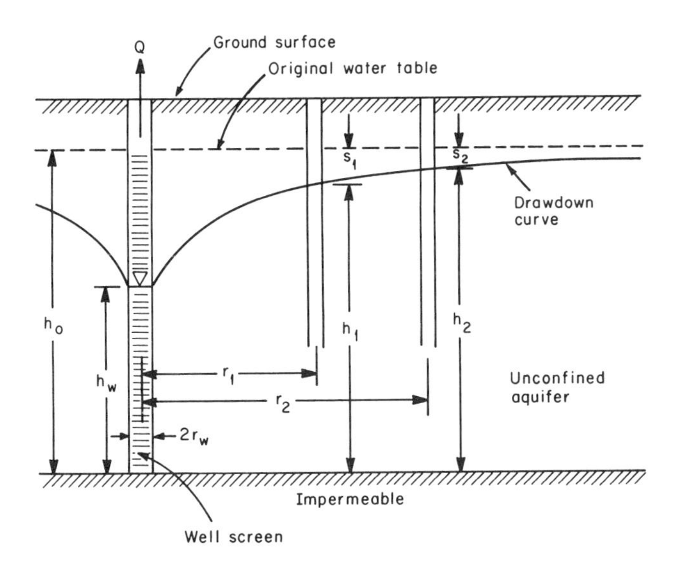

Figure 11-6 Radial flow to a well in an unconfined aquifer infinite in areal extent if the well is screened throughout the depth of the aquifer.

which results in

$$Q = \pi K \left\{ \frac{(h_0)^2 - (h_w)^2}{\ln\left(\frac{r_0}{r_w}\right)} \right\} \tag{11-32}$$

Equation (11-32) may be rewritten for the more general case of measurements made in two observation wells at radial distances r_1 and r_2.

$$Q = \pi K \left\{ \frac{(h_2)^2 - (h_1)^2}{\ln\left(\frac{r_2}{r_1}\right)} \right\} \tag{11-33}$$

In reality a vertical seepage face exists at the well boundary and the height of the water table is underestimated by Eq. (11-33) (see Fig. 11-5). This equation is therefore not an accurate governing equation for the distribution of piezometric head close to the well due to the large components of vertical flow which violate the assumptions required for the Dupuit-Forchheimer equation. Hantush (1964) however has shown that either Eq. (11-32) or (11-33) is accurate for computing discharge to within 1 to 2 percent of the measured value (Mariño and Luthin, 1982). Bear (1979) indicates that at radial distances of $r > 1.5$ times the height of the flow system—that is, $r > 1.5\ h_0$—Eq. (11-33) is also accurate for computing the distribution of the piezometric head.

In practice, the transmissivity of an unconfined aquifer may be computed by multiplying the hydraulic conductivity by the average saturated thickness of the aquifer as determined by measurements in the observation wells (Bear, 1979). The drawdowns should be small in relation to the saturated thickness of the unconfined aquifer to apply this procedure.

Example Problem 11-3

This example problem is similar to problem 11-2 but defined for the case of an unconfined aquifer. The unconfined aquifer of coarse gravel has a hydraulic conductivity of 180 m/d as determined by other well tests in the same aquifer. The static water level before pumping starts is 50.00 m above the impermeable boundary at the bottom of the aquifer. At a pumping rate of 2725 m^3/d, drawdown in the observation well at a radial distance of 120 m from the center of the pumped well is 1.26 m. Determine the expected drawdown at an observation well located at a radial distance of 20.0 m from the pumped well.

Solution Using the form of the discharge equation for unconfined aquifers,

$$Q = \pi K \left\{ \frac{(h_2)^2 - (h_1)^2}{\ln\left(\frac{r_2}{r_1}\right)} \right\}$$

convert the drawdown to piezometric head:

$$h = h_0 - s$$
$$h_2 = h_0 - s_2 = 50.00 \text{ m} - 1.26 \text{ m}$$
$$h_2 = 48.74 \text{ m}$$

Substituting values from the problem,

$$2725 \text{ m}^3/\text{d} = \pi(180 \text{ m/d}) \left\{ \frac{(48.74 \text{ m})^2 - (h_1)^2}{\ln\left(\frac{120 \text{ m}}{20 \text{ m}}\right)} \right\}$$

Rearranging,

$$(h_1)^2 = (48.74 \text{ m})^2 - \left[\frac{2725 \text{ m}^3/\text{d}}{\pi 180 \text{ m/d}}\right] \ln\left(\frac{120 \text{ m}}{20 \text{ m}}\right)$$
$$(h_1)^2 = 2367.95 \text{ m}$$

Taking the square root,

$$h_1 = 48.65 \text{ m}$$

Compute the expected drawdown,

$$s_1 = h_0 - h_1 = 50.00 \text{ m} - 48.65 \text{ m}$$
$$s_1 = 1.35 \text{ m}$$

Comment on Level of Analysis

The hydraulic analysis for flow to wells in confined and unconfined aquifers covered in this section is fundamental and involves only the most straightforward groundwater flow systems. For example, systems which are anisotropic with respect to hydraulic conductivity have not been dealt with. Nor have systems which may interfere with the streamlines of flow towards a well, such as adjacent wells with an overlapping cone of depression or flows from streams which place boundary constraints on analysis of flow towards a well. The preceding equations also deal with the steady state conditions which may take long time periods to reach in confined aquifers.

These topics and others are covered with varying degrees of complexity in other references devoted to hydraulic analysis of groundwater flow systems (Todd, 1980; Mariño and Luthin, 1982; Bear, 1979; Freeze and Cherry, 1979). The material discussed in this section is aimed at developing an understanding of the fundamentals of analysis of flow towards wells so the reader may proceed to apply the concepts to well sizing, well screen design, and well testing in later sections. The concepts and governing equations indicated have been field verified to have application to a wide variety of problems associated with on-farm water supply from groundwater resources and may be applied with confidence to simple flow situations. Analysis of groundwater resources on a larger geographic scale or for more complex or transient state flow systems should be conducted using more sophisticated methods of analysis than those reviewed in this section.

11.3 Well Drilling Methods and Construction

Drilling Methods

Table 11-4 indicates the applicability of various drilling methods with respect to type of geologic formation, normal depths and diameters, and well yields. Historically wells have been dug by hand with simple applications of animal power. In some locations, hand drilling methods continue to be practiced (see Fig. 11-7). When driven by hand, wells tend to be large in diameter to accommodate one or more drillers in the hole and relatively shallow since it is physically impossible to dig very much below the level of the water table. Such wells generally have very limited

Figure 11-7 Large diameter well being dug by hand in central Tunisia.

TABLE 11-4 Water well construction methods and applications. (After U.S. Soil Conservation Service, 1969.)

Method	Materials for Which Best Suited	Water Table Depth for Which Best Suited, m	Usual Maximum Depth, m	Usual Diameter Range, cm	Usual Casing Material	Customary Use	Yield, m^3/d	Remarks
Augering Hand auger	Clay, silt, sand, gravel less than 2 cm	2–9	10	5–20	Sheet metal	Domestic, drainage	15–250	Most effective for penetrating and removing clay. Limited by gravel over 2 cm. Casing required if material is loose.
Power auger	Clay, silt, sand, gravel less than 5 cm	2–15	25	15–90	Concrete, steel or wrought-iron pipe	Domestic, irrigation, drainage	15–500	Limited by gravel over 5 cm, otherwise same as for hand auger.
Driven Wells Hand, air hammer	Silt, sand, gravel less than 5 cm	2–5	15	3–10	Standard weight pipe	Domestic, drainage	15–200	Limited to shallow water table, no large gravel.
Jetted Wells Light, portable rig	Silt, sand, gravel less than 2 cm	2–5	15	4–8	Standard weight pipe	Domestic, drainage	15–150	Limited to shallow water table, no large gravel.
Drilled Wells Cable tool	Unconsolidated and consolidated medium hard and hard rock	Any depth	450	8–60	Steel or wrought-iron pipe	All uses	15–15,000	Effective for water exploration. Requires casing in loose materials. Mudscow and hollow rod bits developed for drilling unconsolidated fine to medium sediments.

Rotary	Silt, sand, gravel less than 2 cm; soft to hard consolidated rock	Any depth	450	8–45	Steel or wrought-iron pipe	All uses	15–15,000	Fastest method for all except hardest rock. Casing usually not required during drilling. Effective for gravel envelope wells.
Reverse-circulation rotary	Silt, sand, gravel, cobble	2–30	60	40–120	Steel or wrought-iron pipe	Irrigation, industrial, municipal	2500–20,000	Effective for large-diameter holes in unconsolidated and partially consolidated deposits. Requires large volume of water for drilling. Effective for gravel envelope wells.
Rotary-percussion	Silt, sand, gravel less than 5 cm; soft to hard consolidated rock	Any depth	600	30–50	Steel or wrought-iron pipe	Irrigation, industrial, municipal	2500–15,000	Now used in oil exploration. Very fast drilling. Combines rotary and percussion methods (air drilling) cuttings removed by air. Would be economical for deep water wells.

pumping capacities both in terms of flow rate and time of pumping. Often hand-driven wells must be left to recover after pumping for just a few hours and pumping cannot be re-initiated until the following day. These wells do not make use of the hydraulic efficiency associated with aquifer properties and potential drawdown of deeper wells.

The predominant methods used for drilling moderate to high capacity irrigation wells are cable-tool, mud rotary, and air rotary. Cable tool is a percussion type method, while mud and air rotary methods both use drill bits in combination with fluids to remove the cuttings from the hole. (The rotary percussion method uses a combination of successive impacts delivered through a rotary drill string. This method is not widely used for irrigation wells.) The cable-tool and rotary drilling methods differ in their recommended applicability to type of geologic formation and well diameter as indicated in Table 11-4. Table 11-5 presents additional information

TABLE 11-5 Performance of drilling methods in various types of geologic formations. (Taken from Speedstar Division)

	Drilling Method		
Type of Formation	Cable-Tool	Rotary	Rotary Percussion[a]
Dune sand	Difficult	Rapid	NR
Loose sand and gravel	Difficult	Rapid	NR
Quicksand	Difficult, except in thin streaks. Requires a string of drive pipe.	Rapid	NR
Loose boulders in alluvial fans or glacial drift	Difficult; slow but generally can be handled by driving pipe	Difficult, frequently impossible	NR
Clay and silt	Slow	Rapid	NR
Firm shale	Rapid	Rapid	NR
Sticky shale	Slow	Rapid	NR
Brittle shale	Rapid	Rapid	NR
Sandstone, poorly cemented	Slow	Slow	NR
Sandstone, well cemented	Slow	Slow	NR
Chert nodules	Rapid	Slow	NR
Limestone	Rapid	Rapid	Very rapid
Limestone with chert nodules	Rapid	Slow	Very rapid
Limestone with small cracks or fractures			Very rapid
Limestone, cavernous	Rapid	Slow to impossible	Difficult
Dolomite	Rapid	Rapid	Very rapid
Basalts, thin layers in sedimentary rocks	Rapid	Slow	Very rapid
Basalts, thick layers	Slow	Slow	Rapid
Metamorphic rocks	Slow	Slow	Rapid
Granite	Slow	Slow	Rapid

[a]NR: not recommended.

regarding recommended drilling methods and relative speed of drilling for a wide variety of geologic formations. Following are basic descriptions of the three predominant methods used for drilling irrigation wells.

Cable-Tool Method

The cable-tool drilling method is a percussion method that relies on the impact of the drill bit to break up the formation material so it may be removed from the bore hole. The impact to the bottom of the hole is provided by what is termed the string line of tools suspended from pulleys by a steel cable which is alternately raised and lowered. A drawing of a typical drill string composed of drill bit, drill stem, jars, and a rope socket for attachment to the drill line is shown in Fig. 11-8. Figure 11-9 is a photo of a typical cable-tool drill rig.

The drill bit which does the actual cutting is a heavy steel bar about 1 to 2.5 m long with varying degrees of sharpness depending on the hardness of the formation to be penetrated. The rope socket attaches the drill string to the drill line using a mandrel within a casing which allows the drill line to turn freely. During the upstroke and downstroke of the drill string, the random swivel action of the rope socket insures that a round hole will be formed by the drill bit. Water, either naturally occurring or pumped into the hole, is required to mix the loosened formation material into a slurry.

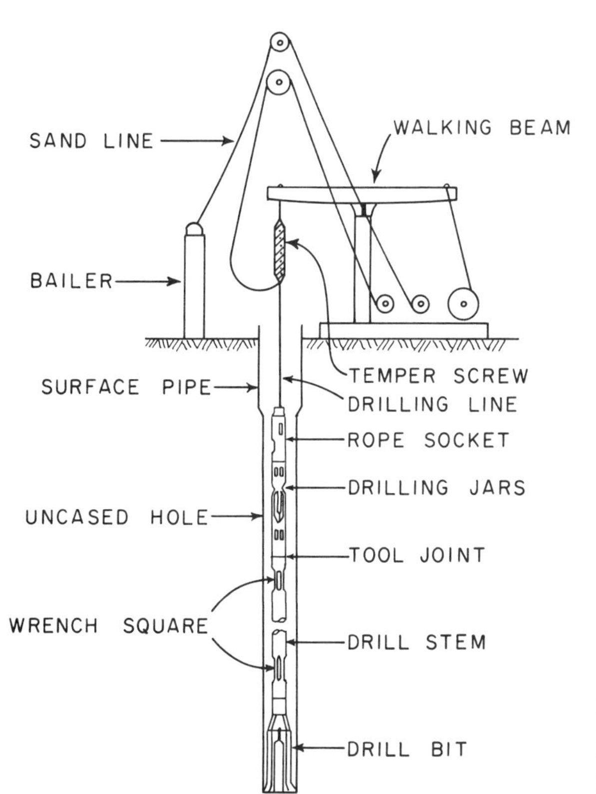

Figure 11-8 Schematic of cable-tool drilling device and equipment required to make up drill string. (Adapted from Mariño and Luthin, 1982.)

Figure 11-9 Cable-tool drilling rig on location in the Willamette Valley, Oregon.

After approximately 1 m penetration into the formation, the cuttings must be removed. The drill string is removed and a bailer or sand pump inserted into the hole using the sand line indicated in Fig. 11-8. A bailer is a pipe about 3 to 9 m in length with a one-way valve at the bottom. When lowered in the hole, the one-way valve opens due to the upward force of the fluid and the bailer fills with the slurry containing the cuttings. It is pulled to the top of the hole by the sand line and emptied. The sand pump is similar to the bailer except it has a piston device which produces a suction which draws in the slurry. The sand pump is more effective in removing cuttings.

Casing is required throughout the length of the hole in unconsolidated formations subject to caving. Blows to the casing using drive clamps attached to the drill stem are used to push lengths of casing down the hole. Casing can be protected by fitting a drive shoe at the bottom and a drive head at the top.

Cable-tool drilling can be used in all formations, although it is relatively slow in poorly consolidated formations, such as sand or sandstone, or extremely hard formations, such as granite or metamorphic rocks. The method can be applied to shallow or deep holes of various diameters, although it is most applicable to relatively shallow holes of less than 50 m depth and of moderate to large diameter starting at 15 cm. Little water is required and the geologic formation can be accurately logged during the drilling operation. Initial equipment and operating costs for many types of formations are modest compared to typical rotary drilling methods.

Rotary Methods

The principles and hydraulics of both mud and air rotary methods are similar. The main difference is the fluid used in the drilling operation is either a synthetic mud or air. The synthetic mud is produced by mixing bentonite with water. The consistency of the mud varies depending on the density of the cuttings to be brought to the surface. Denser particles require mud with a higher concentration of bentonite. Because of similarity of operation, both rotary drilling methods will be described in this section.

The main components for rotary drilling are a pump to pressurize the fluid, hollow drill pipes to transmit the pressurized fluid, a hollow drill bit which grinds the formation and passes the drilling mud, and a settling basin in which the cutting-laden mud is allowed to pond so the cuttings drop to the bottom. In normal rotary drilling, the pressurized drilling fluid is forced through the drill bit and immediately upon exiting the bit comes into contact with the cuttings which are then carried upward in the annular space between the drill pipes and hole wall. After being pumped to the surface, a high fraction of cuttings is removed by passing the drilling mud through a set of settling basins. The drilling mud is then recirculated through the pump. A schematic of the hydraulic rotary drilling method is indicated in Fig. 11-10.

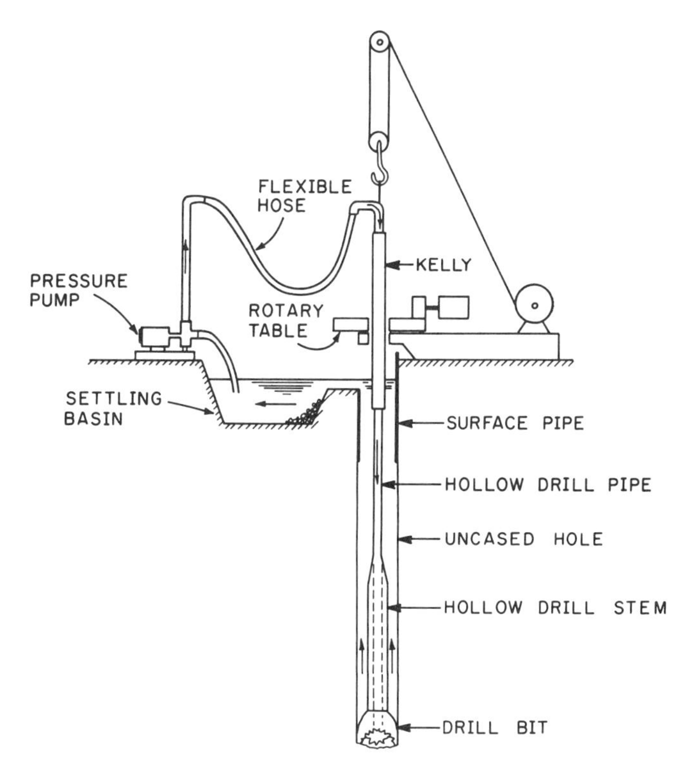

Figure 11-10 Schematic of a mud rotary drilling rig and equipment required for operation. (Adapted from Mariño and Luthin, 1982.)

Figures 11-11 and 11-12 are photographs of a mud rotary drilling rig and a dual settling basin. Wells can be drilled using the mud rotary method without a well casing even in unconsolidated formations due to the hydraulic head built up inside the hole which prevents caving.

Figure 11-11 Mud rotary drilling rig on location in central Tunisia. (From Cuenca, 1986.)

Figure 11-12 Dual settling basin used for mud rotary drilling operation. Water movement is from left to right in photograph. Drilling chips settle out in basin at left. Water laden with drilling mud then flows through opening between basins to the right where it enters recirculating system. (From Cuenca, 1986.)

Choice of drill bit depends on hardness of the geologic formation being drilled. A rotating tri-cone bit is indicated in Fig. 11-13. The drill bit, drill stem, and drill pipe all have threaded fittings. The drill pipe is made of high strength steel typically with a wall thickness of 8.56 mm or larger to insure that the pipe can withstand the forces developed during drilling. The drill string and drill bit are rotated by either a rotating table on the bottom of the drill rig or a rotating head at the top of the drill string. The schematic in Fig. 11-10 has a rotating table while the drill rig shown in Fig. 11-11 has a rotating head.

The rotary drilling method is relatively fast in many types of geologic formations as indicated in Table 11-5. Alternate layers of soft and hard material in the same hole can be handled more effectively by rotary drilling methods than by the cable-tool method. It is difficult to obtain representative formation samples using rotary methods since the material is crushed in the drilling process and cuttings from different layers are mixed in the drilling fluid. Considerable amounts of water are required for the mud rotary method compared to cable-tool, especially in gravel and fissured rock formations. As indicated in Table 11-5, mud rotary drilling in cavernous limestone is slow to impossible due to the loss of fluid down the hole. Air rotary methods, which may use a foam for the drilling fluid, have been successfully applied in formations which were not suitable for the mud rotary method.

Modifications of the normal rotary procedure include reverse rotary and rotary-percussion drilling methods. The reverse rotary method operates on the same basic principles as the normal rotary method except that the drilling fluid circulates down through the annular space between the hole wall and drill pipe and is brought up the drill pipe by a suction pump. The method is particularly well adapted to drilling large diameter wells in unconsolidated formations.

In the rotary-percussion method, an impact is administered by the drill bit to the formation by lifting and dropping the drill string. This method can be used to penetrate particularly hard formations or layers of relatively hard material in forma-

Figure 11-13 Tri-cone drill bit used to grind through strata in mud rotary drilling operation.

tions in which the rotary method would normally be inadequate. Drilling using the rotary-percussion method is not as rapid as normal rotary drilling. The applicability of the rotary-percussion method in different geologic formations is indicated in Table 11-5.

11.4 Criteria for Well Sizing and Screen Selection

Types of Well Installations

There are three basic types of well installations as demonstrated in Fig. 11-14. The first is an uncased hole which may be applied in a consolidated formation in which the material in the aquifer is rigid and will not collapse into the well. The well radius in this case is equal to the radius of the hole.

The second case is one in which the well is installed in an unconsolidated formation and the well must be screened to keep the aquifer material from collapsing into the well. The well radius in this case is also equal to the radius of the hole. As shown in part (b) of Fig. 11-14, for the same well discharge there are additional head losses due to the friction head created when the water flows past the well screen. The amount of loss will depend on the type and design of the well screen.

The third case demonstrated is that of a gravel-packed well in which a blanket of gravel on the order of 5 cm thick is placed surrounding the well screen. The gravel causes increased permeability in the vicinity of the well and results in reduced drawdown for the same well discharge. The gravel pack insures that a material of known diameter exists surrounding the well screen thereby simplifying screen design. The well radius to apply in the case of a gravel-packed well is more difficult to specify. It is common to determine an effective radius by projecting from the level of reduced drawdown horizontally to the intersection with the theoretical drawdown curve [see Fig. 11-14(c)]. An effective radius larger than the well radius often results

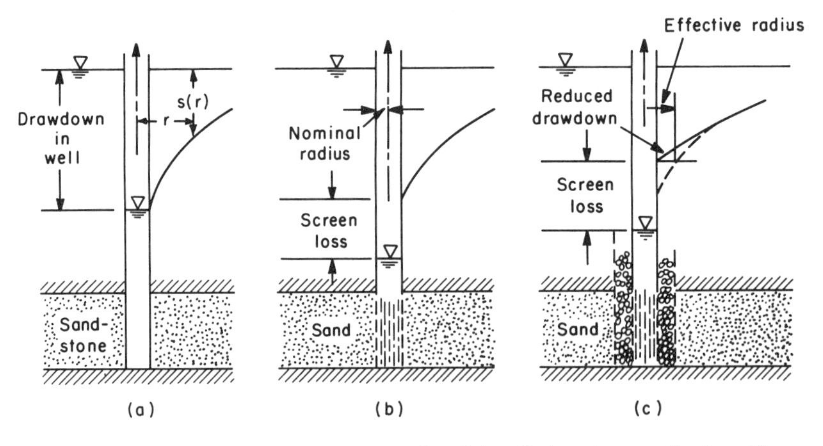

Figure 11-14 Schematic of three types of well installations in a confined aquifer: (a) uncased hole, (b) screened well, and (c) gravel-packed well. (Adapted from Bear, 1979.)

in the case of developed wells in which the permeability of the aquifer adjacent to the well is increased in the development process. The process of well development is discussed later in this section.

Justification for Well Screens

As previously noted, the quality and design of the screen material affects the level of drawdown in the well. The drawdown must be overcome by the pump when pumping water to the surface and therefore there is a cost associated with the drawdown every time the well is operated. It is clear that the operating cost of the well could be reduced if the drawdown is reduced.

The amount of drawdown is directly related to the friction loss across the screen which is a function of the velocity squared of the flow into the well. For the same discharge and screen length and diameter, the velocity of flow into the well is a function of what is called the open area of the screen—that is, that portion of the screen made up of holes which the flow passes through. Increasing the percentage of open area decreases the flow velocity and therefore decreases the head loss if all other factors are constant.

A variety of screen materials have been employed including slots cut into pipe by a welding torch, perforations punched into the casing material by multi-bladed devices, slots machined into the metal casing, louvered screen, and screens developed by winding wire of uniform thickness around a longitudinal wire frame. In any case, the principle described in the previous paragraph is valid—that is, the amount of drawdown decreases as the percentage of open area increases, all other conditions being constant. Figure 11-15 compares the open area of three types of screen material. It can be observed that the wire-wound well screen has almost 7 times the open area of the louvered screen and over 21 times the open area of the slotted pipe. This results in a significant reduction in drawdown for a given discharge rate and aquifer material for the wire-wound screen relative to louvered screen or slotted pipe. The remainder of this section will be directed towards the required specifications of wire-wound screen due to the increased hydraulic and economic efficiency of this type of screen in irrigation wells.

Typical wire-wound screens are fabricated by winding a V-shaped wire around a longitudinal metal skeleton. This so-called V-slot screen has the point of the V fac-

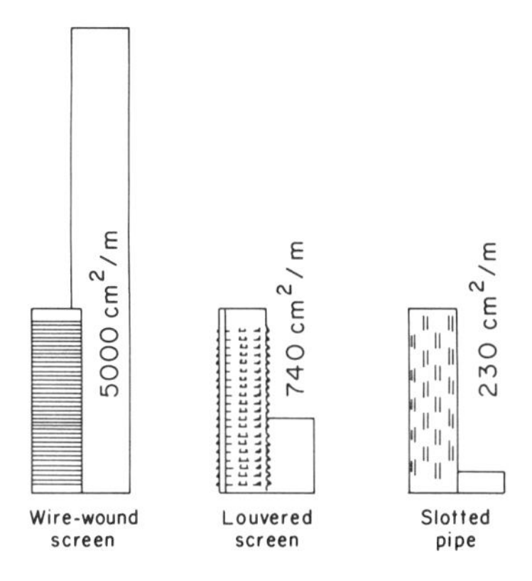

Figure 11-15 Schematic of magnitude of open area for three different types of well screen material.

ing towards the center of the well. This allows all aquifer particles which are close in diameter to the slot opening to pass through the screen and be removed during the development process. The aquifer immediately surrounding the screen thus has enhanced permeability which results in reduced drawdown adjacent to the well.

Specification of Well Casing and Screen Dimensions

The physical dimensions of the well casing and screen are a function of the extent of the aquifer, expected drawdown in unconfined aquifers, particle size distribution of the aquifer, hydraulic conductivity of the aquifer, and required discharge from the well. Often the final choice of screen dimensions will be a compromise between desired discharge, physical characteristics of the aquifer, and screen cost.

The recommended well casing and screen diameter as a function of well discharge is given in Table 11-6. The table also indicates the approximate diameter required to house a pump which could produce the required discharge. Methods of well developments will be described later in this section. It can be noted from Table 11-6 that the screen diameter is less than the casing diameter to allow the screen to be slipped through the casing into position opposite the water-bearing strata of the aquifer.

The screen length is dependent upon acceptable entrance velocities, aquifer thickness, open area per unit length of screen, and the required well discharge. The effective open area can be reduced to about one-half the actual open area due to blocking by sand and gravel, impurities carried to the screen in solution, and bacterial growth. The entrance velocity for water passing the screen boundary is selected to minimize blockage and the encrustation which may occur when impurities precipitate out of solution. The precipitation takes place as the result of chemical reactions related to the increase in velocity and decrease in pressure as the water flows past the screen. Recommended entrance velocities as a function of hydraulic conductivity of the aquifer are indicated in Table 11-7. Other recommended velocities are even more restrictive than those in the table allowing a maximum of 3 cm/s for any value of the hydraulic conductivity equal to or greater than 120 m/day (National Water Well

TABLE 11-6 Recommended minimum diameters for well casings and screens. (Adapted from U.S. Bureau of Reclamation, 1977.)

Well Yield (m^3/d)	Nominal Pump Casing Casing Diameter (cm)	Surface Casing Diameter (cm)		Nominal Screen Diameter (cm)
		Naturally Developed Wells	Gravel-Packed Wells	
<270	15	25	45	5
270–680	20	30	50	10
680–1,900	25	35	55	15
1,900–4,400	30	40	60	20
4,400–7,600	35	45	65	25
7,600–14,000	40	50	70	30
14,000–19,000	50	60	80	35
19,000–27,000	60	70	90	40

TABLE 11-7 Values of entrance velocities through screen openings. (Adapted from Walton, 1962.)

Aquifer Hydraulic Conductivity		Entrance Velocity through Screen Opening	
(m/d)	(gpd/ft²)	(cm/s)	(ft/min)
<20	<500	1	2
20	500	1.5	3
40	1000	2	4
80	2000	3	6
120	3000	4	8
160	4000	4.5	9
200	5000	5	10
240	6000	5.5	11
>240	>6000	6	12

TABLE 11-8 Sample record of cumulative weight by sieve analysis.

Sieve Opening (cm)	Cumulative Weight Retained (gm)	Cumulative Percent Retained
0.165	24	6
0.117	72	18
0.084	140	35
0.058	220	55
0.041	300	75
0.030	344	86
0.020	372	93
Bottom Pan	400	100

Association, 1981). The maximum velocity applied in this section will be 3 cm/s to correspond to these more restrictive standards.

The screen slot size—that is, the minimum width of the opening between successive windings of the screen wire—is determined using the results of sieve analysis of the aquifer material. Table 11-8 and Fig. 11-16 indicate sample tabulated sieve analysis data and the corresponding grain size distribution curve. The effective diameter is used as an index of the fineness of the aquifer material. The grain size corresponding to that for which 90 percent of the aquifer material is coarser is termed the *effective diameter*. This is equivalent to 0.025 cm in the sample data.

The *uniformity coefficient* of an aquifer is defined as the ratio of the grain size for which 40 percent of the aquifer material is coarser to the effective grain size,

$$\mathrm{UC} = \frac{d_{40}}{d_{90}} \tag{11-34}$$

For the sample data,

$$\mathrm{UC} = \frac{0.076 \text{ cm}}{0.025 \text{ cm}} = 3.0 \tag{11-35}$$

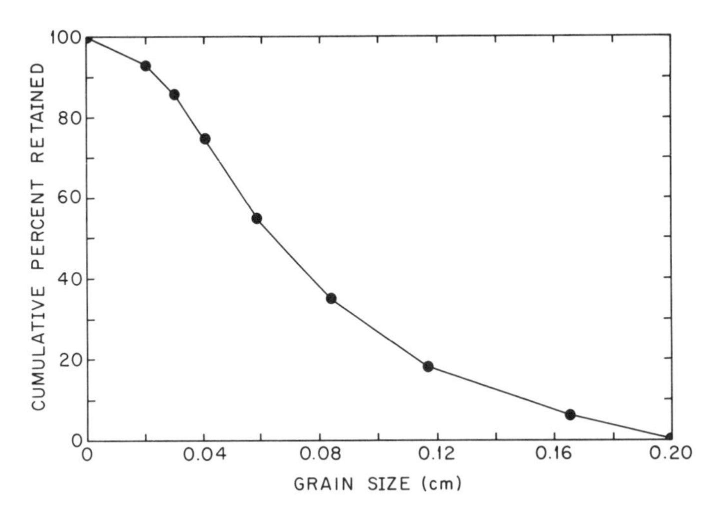

Figure 11-16 Grain size distribution based on data in Table 11-8.

For formations made up of relatively fine and uniform materials—that is, UC $\leq$ 5, the size of the screen opening is selected to correspond to d_{40} if the groundwater is noncorrosive and d_{50} if the groundwater source is corrosive. With a homogeneous formation of coarse sand and gravel, the recommended size of slot openings is from d_{30} to d_{50} of the *sand fraction*—that is, relative to the grain size distribution of only the sand particles.

For nonuniform aquifer materials—that is, UC > 5—the recommended size of screen openings is d_{30} if the material above the aquifer is stable and d_{50} if the overlying material is unstable and subject to caving (i.e., unconsolidated). In nonuniform aquifers, the principle of screen diameter selection promotes entrance of finer material into the well during the well development process at which time it is removed. Coarser material is retained outside the screen, forming a permeable envelope around the well. This is said to result in a *natural gravel envelope* which causes a hydraulic conductivity adjacent to the well which is many times the conductivity of the original aquifer.

For wells with an artificial gravel pack, the d_{90} of the gravel grain size distribution is recommended for the slot size. Gravel packs are commonly applied in sand aquifers in which too small a slot size would have to be specified for practical purposes. The required diameter of the gravel material itself is dependent on the uniformity coefficient of the aquifer. The criteria for recommended gravel diameter as developed by the U.S. Bureau of Reclamation is indicated in tabulated form in Table 11-9.

Once the slot size has been selected, the length of screen required depends on the required well discharge, the percentage of open area for the screen selected, and often on practical considerations of the thickness of water-bearing formations within the aquifer. The percentage of screen open area is a function of the slot size—that is, screen opening—type of screen selected, and screen diameter. This information is available in tabulated form from the screen manufacturer. Table 11-10 is an example of this type of information.

TABLE 11-9 Criteria for selection of gravel pack material. (After U.S. Bureau of Reclamation, 1977.)

Uniformity Coefficient (UC) of Aquifer	Gravel Pack Criteria	Screen Slot Size
<2.5	(a) UC between 1 and 2.5 with the 50% size not greater than 6 times the 50% size of the aquifer (b) If (a) is not available, UC between 2.5 and 5 with 50% size not greater than 9 times the 50% size of the aquifer	≤10% passing size of the gravel pack
2.5–5	(a) UC between 1 and 2.5 with the 50% size not greater than 9 times the 50% size of the formation (b) If (a) is not available, UC between 2.5 and 5 with 50% size not greater than 12 times the 50% size of the aquifer	≤10% passing size of the gravel pack
>5	(a) Multiply the 30% passing size of the aquifer by 6 and 9 and locate the points on the grain-size distribution graph on the same horizontal line (b) Through these points draw two parallel lines representing materials with UC ≤ 2.5 (c) Select gravel pack material that falls between the two lines	≤10% passing size of the gravel pack

TABLE 11-10 Intake areas for telescope V-slot wire-wound well screen. (Areas expressed as square centimeters per lineal meter of screen length.)

Nominal screen size (cm)	10-slot (0.010 in.) (0.25 mm)	20-slot (0.020 in.) (0.50 mm)	40-slot (0.040 in.) (1.0 mm)	60-slot (0.060 in.) (1.5 mm)	80-slot (0.080 in.) (2.0 mm)	100-slot (0.100 in.) (2.5 mm)	150-slot (0.150 in.) (3.7 mm)	250-slot (0.250 in.) (6.2 mm)
7.6	318	550	868	1101	1249	1376	1545	1736
8.9	381	656	1037	1291	1482	1630	1863	2096
10.2	423	741	1207	1503	1715	1863	2138	2434
11.4	487	847	1355	1693	1947	2117	2413	2731
12.7	550	953	1524	1905	2159	2371	2371	2794
14.3	593	1037	1672	2096	2392	2604	2985	3366
15.2	635	1122	1799	2244	2117	2371	2794	3302
20.3	593	1080	1842	2392	2815	3154	3387	4107
25.4	762	1376	2286	2985	3514	3937	4234	5144
30.5	889	1630	2752	3027	3620	4128	5017	5610
35.6	783	1439	2053	2794	3408	3916	4911	6181
40.6	889	1270	2286	3133	3810	4403	5525	6922
45.7	762	1461	2625	3577	4361	5017	6308	7938
50.8	868	1630	2942	4001	4847	5588	5927	7747
61.0	1291	2392	2773	3853	4784	5610	7261	9504
66.0	1334	2498	2921	4043	5017	5885	7620	9970
76.2	1588	2921	3408	4742	5885	6880	8933	11685
91.4	1778	3323	3895	5398	6710	7853	10182	13315

The following relationship has been developed for well screen dimensions as a function of required well discharge and allowable entrance velocity:

$$Q = 0.0864v_s(1 - c)\pi d_s L_s P \tag{11-36}$$

where

Q = well discharge, m^3/d
v_s = screen entrance velocity, cm/s
c = clogging coefficient
d_s = screen diameter, cm
L_s = screen length, m
P = percentage of screen open area

The normal value used for the clogging coefficient is 0.50 which assumes that 50 percent of the screen open area will be blocked by aquifer material following well development.

In practice, the extent of screened area in a well may be constrained by the thickness of water-bearing formations in the aquifer. The following criteria have been specified by the National Water Well Association (1981):

(a) Confined aquifer—at least 80 percent of aquifer thickness is to be screened with the screen centered in the water-bearing formation.
(b) Unconfined aquifer—33 to 50 percent of aquifer thickness is to be screened with the screen being placed at the bottom of the aquifer.

In any case, the screen entrance velocity is not to exceed 3 cm/s.

Example Problem 11-4

A well casing diameter and screen dimensions are to be specified for a well installation in a confined aquifer which is a source of noncorrosive water. The well log and results of pump tests on other wells in the same aquifer indicate that the aquifer thickness is 30 m and the hydraulic conductivity is 180 m/d. The grain size distribution for the aquifer material is that given by the sample data in Table 11-8 and Fig. 11-16. Analysis of those data yields

$$UC = 3.0$$
$$d_{90} = 0.025 \text{ cm}$$
$$d_{40} = 0.076 \text{ cm}$$

No gravel pack is to be used. The required well discharge is 5000 m^3/d. Specify the casing diameter, screen dimensions, and check the feasibility of the installation in this aquifer. Use Table 11-10 for the manufacturer's data on percentage of open area.

Solution Based on the uniformity coefficient criteria—that is, $UC < 5$ for uniform material and noncorrosive water supply—the required slot size is

$$s_s = d_{40} = 0.076 \text{ cm} = 0.030 \text{ in.}$$

Based on the hydraulic conductivity of the aquifer and Table 11-7, the allowable screen velocity is

$$v_s = 4.5 \text{ cm/s}$$

In this case we will apply the more constraining velocity criteria of the National Water Well Association Standards,

$$v_{s-max} = 3.0 \text{ cm/s}$$

Based on the required discharge and Table 11-6, the minimum screen diameter is

$$d_s = 25 \text{ cm}$$

Assume a clogging coefficient,

$$c = 0.50$$

For a telescope type well screen, use the manufacturer's data from Table 11-10 for P. The required slot size, s_s, is 0.76 mm (0.030 in.). From the table select 20-slot size (i.e., s_s = 0.020 in.) as the closest smaller size available. From Table 11-10 for 20-slot screen with 25 cm (approximately 10 inch) diameter,

$$A_{open} = 1376 \text{ cm}^2/\text{m}$$

where per meter refers to per lineal meter of screen length. Computing the total screen area per lineal meter,

$$A_{total} = \pi d_s(100 \text{ cm/m})$$
$$A_{total} = 7854 \text{ cm}^2/\text{m}$$

Compute percent open area:

$$P = \left[\frac{A_{open}}{A_{total}}\right]100$$
$$P = \left[\frac{1376 \text{ cm}^2/\text{m}}{7854 \text{ cm}^2/\text{m}}\right]100$$
$$P = 17.5 \text{ percent}$$

Substituting these values into Eq. (11-36) for Q,

$$Q = 0.0864v_s(1 - c)\pi d_s L_s P$$
$$5000 \text{ m}^3/\text{d} = 0.0864(3.0 \text{ cm/s})(1 - 0.50)\pi(25 \text{ cm})L_s(17.5)$$

Rearranging to solve for screen length,

$$L_s = \frac{5000 \text{ m}^3/\text{d}}{178}$$
$$L_s = 28.1 \text{ m}$$

Since the aquifer thickness b = 30 m,

$$\frac{L_s}{b} = 0.937 > 80 \text{ percent}$$

which checks with the design criteria. As a practical matter, the aquifer would be screened throughout the full aquifer thickness.

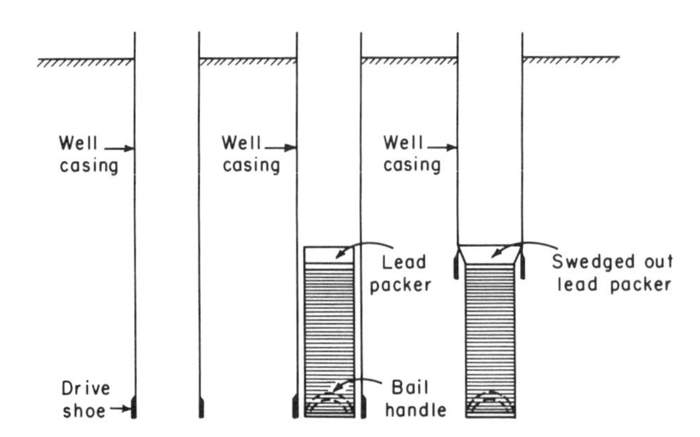

Figure 11-17 Schematic indicating procedure to install well screen using the pull-back method. (Adapted from Mariño and Luthin, 1982.)

Installation of Well Screens

Regardless of the method of well drilling, the simplest method of well screen installation is by the pull-back method. In this method, the casing is extended to the full depth of the drilled hole and the properly sized screen is lowered into the casing. The casing is then pulled back to expose the well screen. What is called a swedge block is then lowered into the soft lead packer which forms the top of the well screen. The swedge block expands the lead packer until it forms a sand-tight fit with the well casing. The excess casing is cut off approximately 60 cm above the surface of the well using a welding torch. The length of casing above the well prevents runoff from entering the well. Figure 11-17 depicts the main steps in the pull-back method. Figure 11-18 is a photograph of a screen being installed by lowering into the well casing.

Well screen can be installed by the bail-down or wash-down method when the pull-back method is not feasible and the aquifer material is unconsolidated and can be removed in a slurry. By this method, the borehole is drilled to the top of the aquifer and the casing is set in its permanent position (see Fig. 11-19). The properly sized screen fitted with a bail-down or wash-down bottom is lowered through the casing to the top of the aquifer. The screen is then lowered into the aquifer by using a bailer or wash line through the screen bottom. The object is to remove unconsolidated material from below the screen so the screen sinks into the desired position in the aquifer. In the wash-down method, the fluid delivered by the wash line circulates around the outside of the screen to bring unconsolidated material to the surface. The bailing pipe or wash line is removed when the desired depth is reached. A screen with a bail-down bottom then requires a lead or weighted wooden plug to seal it. A wash-down bottom is self-sealing. After the bailing pipe or wash line is removed, a swedge block is used to seal the screen against the lead packer.

Well Development

The final step in well construction is well development which is required to promote proper well operation. Well development is a very important step in finishing the construction process and is carried out for two major reasons:

(a) To increase aquifer permeability to the level that existed before drilling and thereby offset the decrease in permeability resulting from compaction and/or lining of the hole with drilling mud during the drilling operation.

(b) To remove material fine enough to pass through the well screen which is in the immediate vicinity of the well to be able to operate the well with sand-free conditions during normal operations.

Well development is accomplished by rapidly moving water through the aquifer formation in the immediate vicinity of the well screen. This movement of water loosens fine material and brings it into the well where it can be removed by

Figure 11-18 Well screen being lowered into well casing using a cable-tool drilling rig. Note the lead packing seal at the top of the screen section.

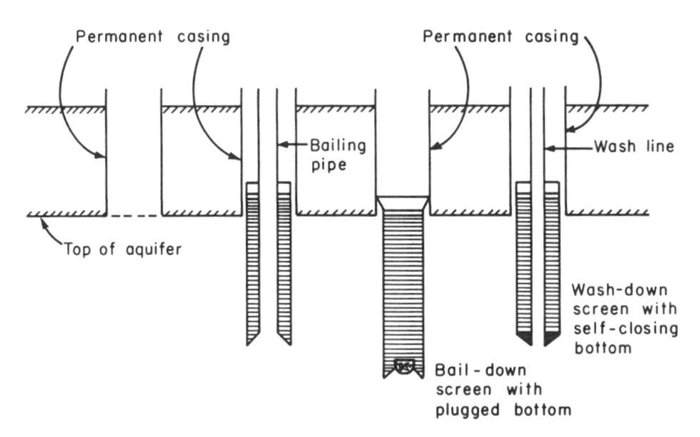

Figure 11-19 Schematic of procedure to install well screen using the bail-down or wash-down method. (Adapted from Mariño and Luthin, 1982.)

pumping or by using the bailer associated with the cable-tool method. The three common well development methods are:

(a) Surge block.
(b) Air displacement.
(c) Jetting.

The main criterion for good development is to vigorously agitate the formation adjacent to the screen and pass water back through the screen into the well at high velocity. In the surge method this is accomplished by oscillating the surge block which is conveniently done using a cable-tool rig. An air compresser at the surface is required for the air displacement method. Jetting is carried out using a high pressure water pump which is temporarily brought to the well site. Water is discharged through nozzles at high speed and passes through the well screen to loosen aquifer material. Jetting is probably the most successful development method considering the objectives, although the other methods have proven to be satisfactory.

The well development process is completed when the well is able to flow sand-free at a discharge rate at least equal to and preferably higher than the maximum design discharge. Practically speaking, sand-free means having no more than a few grains of sand per liter when the well is pumped at maximum or higher flow rates. More specific criteria for establishment of a sand-free condition are given in the Standards of the National Water Well Association (1981).

11.5 Well Testing

Purpose

Contracts for installation of irrigation wells normally require that a well test be conducted as the final step in completing the installation process. Well tests are generally performed for two reasons: (a) to determine the hydraulic characteristics of the aquifer; (b) to determine the operating characteristics of the well. Two common well testing methods are the specific capacity test and the constant-rate test. The most straightforward application of these testing methods will be described in this section. Other types of well tests and analysis of well test data under a wide variety of more complex hydraulic situations are described elsewhere (Kruseman and de Ridder, 1983).

Specific Capacity Test

For this test the well is pumped at different levels of discharge for times equal to the expected time of well operation. It is common to start the test at 20 percent of design discharge and increment the discharge by 20 percent until 120 percent of design discharge is attained. The drawdown corresponding to each level of discharge is measured and used to construct what is termed the well characteristic curve as indicated in Fig. 11-20. The time of pumping is set to correspond to the expected operating time so the results of the specific capacity test may be used to estimate the expected drawdown for a given level of discharge.

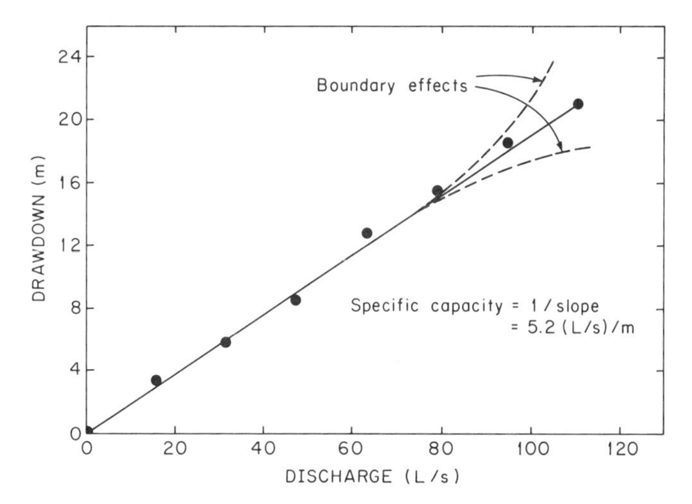

Figure 11-20 Example of a well characteristic curve developed from well test data at different flow rates.

The results of this test yield the specific capacity of the aquifer-well system which is defined as the ratio of the amount of well discharge to drawdown below the static piezometric surface. This parameter can be important in determining the required pump operating characteristics, the potential of interference of cones of depression of wells in the same vicinity, and in performing economic analysis of well operations. The dashed lines indicated in Fig. 11-20 may occur when boundary conditions not described by the straightforward analyses indicated in this chapter are encountered. Examples of such conditions are cones of depression of interfering wells or contribution to aquifer flow by surface streams within the radius of influence of the pumped well.

Constant Rate Test

The constant rate test is performed by measuring the variation in the level of drawdown with time for a given level of discharge. If a single constant rate test is to be run, it should be at the design discharge of the system. If multiple constant rate tests are run, discharge can be varied between 60 and 125 percent of design discharge. The tests should be run for a minimum of 100 minutes and may be run for 24 hours or longer. It is good practice to run the constant rate test for at least the expected operating time of the system.

The principle of the constant-rate test is that drawdown increases at a constant rate of discharge if time of pumping is sufficiently long. This can be demonstrated by referring to the data shown in Fig. 11-21. The hydraulic properties of the aquifer are most conveniently determined by analysis of data over one log-cycle time of pumping. The properties are determined by rearranging the equations used to describe the hydraulics of wells. The aquifer transmissivity is derived from the following relation:

$$T = \frac{2.30\ Q}{4\pi\ \Delta s} \tag{11-37}$$

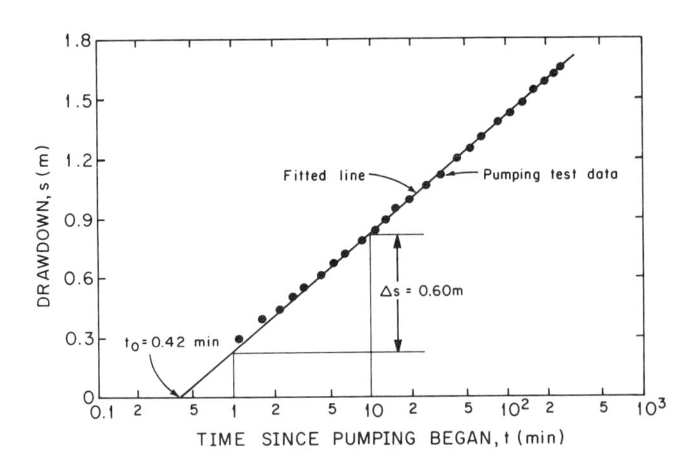

Figure 11-21 Example plot of test results from constant rate well test data.

where

T = transmissivity, m^2/d

Q = well discharge, m^3/d

Δs = drawdown corresponding to one log-cycle of pumping time, m

Another expression is used to compute the aquifer storativity. It is given by the following expression:

$$S = 2.25T\frac{t_0}{r^2} \tag{11-38}$$

where

S = storativity

t_0 = intersection of straight line fit through well test data with time axis at zero drawdown, d

r = radial distance between pumped well and observation well, m

An example plot of data from this type of test is indicated in Fig. 11-21. Since solution for the storativity using Eq. (11-38) requires the transmissivity, Eq. (11-37) must be solved first and the result substituted into the foregoing equation. Application of the equations for transmissivity and storativity of an aquifer is indicated in the following example problem.

Example Problem 11-5

The well test data from Fig. 11-21 are from a constant rate test in a confined aquifer in which drawdown was measured in an observation well 60.0 m radial distance from the pumped well. Discharge from the pumped well was held constant at 2500 m^3/d.

Compute the transmissivity and storativity of the aquifer.

Solution Apply Eq. (11-37) for transmissivity from the constant rate test:

$$T = \frac{2.30\,Q}{4\pi\,\Delta s}$$

$$T = \frac{2.30(2500\ m^3/d)}{4\pi(0.60\ m)}$$

$$T = 763\ m^2/d$$

Apply Eq. (11-38) for storativity,

$$S = \frac{2.25T\,t_0}{r^2}$$

$$S = \frac{2.25(763\ m^2/d)\,(0.42\ min)\left(\frac{1\ h}{60\ min}\right)\left(\frac{1\ d}{24\ h}\right)}{(60\ m)^2}$$

$$S = 0.000139$$

REFERENCES

BEAR, J., *Hydraulics of Groundwater.* New York: McGraw-Hill, 1979.

CUENCA, R. H., "Energy Requirements for Well Installation," in *Energy and Water Management in Western Irrigated Agriculture,* ed. N. K. Whittlesey. Boulder, Colorado: Westview Press, 1986, pp. 223–256.

DARCY, H., *Les Fontaines Publiques de la Ville de Dijon.* Paris: V. Dalmont, 1856.

FREEZE, R. A., and J. A. CHERRY, *Groundwater.* Prentice-Hall, Englewood Cliffs, New Jersey, 1979.

HANTUSH, M. S., "Hydraulics of Wells," in *Advances in Hydroscience,* ed. V. T. Chow. New York: Academic Press, 1964, pp. 281–432.

JOHNSON, A. I., "Specific Yield—Compilation of Specific Yield for Various Materials," *U.S. Geological Survey Water-Supply Paper 1662-D,* 1967.

KRUSEMAN, G. P., and N. A. DE RIDDER, *Analysis and Evaluation of Pumping Test Data.* Wageningen, The Netherlands: International Institute for Land Reclamation and Improvement, 1983.

MARIÑO, M. A., and J. N. Luthin, *Seepage and Groundwater.* New York: Elsevier, 1982.

National Water Well Association, Committee on Water Well Standards, *Water Well Specifications.* Berkeley, California: Premier Press, 1981.

Speedstar Division, *Well Drilling Manual.* Enid, Oklahoma: Koehring Co., (No date).

TODD, D. K., *Groundwater Hydrology.* New York: John Wiley and Sons, 1980.

U.S. Bureau of Reclamation, *Ground Water Manual,* U.S. Dept. of the Interior, 1977.

U.S. Soil Conservation Service, *Engineering Field Manual for Conservation Practices,* Washington, D.C. 1969.

PROBLEMS

11-1. Describe and give examples of the following type of porous media:
(a) Homogeneous and isotropic.
(b) Homogeneous and anisotropic.
(c) Heterogeneous and isotropic.
(d) Heterogeneous and anisotropic.

11-2. An alluvial valley with an unconfined aquifer system has an area of 1000 hectares. The water table is 20 m below land surface. When 25 million m^3 of water is added to the aquifer, the water table is raised 10 m. What is the specific yield of the aquifer material?

11-3. An alluvial valley with an unconfined aquifer system has an area of 1000 hectares. The specific yield of the aquifer material is 0.20.
(a) How much water will be released from storage if the water table is lowered 4 m?
(b) Assume that the aquifer system is bounded by vertical impermeable walls and that the water table is 15 m below land surface. How much water has to be added to the aquifer by artificial recharge to raise the water table 8 m?

11-4. A confined aquifer 20 m thick is composed of uniform sand that has an average grain diameter of 1.2 mm. A 0.3-m well screened throughout the aquifer is pumped at a rate of 10,000 m^3/d. Does the flow near the well obey Darcy's Law?

11-5. A tracer substance was introduced into an aquifer at an upstream location and the time required for it to appear at a downstream point was observed. The average flow velocity was found to be 0.015 m/d. The slope of the piezometric surface is 0.3 m per kilometer. Determine the hydraulic conductivity of the aquifer if the porosity of the medium is 23 percent.

11-6. A well that penetrates a homogeneous-isotropic confined aquifer of uniform thickness is pumped at a constant rate until steady-state conditions are observed. Under these conditions dye travels from an observation well A to observation well B in 40 hours. Compute the time in hours it will take the dye to travel from observation well B to observation well C if A, B, and C are all located on the same radial line out from the pumping well. The distances from the pumped well to A is 18.29 m, to B is 12.19 m, and to C is 7.62 m.

11-7. A 30.00 cm diameter well penetrates an homogeneous-isotropic unconfined aquifer to an impermeable stratum at a depth 18.00 m below the static water table. After pumping at a discharge of 16.7 L/s, steady-state conditions are attained and drawdowns in observation wells 14.00 m and 80.00 m from the pumped well are measured at 2.620 m and 2.567 m, respectively. Compute the hydraulic conductivity of the aquifer in m/d and the transmissivity in m^2/d. Apply the average thickness of the flow system to compute the transmissivity.

11-8. A 30.00 cm diameter well penetrates 32.92 m below the static water table of an homogeneous-isotropic unconfined aquifer to the lower impermeable boundary of the aquifer. After a long enough period of pumping to attain steady-state conditions at a flow rate of 22.10 L/s, the drawdown in observation wells 17.40 m and 45.10 m from the pumped well are 3.658 m and 2.256 m, respectively. Compute the transmissivity of the aquifer in m^2/d based on the average thickness of the flow system.

11-9. A 30.00 cm diameter well produces 3.2 L/s under steady-state conditions when the drawdown in the well is 1.829 m. This well penetrates to an impermeable stratum 32.00 m below the water table of an homogeneous-isotropic unconfined aquifer.

Compute the discharge in L/s from this well for a drawdown of 1.829 m if the diameter was: (a) 20.00 cm; (b) 40.00 cm. Assume that the radius of influence is 760.0 m for both cases.

11-10. Water with a radioactive tracer is injected into a well in a confined aquifer which has a static water level 26 m above a horizontal impermeable layer. Another well 200 m distant from the first in the downstream direction penetrates the same aquifer and has a static water level 19 m above the same datum. The aquifer is a medium-grained sandstone with a transmissivity of 24.8 m^2/d. Estimate the aquifer thickness in meters and compute how many days it will take before the radioactive tracer is picked up in the second well.

11-11. A well is tested in an unconfined aquifer by pumping at a constant rate of 3000 m^3/d. The aquifer is a medium-fine sand and previous aquifer tests indicate a hydraulic conductivity of 8.5 m/d. There are two observation wells at radial distances 40.0 m and 240.0 m from the pumped well. The static water level in both wells is 65.0 m above an impermeable layer. Compute the expected drawdown in meters for the more distant well if the drawdown in the well nearest the pumped well is 2.35 m.

11-12. A well is to be placed in a confined aquifer made up of a fine sand which has a hydraulic conductivity of 2.5 m/d. It is to be placed within a gravel envelope—that is, gravel pack. An analysis of the grain size distribution of the aquifer and gravel material yields the following information:

Percent Retained (by weight)	Fine Sand Sieve Diameter (mm)	Gravel Sieve Diameter (mm)
10	0.400	50.00
30	0.250	28.00
40	0.200	21.00
60	0.180	11.00
90	0.085	0.95

The thickness of the aquifer is 18.0 m and the well is to be screened throughout 80 percent of the aquifer using a telescoping screen. The maximum well screen diameter due to drilling rig limitations is 25 cm. Determine the maximum discharge of the well in m^3/d assuming a 0.5 clogging coefficient.

11-13. A step-drawdown well test results in a linear function of drawdown versus discharge with a slope of 0.00185 $m/(m^3/d)$. Compute the discharge in m^3/d if the well is operating with a total dynamic head of 115 m of which 97 m is the static head, 2.5 m is friction head and 8.5 m the required pressure head. Assume velocity head is negligible.

11-14. A constant rate well test is run at a discharge of 3500 m^3/d for 24 hours. A straight line fit through the data indicates a change in drawdown of 0.55 m for each log-cycle of time and an intercept at zero drawdown of 1.5 minutes. Previous tests indicate a storativity of 0.000076 for the aquifer. Compute the radial distance in meters from the well being pumped to the observation well at which test data were taken.

12

Open Channel Flow

12.1 Open Channel Hydraulics

Flow in open channels can be compared to flow in pipes by inspection of Bernoulli's equation describing the total energy at two points in a flow system:

$$y_1 + z_1 + \frac{(v_1)^2}{2g} = y_2 + z_2 + \frac{(v_2)^2}{2g} + (h_f)_{1-2} \qquad (12\text{-}1)$$

where

y = depth of flow above channel bottom, L

z = position of channel bottom above datum, L

v = flow velocity, L/T

h_f = friction headloss, L

The equivalent application of Eq. (12-1) applied to flow in pipes and to flow in open channels is demonstrated in Fig. 12-1.

Types of Flow

It is helpful to begin our analysis by categorizing different flow regimes. The simplest case is that of *steady flow* which is defined as flow for which the depth is constant with time along the channel. Steady flow can be further divided into uniform

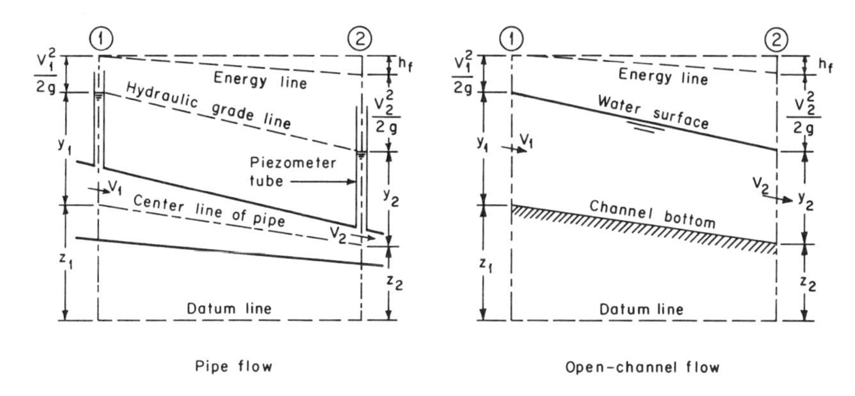

Figure 12-1 Comparison between pipe flow and open-channel flow.

and varied flow regimes. For *uniform flow,* the depth is the same at every section along the channel. In *varied flow,* the depth of flow changes along the channel length. *Gradually varied flow* describes the condition in which the change of depth of flow per unit length of channel is small. *Rapidly varied flow* is used to describe the flow which has abrupt changes in flow depth such as over a weir. Examples of steady flow are indicated in Fig. 12-2.

Unsteady flow is more difficult to handle computationally. It is defined as flow in which the depth changes with time. Unsteady flow may be uniform, in which case the flow depth changes by exactly the same magnitude with time along the entire section of the channel being analyzed. This case is practically impossible to find in nature. Unsteady varied flow can be separated into categories of gradual and rapid variation. An example of the gradually varied case is a flood wave moving along a channel. Rapidly varied flow is demonstrated by tidal bores which may occur in shallow bays and their tributaries. These unsteady varied flow conditions are indicated schematically in Fig. 12-2.

This chapter covers different methods for analysis of steady flow conditions. References for analysis of unsteady flow regimes include the classic work of Chow (1959), the later work of Henderson (1966), and the more recent work of French (1985). The same references were used as sources for the analysis of uniform flow described in this chapter. As just indicated, the occurrence of uniform unsteady flow is difficult to imagine in nature. It will therefore be understood that analysis of uniform flow conditions in this chapter pertains to flow regimes which are steady.

Uniform Flow Equations

The history of the equations to analyze uniform flow conditions derives basically from developments made by hydraulicians in Europe beginning some 200 years ago. More modern developments followed by Bakhmeteff in the Soviet Union (1912) and later in America (1932), as well as others. Chow produced his classical work in the United States in 1959, and it continues to be an important reference today. Many procedures adapted to realistic field situations were developed by governmental agencies in the United States such as the Bureau of Reclamation and Soil Conservation Service. The material which follows briefly explains the development of equa-

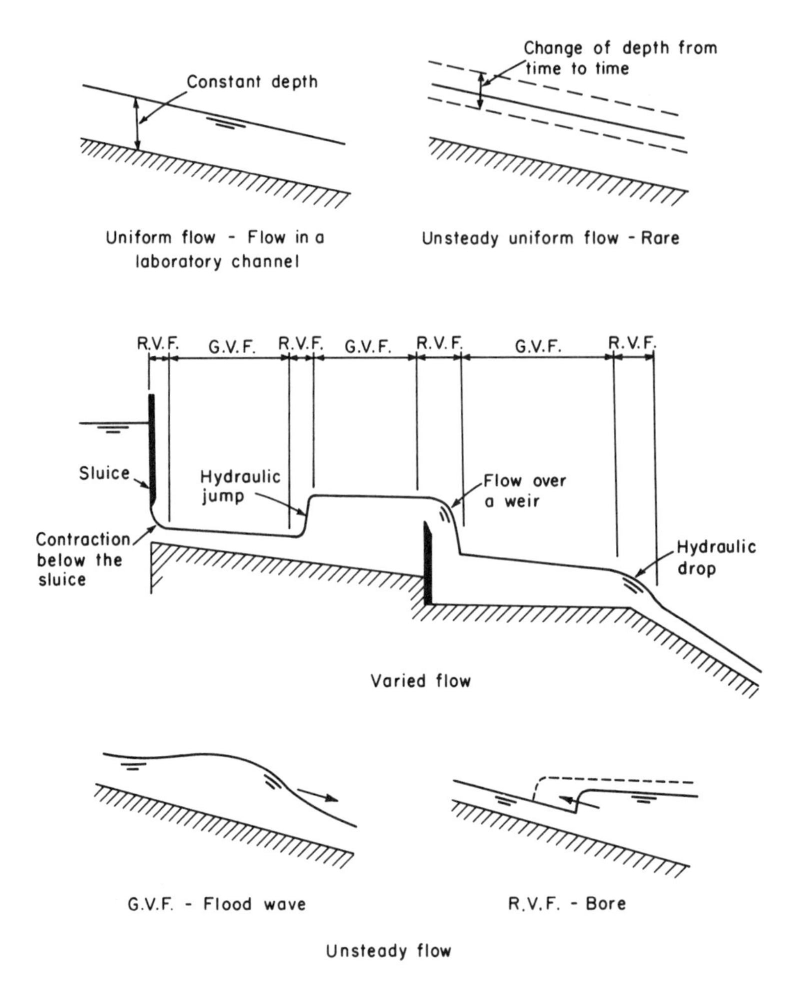

Figure 12-2 Various types of open-channel flow, GVF = gradually varied flow; RVF = rapidly varied flow. (Adapted from Chow, 1959.)

tions to analyze uniform flow conditions up to Manning's equation which is widely applied in the design of open channel systems. This equation will be used as the basis of open channel analysis in this chapter.

The French hydraulician Chézy developed an original relationship between flow velocity, hydraulic characteristics of the flow regime, and the energy gradient from which Manning's equation followed over 100 hundred years later (Chézy, 1769). Chézy's basic work in uniform flow regimes was based on field measurements made on the Courpalet Canal and Seine River in northern France (Chow, 1959). What is called the *Chézy formula* is given as

$$v = C[R_h(S)]^{1/2} \tag{12-2}$$

where

$$v = \text{mean flow velocity, L/T}$$
$$R_h = \text{hydraulic radius of flow system, L}$$
$$S = \text{slope of energy grade line, L/L}$$
$$C = \text{Chézy C} - \text{a flow resistance factor, dimensionless}$$

The hydraulic radius is defined as the cross-sectional flow area divided by the channel wetted perimeter, P_w,

$$R_h = \frac{A}{P_w} \tag{12-3}$$

In uniform flow regimes, the slope of the water surface, channel bottom, and energy grade line are all equal. This condition is demonstrated in Fig. 12-3. Water surface slope will be designated as S_w, channel bed slope as S_0, and slope of the energy grade line as S_f in this text.

Following publication of the Chézy formula, much work centered on evaluating the Chézy C. One hundred years after the work of Chézy, the Swiss hydraulicians Ganguillet and Kutter (1869) developed a complex equation for C as a function of the slope of the energy grade line, hydraulic radius, and a roughness coefficient. Their formula was based on flow measurements in many European rivers, the Mississippi River, and experimental channels operated by Bazin who would later develop his own formula. For English Engineering units of length in feet and time in seconds, the Ganguillet and Kutter formula is given as

$$C = \frac{[a + b + 1.811/n]}{[1 + (a + b)n/(R_h)^{1/2}]} \tag{12-4}$$

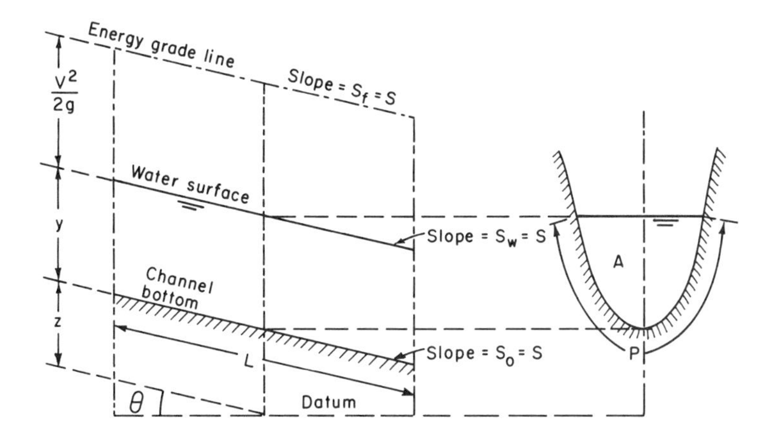

Figure 12-3 Identification of terms for the Chézy formula for uniform flow in open channels.

where

$$a = 41.65$$
$$b = 0.00281/S_f$$
$$n = \text{Kutter's } n$$

Kutter's n is basically equivalent to Manning's n which was developed some 20 years later within a different format but using some of the same experimental data sets.

Publication of the work of Bazin in France (1897) actually followed that of Manning by some eight years. Bazin made measurements which were completed in 1862 after being started by Henry Darcy in 1852 on experimental channels. Bazin's measurements were published in 1865 and applied by virtually all other hydraulicians working on evaluation of the Chézy C in that era, including Manning. Bazin evaluated the Chézy C as a function of the hydraulic radius and roughness factor of the experimental channels. The relationship developed expressed in English Engineering units is given by

$$C = \frac{157.6}{[1 + m/(R_h)^{1/2}]} \tag{12-5}$$

where

$$m = \text{Bazin's } m$$

Bazin's m is also an empirical flow resistance factor. Because it was developed based on measurements in artificial channels, it is not felt to be as accurate as factors developed using more field data from natural channels. Nevertheless, Bazin's method has been widely applied to different flow situations particularly in francophone countries.

The formula of Manning was first presented in Ireland in 1889. It was originally developed based on data made available by Bazin and later verified and modified using 170 other observations. The Manning equation is the most widely applied formula for analysis of open channel flow systems today due to its balance between simplicity and accurate results (Chow, 1959). So many applications have been made of this formula that the empirical values required for the roughness factor are now highly verified.

Considering the Manning formula as an evaluation of the Chézy C, this factor is defined as a function of the hydraulic radius and Manning roughness coefficient as

$$C = \left(\frac{1}{n}\right)(R_h)^{1/6} \tag{12-6}$$

for SI units and as,

$$C = \left(\frac{1.486}{n}\right)(R_h)^{1/6} \tag{12-7}$$

for English Engineering units. The 1.486 is derived from the cube root of the 3.2808 ft/m conversion factor if n is assumed to be constant for both sets of units.

The requirement for the cube root will be evident when the Manning equation is expressed for velocity. Considering the number of significant figures in the empirical roughness coefficient, the constant is expressed with sufficient accuracy as 1.49 (Chow, 1959). Discussion of application of the Manning equation in this book will concentrate on SI units.

Substituting the expression for C in Eq. (12-6) into the Chézy formula in Eq. (12-2) gives the mean channel velocity as

$$v = \left[\frac{1}{n}(R_h)^{1/6}\right][R_h(S_f)]^{1/2} \tag{12-8}$$

which can be simplified to

$$v = \frac{1}{n}(R_h)^{2/3}(S_f)^{1/2} \tag{12-9}$$

where

v = mean flow velocity, m/s
R_h = hydraulic radius, m
S_f = slope of energy grade line, m/m

Note that for uniform flow conditions,

$$S_f = S_w = S_0 \tag{12-10}$$

Chow (1959) presented a table for Manning's roughness coefficients which covers a tremendously diverse range of channel materials. This table is reproduced as Table 12-1 which indicates minimum, normal, and maximum coefficients for each material. Coefficients appearing in bold print are recommended for design applications. Chow (1959) also presents a useful set of photographs giving values of n related to some manmade channels and numerous natural river and stream flow situations.

The volumetric flow rate is equal to the flow velocity times the cross-sectional flow area. Separating the cross-sectional flow area term out of the hydraulic radius and combining terms, Manning's equation can be rearranged to give the volumetric flow rate as

$$Q = \frac{1}{n}\left[\frac{A^{5/3}}{P_w^{2/3}}\right](S_f)^{1/2} \tag{12-11}$$

where

Q = volumetric flow rate, m^3/s
A = cross-sectional flow area, m^2
P_w = wetted perimeter, m

Application of Manning's equation is demonstrated in the following example problem.

TABLE 12-1 Values of the roughness coefficient n. (**Boldface** figures are values generally recommended in design.) (Chow, 1959)

Type of channel and description	Minimum	Normal	Maximum
A. CLOSED CONDUITS FLOWING PARTLY FULL			
A-1. Metal			
a. Brass, smooth	0.009	**0.010**	0.013
b. Steel			
1. Lockbar and welded	0.010	0.012	0.014
2. Riveted and spiral	0.013	0.016	0.017
c. Cast iron			
1. Coated	0.010	0.013	0.014
2. Uncoated	0.011	0.014	0.016
d. Wrought iron			
1. Black	0.012	0.014	0.015
2. Galvanized	0.013	0.016	0.017
e. Corrugated metal			
1. Subdrain	0.017	0.019	0.021
2. Storm drain	0.021	**0.024**	0.030
A-2. Nonmetal			
a. Lucite	0.008	0.009	0.010
b. Glass	0.009	**0.010**	0.013
c. Cement			
1. Neat, surface	0.010	0.011	0.013
2. Mortar	0.011	0.013	0.015
d. Concrete			
1. Culvert, straight and free of debris	0.010	0.011	0.013
2. Culvert with bends, connections, and some debris	0.011	**0.013**	0.014
3. Finished	0.011	0.012	0.014
4. Sewer with manholes, inlet, etc., straight	0.013	0.015	0.017
5. Unfinished, steel form	0.012	0.013	0.014
6. Unfinished, smooth wood form	0.012	**0.014**	0.016
7. Unfinished, rough wood form	0.015	0.017	0.020
e. Wood			
1. Stave	0.010	0.012	0.014
2. Laminated, treated	0.015	0.017	0.020
f. Clay			
1. Common drainage tile	0.011	**0.013**	0.017
2. Vitrified sewer	0.011	0.014	0.017
3. Vitrified sewer with manholes, inlet, etc.	0.013	0.015	0.017
4. Vitrified subdrain with open joint	0.014	0.016	0.018
g. Brickwork			
1. Glazed	0.011	0.013	0.015
2. Lined with cement mortar	0.012	0.015	0.017
h. Sanitary sewers coated with sewage slimes, with bends and connections	0.012	0.013	0.016
i. Paved invert, sewer, smooth bottom	0.016	0.019	0.020
j. Rubble masonry, cemented	0.018	0.025	0.030

TABLE 12-1 Values of the roughness coefficient n. (**Boldface** figures are values generally recommended in design.) (Chow, 1959) (*continued*)

Type of channel and description	Minimum	Normal	Maximum
B. LINED OR BUILT-UP CHANNELS			
B-1. Metal			
a. Smooth steel surface			
1. Unpainted	0.011	**0.012**	0.014
2. Painted	0.012	0.013	0.017
b. Corrugated	0.021	0.025	0.030
B-2. Nonmetal			
a. Cement			
1. Neat, surface	0.010	0.011	0.013
2. Mortar	0.011	0.013	0.015
b. Wood			
1. Planed, untreated	0.010	0.012	0.014
2. Planed, creosoted	0.011	0.012	0.015
3. Unplaned	0.011	0.013	0.015
4. Plank with battens	0.012	0.015	0.018
5. Lined with roofing paper	0.010	0.014	0.017
c. Concrete			
1. Trowel finish	0.011	**0.013**	0.015
2. Float finish	0.013	0.015	0.016
3. Finished, with gravel on bottom	0.015	0.017	0.020
4. Unfinished	0.014	0.017	0.020
5. Gunite, good section	0.016	0.019	0.023
6. Gunite, wavy section	0.018	0.022	0.025
7. On good excavated rock	0.017	0.020	
8. On irregular excavated rock	0.022	0.027	
d. Concrete bottom float finished with sides of			
1. Dressed stone in mortar	0.015	0.017	0.020
2. Random stone in mortar	0.017	0.020	0.024
3. Cement rubble masonry, plastered	0.016	0.020	0.024
4. Cement rubble masonry	0.020	0.025	0.030
5. Dry rubble or riprap	0.020	0.030	0.035
e. Gravel bottom with sides of			
1. Formed concrete	0.017	0.020	0.025
2. Random stone in mortar	0.020	0.023	0.026
3. Dry rubble or riprap	0.023	0.033	0.036
f. Brick			
1. Glazed	0.011	**0.013**	0.015
2. In cement mortar	0.012	**0.015**	0.018
g. Masonry			
1. Cemented rubble	0.017	0.025	0.030
2. Dry rubble	0.023	0.032	0.035
h. Dressed ashlar	0.013	0.015	0.017
i. Asphalt			
1. Smooth	0.013	0.013	
2. Rough	0.016	0.016	
j. Vegetal lining	0.030	—	0.500

TABLE 12-1 Values of the roughness coefficient n. (**Boldface** figures are values generally recommended in design.) (Chow, 1959) (*continued*)

Type of channel and description	Minimum	Normal	Maximum
C. EXCAVATED OR DREDGED			
a. Earth, straight and uniform			
1. Clean, recently completed	0.016	0.018	0.020
2. Clean, after weathering	0.018	**0.022**	0.025
3. Gravel, uniform section, clean	0.022	0.025	0.030
4. With short grass, few weeds	0.022	0.027	0.033
b. Earth, winding and sluggish			
1. No vegetation	0.023	0.025	0.030
2. Grass, some weeds	0.025	0.030	0.033
3. Dense weeds or aquatic plants in deep channels	0.030	0.035	0.040
4. Earth bottom and rubble sides	0.028	0.030	0.035
5. Stony bottom and weedy banks	0.025	0.035	0.040
6. Cobble bottom and clean sides	0.030	0.040	0.050
c. Dragline-excavated or dredged			
1. No vegetation	0.025	0.028	0.033
2. Light brush on banks	0.035	0.050	0.060
d. Rock cuts			
1. Smooth and uniform	0.025	0.035	0.040
2. Jagged and irregular	0.035	0.040	0.050
e. Channels not maintained, weeds and brush uncut			
1. Dense weeds, high as flow depth	0.050	0.080	0.120
2. Clean bottom, brush on sides	0.040	0.050	0.080
3. Same, highest stage of flow	0.045	0.070	0.110
4. Dense brush, high stage	0.080	0.100	0.140
D. NATURAL STREAMS			
D-1. Minor streams (top width at flood stage <100 ft)			
a. Streams on plain			
1. Clean, straight, full stage, no rifts or deep pools	0.025	**0.030**	0.033
2. Same as above, but more stones and weeds	0.030	0.035	0.040
3. Clean, winding, some pools and shoals	0.033	0.040	0.045
4. Same as above, but some weeds and stones	0.035	0.045	0.050
5. Same as above, lower stages, more ineffective slopes and sections	0.040	0.048	0.055
6. Same as 4, but more stones	0.045	0.050	0.060
7. Sluggish reaches, weedy, deep pools	0.050	0.070	0.080
8. Very weedy reaches, deep pools, or floodways with heavy stand of timber and underbrush	0.075	0.100	0.150

TABLE 12-1 Values of the roughness coefficient n. (**Boldface** figures are values generally recommended in design.) (Chow, 1959) (*continued*)

Type of channel and description	Minimum	Normal	Maximum
b. Mountain streams, no vegetation in channel, banks usually steep, trees and brush along banks submerged at high stages			
1. Bottom: gravels, cobbles, and few boulders	0.030	0.040	0.050
2. Bottom: cobbles with large boulders	0.040	0.050	0.070
D-2. Flood plains			
a. Pasture, no brush			
1. Short grass	0.025	0.030	0.035
2. High grass	0.030	0.035	0.050
b. Cultivated areas			
1. No crop	0.020	0.030	0.040
2. Mature row crops	0.025	0.035	0.045
3. Mature field crops	0.030	0.040	0.050
c. Brush			
1. Scattered brush, heavy weeds	0.035	0.050	0.070
2. Light brush and trees, in winter	0.035	0.050	0.060
3. Light brush and trees, in summer	0.040	0.060	0.080
4. Medium to dense brush, in winter	0.045	0.070	0.110
5. Medium to dense brush, in summer	0.070	0.100	0.160
d. Trees			
1. Dense willows, summer, straight	0.110	0.150	0.200
2. Cleared land with tree stumps, no sprouts	0.030	0.040	0.050
3. Same as above, but with heavy growth of sprouts	0.050	0.060	0.080
4. Heavy stand of timber, a few down trees, little undergrowth, flood stage below branches	0.080	0.100	0.120
5. Same as above, but with flood stage reaching branches	0.100	0.120	0.160
D-3. Major streams (top width at flood stage >100 ft). The n value is less than that for minor streams of similar description, because banks offer less effective resistance.			
a. Regular section with no boulders or brush	0.025	—	0.060
b. Irregular and rough section	0.035	—	0.100

Example Problem 12-1

Compute the discharge for uniform flow in a rectangular channel made of unfinished concrete which has a bed slope of 0.0101 m/m, width of 1.80 m, and depth of flow equal to 0.541 m.

Solution From Table 12-1 using the normal range,

$$n = 0.017$$

Compute hydraulic radius:

$$R_h = \frac{A}{P_w} = \frac{1.80\text{ m}(0.541\text{ m})}{1.80\text{ m} + 2(0.541\text{ m})}$$

$$R_h = 0.338\text{ m}$$

Apply Manning's equation:

$$v = \left(\frac{1}{n}\right)(R_h)^{2/3}(S_f)^{1/2}$$

Recognizing that $S_f = S_0$ under uniform flow,

$$v = \left(\frac{1}{0.017}\right)(0.338_m)^{2/3}(0.0101\ m/m)^{1/2}$$

$$v = 2.868\ m/s$$

Compute discharge:

$$Q = vA$$

$$Q = 2.868\ m/s(0.974\ m^2)$$

$$Q = 2.79\ m^3/s$$

State of Flow

With the flow velocity established by application of Manning's equation, determination of the state of flow can be made. In the *laminar* state, viscous forces predominate over inertial forces. In the *turbulent* state, inertial forces predominate over viscous forces. The quantification of the state of flow is made by Reynold's number which is the ratio of inertial and viscous forces:

$$R_N = \frac{\text{inertial forces}}{\text{viscous forces}} \tag{12-12}$$

and which is expressed as

$$R_N = \rho v \frac{1}{\mu} \tag{12-13}$$

where

R_N = Reynold's number, dimensionless
ρ = fluid density, kg/m^3
1 = characteristic length, m
μ = dynamic viscosity, N (s)/m^2

Equation (12-13) can be rearranged by substitution of the kinematic viscosity which equals the dynamic viscosity divided by the fluid density:

$$\nu = \frac{\mu}{\rho} \tag{12-14}$$

where

ν = kinematic viscosity, m^2/s

Rewriting the Reynold's number,

$$R_N = v\frac{1}{\nu} \tag{12-15}$$

It was previously shown that for flow in pipes the characteristic length is the diameter. For open channel flow, the characteristic length is the hydraulic radius. The differentiation between laminar and turbulent flow in pipes was made at a Reynold's number of approximately 2000 so for $R_N < 2000$ flow was laminar and for $R_N > 2000$ flow was turbulent. In reality there is a critical zone between a Reynold's number of 2000 and 4000 in which the state of flow is uncertain. However, in normal engineering applications, flow in this zone is not common (Mott, 1982). An analagous situation exists for flow in open channels. In this case the point of differentiation is 500 so flow is laminar for $R_N < 500$ and turbulent for $R_N > 500$. A transition range lies between a Reynolds number of 500 and 4000, but flow in this range is not common in engineering applications.

Example Problem 12-2

Classify the flow regime for the channel made of unfinished concrete in example problem 12-1. Assume 20°C for water temperature.

Solution From the solution of example problem 12-1,

$$R_h = 0.338 \text{ m}$$

and

$$v = 2.868 \text{ m/s}$$

For standard temperature of 20°C,

$$\rho = 998 \text{ kg/m}^3$$

and

$$\mu = 1.0 \times 10^{-3}\text{N s/m}^2$$

Compute Reynold's number:

$$R_N = \frac{\rho v R_h}{\mu}$$

$$R_N = 998 \text{ kg/m}^3(2.868 \text{ m/s})\frac{0.338 \text{ m}}{1.0 \times 10^{-3}\text{N s/m}^2}$$

$$R_N = 9.67 \times 10^5$$

State of flow is therefore fully turbulent.

Specific Energy Principle

It is convenient in open channel flow analysis to make use of the specific energy principle. The specific energy at any point in a channel is defined as the sum of the flow depth and velocity head:

$$E = y + \frac{v^2}{2g} \tag{12-16}$$

where

$$E = \text{specific energy, m}$$

The concept of discharge per unit width will be applied to aid in visualizing the specific energy relations. For rectangular channels of width b and discharge Q, the discharge per unit width is

$$q = \frac{Q}{b} \tag{12-17}$$

where

$$q = \text{discharge per unit width, m}^2/\text{s}$$

The flow velocity may now be expressed using q as

$$v = \frac{Q}{A} = \frac{q(b)}{y(b)} = \frac{q}{y} \tag{12-18}$$

which may be substituted into Eq. (12-16) to give an expression for the specific energy in terms of discharge per unit width:

$$E = y + \frac{q^2}{2gy^2} \tag{12-19}$$

Equation (12-19) may be rearranged as

$$(E - y)y^2 = \frac{q^2}{2g} \tag{12-20}$$

in which the term on the right-hand side is constant for steady-state flow. Plotting the foregoing equation for a given value of q, it is seen to be asymptotic at $(E - y) = 0$ which is equivalent to $E = y$, and at $y = 0$. The plot of Eq. (12-20) for a given value of q results in the specific energy curve demonstrated in Fig. 12-4.

At a given value of specific energy E, the depths y_{sup} and y_{sub} on the two limbs of the curve are termed alternate depths. The concepts of supercritical and subcritical flow depths will be explained in a later section. The minimum specific energy for a given value of q occurs at what is called the critical flow depth. Critical flow will be investigated in the following section.

Considerable useful information for the analysis of open channel systems can be obtained from the specific energy curve. Let us apply the specific energy concept to analysis of the flow situation indicated in Fig. 12-5. In the figure, flow approaches from the right in a rectangular channel with a horizontal bed at discharge q.

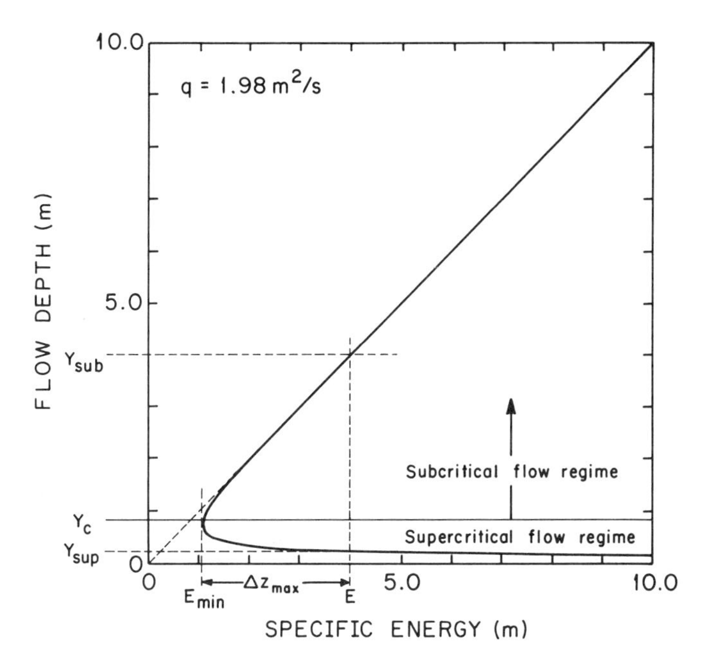

Figure 12-4 Specific energy curve for given discharge per unit width q.

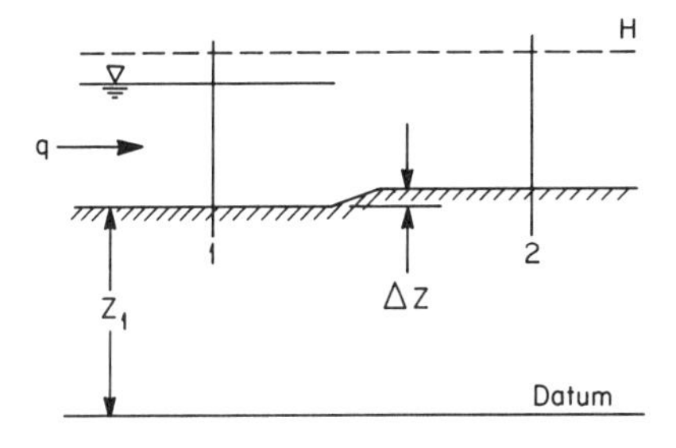

Figure 12-5 Example for application of specific energy principle to flow in a horizontal rectangular channel with a step.

There is a step flow constriction in the channel of height Δz between flow sections 1 and 2. Friction losses across the step and along the channel will be assumed negligible resulting in a constant total head H.

Assume that flow conditions are such that we are operating on the upper portion of the specific energy curve—that is, at a depth greater than the critical depth. Applying Bernoulli's equation to flow sections 1 and 2 with no loss in head,

$$z_1 + y_1 + \frac{(v_1)^2}{2g} = (z_1 + \Delta z) + y_2 + \frac{(v_2)^2}{2g} \tag{12-21}$$

Substituting for v from Eq. (12-18) in terms of discharge per unit width,

$$z_1 + y_1 + \frac{q^2}{2g(y_1)^2} = (z_1 + \Delta z) + y_2 + \frac{q^2}{2g(y_2)^2} \tag{12-22}$$

which reduces to

$$H = z_1 + E_1 = (z_1 + \Delta z) + E_2 \tag{12-23}$$

Solving for the specific energy in section 2,

$$E_2 = E_1 - \Delta z \tag{12-24}$$

Since we are operating on the upper limb of the specific energy curve,

$$y_2 < y_1 \tag{12-25}$$

so the depth of flow is seen to decrease following the step. The new depth of flow y_2 is solved from a cubic equation. A numerical technique to solve such a flow situation is demonstrated in the following example problem.

Example Problem 12-3

Water with a velocity of 1.1 m/s flows at a depth of 1.8 m in a rectangular channel with a horizontal bed. There is a smooth upward step of 0.15 m in the channel bed. Compute the depth of flow after the step and the absolute water level—that is, the depth of flow with reference to the upstream channel section. Assume negligible losses due to friction.

Solution

$$H = z_1 + E_1 = (z_1 + \Delta z) + E_2$$

$$E_2 = E_1 - 0.15 \text{ m}$$

$$E_1 = y_1 + \frac{(v_1)^2}{2g}$$

$$E_1 = 1.8 \text{ m} + \frac{(1.1 \text{ m/s})^2}{2(9.81 \text{ m/s}^2)}$$

$$E_1 = 1.86 \text{ m}$$

$$E_2 = 1.86 \text{ m} - 0.15 \text{ m} = 1.71 \text{ m}$$

Applying the concept of discharge per unit width,

$$E_2 = y_2 + \frac{q^2}{2g(y_2)^2}$$

$$E_2 = y_2 + \frac{[1.1 \text{ m/s}(1.8 \text{ m})]^2}{2(9.81 \text{ m/s}^2)(y_2)^2}$$

$$E_2 = y_2 + \frac{0.200}{(y_2)^2} = 1.71 \text{ m}$$

Rearranging as a cubic equation,

$$(y_2)^3 - 1.71(y_2)^2 + 0.200 = 0$$

Newton's method to solve nonlinear equations will be used to compute the depth in section 2. Setting up the function of depth and its first derivative for solution by Newton's method,

$$f(y) = 0 = (y_2)^3 - 1.71(y_2)^2 + 0.200$$

$$f'(y) = 3(y_2)^2 - 3.42y_2$$

For Newton's method, let the value of the initial estimate or previous iteration be y_n and the value for the current iteration be y_{n+1}. The iterative scheme for Newton's method is

$$y_{n+1} = y_n - \frac{f(y_n)}{f'(y_n)}$$

which for this problem becomes

$$y_{n+1} = y_n - \frac{(y_n)^3 - 1.71(y_n)^2 + 0.200}{3(y_n)^2 - 3.42y_n}$$

The tabulated solution using 1.500 m as the initial estimate for y_n is

y_n	y_{n+1}
1.500	1.668
1.668	1.637
1.637	1.635
1.635	1.635

where the final result is unchanged to three decimal places. Therefore,

$$y_2 = 1.635 \text{ m}$$

The absolute level of the water surface in section 2 above the upstream channel bed is calculated from

$$y_{ab-2} = y_2 + \Delta z$$

$$y_{ab-2} = 1.635 \text{ m} + 0.15 \text{ m}$$

$$y_{ab-2} = 1.785 \text{ m}$$

The flow depth has decreased from the initial absolute depth of 1.800 m.

Critical State of Flow

Rectangular sections

Critical flow is defined as the state of flow for which the specific energy is a minimum for a given discharge. If we refer to the specific energy curve in Fig. 12-4, we see that critical flow occurs at the critical depth, y_c. We will proceed with derivation of the magnitude of the critical depth using a rectangular channel. We noted in Eq. (12-19) that the specific energy was given as

$$E = y + \frac{q^2}{2gy^2} \tag{12-19}$$

By observation of the specific energy curve at E_{min}, the derivative of the specific energy with respect to depth equals zero and we can write

$$\frac{dE}{dy} = 1 - \frac{q^2}{gy^3} = 0 \tag{12-26}$$

which can be rearranged as

$$y_c = \left[\frac{q^2}{g}\right]^{1/3} \tag{12-27}$$

The velocity at critical depth, v_c, can be derived using Eq. (12-19) as

$$v_c = [gy_c]^{1/2} \tag{12-28}$$

To evaluate the critical depth in terms of the minimum specific energy, we rearrange Eq. (12-28) as

$$\frac{(v_c)^2}{2g} = \frac{y_c}{2} \tag{12-29}$$

Expressing the minimum specific energy,

$$E_{min} = y_c + \frac{(v_c)^2}{2g} = \frac{3}{2}y_c \tag{12-30}$$

which can be rearranged as

$$y_c = \frac{2}{3}E_{min} \tag{12-31}$$

An alternate derivation (Henderson, 1966) allows us to state that the critical state of flow is the maximum discharge per unit width for a given specific energy. Application of the critical depth is indicated in the following example problem.

Example Problem 12-4

Water flows at a velocity of 3.0 m/s and a depth of 3.0 m in a smooth channel of rectangular section. Compute the maximum possible size of an upward step while maintaining the flow conditions indicated. Assume friction losses are negligible across the step.

Solution Compute specific energy upstream of step:

$$E_1 = y_1 + \frac{(v_1)^2}{2g}$$

$$E_1 = 3.0 \text{ m} + \frac{(3.0 \text{ m/s})^2}{2(9.81 \text{ m/s}^2)}$$

$$E_1 = 3.459 \text{ m}$$

Maximum allowable step size is equal to the difference between upstream specific energy and minimum specific energy which is that which occurs at critical depth. Compute the critical depth using Eq. (12-27),

$$y_c = \left[\frac{q^2}{g}\right]^{1/3}$$

$$y_c = \left\{\frac{[3.0 \text{ m/s}(3.0 \text{ m})]^2}{9.81 \text{ m/s}^2}\right\}^{1/3}$$

$$y_c = 2.021 \text{ m}$$

Applying Eq. (12-31),

$$y_c = \frac{2}{3}E_{min}$$

which can be rearranged to

$$E_{min} = \frac{3}{2}y_c$$

$$E_{min} = \frac{3}{2}(2.02 \text{ m})$$

$$E_{min} = 3.032 \text{ m}$$

Therefore maximum step size,

$$\Delta z_{max} = E_1 - E_{min}$$

$$\Delta z_{max} = 3.459 \text{ m} - 3.032 \text{ m}$$

$$\Delta z_{max} = 0.427 \text{ m}$$

If a higher step is placed in the channel, flow will be restricted or choked because the same level of discharge can no longer be maintained. A higher step will force the channel to operate on a specific energy curve shifted to the left in Fig. 12-4; that is, a specific energy curve for a lower discharge per unit width.

Nonrectangular sections

The specific energy is defined using Eq. (12-16) for both rectangular and nonrectangular sections. For the more general case of nonrectangular sections, the specific energy may be written as

$$E = y + \frac{Q^2}{2gA^2} \tag{12-32}$$

where

A = the cross-sectional flow area

Differentiating Eq. (12-32),

$$\frac{dE}{dy} = 1 - \frac{Q^2}{gA^3}\left(\frac{dA}{dy}\right) \tag{12-33}$$

It is now convenient to define the derivative of the area as

$$dA = T\,dy \tag{12-34}$$

where

T = width of the free water surface

Substituting into Eq. (12-33),

$$\frac{dE}{dy} = 1 - \frac{Q^2T}{gA^3} \tag{12-35}$$

For critical flow, $E = E_{min}$ and $dE/dy = 0$ so Eq. (12-35) becomes

$$(Q_c)^2 T_c = g(A_c)^3 \tag{12-36}$$

which can be reduced to

$$v_c = \left[g\left(\frac{A_c}{T_c}\right)\right]^{1/2} \tag{12-37}$$

Equation (12-37) indicates that A_c/T_c is equivalent to the critical depth term for nonrectangular channels.

Froude Number

Rectangular sections

The Froude number is the ratio of inertial to gravity forces:

$$F_N = \frac{\text{inertial forces}}{\text{gravity forces}} \tag{12-38}$$

and is defined as

$$F_N = \frac{v}{(gy)^{1/2}} \tag{12-39}$$

By squaring both sides of Eq. (12-39), it may be rewritten as

$$(F_N)^2 = \frac{q^2}{gy^3} \tag{12-40}$$

The right-hand side of Eq. (12-40) expresses the same condition as given by Eq. (12-26) for critical flow if F_N is equal to 1. The following states of flow can therefore be designated with respect to the Froude number:

(a) $F_N = 1$ represents critical flow.
(b) $F_N > 1$ is termed supercritical flow.
(c) $F_N < 1$ is termed subcritical flow.

As shown in Fig. 12-4, supercritical flow occurs in the lower limb of the specific energy curve, that is at depths less than the critical depth, and subcritical flow occurs in the upper limb. The Froude number for gravity waves in water shares many physical analogies with the Mach number for sonic waves in air.

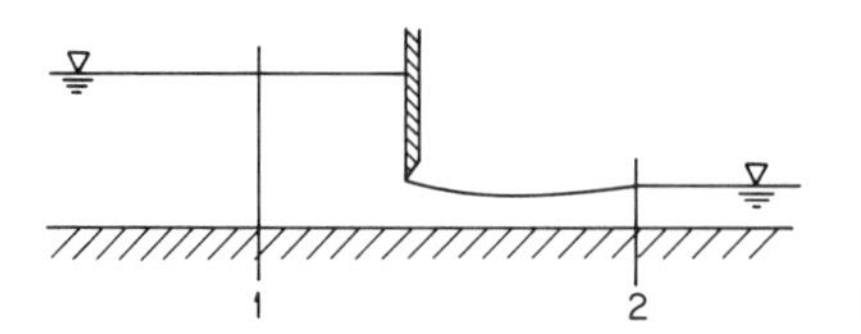

Figure 12-6 Flow across a sluice gate.

Example Problem 12-5

Discharge in a smooth rectangular channel 1.0 m in width is controlled by a sluice gate as shown in Fig. 12-6. The depth of flow is 0.95 m in section 1 and 0.25 m in section 2. Compute the Froude number and flow velocity in section 1 and the channel discharge. Assume the energy loss across the gate is negligible.

Solution Assuming no energy loss across the gate,

$$E_1 = E_2$$

By continuity,

$$q_1 = v_1 y_1 = q_2 = v_2 y_2$$

Set up the solution using the following equations:

$$E_1 = y_1 + \frac{q^2}{2g(y_1)^2}$$

$$E_2 = y_2 + \frac{q^2}{2g(y_2)^2}$$

$$y_1 - y_2 = q^2\left[\frac{1}{2g(y_2)^2} - \frac{1}{2g(y_1)^2}\right]$$

$$2g(y_1 - y_2) = q^2\left[\frac{(y_1)^2 - (y_2)^2}{(y_2)^2(y_1)^2}\right]$$

Solving for q^2,

$$q^2 = 2g(y_1 - y_2)\left[\frac{(y_2)^2(y_1)^2}{(y_1 - y_2)(y_1 + y_2)}\right]$$

$$q^2 = 2g\left[\frac{(y_2)^2(y_1)^2}{(y_1 + y_2)}\right]$$

Defining $(F_{N1})^2$ and $(v_1)^2$,

$$(F_{N1})^2 = \frac{(v_1)^2}{gy_1} \qquad (v_1)^2 = \frac{q^2}{(y_1)^2}$$

$$(F_{N1})^2 = \frac{q^2}{g(y_1)^3}$$

$$(F_{N1})^2 = \frac{2g(y_2)^2(y_1)^2}{g(y_1)^3(y_1 + y_2)}$$

$$(F_{N1})^2 = \frac{2(y_2)^2}{y_1(y_1 + y_2)}$$

$$(F_{N1})^2 = \frac{2}{\frac{y_1}{(y_2)^2}(y_1 + y_2)}$$

$$(F_{N1})^2 = \frac{2}{\frac{y_1}{y_2}\left[\frac{y_1}{y_2} + 1\right]}$$

Substituting in the values given for the problem,

$$(F_{N1})^2 = \frac{2}{\dfrac{0.95}{0.25}\left(\dfrac{0.95}{0.25} + 1\right)} = 0.1096$$

$$F_{N1} = 0.331$$

$$v_1 = F_{N1}(gy_1)^{1/2}$$

$$v_1 = 0.331[(9.81 \text{ m/s}^2)(0.95 \text{ m})]^{1/2}$$

$$v_1 = 1.011 \text{ m/s}$$

$$Q = v_1 y_1 b = 1.011 \text{ m/s}(0.95 \text{ m})(1.00 \text{ m})$$

$$Q = 0.960 \text{ m}^3/\text{s}$$

Nonrectangular sections

Combining Eq. (12-39) for the definition of the Froude number with the concept of the hydraulic depth, A/T, for general channel shapes, the Froude number can be written as

$$F_N = \frac{v}{\left[g\left(\dfrac{A}{T}\right)\right]^{1/2}} \tag{12-41}$$

Squaring both sides of the preceding equation, we can write,

$$(F_N)^2 = \frac{v^2T}{gA} = \frac{Q^2T}{gA^3} \tag{12-42}$$

Substituting Eq. (12-42) into Eq. (12-35) we can develop an expression for the derivative of the specific energy as a function of depth which is valid for nonrectangular as well as rectangular channels. It is given as

$$\frac{dE}{dy} = 1 - (F_N)^2 \tag{12-43}$$

The general relationship given in Eq. (12-43) can be applied to the analysis of flow depth profiles.

Model-prototype relations

In general, the hydraulic depth in open channel flow is given as

$$D = \frac{A}{T} \tag{12-44}$$

which is equal to the depth of flow y for the special case of rectangular channels. For the critical state of flow, the Froude number is equal to 1 and Eq. (12-39) may be expressed as,

$$v_c = [gD_c]^{1/2} \tag{12-45}$$

This critical velocity is the celerity of small gravity waves that occur in shallow channels. These waves are due to obstacles in the channel that cause a displacement of water above or below the mean surface level. These displacements create waves that ultimately control flow in most channels. A laboratory model used to simulate effects in a full-scale prototype must be designed to create the same effects. This is accomplished by having the Froude number of the flow in the model channel equal the Froude number of flow in the prototype. This is given by the expression

$$\frac{v_m}{[gD_m]^{1/2}} = \frac{v_p}{[gD_p]^{1/2}} \tag{12-46}$$

where the subscript m denotes properties in the model channel and p denotes properties in the prototype channel. Application of this relationship is given in the following example problem.

Example Problem 12-6

A prototype rectangular channel 1.0 m in width is to be constructed of lined concrete with $n = 0.015$ on a slope of 0.002 m/m with a discharge capacity of 2.0 m^3/s under uniform flow conditions. The channel design is to be tested on a model rectangular flume 20 cm in width made of plexiglass with $n = 0.010$ and set to the same slope.

Calculate the required discharge in the model channel.

Solution For uniform flow in the prototype channel,

$$Q = \left(\frac{1}{n}\right)A(R_h)^{2/3}(S_0)^{1/2}$$

$$2.0\ m^3/s = \frac{1}{0.015}(1.0\ m)y\left[\frac{1.0\ m(y)}{1.0\ m + 2(y)}\right]^{2/3}(0.002\ m/m)^{1/2}$$

$$0.6708 = y\left[\frac{y}{1.0 + 2y}\right]^{2/3}$$

$$(0.6708)^{3/2} = y^{3/2}\left[\frac{y}{1.0 + 2y}\right]$$

$$0.5494(1.0 + 2y) = y^{5/2}$$

or

$$f(y) = 0 = y^{5/2} - 1.0989y - 0.5494$$

Using Newton's method, the solution to the foregoing equation for the prototype is given by

$$y_p = 1.319\ m$$

$$v_p = \frac{Q}{A} = \frac{2.0\ m^3/s}{1.319\ m(1.0\ m)}$$

$$v_p = 1.516\ m/s$$

$$F_{Np} = \frac{v_p}{[gy_p]^{1/2}}$$

$$F_{Np} = \frac{1.516\ m/s}{[(9.81\ m/s)(1.319\ m)]^{1/2}}$$

$$F_{Np} = 0.4214$$

Setting the same conditions for the Froude number in the model as in the prototype,

$$F_{Np} = F_{Nm} = \frac{v_m}{[gy_m]^{1/2}}$$

For uniform flow in the model,

$$v = \frac{1}{n}(R_n)^{2/3}(S_0)^{1/2}$$

$$v_m = \frac{1}{0.010}\left[\frac{0.20\ m(y)}{0.20\ m + 2y}\right]^{2/3}(0.002\ m/m)^{1/2}$$

But

$$v_m = F_{Nm}[gy_m]^{1/2} = F_{Np}[gy_m]^{1/2}$$

$$v_m = 0.4214(9.81\ m/s^2)^{1/2}(y_m)^{1/2}$$

$$v_m = 1.320(y_m)^{1/2}$$

Substituting v_m into the uniform flow equation,

$$1.320(y_m)^{1/2} = \frac{1}{0.010}\left[\frac{0.20\ m(y_m)}{0.20\ m + 2y_m}\right]^{2/3}(0.002)^{1/2}$$

By algebraic manipulation, this equation can be converted to

$$f(y) = 0 = 0.3207(y_m)^{7/4} + 0.0321(y_m)^{3/4} - 0.20y_m$$

Solving the preceding equation using Newton's method,

$$y_m = 0.3966\ m$$

For other model channel parameters,

$$v_m = F_{Np}[gy_m]^{1/2}$$

$$v_m = 0.4214[(9.81\ m/s^2)(0.3966\ m)]^{1/2}$$

$$v_m = 0.8312\ m/s$$

$$A_m = 0.20\ m(y_m) = 0.20\ m(0.3966\ m)$$

$$A_m = 0.0793\ m^2$$

$$Q_m = v_m A_m = 0.8312\ m/s(0.0793\ m^2)$$

$$Q_m = 0.0659\ m^3/s$$

Control Sections for Flow

The relationship between depth of flow and discharge, or stage-discharge relationship, is determined at a control section. Such a stage-discharge relationship is often termed a rating curve. It can only be defined at a control section. Control sections can be designated for uniform flow, critical flow, and nonuniform flow. These three cases will be investigated in this section. It is important to recognize that the location of the control section is dependent on the state of flow. Supercritical flow with $F_N > 1$ moves at a higher velocity downstream than any gravity wave can move upstream. In this respect, supercritical flow cannot anticipate obstructions or controls to flow downstream. Control for flow in a supercritical section must therefore be upstream of the section. The opposite conditions hold for subcritical flow. Flow in a subcritical section does not move fast enough to erase the effects of gravity waves moving upstream. Therefore a subcritical flow section is fully controlled by conditions and obstacles downstream of the section.

Uniform flow

Uniform flow is the state which flow tends to assume in a long channel of uniform cross section and constant slope when no other controls or obstructions are present. An equation which balances resistance with gravitational forces such as Manning's may be applied for uniform flow to define the depth-discharge relationship. A uniform flow section is not associated with any localized features. It is associated with long, uniform channels which have no local obstructions to flow.

Under conditions of uniform flow the depth of flow is termed the normal depth. The normal depth may be greater than, equal to, or less than the critical depth depending on the slope of the channel bottom. It should be recalled that under uniform flow conditions the channel bed slope, surface water slope, and slope of the energy grade line are all equal. In what is termed a *subcritical channel,* the flow is subcritical and the flow depth must therefore be greater than the critical depth. In this case we have

$$y_n > y_c \tag{12-47}$$

where

$$y_n = \text{normal depth}$$

In a *critical channel* the normal depth is equal to the critical depth:

$$y_n = y_c \tag{12-48}$$

In a *supercritical channel* the normal depth is less than the critical depth:

$$y_n < y_c \tag{12-49}$$

The state of flow in a channel, if all other conditions such as discharge, channel width, and Manning's roughness coefficient remain constant, can be controlled by the slope of the channel bed. Channels are therefore categorized as having subcritical, critical, or supercritical slopes depending on whether the normal depth is greater than, equal to, or less than the critical depth, respectively. Figure 12-7 indicates these three flow configurations. Notice that flow at critical depth is not stable because it has the tendency to oscillate between subcritical and supercritical flow.

Critical flow

Critical flow may serve as a channel control because a definitive stage-discharge relationship exits at critical depth. Critical flow serves as a control at sections such as at the free outfall from a lake over a spillway. This may be demonstrated by writing Bernoulli's equation for flow over a spillway in which energy losses are as-

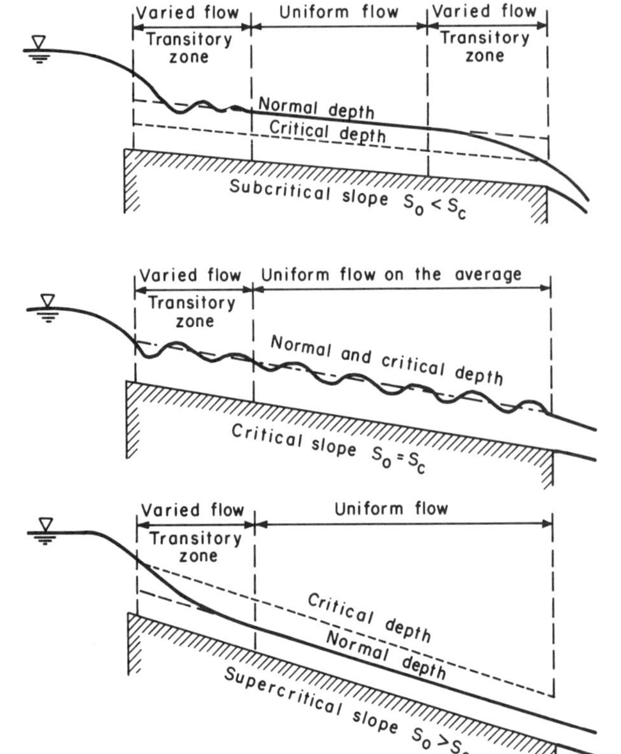

Figure 12-7 Establishment of uniform flow in a long channel. (Adapted from Chow, 1959.)

sumed negligible over the crest. For a rectangular channel of constant width, H and q are constant so

$$H = y + z + \frac{q^2}{2gy^2} = E + z = \text{constant} \tag{12-50}$$

Differentiating with respect to the distance x along the channel,

$$\frac{dE}{dx} + \frac{dz}{dx} = 0 \tag{12-51}$$

which may be written as

$$\frac{dy}{dx}\frac{dE}{dy} + \frac{dz}{dx} = 0 \tag{12-52}$$

It was shown in Eq. (12-43) that

$$\frac{dE}{dy} = 1 - (F_N)^2 \tag{12-43}$$

Substituting this relation into Eq. (12-52),

$$\frac{dy}{dx}[1 - (F_N)^2] + \frac{dz}{dx} = 0 \tag{12-53}$$

Equation (12-53) can be used to analyze the condition of a step in a horizontal channel and other situations. Application of this equation for analysis of flow over a spillway crest is demonstrated in the following example problem.

Example Problem 12-7

Water flows from a lake into a steep rectangular channel 3.0 m in width. The level of the lake is 3.0 m above the channel bed at the outfall. Assuming the flow is over a smooth crest with no energy loss, compute the discharge from the lake into the channel. This situation is demonstrated in Fig. 12-8.

Solution At the crest,

$$\frac{dz}{dx} = 0 \qquad \frac{dy}{dx} \neq 0$$

Substituting these conditions into Eq. (12-53), the term in brackets must be zero—that is,

$$1 - (F_N)^2 = 0$$

or

$$(F_N)^2 = 1$$

Therefore at the crest $F_N = 1$ and $y = y_c$

$$y_c = \frac{2}{3}E_1 = \frac{2}{3}(3\text{ m}) = 2\text{ m}$$

It was previously shown that

$$y_c = \left[\frac{q^2}{g}\right]^{1/3}$$

Therefore,

$$q^2 = (y_c)^3 g = (2\text{ m})^3(9.81\text{ m/s}^2)$$
$$q^2 = 78.48\text{ m}^4/\text{s}^2$$

or

$$q = 8.86\text{ m}^2/\text{s}$$

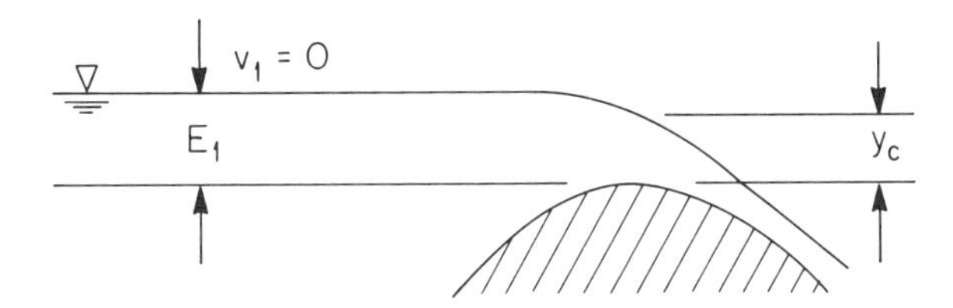

Figure 12-8 Flow at the outfall of a lake over a smooth-crested spillway.

Compute the discharge:

$$Q = qb = 8.86 \text{ m}^2/\text{s}(3.0 \text{ m})$$

$$Q = 26.58 \text{ m}^3/\text{s}$$

Nonuniform flow

Physical obstacles to flow in a channel tend to pull the flow away from the uniform condition. The transition between uniform flow and some other control section for flow will exhibit gradual or rapidly varied nonuniform flow. The nonuniform flow case can be analyzed by letting the slope of the total energy line be designated as S_f. Differentiating the Bernoulli equation for total energy with respect to horizontal distance x along the channel,

$$\frac{dH}{dx} = \frac{d}{dx}\left[z + y + \frac{v^2}{2g}\right] = -S_f \tag{12-54}$$

where the slope of the total energy line is negative because it is decreasing in the positive x-direction. Rearranging Eq. (12-54),

$$\frac{d}{dx}\left[y + \frac{v^2}{2g}\right] = -\frac{dz}{dx} - S_f \tag{12-55}$$

which reduces to

$$\frac{dE}{dx} = S_0 - S_f \tag{12-56}$$

where

$$S_0 = \text{channel bed slope}$$

As previously shown in Eq. (12-43) for any channel cross-sectional shape,

$$\frac{dE}{dy} = 1 - (F_N)^2 \tag{12-43}$$

Equations (12-56) and (12-43) can be rearranged as

$$dE = [S_0 - S_f]\,dx \tag{12-57}$$

and

$$dE = [1 - (F_N)^2]\,dy \tag{12-58}$$

These two equations can be combined as

$$\frac{dy}{dx} = \frac{S_0 - S_f}{1 - (F_N)^2} \tag{12-59}$$

This equation defines the rate of change of the depth of flow with horizontal distance, dy/dx, as a function of the bed slope, friction slope, and Froude number. Such a relationship is applied in the analysis of the flow profile in the following example problem.

Example Problem 12-8

A concrete lined rectangular channel with a Manning's roughness coefficient of 0.015 and 0.60 m in width has a bed slope of 0.007 m/m. The channel is to carry a discharge of 0.80 m^3/s.

(a) Designate the channel as subcritical, critical, or supercritical.

(b) If at some point r along the channel the flow depth is equal to 0.75 m, compute the flow depth at channel station r + 10 m where r is measured along the horizontal.

Solution

(a) Does the channel have a subcritical, critical, or supercritical slope?

$$q = \frac{Q}{b} = \frac{0.80\ m^3/s}{0.60\ m}$$

$$q = 1.333\ m^2/s$$

$$y_c = \left[\frac{q^2}{g}\right]^{1/3} = \left[\frac{(1.333\ m^2/s)^2}{9.81\ m/s^2}\right]^{1/3}$$

$$y_c = 0.566\ m$$

Normal depth is depth attained for uniform flow. Applying Manning's equation,

$$Q = \frac{A}{n}(R_h)^{2/3}(S_0)^{1/2}$$

$$0.80\ m^3/s = \left[\frac{(0.60\ m)y_n}{0.015}\right]\left\{\frac{0.60\ m(y_n)}{0.60\ m + 2y_n}\right\}^{2/3}(0.007\ m/m)^{1/2}$$

$$0.2390 = y_n\left\{\frac{0.60\ m(y_n)}{0.60\ m + 2y_n}\right\}^{2/3}$$

$$0.3360 = \frac{(y_n)^{5/3}}{[0.60 + 2y_n]^{2/3}}$$

Set up the solution using Newton's method.

$$f(y) = 0 = \frac{(y_n)^{5/3}}{[0.60 + 2y_n]^{2/3}} - 0.3360$$

$$f'(y) = \frac{5/3\,(y_n)^{2/3}}{[0.60 + 2y_n]^{2/3}} - \frac{4/3\,(y_n)^{5/3}}{[0.60 + 2y_n]^{5/3}}$$

Set up the solution as

$$y_{n+1} = y_n - \frac{f(y_n)}{f'(y_n)}$$

Table of iterations,

y_n	$f(y_n)$	$f'(y_n)$	y_{n+1}
1.0	0.1928	1.0621	0.8185
0.8185	0.0827	0.7053	0.7012
0.7012	0.0123	0.5060	0.6769
0.6769	−0.0021	0.4679	0.6814
0.6814	0.0006	0.5946	0.6804

Therefore,

$$y_n = 0.680 \text{ m}$$

Since $y_n > y_c$ the channel has a subcritical slope.

(b) If at some point r the flow depth y = 0.75 m, what is the flow depth at channel station r + 10 m?

$$A = (0.75 \text{ m})(0.60 \text{ m}) = 0.45 \text{ m}^2$$

$$v = \frac{Q}{A} = \frac{0.80 \text{ m}^3/\text{s}}{0.45 \text{ m}^2} = 1.778 \text{ m/s}$$

$$R_h = \frac{A}{P_w} = \frac{0.45 \text{ m}^2}{0.60 \text{ m} + 2(0.75 \text{ m})} = 0.2143 \text{ m}$$

Applying Manning's equation and using S_f for S,

$$v = \left(\frac{1}{n}\right)(R_h)^{2/3}(S_f)^{1/2}$$

$$(S_f)^{1/2} = \frac{v(n)}{(R_h)^{2/3}} = \frac{1.778(0.015)}{(0.2143)^{2/3}}$$

$$(S_f)^{1/2} = 0.0745$$

$$S_f = 0.0055 \text{ m/m}$$

Applying the governing equation,

$$\frac{dy}{dx} = \frac{S_0 - S_f}{1 - (F_N)^2}$$

$$(F_N)^2 = \frac{v^2}{gy}$$

$$(F_N)^2 = \frac{(1.778 \text{ m/s})^2}{9.81 \text{ m/s}^2(0.75 \text{ m})}$$

$$(F_N)^2 = 0.4296$$

$$\frac{dy}{dx} = \frac{0.007 - 0.0055}{1 - 0.4296} = +0.0026 \text{ m/m}$$

At channel station r + 10 m,

$$dy = \left(\frac{dy}{dx}\right)\Delta x$$

$$dy = +0.0026 \text{ m/m}(10 \text{ m}) = 0.026 \text{ m}$$

$$y_{r+10} = y_r + dy = 0.75 \text{ m} + 0.026 \text{ m}$$

$$y_{r+10} = 0.776 \text{ m}$$

Application of Momentum Principle

It is necessary to apply the momentum principle to analysis of open channel flow systems when the specific energy principle cannot be applied due to unknown energy loss in the flow section. The momentum principle may also be applied to sections which experience no energy loss, but it is felt that the energy principle is more direct in such cases (Chow, 1959). Unknown losses could be caused by an obstruction within the channel, such as a broad-crested weir, or a hydraulic jump which has high internal energy changes as the flow regime goes from supercritical to subcritical. In such cases there is an unknown and non-negligible energy loss in the channel section which precludes the use of the specific energy principle.

A description of application of the momentum principle is beyond the scope of this book. Other references dedicated to the hydraulics of open channel flow are recommended for discussion of this important topic (Chow, 1959; Henderson, 1966; French, 1985).

12.2 Channel Design for Uniform Flow

Nonerodible Channels

Channels which are constructed or lined with concrete or other durable material or excavated out of bedrock are considered nonerodible. Dimensions for such channels are determined using uniform flow formulas and considering hydraulic efficiency, empirical relations which have proven effective, practicability of construction, and economic considerations. Figures 12-9 and 12-10 are examples of two nonerodible channels with different cross sections used for delivery of irrigation water.

Minimum permissible velocity

The minimum permissible velocity is the velocity which will maintain sediment in suspension in the channel and not allow deposition of silt on the channel bed. It is also the velocity which will not induce growth of aquatic plants and moss in the channel. These conditions alter the shape of the channel cross section and cause the flow to deviate from conditions predicted by the uniform flow equation. A value of 0.6 to 0.9 m/s (2.0 to 3.0 ft/s) is normally accepted as the minimum permissible velocity. Applying a uniform flow equation, such as Manning's, indicates that the flow velocity is dependent on the slope of the channel bed. Practically speaking, the slope of the channel bed is dependent on the topography of the surface over which the channel is to pass. The slope should be checked in uniform sections to see that it allows for the minimum permissible velocity.

Best hydraulic section

Empirical equations for the design depth in open channels, such as Manning's, are functional representations of the balance between gravitational forces moving water down the channel and resistance forces acting against the water. The channel section which has the least wetted perimeter for a given cross-sectional flow area

Figure 12-9 Lined trapezoidal channel used to supply irrigation water in Nile River Valley, Egypt.

Figure 12-10 Hydraulically efficient semicircular channel section set to grade for irrigation distribution system in central Tunisia.

will be the most efficient in transporting water from the hydraulic point of view. For any given cross-sectional shape, the most efficient water transporting geometry is called the best hydraulic section. Table 12-2 indicates different geometric parameters for best hydraulic sections as a function of the design depth.

TABLE 12-2 Best hydraulic sections. (Adapted from Chow, 1959.)

Cross Section	Area A	Wetted Perimeter P	Hydraulic Radius R	Top Width T	Hydraulic Depth D
Trapezoid, half of a hexagon	$\sqrt{3}\, y^2$	$2\sqrt{3}\, y$	$\frac{1}{2}y$	$\frac{4}{3}\sqrt{3}\, y$	$\frac{3}{4}y$
Rectangle, half of a square	$2y^2$	$4y$	$\frac{1}{2}y$	$2y$	y
Triangle, half of a square	y^2	$2\sqrt{2}\, y$	$\frac{1}{4}\sqrt{2}\, y$	$2y$	$\frac{1}{2}y$
Semicircle	$\frac{\pi}{2} y^2$	πy	$\frac{1}{2}y$	$2y$	$\frac{\pi}{4}y$
Parabola, $T = 2\sqrt{2}\, y$	$\frac{4}{3}\sqrt{2}\, y^2$	$\frac{8}{3}\sqrt{2}\, y$	$\frac{1}{2}y$	$2\sqrt{2}\, y$	$\frac{2}{3}y$
Hydrostatic catenary	$1.39586y^2$	$2.9836y$	$0.46784y$	$1.917532y$	$0.72795y$

The semicircle is the most hydraulically efficient of all the channel shapes. In certain parts of the world, semicircular sections are the norm and are the key feature of large-scale irrigation projects. Trapezoidal sections tend to be used in the western United States because of ease of construction. Typically channel sections are designed based on the best hydraulic section because that is the cross-sectional shape which will transport the greatest amount of water for the available head. The design or normal depth is computed by substituting the dimensions given for the best hydraulic section into the governing equation for channel flow. The design depth may then require alteration based on the practical considerations of terrain, expense, and construction techniques.

The principle of basing the design on the best hydraulic section applies only to nonerodible channels. Designs for channels constructed of erodible materials must consider maximum recommended flow velocities and the distribution of forces necessary to maintain channel stability.

Freeboard

Freeboard is the vertical distance from the top of the water surface to the top of the lined channel section at design flow conditions. It is a safety factor which is required to insure the physical integrity of the channel even if the channel must temporarily pass flows greater than the design discharge. The amount of freeboard required is sufficient to prevent waves or fluctuations in the water surface from overflowing the channel sides and weakening channel supports.

A common method used to initially estimate the required height of freeboard is that developed by the U.S. Bureau of Reclamation. Converted to SI, this equation is expressed as,

$$f = [0.4572\ y_n]^{1/2} \quad \text{for } Q = 0.57\ \text{m}^3/\text{s (20 cfs)} \tag{12-60}$$

$$f = [0.7620\ y_n]^{1/2} \quad \text{for } Q = 85\ \text{m}^3/\text{s (3000 cfs)} \tag{12-61}$$

where

f = freeboard, m

y_n = normal depth under uniform flow, m

Q = design discharge

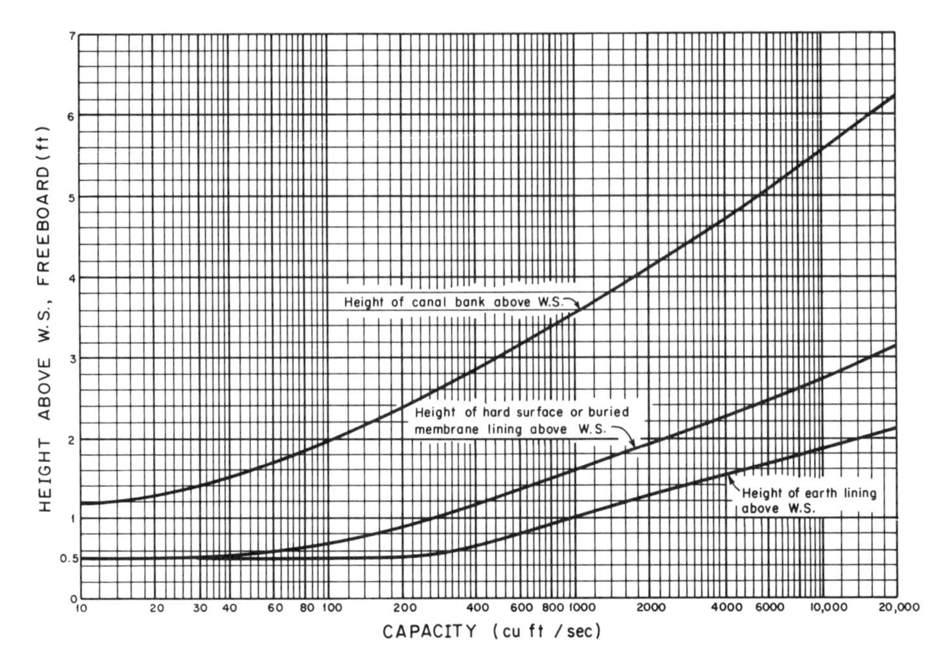

Figure 12-11 Recommended average freeboard and height of bank of lined channels as function of capacity. (Aisenbrey et al., 1974.)

Intermediate values are taken from Figure 12-11 which indicates recommended values of freeboard and height of unlined bank above the level of the water surface at the design discharge.

Calculation of Section Dimensions

Determination of section dimensions will be explained in this section with reference to Manning's equation. The procedures described are a systematic method for determining the design flow depth and other geometric characteristics of the channel section. Certain empirical relations for channel design which have proven reasonable in numerous applications will also be demonstrated.

The first step is to estimate Manning's n based on channel material, select the channel bed slope, S_0, based on topography and practical considerations, and determine the design discharge Q dependent on channel delivery requirements. The combination of these terms is called the section factor, SF, given as

$$SF = A(R_h)^{2/3} = \frac{nQ}{(S_0)^{1/2}} \tag{12-62}$$

The section factor is comprised of the geometric elements of the channel section. The equations for these geometric elements may be taken from Table 12-3 for the general case or from Table 12-2 for the best hydraulic section. The result of solving Eq. (12-62) is the design depth of flow, y_n. This will often require a trial and error solution. The solution can be conveniently obtained using Newton's method as

TABLE 12-3 Geometric elements of channel sections. (Taken from Chow, 1959.)

Section	Area A	Wetted perimeter P_w	Hydraulic radius R	Top width T	Hydraulic depth D
Rectangle	by	$b + 2y$	$\frac{by}{b + 2y}$	b	y
Trapezoid	$(b + zy)y$	$b + 2y\sqrt{1 + z^2}$	$\frac{(b + zy)y}{b + 2y\sqrt{1 + z^2}}$	$b + 2zy$	$\frac{(b + zy)y}{b + 2zy}$
Triangle	zy^2	$2y\sqrt{1 + z^2}$	$\frac{zy}{2\sqrt{1 + z^2}}$	$2zy$	$\frac{1}{2}y$
Circle	$\frac{1}{8}(\theta - \sin\theta)\,d_0^2$	$\frac{1}{2}\theta\, d_0$	$\frac{1}{4}\left(1 - \frac{\sin\theta}{\theta}\right) d_0$	$(\sin \frac{1}{2}\theta)\, d_0$ or $2\sqrt{y(d_0 - y)}$	$\frac{1}{8}\left(\frac{\theta - \sin\theta}{\sin \frac{1}{2}\theta}\right) d_0$
Parabola	$\frac{2}{3}Ty$	$T + \frac{8}{3}\frac{y^2}{T}$ *	$\frac{2T^2y}{3T^2 + 8y^2}$ *	$\frac{3}{2}\frac{A}{y}$	$\frac{2}{3}y$
Round-cornered rectangle ($y > r$)	$\left(\frac{\pi}{2} - 2\right)r^2 + (b + 2r)y$	$(\pi - 2)r + b + 2y$	$\frac{(\pi/2 - 2)r^2 + (b + 2r)y}{(\pi - 2)r + b + 2y}$	$b + 2r$	$\frac{(\pi/2 - 2)r^2}{b + 2r} + y$
Round-bottomed triangle	$\frac{T^2}{4z} - \frac{r^2}{z}(1 - z\cot^{-1}z)$	$\frac{T}{z}\sqrt{1 + z^2} - \frac{2r}{z}(1 - z\cot^{-1}z)$	$\frac{A}{P}$	$2[z(y - r) + r\sqrt{1 + z^2}]$	$\frac{A}{T}$

* Satisfactory approximation for the interval $0 < x \leqq 1$, where $x = 4y/T$. When $x > 1$, use the exact expression $P = (T/2)[\sqrt{1 + x^2} + 1/x \ln(x + \sqrt{1 + z^2})]$.

shown in previous example problems. This method is commonly written within a computerized solution for open channel flow. The computer problem at the end of the chapter requires such a solution.

Trapezoidal sections have various possible solutions depending on the side-slope ratio and bottom width selected. A side-slope ratio of 0.5774 horizontal to 1.0 vertical results in the best hydraulic section; however, ratios of 2.0 or 3.0 to 1.0 are more convenient to construct and more common. Figure 12-12 indicates bottom widths which have proven practical in numerous design applications. This information can be used for an initial estimate of channel dimensions.

Once a consistent set of channel dimensions which satisfy Eq. (12-62) have been determined, the channel cross section must be tested to see that it conforms to velocity restrictions. The minimum permissible velocity to avoid channel siltation has previously been discussed. In addition, unless there is a specific reason for doing so, the design depth is fixed at greater than the critical depth so the flow will be subcritical. This is done by checking that the Froude number is less than some specified value. The value chosen for typical design applications is $F_N < 0.8$. For channel sections which serve as approaches for a broad-crested weir to gauge channel discharge, a more restrictive value of 0.5 is recommended (Bos et al., 1984). It must be recognized that the faster the channel flow, the greater the amount of discharge for a given cross-sectional area. The most cost-effective channel is therefore that which transmits the design discharge with the minimum cross-sectional area. The cost can be minimized by designing the channel up to the limiting value of the Froude number.

The final step is to add the freeboard and height of bank, if applicable, to the design channel depth. Application of the procedure to determine section dimensions is indicated in the following example problem.

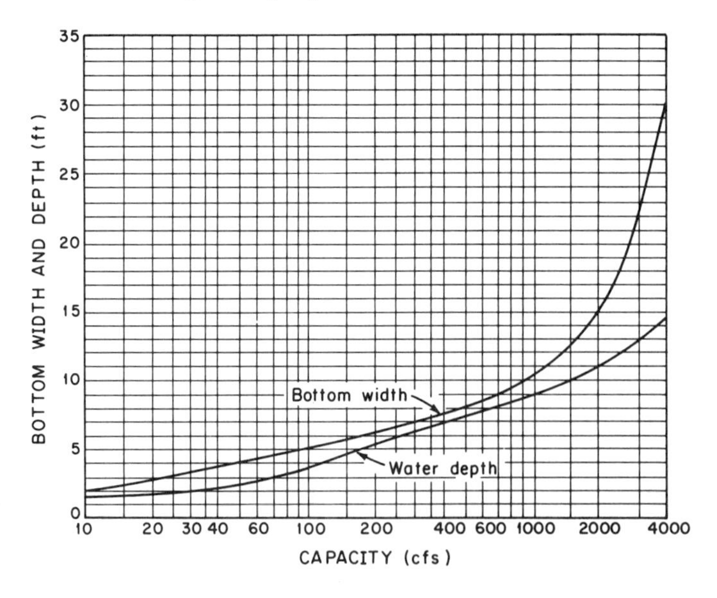

Figure 12-12 Experience curves showing bottom width and depth of lined trapezoidal channels. (U.S. Bureau of Reclamation, 1952.)

Example Problem 12-9

A concrete lined channel with a Manning's n of 0.015 has a required discharge of 1.7 m^3/s and is to be built on a bed-slope of 0.002 m/m. The channel is to be trapezoidal in cross section.

(a) Solve for the trapezoidal section dimensions using Fig. 12-12 for the initial estimate of the channel width and assuming a 2 to 1 side-slope ratio.

(b) Solve for the trapezoidal section dimensions using the parameters for the best hydraulic section given in Table 12-2.

Solution From Fig. 12-12 for Q = 1.7 m^3/s (60 cfs),

$$b = 4.4 \text{ ft} = 1.34 \text{ m}$$

Compute the section factor:

$$SF = A(R_h)^{2/3} = \left[n\frac{Q}{(S_0)^{1/2}}\right]$$

$$A(R_h)^{2/3} = 0.015\frac{1.7 \text{ m}^3/\text{s}}{(0.002 \text{ m/m})^{1/2}}$$

$$A(R_h)^{2/3} = 0.5702$$

From Table 12-3 for channel geometric elements,

$$A(R_h)^{2/3} = (b + zy)\,y\left\{\frac{(b + zy)y}{b + 2y(1 + z^2)^{1/2}}\right\}^{2/3}$$

Substituting z = 2 and b = 1.34 and rearranging,

$$[A(R_h)^{2/3}]^{3/2} = [1.34y + 2y^2]^{3/2}\frac{1.34y + 2y^2}{1.34 + 4.472y}$$

$$(0.5702)^{3/2}[1.34 + 4.472y] = [1.34y + 2y^2]^{5/2}$$

$$0.5770 + 1.9255y = [1.34y + 2y^2]^{5/2}$$

Set up for numerical solution using Newton's method:

$$f(y) = 0 = [1.34y + 2y^2]^{5/2} - 1.9255y - 0.5770$$

The solution to the preceding equation is

$$y_n = 0.507 \text{ m}$$

Check the minimum permissible velocity:

$$A = (b + zy)y = 1.34 \text{ m}(0.507 \text{ m}) + 2(0.507 \text{ m})^2$$

$$A = 1.1935 \text{ m}^2$$

$$v_{avg} = \frac{Q}{A} = \frac{1.7 \text{ m}^3/\text{s}}{1.1935 \text{ m}^2} = 1.424 \text{ m/s}$$

$$v_{avg} > 0.6 \text{ m/s minimum}$$

Check the maximum velocity by checking the Froude number:

For general cross-sectional shape,

$$F_N = \frac{V_{avg}}{\left[g\left(\frac{A}{T}\right)\right]^{1/2}}$$

Compute the top width:

$$T = b + 2z(y)$$
$$T = 1.34 \text{ m} + 2(2)(0.507 \text{ m})$$
$$T = 3.368 \text{ m}$$

$$F_N = \frac{1.424 \text{ m/s}}{9.81 \text{ m/s}^2 \left(\frac{1.1935 \text{ m}^2}{3.368 \text{ m}}\right)^{1/2}}$$

$$F_N = 0.764 < 0.8 \text{ maximum}$$

Add freeboard, f, using Fig. 12-11 for $Q = 1.7 \text{ m}^3/\text{s}$ (60 cfs):

$$f = 0.6 \text{ ft} = 0.182 \text{ m}$$

Total depth of lined trapezoidal channel:

$$d_T = y_n + f$$
$$d_T = 0.507 \text{ m} + 0.182 \text{ m} = 0.689 \text{ m}$$

Erodible Channels

Channels constructed in noncohesive soils or lined with grass are considered erodible. Figure 12-13 shows an example of such a channel. The stability of these channels must be maintained to insure that they perform to design standards. Channel stability is typically maintained by limiting flow velocities or by limiting the force exerted on the sides and bottom of the channel by the fluid volume. The latter procedure is termed the tractive force method and is described in other references on open channel flow (Chow, 1959; Henderson, 1966; French, 1985). This section will discuss design of erodible channels by application of maximum velocity criteria.

Maximum permissible velocity

The maximum permissible velocity is that velocity which can be maintained in open channel flow without eroding the channel cross-sectional area or otherwise damaging the channel flow characteristics. Maximum permissible velocities for different soil types along with their respective Manning's roughness coefficients are given in Table 12-4. This table is divided into one column for clear water and one for silt laden water. Due to the dynamic equilibrium of silt concentration in the flow stream, channels which carry silt-laden water are less likely to cause erosion than clear water channels. For this reason the maximum permissible velocity in silt-laden water is higher than for clear water.

Figure 12-13 Large unlined delivery channel used for irrigation as well as domestic water source in Nile River Valley, Egypt.

TABLE 12-4 Maximum permissible velocities (v) recommended by Fortier and Scobey and the corresponding unit-tractive force values (τ) converted by the U.S. Bureau of Reclamation.* (For straight channels with small slope, after aging.)

		Clear water				Water transporting colloidal silts			
		v		τ		v		τ	
Material	Manning n	ft/s	m/s	lb/ft^2	N/m^2	ft/s	m/s	lb/ft^2	N/m^2
Fine sand, colloidal	0.020	1.50	0.46	0.027	1.293	2.50	0.76	0.075	3.591
Sandy loam, noncolloidal	0.020	1.75	0.53	0.037	1.772	2.50	0.76	0.075	3.591
Silt loam, noncolloidal	0.020	2.00	0.61	0.048	2.298	3.00	0.91	0.110	5.267
Alluvial silts, noncolloidal	0.020	2.00	0.61	0.048	2.298	3.50	1.07	0.150	7.182
Ordinary firm loam	0.020	2.50	0.76	0.075	3.591	3.50	1.07	0.150	7.182
Volcanic ash	0.020	2.50	0.76	0.075	3.591	3.50	1.07	0.150	7.182
Stiff clay, very colloidal	0.025	3.75	1.14	0.260	12.449	5.00	1.52	0.460	22.025
Alluvial silts, colloidal	0.025	3.75	1.14	0.260	12.449	5.00	1.52	0.460	22.025
Shales and hardpans	0.025	6.00	1.83	0.670	32.080	6.00	1.83	0.670	32.080
Fine gravel	0.020	2.50	0.76	0.075	3.591	5.00	1.52	0.320	15.322
Graded loam to cobbles when noncolloidal	0.030	3.75	1.14	0.380	18.194	5.00	1.52	0.660	31.601
Graded silts to cobbles when noncolloidal	0.030	4.00	1.22	0.430	20.588	5.50	1.68	0.800	38.304
Coarse gravel, noncolloidal	0.025	4.00	1.22	0.300	14.364	6.00	1.83	0.670	32.080
Cobbles and shingles	0.035	5.00	1.52	0.910	43.571	5.50	1.68	1.100	52.668

*The Fortier and Scobey values were recommended for use in 1926 by the Special Committee on Irrigation Research of the American Society of Civil Engineers.

The procedure to design a channel for erodible soils is to choose a value of n and the maximum velocity for the particular soil type and substitute these values into Manning's equation. A section factor relative to the velocity of flow, SF_v, can then be calculated as

$$SF_v = (R_h)^{2/3} = \frac{nv}{(S_0)^{1/2}} \tag{12-63}$$

Application of Eq. (12-63) to solve for geometric channel parameters is illustrated in the following example problem.

Example Problem 12-10

A trapezoidal channel of bottom width 0.60 m and side-slope ratio 2.0 horizontal to 1.0 vertical is to be constructed in a noncolloidal silty loam. Compute the maximum channel discharge in m^3/s if the bed-slope is 0.002 m/m and the channel is to transport clear water.

Solution From Table 12-4 for noncolloidal silty loam,

$$n = 0.020$$

$$v_{max} = 0.61 \text{ m/s for clear water}$$

Substituting into Eq. (12-63),

$$SF_v = \frac{nv_{max}}{(S_0)^{1/2}} = (R_h)^{2/3}$$

$$SF_v = \frac{0.020(0.61 \text{ m/s})}{(0.002 \text{ m/m})^{1/2}}$$

$$SF_v = 0.2728$$

From Table 12-3 for trapezoidal channels,

$$R_h = \frac{(b + zy)y}{b + 2y(1 + z^2)^{1/2}}$$

$$R_h = \frac{(0.60 + 2y)y}{0.60 + 2y(5)^{1/2}}$$

$$R_h = \frac{0.60y + 2y^2}{0.60 + 4.4721y}$$

$$R_h = (SF_v)^{3/2} = (0.2728)^{3/2}$$

$$R_h = 0.1425$$

$$0.1425(0.60 + 4.4721y) = 0.60y + 2y^2$$

which reduces to

$$0 = 2(y_n)^2 - 0.0372y_n - 0.0855$$

Solve for y_n using the quadratic formula

$$y_n = \frac{+0.0372 \pm [(0.0372)^2 + 4(2)(0.0855)]^{1/2}}{2(2)}$$

in which only the + operation yields a feasible solution:

$$y_n = 0.2163 \text{ m}$$

From Table 12-3,

$$A = (b + zy)y = [0.60 + 2(0.2163)]0.2163$$
$$A = 0.2233 \text{ m}^2$$

Compute Q_{max}:

$$Q_{max} = v_{max}A$$
$$Q_{max} = 0.61 \text{ m/s}(0.2233 \text{ m}^2) = 0.136 \text{ m}^3/\text{s}$$

Grass-lined channels can also be sized using the maximum permissible velocity criteria. For grass-lined channels, the equivalent of Manning's roughness coefficient is referred to as the *retardance coefficient*. The difficulty in solving for the normal depth using Manning's equation for grass-lined channels is that the retardance coefficient is a function of grass variety, stand density, and grass length. Additionally, since grass is flexible, the retardance coefficient is nonlinearly related to the product of flow velocity times the hydraulic radius (Chow, 1959).

Because of the nonlinearity of the retardance coefficient, a trial and error solution is required using graphical or computerized functions of the retardance coefficient versus the $v(R_h)$ product for different categories of vegetal retardance. The application of the principles required to correctly solve for grass-lined channel flow sections is beyond the scope of this book but is described in Chow (1959). Table 12-5 indicates maximum permissible velocities for different grass varieties as a function of soil erodibility and slope. These maximum velocities may be applied with the Manning's roughness coefficients given in Table 12-1 as an initial approximation of

TABLE 12-5 Permissible velocities for channels lined with grass. (Adapted from U.S.-SCS, 1954.)

Cover	Slope range, Percent	Permissible velocity: Erosion-resistant soils, m/s	Erosion-resistant soils, ft/s	Easily-eroded soils, m/s	Easily-eroded soils, ft/s
Bermuda grass	0–5	2.4	8	1.8	6
	5–10	2.1	7	1.5	5
	>10	1.8	6	1.2	4
Buffalo grass, Kentucky bluegrass, smooth brome, blue grama	0–5	2.1	7	1.5	5
	5–10	1.8	6	1.2	4
	>10	1.5	5	0.9	3
Grass mixture	0–5	1.5	5	1.2	4
	5–10	1.2	4	0.9	3
	Do not use on slopes steeper than 10%				
Lespedeza sericea, weeping love grass, ischaemum (yellow bluestem), kudzu, alfalfa, crabgrass	0–5	1.1	3.5	0.8	2.5
	Do not use on slopes steeper than 5%, except for side slopes in a combination channel				
Annuals—used on mild slopes or as temporary protection until permanent covers are established, common lespedeza, Sudan grass	0–5	1.1	3.5	0.8	2.5
	Use on slopes steeper than 5% is not recommended				

REMARKS. The values apply to average, uniform stands of each type of cover. Use velocities exceeding 1.5 m/s (5 ft/s) only where good covers and proper maintenance can be obtained.

the maximum capacity of a grass channel. The solution follows the pattern indicated in example problem 12-10. For more accurate design of grass-lined channels, the procedure which accounts for the nonlinearity of the retardance coefficient should be applied.

12.3 Discharge Measurement and Channel Controls

Weirs

General requirements

Weirs are structures which are inserted into the flow stream in an open channel for purposes of measuring discharge. They have an opening with a sharp edge in the upstream direction through which the flow passes for measurement. Various shapes have been used for the opening. This section will discuss rectangular, Cipolletti or trapezoidal, and 90 degree V-notch weirs. The amount of discharge flowing through the opening is nonlinearly related to the width of the opening and the depth of the water level in the approach section above the height of the weir crest.

The fall of water over the opening is called the nappe. Weirs are classified as being *contracted* or *suppressed* depending on whether or not the nappe is constrained by the edges of the channel. If the nappe is open to the atmosphere at the edges, it is said to be contracted because the flow contracts as it passes through the flow section and the width of the nappe is slightly less than the width of the weir crest. This type of weir is demonstrated in Figs. 12-14 and 12-15. If the sides of the

Figure 12-14 Contracted sharp-crested weir installed on an irrigation delivery ditch. Measurement of head above the weir crest is made in the pool upstream of the weir where the water level is not affected by flow over the weir. (Photo courtesy of Marvin Shearer.)

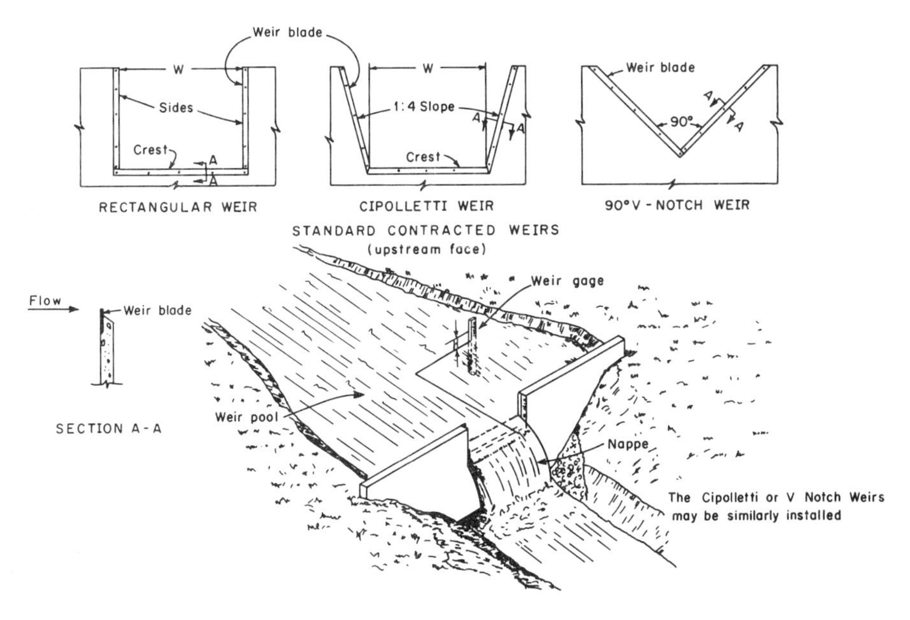

Figure 12-15 Standard contracted weir discharging at free flow. (Adapted from Aisenbrey et al., 1974.)

channel are also the sides of the weir opening, the streamlines of flow are parallel to the walls of the channel and there is no contraction of flow. In this case the weir is said to be suppressed. Some type of air vent must be installed in a suppressed weir so air at atmospheric pressure is free to circulate beneath the nappe. Serious error can result in the estimated discharge if the flow is not ventilated in a suppressed weir. Figure 12-16 indicates suppressed weir installation.

The flow may also be classified as free or submerged flow. Under free flow conditions, the water surface in the channel downstream of the weir is lower than the level of the weir crest. If the water level downstream is higher than the level of the crest, the flow is said to be submerged. The discharge equations in this section assume free flow over the crest which is the condition for the most accurate estimation of discharge.

Weirs can be placed in many positions in a channel distribution system, including at division structures and turnouts where the flow is directed to specific locations. Such an installation in an irrigation district is demonstrated in Fig. 12-17. The weirs discussed in this section are appropriate for discharges in the range of approximately 0.03 to 2.8 m^3/s (1 to 100 cfs). The contracted rectangular and Cipolletti weir shapes are equally applicable for the higher flow rates. The V-notch weir is the most accurate of the shapes presented for discharge rates below 0.12 m^3/s (4.3 cfs).

Certain requirements for weir construction and installation must be met for accurate estimates of discharge. The weir blade which surrounds the weir opening should be made of metal and have a sharp 90 degree edge in the upstream direction. Care should be taken to insure that the weir blade is installed vertically. The weir blade should be kept free from nicks and rust.

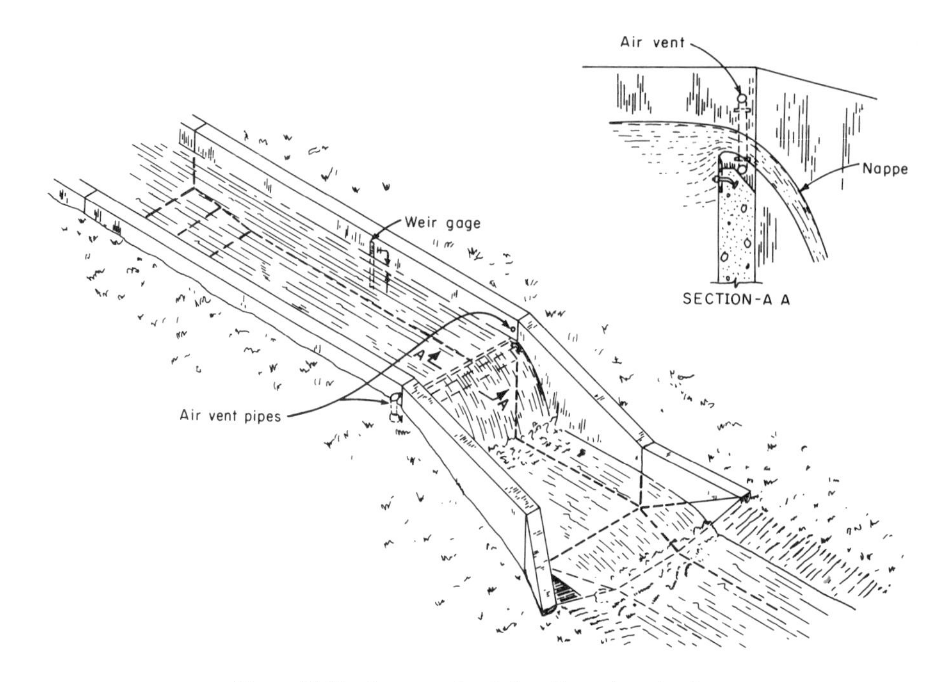

Figure 12-16 Suppressed weir in a flume drop structure.

Figure 12-17 Flow diversion system on open channel network using movable gates as weirs to control discharge.

Nappes should be fully ventilated—that is, air should be able to circulate freely—and suppressed weirs should be vented on both sides. The nappe should flow free of the weir bottom and sides on the downstream side under normal operating conditions.

One of the key features in correct estimation of discharge using weirs is that the measurement of head above the weir crest be made according to standardized procedures. One of the criteria required is that the flow velocity in the approach section be less than or equal to about 0.15 m/s (0.5 ft/s). This can normally be maintained by establishing a pool of water behind the weir which extends upstream a distance of 15 to 20 times the head on the weir crest. The head on the weir crest is equal to the difference between the height of the weir crest and the upstream water level. If the approach velocity cannot be maintained within this criterion, discharge correction coefficients must be applied (Aisenbrey et al., 1974).

The depth of water below the crest in the upstream weir pool should be at least two times the head on the weir and not less than 0.3 m (1 ft). The lateral distance between the edges of the weir and the sides of the approach channel for a contracted flow weir should also be at least two times the head on the weir and not less than 0.3 m (1 ft). The head on the weir should be measured at a distance of not less than four times the maximum head upstream from the crest. The head is usually measured on a staff gauge installed at this position in the weir pool or may be measured using a stilling well which measures the water level at the same position (Aisenbrey et al., 1974).

Discharge formulae

The type of weir to be selected and length of the weir crest are a function of the design discharge. 90 degree V-notch weirs can be accurately applied for measuring low flow volumes up to a maximum of about 0.12 m^3/s (4.3 cfs). This type of weir is most accurate for low levels of discharge. Suppressed rectangular weirs are normally employed for measuring discharge up to 1.0 m^3/s (35.6 cfs). Both contracted rectangular weirs and Cipolletti weirs can be used for measuring discharge from a minimum of about 0.03 m^3/s (1 cfs) up to approximately 2.8 m^3/s (100 cfs). A summary table of maximum weir discharges and recommended lengths of weir crests is given in Table 12-6. This table can be used to select the proper weir size for a given design discharge.

The discharge formulae for different weir cross sections are given next. These formulae are derived from procedures and field tests results developed by the Bureau of Reclamation (Aisenbrey et al., 1974). They assume that proper consideration has been made of approach velocity in the weir pool, lateral distance from the weir opening to side of the channel, and upstream distance from the weir crest for measurement of head.

90 degree V-notch:

$$Q = 1.3424(h)^{2.48} \tag{12-64a}$$

where

$$Q = \text{discharge, m}^3/\text{s}$$

$$h = \text{head above invert of V-notch, m}$$

For the English system the formula is

$$Q = 2.49(h)^{2.48} \tag{12-64b}$$

TABLE 12-6 Maximum discharges and recommended widths of weir crest.

Rectangular Contracted		Rectangular Suppressed		Cipolletti	
W (m)	Qmax (m^3/s)	W (m)	Qmax (m^3/s)	W (m)	Qmax (m^3/s)
0.30	0.03	0.30	0.02	0.30	0.03
0.60	0.18	0.46	0.05	0.60	0.20
0.90	0.26	0.60	0.10	0.90	0.39
1.22	0.54	0.90	0.28	1.22	0.59
1.52	0.94	1.22	0.58	1.52	1.02
1.83	1.49	1.52	1.01	1.83	1.62
2.13	1.98			2.13	1.99
2.44	1.98			2.44	1.99
2.74	1.98			2.74	1.99
3.05	1.98			3.05	1.99
3.66	2.03			3.35	1.99
4.57	2.54			3.66	2.10
5.49	2.85			3.96	2.28
6.10	2.85			4.27	2.45
				4.57	2.63
				4.88	2.80

where

$$Q = \text{discharge, cfs}$$
$$h = \text{head above invert of V-notch, ft}$$

Rectangular suppressed:

$$Q = 1.838W(h)^{1.5} \tag{12-65a}$$

where

$$W = \text{width of weir crest, m}$$
$$h = \text{head above weir crest, m}$$

which is given in the English system as

$$Q = 3.33W(h)^{1.5} \tag{12-65b}$$

where

$$W = \text{width of weir crest, ft}$$
$$h = \text{head above weir crest, ft}$$

Rectangular contracted:

$$Q = 1.838(W - 0.2h)(h)^{1.5} \tag{12-66a}$$

for the SI units previously given and as follows for English units,

$$Q = 3.33(W - 0.2h)(h)^{1.5} \tag{12-66b}$$

Cipolletti:

$$Q = 1.859W(h)^{1.5} \tag{12-67a}$$

for SI units and for English units,

$$Q = 3.367W(h)^{1.5} \tag{12-67b}$$

Application of the tables for weir size and discharge formulae are indicated in the following example problem.

Example Problem 12-11

Design discharge for an irrigation channel is 1.560 m^3/s (55.1 cfs) and the discharge is to be gauged using a Cipolletti weir. Indicate the appropriate weir length, height of weir crest above the channel invert, lateral distance from the edge of the weir to the side of the approach section, and required distance upstream for measurement of head above the weir crest.

Solution Calculate the required weir height using the discharge formula given by Eq. (12-67),

$$Q = 1.859W(h)^{1.5}$$

From Table 12-6,

$$W = 1.83 \text{ m}$$

$$(h)^{1.5} = \frac{1.560 \text{ m}^3/\text{s}}{1.859(1.83 \text{ m})}$$

$$h = 0.595 \text{ m}$$

Required height of weir crest above bottom of approach channel:

$$h_c \geq 2(h) \geq 1.190 \text{ m}$$

Required lateral distance from weir edge to side of channel:

$$d_1 \geq h_c \geq 1.190 \text{ m}$$

Required distance upstream from weir to measure h:

$$d_h \geq 4(h) \geq 2.380 \text{ m}$$

Portable weirs can be constructed out of adequately strong but flexible material, such as a reinforced, waterproof cloth. A contracted rectangular flow section using metal strips riveted to the tear-resistant material is commonly used for such portable weirs. These weirs are suspended across the channel by a lightweight metal bar inserted into a tube of material sown along the top edge of the weir. They can be anchored into the channel by burying the excess weir material on the banks and bottom of the upstream channel section. Such weirs can be temporarily installed in channels in which it would otherwise not be feasible to determine the discharge. A weir installation of this type is depicted in Fig. 12-18. Details of construction requirements for the more permanent weirs described in this section can be found in the reference by the Bureau of Reclamation (Aisenbrey et al., 1974).

Figure 12-18 Portable sharp-crested weir installed for irrigation efficiency evaluation in central Tunisia. Aluminum pipe is fit through loop sewn into top of waterproof material for rigidity.

Parshall Flumes

The Parshall flume is another discharge measuring device. Parshall flumes can accurately measure channel discharge over an extremely wide range of flow rates from as low as 0.0003 m^3/s (0.01 cfs) to as high as 85 m^3/s (3000 cfs). This section will focus on application of Parshall flumes for measurement of flow from 0.0014 m^3/s (0.05 cfs) to 3.95 m^3/s (139.5 cfs). Due to the wide range of discharges which can be measured, Parshall flumes are applicable on channels as small as individual furrows in a surface irrigated field on up to major distribution canals.

The required geometry of a Parshall flume must meet specific criteria as indicated in the plan and profile views of the flume shape in Fig. 12-19 and the accompanying Table 12-7. The throat width of flumes corresponding to the range of discharges to be covered in this section varies from 0.15 m (0.5 ft) to 2.44 m (8.0 ft). The wide range of discharges which may be accurately measured with such a flume have led to its widespread application on irrigation projects. Although construction of such a flume must meet the specifications to enable application of the discharge equations, some firms have specialized in fabrication of such flumes and significantly reduced the unit cost. Figure 12-20 demonstrates this type of flume.

Parshall flumes are operated under what are called free-flow or submerged conditions. Under free-flow conditions, the tailwater level does not affect flow through the convergent crest section. The flow passes through critical depth at the crest and only one depth measurement at point H_a in Fig. 12-19 is required to evaluate the discharge. Under submerged-flow conditions, the tailwater depth is high

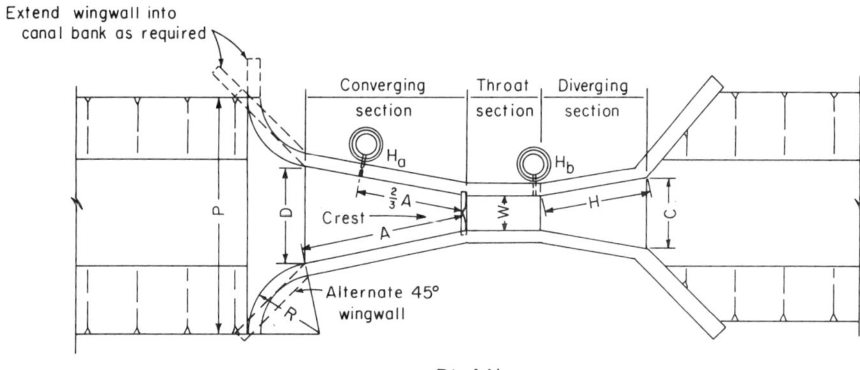

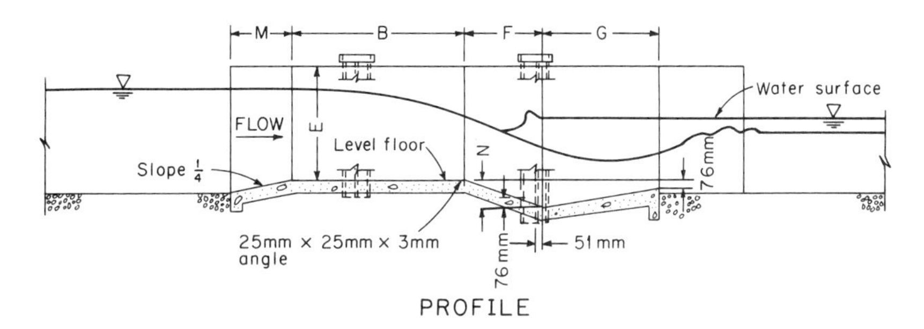

Figure 12-19 Plan and profile views of a Parshall flume. (Adapted from Aisenbrey et al., 1974.)

Figure 12-20 Prefabricated Parshall flume fitted with staff gauge used to measure discharge from upstream pond.

TABLE 12-7 Standard dimensions and capabilities of the Parshall flume for various throat widths (W) for free flow.

Throat Width W	A	B	C	D	E	F	G	K	N	X	Y	Free-Flow Capacity Minimum	Free-Flow Capacity Maximum
1. English units 2. SI units	ft in cm	ft in cm	ft in cm	ft in cm	ft in cm	ft in cm	ft in cm	in cm	ft in cm	in cm	in cm	1. ft^3/s 2. L/s	1. ft^3/s 2. L/s
6 in	1-4$\frac{5}{16}$	2-0	1-3$\frac{1}{2}$	1-3$\frac{5}{8}$	2-0	1-0	2-0	3	0-4$\frac{1}{2}$	2	3	0.05	3.9
15.2 cm	41.5	63.0	50.8	44.3	61.0	30.5	61.0	7.6	11.4	5.1	7.6	1.42	110.4
9 in	1-11$\frac{1}{8}$	2-10	1-3	1-10$\frac{5}{8}$	2-6	1-0	2-6	3	0-4$\frac{1}{2}$	2	3	0.09	8.9
22.9 cm	58.8	86.4	38.1	57.5	76.3	30.5	76.2	7.6	11.4	5.1	7.6	2.55	251.8
1 ft	3-0	4-4$\frac{7}{8}$	2-0	2-9$\frac{1}{4}$	3-0	2-0	3-0	3	0-9	2	3	0.11	16.1
30.5 cm	91.5	134.4	61.0	84.5	91.5	61.0	91.5	7.6	22.9	5.1	7.6	3.11	455.6
1$\frac{1}{2}$ ft	3-2	4-7$\frac{7}{8}$	2-6	3-4$\frac{3}{8}$	3-0	2-0	3-0	3	0-9	2	3	0.15	24.6
45.8 cm	96.6	142.3	76.2	102	91.5	61.0	91.5	7.6	22.9	5.1	7.6	4.29	696.2
2 ft	3-4	4-10$\frac{7}{8}$	3-0	3-11$\frac{1}{2}$	3-0	2-0	3-0	3	0-9	2	3	0.42	33.1
61 cm	101.7	149.6	91.5	120.71	91.5	61.0	91.5	7.6	22.9	5.1	7.6	11.89	936.7
3 ft	3-8	5-4$\frac{3}{4}$	4-0	5-1$\frac{7}{8}$	3-0	2-0	3-0	3	0-9	2	3	0.61	50.4
91.5 cm	111.8	164.6	122.0	157.3	91.5	61.0	91.5	7.6	22.9	5.1	7.6	17.26	1426
4 ft	4-0	5-10$\frac{5}{8}$	5-0	6-4$\frac{1}{4}$	3-0	2-0	3-0	3	0-9	2	3	1.3	67.9
122.0 cm	122.0	179.5	152.5	193.8	91.5	61.0	91.5	7.6	22.9	5.1	7.6	36.79	1922
5 ft	4-4	6-4$\frac{1}{2}$	6-0	7-6$\frac{5}{8}$	3-0	2-0	3-0	3	0-9	2	3	1.6	85.6
152.5 cm	132.2	194.4	183.0	230.3	91.5	61.0	91.5	7.6	22.9	5.1	7.6	45.28	2422
6 ft	4-8	6-10$\frac{3}{8}$	7-0	8-9	3-0	2-0	3-0	3	0-9	2	3	2.6	103.5
183.0 cm	142.3	209.4	213.5	266.9	91.5	61.0	91.5	7.6	22.9	5.1	7.6	73.58	2929

enough to affect flow through the crest section. In this case two depth measurements at H_a and H_b are required to determine the discharge.

The degree of submergence is specified by the ratio of the depths H_b/H_a. For flumes with throat widths less than 0.23 m (0.75 ft), free-flow conditions are assumed to exist for

$$\frac{H_b}{H_a} \leq 0.60 \tag{12-68}$$

For flumes with throat widths greater than or equal to 0.23 m (0.75 ft), free-flow conditions are assumed to be maintained up to the limit:

$$\frac{H_b}{H_a} \leq 0.70 \tag{12-69}$$

The maximum degree of submergence of flow through the crest section of a Parshall flume while still maintaining accurate discharge estimation is given by

$$\frac{H_b}{H_a} \leq 0.95 \tag{12-70}$$

The accuracy of the discharge estimates using the equations which follow is in the range of two percent for free-flow conditions and five percent for submerged flows.

Solution of discharge for submerged-flow conditions requires application of graphical and tabulated material not provided in this chapter. Other design manuals and publications can be referred to for the necessary procedure (French, 1985; Kraatz and Mahajan, 1975; Aisenbrey et al., 1974). The following discharge equations are only applicable for free-flow conditions.

The discharge equation for a Parshall flume with a throat width of 0.1524 m (0.5 ft) is

$$Q = 0.3812[H_a]^{1.58} \tag{12-71}$$

where

Q = discharge, m^3/s

H_a = depth of flow measured in convergent section (see Fig. 12-19), m

For a Parshall flume with a throat width of 0.2286 m (0.75 ft), the discharge equation is

$$Q = 0.5354[H_a]^{1.53} \tag{12-72}$$

where the units are as previously defined. For flumes with throat widths in the range of 0.3048 m (1.0 ft) to 2.4384 m (8.0 ft), the governing discharge equation is

$$Q = 0.3716(W)[3.2801(H_a)]^{1.5697(W)^{0.026}} \tag{12-73}$$

where

W = throat width, m

and the other units are as previously given. Equation (12-73) owes its awkward form to the fact that the discharge relationship was originally derived in English engineering units and the nonlinear exponential on the depth term H_a requires use of the original regression coefficients. Application of the discharge equations for free-flow conditions in a Parshall flume are demonstrated in the following example problem.

Example Problem 12-12

A Parshall flume is to be constructed to measure a design discharge of 2.70 m^3/s (95.35 cfs). Specify the required width and length of the throat section, lengths of the convergent and divergent sections, and flow depth measured in the convergent section at design discharge. Assume free-flow conditions. Compute the maximum tailwater depth for the assumption of free-flow conditions to be valid.

Solution From Table 12-7,

$$W = 1.83 \text{ m}$$

$$\text{Length of throat section} = F = 0.610 \text{ m}$$

$$\text{Length of convergent section} = B = 2.094 \text{ m}$$

$$\text{Length of divergent section} = G = 0.915 \text{ m}$$

To compute the design depth of flow in the convergent section, apply Eq. (12-73) for W = 1.83 m:

$$Q = 0.3716(1.83)\,[3.2801(H_a)]^{1.5697(1.83)^{0.026}}$$

$$Q = 0.6800[3.2801(H_a)]^{1.5946}$$

$$\left[\frac{2.70}{0.6800}\right]^{1/1.5946} = 3.2801(H_a)$$

$$H_a = (0.3048)2.3744$$

$$H_a = 0.724 \text{ m}$$

For the maximum tailwater depth,

$$H_{b-max} = 0.95(H_a)$$

$$H_{b-max} = 0.688 \text{ m}$$

Lack of space precludes the description of a number of alternative methods to gauge channel discharge. Among other methods which have proved useful and practical are the cut-throat flume described by Kraatz and Mahajan (1975) and the broad-crested weir described by Bos et al., (1985). The reference by Bos et al., (1985) includes a complete description of the hydraulics of flow over a broad-crested weir and a computer program for discharge calculation. Sluice gates have also been used in many installations to control discharge at predetermined levels as demonstrated previously using the specific energy principle. Figure 12-21 indicates application of sluice gates to control channel discharge on an irrigation project.

Drop Structures

Drop structures, typically constructed out of concrete, can accommodate a sudden change in elevation of the channel bottom while maintaining control of the flow. They are a basic type of flow control structure and will be the only structure discussed in this section. [More elaborate structures which may be found as part of

Figure 12-21 Effective channel flow control using sluice gates with different openings in central Tunisia. Reservoir backed up behind gates seen in foreground.

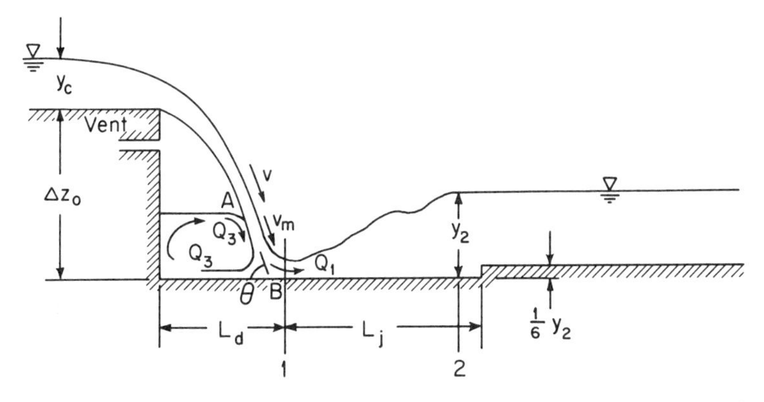

Figure 12-22 The drop structure used as an energy dissipating device in open channels.

large irrigation project distribution systems but not typically in on-farm systems include stilling basins and spillways. These are described in other references including Asenbrey et al., (1975) and Henderson, (1966).] Drop structures are used in channels which must be laid along relatively steep gradients to allow for dissipation of energy without causing scour in the channel itself. In such applications the drop structure allows the main channel to be laid on subcritical slope while the excess potential energy of the flow due to the steep topography is absorbed in the drop structure. Figure 12-22 depicts a drop structure. The following example problem indicates application of drop structures as an energy dissipating devicc in relatively steep terrain.

Example Problem 12-13

An irrigation channel with a design discharge of 2.265 m^3/s (80.0 cfs) is to be laid along terrain having an average slope of 0.005 m/m. To maintain subcritical flow in the channel section, the bottom slope of the channel must be limited to 0.001 m/m. The extra fall is to be absorbed by drop structures such as indicated in Fig. 12-22 having a width of 3.048 m (10.0 ft). Compute the number of structures required in a 16.09 km length of line if the drop height Δz is equal to 1.829 m (3.0 ft).

Solution Compute total drop to be absorbed by structures:

$$Z_T = (S_t - S_o)L$$

where

$$S_t = \text{terrain slope}$$
$$L = \text{distance}$$
$$Z_T = (0.005\ m/m - 0.001\ m/m)16.09\ km$$
$$Z_T = 64.36\ m$$

Compute number of structures required:

$$N_s = \frac{Z_t}{\Delta z} = \frac{64.36\ m}{1.829\ m}$$
$$N_s = 36 \text{ structures}$$

where the solution must be rounded up to next whole integer to maintain the channel slope at $\leq$0.001 m/m.

The various dimensions of the drop structure are illustrated in Fig. 12-22. The following relations have been found to predict the hydraulic conditions existing in a drop structure to within five percent (Hendersen, 1966). These relations assume that the channel slope upstream of the drop is subcritical so flow at the brink of the overfall is critical. The design relationships are given by

$$\frac{y_1}{\Delta z} = 0.54\left[\frac{y_c}{\Delta z}\right]^{1.275} \tag{12-74}$$

where

$$y_c = \text{critical depth in upstream channel}$$

Equation (12-74) can be rewritten as

$$\frac{y_1}{y_c} = 0.54\left[\frac{y_c}{\Delta z}\right]^{0.275} \tag{12-75}$$

Other relations are

$$\frac{y_2}{\Delta z} = 1.66\left[\frac{y_c}{\Delta z}\right]^{0.81} \tag{12-76}$$

$$\frac{L_d}{\Delta z} = 4.30\left[\frac{y_c}{\Delta z}\right]^{0.09} \tag{12-77}$$

and,

$$L_j = 6.9[y_2 - y_1] \tag{12-78}$$

Application of these design relationships is indicated in the following example problem.

Example Problem 12-14

Compute the total required length of the drop structure for the flow conditions specified in example problem 12-13. Assume a mild channel slope upstream of the drop.

Solution Compute critical depth for rectangular channel:

$$y_c = \left[\frac{q^2}{g}\right]^{1/3}$$

$$y_c = \left(\frac{[(2.265\ m^3/s)/3.048\ m]^2}{9.81\ m/s^2}\right)^{1/3}$$

$$y_c = 0.383\ m$$

Apply Eqs. (12-75) through (12-78) to compute the drop structure dimensions:

$$y_1 = 0.54\left(\frac{y_c}{\Delta z}\right)^{0.275} y_c$$

$$y_1 = \frac{0.54(y_c)^{1.275}}{(\Delta z)^{0.275}}$$

$$y_1 = 0.1347\ m$$

$$y_2 = 1.66\left(\frac{y_c}{\Delta z}\right)^{0.81} \Delta z$$

$$y_2 = 1.66(y_c)^{0.81}(\Delta z)^{0.19}$$

$$y_2 = 0.856\ m$$

$$L_d = 4.30\left(\frac{y_c}{\Delta z}\right)^{0.09} \Delta z$$

$$L_d = 4.30(y_c)^{0.09}(\Delta z)^{0.91}$$

$$L_d = 6.832\ m$$

$$L_j = 6.9(y_2 - y_1) = 4.977\ m$$

$$L_T = L_d + L_j = 11.81\ m$$

Other structures which have been employed by the U.S. Bureau of Reclamation on irrigation projects to regulate and control flow are described by Aisenbrey et al., (1974). This reference indicates design dimensions and construction details for numerous structures.

REFERENCES

AISENBREY, A. J., Jr., R. B. HAYES, H. J. WARREN, D. L. WINSETT, and R. B. YOUNG, *Design of Small Canal Structures*. Denver, Colorado: U.S. Department of the Interior, Bureau of Reclamation, 1974.

BAKHMETEFF, B. A., *O Neravnomernom Dvizhenii Zhidkosti v Otkrytom Rusle (Varied Flow in Open Channels)*. St. Petersburg, Russia: 1912.

BAKHMETEFF, B. A., *Hydraulics of Open Channels*. New York: McGraw-Hill, Inc., 1932.

BAZIN, H., "Etude d'une Nouvelle Formule Pour Calculer le Debit des Canaux Decouverts" ("Study of a New Formula to Calculate the Discharge of Open Channels"), Memoire no. 41, Annales de Ponts et Chaussees, vol. 14, Serie 7, 4me Trimestre, 1897, pp. 20–70.

BOS, M. G., J. A. REPLOGLE, and A. J. CLEMMENS, *Flow Measuring Flumes for Open Channel Systems*. New York: John Wiley and Sons, 1984.

CHÉZY, A., "Report on Hydraulic Experiments Made on the Courpalet Canal, River Seine, Paris," 1769. Referenced in V. T. Chow, 1959.

CHOW, V. T., *Open-Channel Hydraulics*. New York: McGraw-Hill Co., 1959.

FRENCH, R. H., *Open-Channel Hydraulics*. New York: McGraw-Hill Co., 1985.

GANGUILLET, E. and W. R. KUTTER, "Versuch zur Aufstellung einer neuen Allegemeinen Formel fur die Gleighformige Bewegung des Wassers in Canalan und Flussen" ("An Investigation to Establish a New General Formula for Uniform Flow of Water in Canals and Rivers"), *Zeitschrift des Oesterreichischen Ingenieur- und Architekten Vereines,* vol. 21, no. 1, Vienna, 1869, pp. 6–25; no. 2–3, Vienna, 1869, pp. 46–59.

HENDERSON, F. M., *Open Channel Flow*. New York: Macmillan Co., 1966.

KRAATZ, D. B. and I. K. MAHAJAN, *Small Hydraulic Structures*. Rome: Food and Agricultural Organization of the United Nations, Irrigation and Drainage Paper no. 26, vol. 1 and vol. 2, 1975.

MANNING, R., "On the Flow of Water in Open Channels and Pipes," paper read at meeting of Institution of Civil Engineers of Ireland, Transactions, Institution of Civil Engineers of Ireland, vol. 20, Dublin, 1889, pp. 161–207.

MOTT, R. L., *Applied Fluid Mechanics*. Columbus, Ohio: Merrill Publishing Co., 1982.

U.S. Bureau of Reclamation, "Linings for Irrigation Canals," 1952.

U.S. Soil Conservation Service, "Handbook of Channel Design for Soil and Water Conservation," SCS-TP-61, Stillwater Outdoor Hydraulic Laboratory, 1954.

PROBLEMS

12-1. Water is flowing at a velocity of 5.0 m/s and a depth of 3.5 m in a rectangular channel.

(a) Find the maximum allowable size of an upward step in the channel while maintaining the same channel discharge.

(b) Find the change in flow depth and value of the absolute water level for a smooth downward step of 35 cm in the channel bed. Use the upstream channel bottom as the reference datum for the absolute water level.

12-2. Assume the same upstream conditions as in problem 12-1 and the width of the channel equal to 3.0 m.

(a) Find the maximum allowable contraction in width while maintaining the same channel discharge.

(b) Find the change in flow depth and value of the absolute water level produced by a smooth expansion to a width of 3.3 m. Use the upstream channel bottom as the reference datum for absolute water level.

12-3. The flow depths a short distance upstream and downstream of a sluice gate in a horizontal channel are 2.440 m and 0.610 m, respectively. The channel is rectangular in section with a width of 3.050 m. Compute the discharge under the sluice gate in m^3/s.

12-4. Water has a flow velocity of 0.90 m/s and a depth of 1.525 m in a rectangular channel. There is a smooth upward step of 0.152 m in the channel bed. Calculate the flow depth over the step and the change in absolute water level using the upstream channel bottom as a reference.

12-5. Compute the critical depth and critical flow velocity of a trapezoidal channel with 2.0 side-slope ratio and width of 6.100 m if the channel discharge is 11.33 m^3/s.

12-6. For the trapezoidal channel section in problem 12-5, let the channel bed slope equal 0.0016 m/m, Manning's roughness coefficient equal 0.025, and discharge equal to 11.33 m^3/s. Compute the normal depth of flow in m for a long, uniform channel and the corresponding flow velocity in m/s.

12-7. Determine the discharge in m^3/s for channels having a normal flow depth equal to 1.830 m, Manning's roughness coefficient equal to 0.015, channel bottom slope equal to 0.0020 m/m, and the following cross sections:

(a) Rectangular section 6.100 m in width.

(b) Trapezoidal section with side-slope ratio equal to 2.0 and bottom width equal to 6.100 m.

(c) Circular section 4.570 m in diameter.

12-8. A long uniform rectangular channel with bottom width of 2.0 m is concrete lined with a Manning roughness coefficient of 0.015. Maximum channel discharge is 1.5 m^3/s and the channel bottom slope is 0.0030 m/m. Flow in the channel is to pass over a smooth broad-crested weir in a horizontal channel section at critical depth to measure the discharge Fig. P12-8. Compute the maximum weir height.

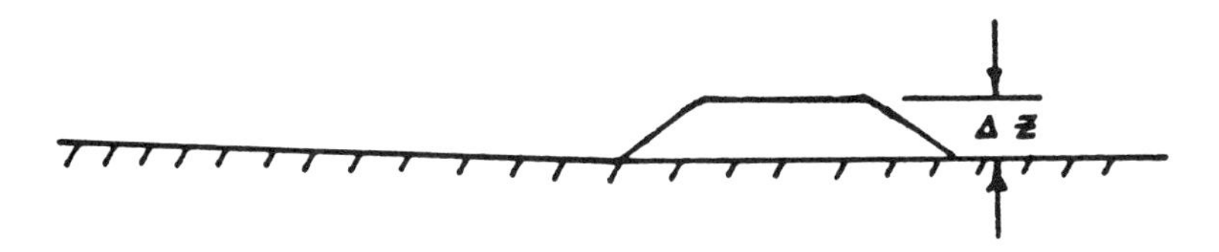

12-9. A rectangular spillway 8.00 m in width is designed to drain a reservoir at a rate of 60.00 m^3/s with a depth of flow equal to 0.300 m at the bottom of the spillway. Following the spillway is a horizontal section in which a hydraulic jump occurs before the flow enters a natural river channel. Compute the depth of flow in meters as the water enters the natural river channel.

12-10. A concrete-lined trapezoidal channel with Manning's roughness coefficient of 0.013, side-slope ratio of 2 to 1, and bottom width of 1.00 m has a design discharge of 1.2 m^3/s. The channel is to be designed so that under uniform flow conditions, the Froude number is less than or equal to 0.85. Compute the maximum allowable channel bottom slope.

12-11. A model stilling basin is developed in a plexiglass lab flume which has a Manning's roughness coefficient of 0.010. The flume is rectangular and the discharge per unit width is 0.2 m^2/s. Under satisfactory operating conditions in the lab, a hydraulic jump is properly located in the stilling basin when the depth of flow before the jump is 0.05 m and the height of an end-sill is 0.09 times the depth of the flow after the jump. Compute the required height of the end-sill on a concrete-lined prototype stilling basin which is 0.8 m in width and has a design discharge of 1.6 m^3/s.

12-12. The design discharge for a rectangular irrigation channel is 2.83 m^3/s. Select the appropriate rectangular weir to measure the design discharge and specify required weir width, height of weir crest above channel invert, lateral distance from edge of weir to side approach section, and required distance upstream for measurement of head above the weir crest.

12-13. A trapezoidal channel is to be constructed in a sandy loam soil along a bottom slope of 0.003 m/m and deliver a maximum discharge of 2.0 m^3/s of water carrying little or no silt.

(a) Compute the wetted perimeter of the channel in meters.

(b) If the depth of uniform flow for this channel is 0.06 m, compute the required channel bottom width.

COMPUTER PROBLEM

Develop a computer program to calculate the flow parameters for a trapezoidal channel. The program should be interactive and accept the following input parameters:

(a) Channel discharge, m^3/s.
(b) Manning's roughness coefficient.
(c) Channel bed slope, m/m.
(d) Channel side-slope ratio.
(e) Channel bottom width, m.
(f) Initial estimate of flow depth, m.

The program should compute the normal depth for uniform flow conditions using the Manning equation. The depth will have to be solved for using an iterative procedure. Newton's method is the recommended procedure but other methods are acceptable. The program will compute two iterative solutions until they are the same within 1 mm. For each iteration the program should print the iteration number, initial depth estimate, and revised depth estimate. When the normal depth is determined within the convergence criteria, the program should print out the input information and the following results:

(a) Flow depth, m.
(b) Cross-sectional flow area, m^2.
(c) Channel top width, m.
(d) Flow velocity, m/s.
(e) Froude number.

Run the program for the following test cases:
Case #1

$Q = 1.7\ m^3/s$

$n = 0.015$

side-slope ratio = z = 2.0:1

$S_0 = 0.004$ m/m

channel bottom width = b = 1.34 m

Case #2

Same as Case #1 except $S_0 = 0.002$ m/m.

Case #3

Same as Case #2 except $Q = 2.0\ m^3/s$ and b = 1.4 m.

Solution The computer program listing which follows performs the tasks outlined in this problem. It is written in BASIC and writes only to the screen. It makes a maximum of 10 iterations of Newton's method in search of a convergent solution based on the specified error criterion. As written, the problem stops at line 360 after achieving a convergent solution. The user must then push the special function key 5 (CONTINUE) to write the final results to the screen.

Numerous improvements to this program are possible. Some of these have been described in other Computer Problem sections of this book. These include addition of a default data set, the option of writing results to the printer after they are displayed on the screen, and the division of the program into subroutines. This last modification would be particularly important if subroutines for other channel geometries were to be added to the program. A message could also be printed out if the Froude number were supercritical or greater than the design criterion described in this chapter. The user may take the simple program listed and make modifications including the addition of subroutines for other channel geometries.

```
10 REM ****************************************************************
20 REM
30 REM      *****   PROGRAM TO COMPUTE DIMENSIONS OF NONERODIBLE   *****
40 REM      *****             TRAPEZIODAL CHANNEL                  *****
50 REM
60 REM         Richard H. Cuenca                         Winter 1985
70 REM
80 REM ****************************************************************
90 REM
100 CLS
110 INPUT "CHANNEL DISCHARGE (m^3/S) = "; Q
120 INPUT "MANNING'S ROUGHNESS COEFFICIENT = "; N
130 INPUT "CHANNEL BOTTOM SLOPE (m/m) = "; S
140 INPUT "SIDE-SLOPE RATIO = "; Z
150 REM   **  COMPUTE SECTION FACTOR  **
160 REM
170 SECFAC = N * Q/S^.5
180 INPUT "BOTTOM WIDTH (m) = "; B
190 REM
200 REM   **  GENERAL FORM OF NEWTON'S METHOD  **
210 REM
220 DIM Y(15)
230 INPUT "INITIAL ESTIMATE OF DEPTH (m) = "; Y(1)
```

```
240 FOR I = 2 TO 10
250 F = ((B + Z * Y(I-1)) * Y(I-1))^2.5 - (SECFAC)^1.5 * (B + 2! * Y(I-1) *     (
1! + Z^2!)^.5)
260 FF = 2.5 * ((B + Z * Y(I-1)) * Y(I-1))^1.5 * (B + 2! * Z * Y(I-1)) -        (
SECFAC)^1.5 * (2! * (1! + Z^2!)^.5)
270 Y(I) = Y(I-1) - F/FF
280 DY = ABS(Y(I) - Y(I-1))
290 IF DY < .001 THEN 310
300 NEXT I
310 ILAST = I - 1
320 FOR I = 1 TO ILAST
330 PRINT
340 PRINT USING "     ##     ###.###     ###.###     "; I, Y(I), Y(I+1)
350 NEXT I
360 STOP
370 YF = Y(ILAST + 1)
380 A = (B + Z * YF) * YF
390 T = B + 2! * Z * YF
400 VAVG = Q/A
410 FR = VAVG/(9.80665 * A/T)^.5
420 REM
430 REM   **  PRINT OUT RESULTS  **
440 REM
450 PRINT "DISCHARGE (m^3/S) = "; Q
460 PRINT "MANNING'S ROUGHNESS COEFFICIENT = "; N
470 PRINT "CHANNEL BOTTOM SLOPE (m/m) = "; S
480 PRINT "SIDE-SLOPE RATIO = "; Z
490 PRINT "CHANNEL BOTTOM WIDTH (m) = "; B
500 PRINT "FLOW DEPTH (m) = "; YF
510 PRINT "FLOW AREA (m^2) = "; A
520 PRINT "TOP WIDTH (m) = "; T
530 PRINT "AVERAGE VELOCITY (m/s) = "; VAVG
540 PRINT "FROUDE NUMBER = "; FR
550 END
```

A

Use of SI (Système International) Units

A.1 Purpose

This appendix is a guide for application of SI units to quantities typically encountered in the fields of irrigation, water resources, and energy analysis. Material for this section is basically taken from American Society of Agricultural Engineers (ASAE) Engineering Practice EP285.6 found in the 1982 Agricultural Engineers Yearbook on pp. 100–107. This Engineering Practice should be referred to for further details.

A.2 SI Units of Measure

SI consists of seven base units, two supplementary units, and a series of derived units consistent with the base and supplementary units. There is also a series of approved prefixes for the formation of multiples and submultiples of various units shown in Table A-1. A number of derived units are listed in Table A-2.

BASE UNITS

meter (m)—unit of length

second (s)—unit of time

kilogram (kg)–unit of mass

TABLE A-1 SI unit prefixes.

Multiples & Submultiples	Prefix	SI Symbol
10^{18}	exa	E
10^{15}	peta	P
10^{12}	tera	T
10^{9}	giga	G
10^{6}	mega	M
10^{3}	kilo	k
10^{2}	hecto	h
10^{1}	deka	da
10^{-1}	deci	d
10^{-2}	centi	c
10^{-3}	milli	m
10^{-6}	micro	μ
10^{-9}	nano	n
10^{-12}	pico	p
10^{-15}	femto	f
10^{-18}	atto	a

kelvin (K)–unit of thermodynamic temperature
ampere (A)—unit of electric current
candela (cd)—luminous intensity
mole (mol)—the amount of a substance

SUPPLEMENTARY UNITS

radian (rad)—plane angle
steradian (sr)—solid angle

A.3 Rules for SI Usage

Capitalization

Symbols for SI units are only capitalized when the unit is derived from a proper name; for example, N for Isaac Newton (except liter, L). Unabbreviated units are not capitalized; for example kelvin and newton. Numerical prefixes given in Table A-1 and their symbols are not capitalized; except for the symbols M (mega), G (giga), T (tera), P (peta), and E (exa).

Plurals

Unabbreviated SI units form their plurals in the usual manner. SI symbols are always written in singular form. For example,

50 newtons or 50 N
25 millimeters or 25 mm

TABLE A-2 Derived units.

Quantity	Unit	SI Symbol	Formula
acceleration	meter per second squared	—	m/s^2
angular acceleration	radian per second squared	—	rad/s^2
angular velocity	radian per second	—	rad/s
area	square meter	—	m^2
density	kilogram per cubic meter	—	kg/m^3
electrical capacitance	farad	F	A · s/V
electrical conductance	siemens	S	A/V
electrical field strength	volt per meter	—	V/m
electrical inductance	henry	H	V · s/A
electrical potential difference	volt	V	W/A
electrical resistance	ohm	Ω	V/A
electromotive force	volt	V	W/A
energy	joule	J	N · m
entropy	joule per kelvin	—	J/K
force	newton	N	$kg \cdot m/s^2$
frequency	hertz	Hz	(cycle)/s
power	watt	W	J/s
pressure	pascal	Pa	N/m^2
quantity of electricity	coulomb	C	A · s
quantity of heat	joule	J	N · m
specific heat	joule per kilogram-kelvin	—	J/kg · K
stress	pascal	Pa	N/m^2
thermal conductivity	watt per meter-kelvin	—	W/m · K
velocity	meter per second	—	m/s
viscosity, dynamic	pascal-second	—	Pa · s
viscosity, kinematic	square meter per second	—	m^2/s
voltage	volt	V	W/A
volume	cubic meter	—	m^3
work	joule	J	N · m

Punctuation

Whenever a numerical value is less than one, a zero should precede the decimal point. Periods are not used after any SI unit symbol. English speaking countries use a dot for the decimal point, others use a comma. Use spaces instead of commas for grouping numbers into threes (thousands). For example,

6 357 831.376 88
not 6,357,831.376,88

Derived Units

The product of two or more units in symbolic form is preferably indicated by a dot midway in relation to unit symbol height. For example,

Use N · m but not mN

A solidus (oblique stroke, /), a horizontal line, or negative powers may be used to express a derived unit formed from two others by division. For example,

$$\mathrm{m/s},\ \frac{\mathrm{m}}{\mathrm{s}},\ \text{or}\ \mathrm{m \cdot s^{-1}}$$

A.4 Non-SI Units

Certain units outside the SI are recognized by the International Organization for Standardization (ISO) because of their practical importance in specialized fields. These include units for temperature, time, and angle. Also included are names for some multiples of units such as liter (L) for volume, hectare (ha) for land measure, and metric ton (t) for mass.

Temperature

The SI base unit for thermodynamic temperature is kelvin (K). Because of the wide usage of the degree Celsius, the Celsius scale (formerly called the centigrade scale) may be used when expressing temperature. The Celsius scale is related directly to the kelvin scale as follows:

one degree Celsius (1°C) equals one kelvin (1 K), exactly

A Celsius temperature (t) is related to a kelvin temperature (T) as follows:

$$t = T - 273.15$$

Time

The SI unit for time is the second. This unit is preferred and should be used when technical calculations are involved. In other cases use of the minute (min), hour (h), day (d), and so on, is permissible.

Angles

The SI unit for plane angle is the radian. The use of arc degrees (°) and their decimal or minute (′), second (″) submultiples is permissible when the radian is not a convenient unit.

A.5 Preferred Units and Conversion Factors

Preferred units for expressing physical quantities commonly encountered in water resources engineering are listed in Table A-3. These are presented as an aid to selecting proper units for given applications and to promote consistency where interpretation of the general rules of SI may not produce consistent results.

TABLE A-3 Preferred units for expressing physical quantities.

1. Quantities are arranged in alphabetical order by principal nouns. For example, surface tension is listed as tension, surface.
2. All possible applications are not listed, but others such as rates can be readily derived. For example, from the preferred units for energy and volume the units for heat energy per unit volume, kJ/m^3, may be derived.
3. Conversion factors are shown to seven significant digits, unless the precision with which the factor is known does not warrant seven digits.

Quantity	From: Former Units	To: SI Units	Multiply By:
Acceleration, angular	rad/s^2	rad/s^2	
Acceleration, linear	(mile/h)/s	(km/h)/s	1.609 344*
	ft/s^2	m/s^2	0.304 8*
Angle, plane	r (revolution)	r (revolution)	
	rad	rad	
	° (deg)	°	
	′ (min)	° (decimalized)	1/60*
	′ (min)	′	
	″ (sec)	° (decimalized)	1/3600*
	″ (sec)	″	
Area	$in.^2$	m^2	0.000 645 16*
	ft^2	m^2	0.092 903 04*
	$in.^2$	mm^2	645.16*
	$in.^2$	cm^2	6.451 6*
	acre	ha	0.404 687 3
	$mile^2$	km^2	2.589 998
Area per time	acre/h	ha/h	0.404 687 3
	ft^2/s	m^2/s	0.092 903 04*
Capacitance, electric	μF	μF	
Capacity, electric	A · h	A · h	
Capacity, heat	Btu/°F	$kJ/K^{\dagger}$	1.899 101
Capacity, heat, specific	Btu/(lb.°F)	$kJ/(kg \cdot K)^{\dagger}$	4.186 8*
Coefficient of Heat Transfer	$Btu/(h \cdot ft^2 \cdot °F)$	$W/(m^2 \cdot K)^{\dagger}$	5.678 263
Coefficient of Linear Expansion	$°F^{-1}$, (1/°F)	K^{-1}, $(1/K)^{\dagger}$	1.8*
Conductance, electric	mho	S	1*
Conductivity, electric	mho/ft	S/m	3.280 840
Conductivity, thermal	$Btu \cdot ft/(h \cdot ft^2 \cdot °F)$	$W/(m \cdot K)^{\dagger}$	1.730 735
Consumption, fuel	gal/h	L/h	3.785 412
Consumption, oil	qt/(1000 miles)	L/(1000 km)	0.588 036 4
Consumption, specific, oil	lb/(hp · h)	g/(kW · h)	608.277 4
	lb/(hp · h)	g/MJ	168.965 9
Current, electric	A	A	
Density, current	$A/in.^2$	kA/m^2	1.550 003
	A/ft^2	A/m^2	10.763 91
Density, (mass)	lb/yd^3	kg/m^3	0.593 276 3
	$lb/in.^3$	kg/m^3	27 679.90
	lb/ft^3	kg/m^3	16.018 46
	lb/gal	kg/L	0.119 826 4
	—	g/m^3, mg/L	—
Density of heat flow rate	$Btu/(h \cdot ft^2)$	W/m^2	3.154 591††

TABLE A-3 Preferred units for expressing physical quantities. (*continued*)

Quantity	From: Former Units	To: SI Units	Multiply By:
Efficiency, fuel	mile/gal	km/L	0.415 143 7
	—	L/(100 km)	§
	lb/(hp · h)	g/MJ	168.965 9
	hp · h/gal	kW · h/L	0.196 993 1
	lb/(hp · h)	kg/(kW · h)§§	0.608 277 4
Energy work, enthalpy, quantity of heat	ft · lbf	J	1.355 818
	Btu	kJ	1.055 056
	kcal	kJ	4.186 8*
	kW · h	kW · h	
	kW · h	MJ	3.6
	hp · h	MJ	2.684 520
	hp · h	kW · h	0.745 699 9
Energy, specific	cal/g‡	J/g	4.186 8*
	Btu/lb	kJ/kg	2.326*
Flow, mass, rate	lb/min	kg/min	0.453 592 4
	lb/s	kg/s	0.453 592 4
	g/min	g/min	
	ton (short)/h	t/h, Mg/h**	0.907 184 7
Flow, volume	ft^3/s	m^3/s	0.028 316 85
	ft^3/s	m^3/min	1.699 011
	gal/s (gps)	L/s	3.785 412
	gal/s (gps)	m^3/s	0.003 785 412
	gal/min (gpm)	L/min	3.785 412
	gal/min (gpm)	L/s	0.063 090 20
	oz/s	mL/s	29.573 53
	oz/min	mL/min	29.573 53
	gal/h	L/h	3.785 412
	ft^3/s	m^3/s	0.028 316 85
Force, thrust, drag	lbf	N	4.448 222
	ozf	N	0.278 013 9
	kgf	N	9.806 650
	dyne	N	0.000 01*
	lbf	kN	0.004 448 222
Force per length	lbf/ft	N/m	14.593 90
	lbf/in.	N/mm	0.175 126 8
Frequency	Mc/s	MHz	1*
	kc/s	kHz	1*
	Hz, c/s	Hz	1*
	r/s (rps)	s^{-1}, r/s	1*
	r/min (rpm)	min^{-1}, r/min	1*
	rad/s	rad/s	
Heat, specific	cal/g‡	kJ/kg	4.186 8*
	Btu/lb	kJ/kg	2.326*
Impedance, mechanical	lbf · s/ft	N · s/m	14.593 90
Inductance, electric	H	H	

TABLE A-3 Preferred units for expressing physical quantities. (*continued*)

Quantity	From: Former Units	To: SI Units	Multiply By:
Length	mile	km	1.609 344*‖
	rod	m	5.029 210‖
	yd	m	0.914 4
	ft	m	0.304 8*
	in.	cm	2.54*
	in.	mm	25.4*
	mil	μm	25.4*
	μin.	μm	0.025 4*
	micron	μm	1*
	μin.	nm	25.4*
Length per time	in./h	mm/h	25.4*
	in./h	cm/h	2.54*
Mass	ton (long)	t, Mg**	1.016 047
	ton (short)	t, Mg**	0.907 184 7
	lb	kg	0.453 592 4
	slug	kg	14.593 90
	oz	g	28.349 52
Mass per area	oz/yd^2	g/m^2	33.905 75
	lb/ft^2	kg/m^2	4.882 428
	oz/ft^2	g/m^2	305.151 7
	lb/acre	kg/ha	1.120 851
	ton (short)/acre	t/ha**	2.241 702
Mass per length	lb/ft	kg/m	1.488 164
	lb/yd	kg/m	0.496 054 7
Mass per time	ton (short)/h	t/h, Mg/h**	0.907 184 7
Modulus of elasticity	lbf/in.^{2}	MPa	0.006 894 757
Modulus, bulk	psi	kPa	6.894 757
Momentum, linear	lb · ft/s	kg · m/s	0.138 255 0
Momentum, angular	lb · ft^2/s	kg · m^2/s	0.042 140 11
Power	W	W	
	Btu/min	W	17.584 27
	Btu/h	W	0.293 071 1
	hp (550 ft · lbf/s)	kW	0.745 699 9
Pressure	lbf/in.^{2} (psi)	kPa	6.894 757
	in. Hg (60°F)	kPa	3.376 85
	in. H_2O (60°F)	kPa	0.248 84
	mmHg (0°C)	kPa	0.133 322
	kgf/cm^2	kPa	98.066 5
	bar	kPa	100.0*
	lbf/ft^2	kPa	0.047 880 26
	atm (normal = 760 torr)	kPa	101.325*
	lbf/in.^{2} (psi)	Pa	6 894.757
Radiation, solar	Btu	J	1 055.056
	cal	J	4.1868*
	langley	J/m^2	41 868*
	Btu/ft^2	J/m^2	11 356.53
	langley/min	W/m^2	697.8*
	Btu/ft^2/min	W/m^2	189.275 5
Resistance, electric	Ω	Ω	
Resistivity, electric	Ω · ft	Ω · m	0.304 8*
	Ω · ft	Ω · cm	30.48*

TABLE A-3 Preferred units for expressing physical quantities (continued).

Quantity	From: Former Units	To: SI Units	Multiply By:
Stress	lbf/in.^{2}	MPa	0.006 894 757
Temperature	°F	°C	$t_{°C} = (t_{°F} - 32)/1.8$*
	°R	K	$T_K = T_{°R}/1.8$*
Temperature interval	°F	K†	1 K = 1°C = 1.8°F*
Tension, surface	lbf/in.	mN/m	175 126.8
	dyne/cm	mN/m	1*
Thermal diffusivity	ft^2/h	m^2/h	0.092 903 04
Time	s	s	
	h	h	
	min	min	
Velocity, linear	mile/h	km/h	1.609 344*
	ft/s	m/s	0.304 8*
	in./s	mm/s	25.4*
	ft/min	m/min	0.304 8*
Viscosity, dynamic	centipoise	mPa · s	1*
Viscosity, kinematic	centistokes	mm^2/s	1*
Volume	yd^3	m^3	0.764 554 9
	ft^3	m^3	0.028 316 85
	bushel	L	35.239 07
	ft^3	L	28.316 85
	in.^{3}	L	0.016 387 06
	in.^{3}	cm^3	16.387 06
	gal	L	3.785 412
	qt	L	0.946 352 9
	pt	L	0.473 176 5
	oz	mL	29.573 53
	acre · ft	m^3	1 233.489‖
		dam^3	1.233 489‖
	bushel (U.S.)	m^3	0.035 239 07
Volume per area	gal/acre	L/ha	9.353 958
Volume per time	gal/h	L/h	3.785 412
Weight	May mean either mass or force—avoid use of weight		

*Indicates exact conversion factor.

†In these expressions K indicates temperature intervals. Therefore K may be replaced with °C if desired without changing the value or affecting the conversion factor. kJ/(kg · K) = kJ/(kg · °C)

‡Not to be confused with kcal/g. kcal often called calorie.

§Convenient conversion: 235.215 ÷ (mile per gal) = L/(100 km).

‖Official use in surveys and cartography involves the U.S. survey mile based on the U.S. survey foot, which is longer than the international foot by two parts per million. The factors used in this standard for acre, acre foot, rod are based on the U.S. survey foot. Factors for all other former length units are based on the international foot. (See ANSI/ASTM Standard E380-76, Metric Practice.)

#Standard acceleration of gravity is 9.806 650 m/s^2 exactly. (Adopted by the General Conference on Weights and Measures.)

**The symbol t is used to designate metric ton. The unit metric ton (exactly 1 MG) is in wide use but should be limited to commercial description of vehicle mass, freight mass, and agricultural commodities. No prefix is permitted.

††Conversions of Btu are based on the International Table Btu.

§§ASAE S209 and SAE J708, Agricultural Tractor Test Code, specify kg/(kW · h). It should be noted that there is a trend toward use of g/MJ as specified for highway vehicles.

A.6 Conversion Techniques

Conversion of quantities between systems of units involves careful determination of the number of significant digits to be retained. All conversions, to be logically established, must depend upon an intended precision of the original quantity—either implied by a specific tolerance, or by the nature of the quantity. The implied precision of a value should relate to the number of significant digits shown. The implied precision is plus or minus one-half unit of the last significant digit in which the value is stated. Whether rounded or not, a quantity should always be expressed with this implication of precision in mind.

Quantities should be expressed in digits which are intended to be significant. Quantities should not be expressed with significant zeros omitted. The dimension 2 in. may mean "about 2 in.," or it may, in fact, mean a very accurate expression which should be written 2.0000 in. In the latter case, while the added zeros are not significant in establishing the value, they are very significant in expressing the proper intended precision.

A.7 Rules for Round-Off

When a number is to be rounded to fewer decimal places the procedure shall be as follows:

> When the first digit discarded is less than 5, the last digit retained shall not be changed. For example, 3.463 25, if rounded to three decimal places, would be 3.463; if rounded to two decimal places, would be 3.46.
>
> When the first digit discarded is greater than 5, or if it is a 5 followed by at least one digit other than 0, the last figure retained shall be increased by one unit. For example, 8.376 52, if rounded to three decimal places, would be 8.377; if rounded to two decimal places, would be 8.38.
>
> Round to closest even number when first digit discarded is 5, followed only by zeros.
>
> Numbers are rounded directly to the nearest value having the desired number of decimal places. Rounding must not be done in successive steps to fewer places.
>
> For example, 27.46 rounded to a whole number = 27. This is correct because the ".46" is less than one-half. 27.46 rounded to one decimal place is 27.5. This is a correct value. But, if the 27.5 is in turn rounded to a whole number, this is successive rounding and the result, 28, is incorrect.

Inch-millimeter Linear Dimensioning Conversion.

1 inch (in.) = 25.4 millimeters (mm) exactly. The term "exactly" has been used with all exact conversion factors. Conversion factors not so labeled have been rounded in accordance with these rounding procedures. To maintain intended preci-

sion during conversion without retaining an unnecessary number of digits, the millimeter equivalent shall be carried to one decimal place more than the inch value being converted and then rounded to the appropriate significant figure in the last decimal place.

A.8 Cited Standards

ASAE S209, Agricultural Tractor Test Code

ANSI/ASTM E380-76, Metric Practice

ISO 1000, SI Units and Recommendations for the Use of Their Multiples and of Certain Other Units

B

Physical Properties of Liquid Water

TABLE B-1 Physical properties of liquid water.

Temperature °C	Vapor Pressure		Specific gravity	Specific heat J/g °C	Latent Heat of Vaporization kJ/g	Surface Tension g/s^2	Thermal Conductivity $J \times 10^{-3}/cm \cdot s \cdot °C$	Viscosity $g \times 10^{-2}/cm \cdot s$	Kinematic Viscosity cm^2/s
	kPa	cm water							
−10	—	—	0.99794	4.271	2.526	—	—	—	—
−05	—	—	0.99918	4.229	2.514	76.4	—	—	—
0	0.6109	6.23	0.99987	4.216	2.503	75.6	5.61	1.787	0.0179
4	0.8131	8.29	1.00000	4.208	2.493	75.0	5.70	1.567	0.0157
5	0.8721	8.89	0.99999	4.204	2.491	74.8	5.74	1.519	0.0152
10	1.2276	12.51	0.99973	4.191	2.479	74.2	5.87	1.307	0.0131
15	1.7051	17.37	0.99913	4.187	2.467	73.4	5.95	1.139	0.0114
20	2.339	23.81	0.99823	4.183	2.455	72.7	6.03	1.002	0.01007
25	3.169	32.31	0.99708	4.178	2.443	71.9	6.12	0.890	0.00897
30	4.246	43.11	0.99568	4.178	2.432	71.1	6.20	0.798	0.00804
35	5.628	57.05	0.99406	4.178	2.420	70.3	6.28	0.719	0.00733
40	7.384	74.71	0.99225	4.178	2.408	69.5	6.33	0.653	0.00661
45	9.593	96.86	0.99024	4.178	2.396	68.7	6.41	0.596	0.00609
50	12.349	124.42	0.98807	4.183	2.384	67.9	6.45	0.547	0.00556

Adapted from: Hillel, D., *Fundamentals of Soil Physics*. New York: Academic Press, 1980.
Colt Industries, *Hydraulic Handbook*. Kansas City, Kansas: Fairbanks Morse Pump Division, 1980.
Keenan, J. H., F. G. Keyes, P. G. Hill, and J. G. Moore, *Steam Tables: Thermodynamic Properties of Water Including Vapor, Liquid, and Solid Phases*. New York: John Wiley & Sons, 1978.

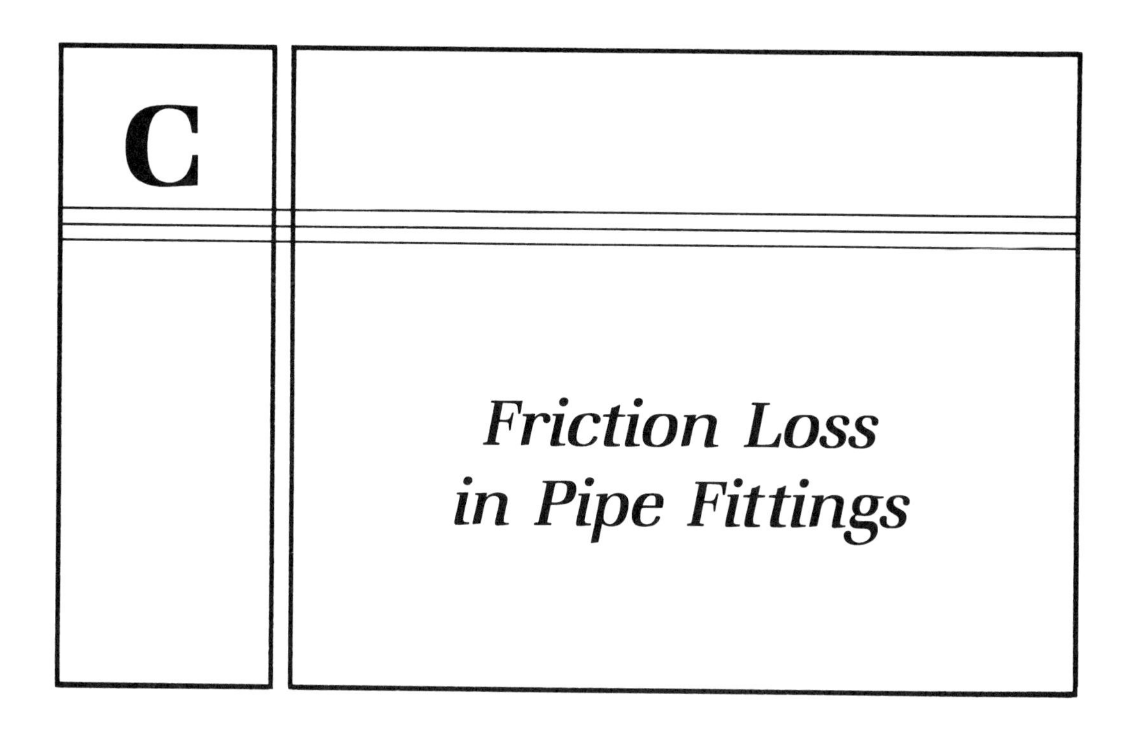

C Friction Loss in Pipe Fittings

[Reproduced with permission from Ingersoll-Rand. Taken from Cameron Hydraulic Data (Westaway and Loomis, 1979). Nominal pipe sizes indicated are in inches.]

Friction of Water
Friction Loss in Pipe Fittings

Resistance coefficient K (use in formula $h_f = K \frac{V^2}{2g}$)

Note: Fittings are standard with full openings.

Fitting	L/D	Nominal pipe size ½	¾	1	1¼	1½	2	2½–3	4	6	8–10	12–16	18–24
		K value											
Gate Valves	8	0.22	0.20	0.18	0.18	0.15	0.15	0.14	0.14	0.12	0.11	0.10	**0.10**
Globe Valves	340	9.2	8.5	7.8	7.5	7.1	6.5	6.1	5.8	5.1	4.8	4.4	**4.1**
Angle Valves	55	1.48	1.38	1.27	1.21	1.16	1.05	0.99	0.94	0.83	0.77	0.72	**0.66**
Angle Valves	150	4.05	3.75	3.45	3.30	3.15	2.85	2.70	2.55	2.25	2.10	1.95	**1.80**
Ball Valves	3	0.08	0.08	0.07	0.07	0.06	0.06	0.05	0.05	0.05	0.04	0.04	**0.04**

Calculated from data in Crane Co. Technical Paper No. 410.

Friction of Water
Friction Losses in Pipe Fittings

Resistance coefficient K $\left(\text{use in formula } h_f = K \frac{V^2}{2g}\right)$

Note: Fittings are standard with full openings.

Fitting		L/D	Nominal pipe size ½	¾	1	1¼	1½	2	2½–3	4	6	8–10	12–16	18–24
			K value											
Butterfly Valve								0.86	0.81	0.77	0.68	0.63	0.35	**0.30**
Plug Valve straightway		18	0.49	0.45	0.41	0.40	0.38	0.34	0.32	0.31	0.27	0.25	0.23	**0.22**
Plug Valve 3-way thru-flo		30	0.81	0.75	0.69	0.66	0.63	0.57	0.54	0.51	0.45	0.42	0.39	**0.36**
Plug Valve branch-flo		90	2.43	2.25	2.07	1.98	1.89	1.71	1.62	1.53	1.35	1.26	1.17	**1.08**
Standard elbow	90°	30	0.81	0.75	0.69	0.66	0.63	0.57	0.54	0.51	0.45	0.42	0.39	**0.36**
	45°	16	0.43	0.40	0.37	0.35	0.34	0.30	0.29	0.27	0.24	0.22	0.21	**0.19**
	long radius 90°	16	0.43	0.40	0.37	0.35	0.34	0.30	0.29	0.27	0.24	0.22	0.21	**0.19**

Calculated from data in Crane Co., Technical Paper No. 410.

Friction of Water
Friction Losses in Pipe Fittings

Resistance coefficient K $\left(\text{use in formula } h_f = K\dfrac{V^2}{2g}\right)$

Note: Fittings are standard with full openings.

Fitting	Type of bend	L/D	½	¾	1	1¼	1½	2	2½–3	4	6	8–10	12–16	18–24
			Nominal pipe size – K value											
Close Return Bend		50	1.35	1.25	1.15	1.10	1.05	0.95	0.90	0.85	0.75	0.70	0.65	0.60
Standard Tee	thru flo	20	0.54	0.50	0.46	0.44	0.42	0.38	0.36	0.34	0.30	0.28	0.26	0.24
	thru branch	60	1.62	1.50	1.38	1.32	1.26	1.14	1.08	1.02	0.90	0.84	0.78	0.72
90° Bends. Pipe bends, flanged elbows, butt welded elbows	r/d = 1	20	0.54	0.50	0.46	0.44	0.42	0.38	0.36	0.34	0.30	0.28	0.26	0.24
	r/d = 2	12	0.32	0.30	0.28	0.26	0.25	0.23	0.22	0.20	0.18	0.17	0.16	0.14
	r/d = 3	12	0.32	0.30	0.28	0.26	0.25	0.23	0.22	0.20	0.18	0.17	0.16	0.14
	r/d = 4	14	0.38	0.35	0.32	0.31	0.29	0.27	0.25	0.24	0.21	0.20	0.18	0.17
	r/d = 6	17	0.46	0.43	0.39	0.37	0.36	0.32	0.31	0.29	0.26	0.24	0.22	0.20
	r/d = 8	24	0.65	0.60	0.55	0.53	0.50	0.46	0.43	0.41	0.36	0.34	0.31	0.29
	r/d = 10	30	0.81	0.75	0.69	0.66	0.63	0.57	0.54	0.51	0.45	0.42	0.39	0.36
	r/d = 12	34	0.92	0.85	0.78	0.75	0.71	0.65	0.61	0.58	0.51	0.48	0.44	0.41
	r/d = 14	38	1.03	0.95	0.87	0.84	0.80	0.72	0.68	0.65	0.57	0.53	0.49	0.46
	r/d = 16	42	1.13	1.05	0.97	0.92	0.88	0.80	0.76	0.71	0.63	0.59	0.55	0.50
	r/d = 18	46	1.24	1.15	1.06	1.01	0.97	0.87	0.83	0.78	0.69	0.64	0.60	0.55
	r/d = 20	50	1.35	1.25	1.15	1.10	1.05	0.95	0.90	0.85	0.75	0.70	0.65	0.60
Mitre Bends	α = 0°	2	0.05	0.05	0.05	0.04	0.04	0.04	0.04	0.03	0.03	0.03	0.03	0.02
	α = 15°	4	0.11	0.10	0.09	0.09	0.08	0.08	0.07	0.07	0.06	0.06	0.05	
	α = 30°	8	0.22	0.20	0.18	0.18	0.17	0.15	0.14	0.14	0.12	0.11	0.10	0.10
	α = 45°	15	0.41	0.38	0.35	0.33	0.32	0.29	0.27	0.26	0.23	0.21	0.20	0.18
	α = 60°	25	0.68	0.63	0.58	0.55	0.53	0.48	0.45	0.43	0.38	0.35	0.33	0.30
	α = 75°	40	1.09	1.00	0.92	0.88	0.84	0.76	0.72	0.68	0.60	0.56	0.52	0.48
	α = 90°	60	1.62	1.50	1.38	1.32	1.26	1.14	1.08	1.02	0.90	0.84	0.78	0.72

Calculated from data in Crane Co. Technical Paper No. 410.

Friction of Water
Friction Losses in Pipe Fittings

Resistance coefficient K $\left(\text{use in formula } h_f = K\frac{V^2}{2g}\right)$

Note: Fittings are standard with full port openings.

Fitting stop-check valves	L/D	Minimum velocity for full disc lift		Nominal pipe size											
		general ft/sec†	water ft/sec	½	¾	1	1¼	1½	2	2½–3	4	6	8–10	12–16	18–24
				K value*											
	400	$55\sqrt{\bar{V}}$	6.96	10.8	10	9.2	8.8	8.4	7.5	7.2	6.8	6.0	5.6	5.2	4.8
	200	$75\sqrt{\bar{V}}$	9.49	5.4	5	4.6	4.4	4.2	3.8	3.6	3.4	3.0	2.8	2.6	2.4
	350	$60\sqrt{\bar{V}}$	7.59	9.5	8.8	8.1	7.7	7.4	6.7	6.3	6.0	5.3	4.9	4.6	4.2
	300	$60\sqrt{\bar{V}}$	7.59	8.1	7.5	6.9	6.6	6.3	5.7	5.4	5.1	4.5	4.2	3.9	3.6
	55	$140\sqrt{\bar{V}}$	17.7	1.5	1.4	1.3	1.2	1.2	1.1	1.0	.94	.83	.77	.72	.66

Calculated from data in Crane Co. Technical Paper No. 410.
* These K values for flow giving full disc lift. K values are higher for low flows giving partial disc lift.
† In these formulas, $\bar{V}$, is specific volume—ft^3/lb.

Friction of Water

Friction Loss in Pipe Fittings

Resistance coefficient K (use in formula $h_f = K \frac{V^2}{2g}$)

Note: Fittings are standard with full port openings.

Fitting	L/D	Minimum velocity for full disc lift: general ft/sec†	Minimum velocity for full disc lift: water ft/sec	½	¾	1	1¼	1½	2	2½–3	4	6	8–10	12–16	18–24
				Nominal pipe size — K value*											
Swing check valve	100	$35\sqrt{\bar{V}}$	4.43	2.7	2.5	2.3	2.2	2.1	1.9	1.8	1.7	1.5	1.4	1.3	1.2
	50	$48\sqrt{\bar{V}}$	6.08	1.4	1.3	1.2	1.1	1.1	1.0	0.9	0.9	.75	.70	.65	.6
Lift check valve	600	$40\sqrt{\bar{V}}$	5.06	16.2	15	13.8	13.2	12.6	11.4	10.8	10.2	9.0	8.4	7.8	7.2
	55	$140\sqrt{\bar{V}}$	17.7	1.5	1.4	1.3	1.2	1.2	1.1	1.0	.94	.83	.77	.72	.66
Tilting disc check valve	5°	$80\sqrt{\bar{V}}$	10.13						.76	.72	.68	.60	.56	.39	.24
	15°	$30\sqrt{\bar{V}}$	3.80						2.3	2.2	2.0	1.8	1.7	1.2	.72
Foot valve with strainer poppet disc	420	$15\sqrt{\bar{V}}$	1.90	11.3	10.5	9.7	9.3	8.8	8.0	7.6	7.1	6.3	5.9	5.5	5.0
Foot valve with strainer hinged disc	75	$35\sqrt{\bar{V}}$	4.43	2.0	1.9	1.7	1.7	1.7	1.4	1.4	1.3	1.1	1.1	1.0	.90

Calculated from data in Crane Co. Technical Paper No. 410.

* These K values for flow giving full disc lift. K values are higher for low flows giving partial disc lift.

† In these formulas, $\bar{V}$, is specific volume—ft^3/lb.

Friction of Water
Friction Loss in Pipe Fittings

Resistance coefficient $\left(\text{use in formula } h_f = K \frac{V^2}{2g}\right)$

Fitting	Description	All pipe sizes K value
Pipe exit	projecting sharp edged rounded	1.0
Pipe entrance	inward projecting	0.78
Pipe entrance flush	sharp edged	0.5
	r/d = 0.02	0.28
	r/d = 0.04	0.24
	r/d = 0.06	0.15
	r/d = 0.10	0.09
	r/d = 0.15 & up	0.04

From Crane Co. Technical Paper 410.

Friction of Water

Friction Loss Due to Change in Pipe Size—Feet of Liquid

Loss of head in ft of liquid — Based on velocity in smaller pipe

Velocity of d_1 fps	Sudden Enlargements d_1/d_2									Sudden Contractions d_1/d_2									Velocity of d_1 fps
	0.9	0.8	0.7	0.6	0.5	0.4	0.3	0.2	0.1	0.9	0.8	0.7	0.6	0.5	0.4	0.3	0.2	0.1	
2	.00	.01	.02	.03	.03	.04	.05	.06	.06	.01	.01	.02	.02	.02	.03	.03	.03	.03	2
3	.01	.02	.04	.06	.08	.10	.12	.13	.14	.01	.03	.03	.04	.05	.06	.06	.07	.07	3
4	.01	.03	.06	.10	.14	.17	.21	.23	.24	.02	.04	.06	.08	.09	.10	.11	.12	.12	4
5	.01	.05	.10	.16	.22	.27	.32	.36	.38	.04	.07	.10	.12	.15	.16	.18	.19	.19	5
6	.02	.07	.15	.23	.31	.39	.46	.51	.55	.05	.10	.14	.18	.21	.23	.26	.27	.28	6
7	.03	.10	.20	.31	.43	.53	.63	.70	.75	.07	.14	.19	.24	.29	.32	.35	.37	.38	7
8	.04	.13	.26	.41	.56	.70	.83	.92	.97	.09	.18	.25	.32	.38	.42	.46	.48	.50	8
9	.05	.16	.33	.52	.70	.88	1.04	1.16	1.23	.12	.23	.31	.40	.48	.53	.56	.60	.63	9
10	.06	.20	.40	.64	.87	1.09	1.29	1.43	1.52	.15	.28	.38	.50	.59	.65	.71	.75	.77	10
12	.08	.29	.58	.92	1.25	1.57	1.86	2.06	2.19	.21	.40	.56	.72	.85	.94	1.03	1.07	1.12	12
15	.13	.45	.91	1.43	1.96	2.45	2.90	3.22	3.43	.33	.63	.87	1.12	1.33	1.47	1.61	1.68	1.75	15
20	.22	.80	1.62	2.55	3.48	4.35	5.16	5.72	6.09	.59	1.12	1.55	1.99	2.36	2.61	2.86	2.98	3.10	20
25	.35	1.26	2.53	3.98	5.44	6.80	8.06	8.94	9.52	.92	1.75	2.43	3.11	3.69	4.08	4.47	4.66	4.86	25
30	.50	1.82	3.64	5.73	7.83	9.79	11.6	12.9	13.7	1.32	2.52	3.50	4.48	5.31	5.87	6.43	6.71	6.99	30
40	.90	3.23	6.46	10.2	13.9	17.4	20.6	22.9	24.4	2.36	4.48	6.22	7.96	9.45	10.4	11.4	11.9	12.4	40
K value	.036	0.13	0.26	0.41	0.56	0.70	0.83	0.92	0.98	.095	0.18	0.25	0.32	0.38	0.42	0.46	0.48	0.50	K value

Calculated from formula $h_f = K\frac{V^2}{2g}$ For sudden enlargements $K = \left(1 - \frac{d_1^2}{d_2^2}\right)^2$ For sudden contractions $K = 0.5\left(1 - \frac{d_1^2}{d_2^2}\right)$

Example: Assume $d_1 = 6''$; $d_2 = 10''$; velocity $d_1 = 10$ fps. $\frac{d_1}{d_2} = 0.60$

From chart: for sudden enlargements: $h_f = 0.64$ feet
from sudden contractions: $h_f = 0.50$ feet

Friction of Water
Friction Loss due to Change in Pipe Size
Head loss in feet of liquid

(K values are for velocity in the small pipe)

Gradual Contraction (Based on velocity in small pipe)

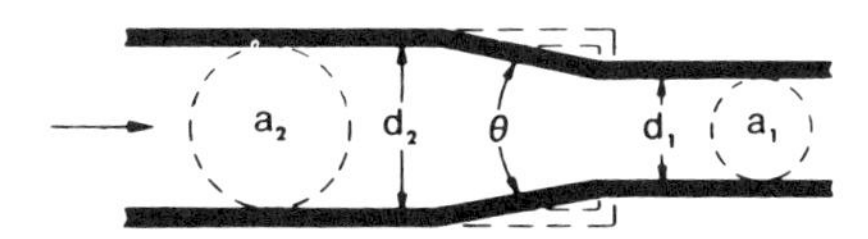

$\theta < 45°$ $$K = 0.8 \sin\frac{\theta}{2}\left(1 - \frac{d_1^2}{d_2^2}\right)$$

$\theta > 45° < 180°$ $$K = 0.5\left(1 - \frac{d_1^2}{d_2^2}\right)\sqrt{\sin\frac{\theta}{2}}$$

Gradual Enlargement (Based on velocity in small pipe)

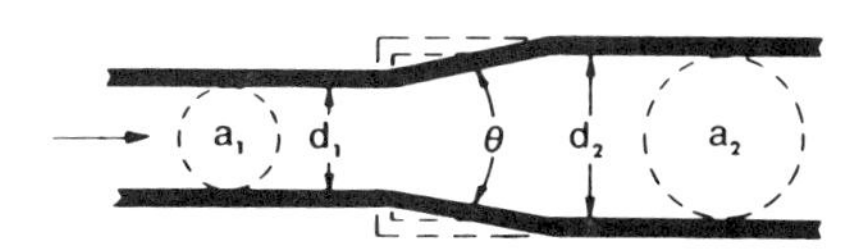

$\theta < 45°$ $$K = 2.6 \sin\frac{\theta}{2}\left(1 - \frac{d_1^2}{d_2^2}\right)^2$$

$\theta > 45° < 180°$ $$K = \left(1 - \frac{d_1^2}{d_2^2}\right)^2$$

Substitute above values of K in formula $h_f = K\dfrac{V^2}{2g}$ If desired, areas can be used instead of diameters in which case substitute

$$\frac{a_1}{a_2} \text{ for } \frac{d_1^2}{d_2^2} \qquad \text{and} \qquad \left(\frac{a_1}{a_2}\right)^2 \text{ for } \left(\frac{d_1}{d_2}\right)^4$$

D

Minimum Standards for Aluminum Sprinkler Irrigation Tubing

This Appendix was developed from ASAE Standard S263.3 (ASAE, 1987) based on the minimum requirement for "Class 150" aluminum tubing for which the maximum specified operating pressure is 1.0 MPa (145 psi). Further details may be found in that reference.

Hydrostatic Test

At a maximum operating pressure of 1.0 MPa (145 psi) all tubing must be capable of withstanding an internal hydrostatic test pressure of 3.0 MPa (435 psi) for 2 min without leaking.

Denting Factor

The tubing must have a denting factor equal or superior to the denting factor in Table D-1 which is calculated by:

$$D_t = F_{ty}t^2$$

where

F_{ty} = minimum tensile yield strength of material in MPa (from Table D-2)

t = specified wall thickness, mm

The wall thickness may be computed by application of the foregoing equation with the Minimum Denting Factor from Table D-1.

Bending Stresses

The tubing must be capable of spanning 9 m (29.5 ft) as a simple beam without permanent deflection or local buckling when filled with water at atmospheric pressure. The bending stresses, f, shall not exceed the smaller of:

$$f = 90\% \text{ of minimum tensile yield strength}$$

or

$$f = 1.57F_{ty} - 2.47 \times 10^{-5} F_{ty}^2 \frac{D}{t}$$

where

D = outside diameter, mm

t = wall thickness, mm

Torque Strength

It is recommended that manufacturers of mechanical move systems use a safety factor of 2 for protection due to torque loading. The torque strength is computed using

$$T = 79.3KD^{1/2}t^{5/2}$$

TABLE D-1 Denting factors for aluminum irrigation tubing.

Outside Diameter		Minimum Denting Factor
mm	in.	N
51	2	280
76	3	280
102	4	280
127	5	300
152	6	370
178	7	450
203	8	580
229	9	750
254	10	980

where

$$T = \text{torque strength, N(m)}$$
$$K = \text{stiffening factor}$$

K equals 1 for extruded tubing and equals 1 or value specified from physical testing for welded tubing.

Theoretical Bursting Pressure

The theoretical bursting pressure is computed from

$$P = 2F_{tu}\frac{t}{D}$$

where

P = bursting pressure, MPa

F_{tu} = ultimate tensile strength from Table D-2 or value guaranteed by manufacturer

TABLE D-2 Mechanical properties.

Aluminum alloys and tempers ASTM No.	Minimum Tensile Yield, F_{ty}		Ultimate Tensile Strength, F_{tu}			
			Welded tubing		Seamless tubing	
	MPa	psi	MPa	psi	MPa	psi
3003-H14	117	17,000	97	14,000	138	20,000
3003-H16	145	21,000	97	14,000	165	24,000
3003-H18	165	24,000	97	14,000	186	27,000
3004-H32	145	21,000	159	23,000	193	28,000
3004-H34	172	25,000	159	23,000	221	32,000
5050-H34	138	20,000	124	18,000	172	25,000
5050-H36	152	22,000	124	18,000	186	27,000
5050-H38	165	24,000	124	18,000	200	29,000
5052-H32	159	23,000	172	25,000	214	31,000
5052-H34	179	26,000	172	25,000	234	34,000
5052-H36	200	29,000	172	25,000	255	37,000
5052-H38	214	31,000	172	25,000	269	39,000
5086-H32	193	28,000	241	35,000	276	40,000
5086-H34	234	34,000	241	35,000	303	44,000
5086-H36	262	38,000	241	35,000	324	47,000
5154-H32	179	26,000	207	30,000	248	36,000
5154-H34	200	29,000	207	30,000	269	39,000
5154-H36	221	32,000	207	30,000	290	42,000
5154-H38	234	34,000	207	30,000	310	45,000
6061-T6	241	35,000	165	24,000	262	38,000
6063-T6	172	25,000	117	17,000	207	30,000
6063-T31	193	28,000			207	30,000

Index